Lecture Notes in Mathematics

For information about Vols. 1–226, please contact your bookseller or Springer-Verlag.

Vol. 277: Séminaire Banach. [illegible] par C. Houzel. VII, 229 pages. 1972.

Vol. 278: H. Jacquet, Automorphic Forms on GL(2). Part II. XIII, 142 pages. 1972.

Vol. 279: R. Bott, S. Gitler and I. M. James, Lectures on Algebraic and Differential Topology. V, 174 pages. 1972.

Vol. 280: Conference on the Theory of Ordinary and Partial Differential Equations. Edited by W. N. Everitt and B. D. Sleeman. XV, 367 pages. 1972.

Vol. 281: Coherent [illegible]. [illegible] pages. 1972.

Vol. 282: W. Klingenberg und P. Flaschel, Riemannsche Hilbertmannigfaltigkeiten. Periodische Geodätische. VII, 211 pages. 1972.

Vol. 283: L. Illusie, Complexe Cotangent et Déformations II. VII, 304 pages. 1972.

Vol. 284: P. A. Meyer, Martingales and Stochastic Integrals I. VI, 89 pages. 1972.

Vol. 285: P. de la Harpe, Classical Banach-Lie Algebras and Banach-Lie Groups of Operators in Hilbert Space. III, 160 pages. 1972.

Vol. 286: S. Murakami, On Automorphic Forms on Siegel Domains. V, 85 pages. 1972.

Vol. 287: Hyperfunctions and Pseudo-Differential Equations. Edited by H. Komatsu. VII, 529 pages. 1973.

Vol. 288: Groupes de Monodromie en Géométrie Algébrique. (SGA 7 I). Dirigé par A. Grothendieck. IX, 523 pages. 1972.

Vol. 289: B. Fuglede, Finely Harmonic Functions. III, 188. 1972.

Vol. 290: D. B. Zagier, Equivariant Pontrjagin Classes and Applications to Orbit Spaces. IX, 130 pages. 1972.

Vol. 291: P. Orlik, Seifert Manifolds. VIII, 155 pages. 1972.

Vol. 292: W. D. Wallis, A. P. Street and J. S. Wallis, Combinatorics: Room Squares, Sum-Free Sets, Hadamard Matrices. V, 508 pages. 1972.

Vol. 293: R. A. DeVore, The Approximation of Continuous Functions by Positive Linear Operators. VIII, 289 pages. 1972.

Vol. 294: Stability of Stochastic Dynamical Systems. Edited by R. F. Curtain. IX, 332 pages. 1972.

Vol. 295: C. Dellacherie, Ensembles Analytiques, Capacités, Mesures de Hausdorff. XII, 123 pages. 1972.

Vol. 296: Probability and Information Theory II. Edited by M. Behara, K. Krickeberg and J. Wolfowitz. V, 223 pages. 1973.

Vol. 297: J. Garnett, Analytic Capacity and Measure. IV, 138 pages. 1972.

Vol. 298: Proceedings of the Second Conference on Compact Transformation Groups. Part 1. XIII, 453 pages. 1972.

Vol. 299: Proceedings of the Second Conference on Compact Transformation Groups. Part 2. XIV, 327 pages. 1972.

Vol. 300: P. Eymard, Moyennes Invariantes et Représentations Unitaires. II, 113 pages. 1972.

Vol. 301: F. Pittnauer, Vorlesungen über asymptotische Reihen. VI, 186 Seiten. 1972.

Vol. 302: M. Demazure, Lectures on p-Divisible Groups. V, 98 pages. 1972.

Vol. 303: Graph Theory and Applications. Edited by Y. Alavi, D. R. Lick and A. T. White. IX, 329 pages. 1972.

Vol. 304: A. K. Bousfield and D. M. Kan, Homotopy Limits, Completions and Localizations. V, 348 pages. 1972.

Vol. 305: Théorie des Topos et Cohomologie Etale des Schémas. Tome 3. (SGA 4). Dirigé par M. Artin, A. Grothendieck et J. L. Verdier. VI, 640 pages. 1973.

Vol. 306: H. Luckhardt, Extensional Gödel Functional Interpretation. VI, 161 pages. 1973.

Vol. 307: J. L. Lions, G. Papanicolaou, S. D. Chatterji, Ecole d'Eté de Probabilités: Processus Stochastiques. VI, 198 pages. 1973.

Vol. 308: D. Knutson, λ-Rings and the Representation Theory of the Symmetric Group. IV, 203 pages. 1973.

Vol. 309: D. H. Sattinger, Topics in Stability and Bifurcation Theory. VI, 190 pages. 1973.

Vol. 310: B. Iversen, Generic Local Structure of the Morphisms in Commutative Algebra. IV, 108 pages. 1973.

Vol. 311: Conference on Commutative Algebra. Edited by J. W. Brewer and E. A. Rutter. VII, 251 pages. 1973.

Vol. 312: Symposium on Ordinary Differential Equations. Edited by W. A. Harris, Jr. and Y. Sibuya. VIII, 204 pages. 1973.

Vol. 313: K. Jörgens and J. Weidmann, Spectral Properties of Hamiltonian Operators. III, 140 pages. 1973.

Vol. 314: M. Deuring, Lectures on the Theory of Algebraic Functions of One Variable. VI, 151 pages. 1973.

Vol. 315: K. Bichteler, Integration Theory (with Special Attention to Vector Measures). VI, 357 pages. 1973.

Vol. 316: Symposium on Non-Well-Posed Problems and Logarithmic Convexity. Edited by R. J. Knops. V, 176 pages. 1973.

Vol. 317: Séminaire Bourbaki – vol. 1971/72. Exposés 400–417. IV, 361 pages. 1973.

Vol. 318: Recent Advances in Topological Dynamics. Edited by A. Beck. VIII, 285 pages. 1973.

Vol. 319: Conference on Group Theory. Edited by R. W. Gatterdam and K. W. Weston. V, 188 pages. 1973.

Vol. 320: Modular Functions of One Variable I. Edited by W. Kuyk. V, 195 pages. 1973.

Vol. 321: Séminaire de Probabilités VII. Edité par P. A. Meyer. VI, 322 pages. 1973.

Vol. 322: Nonlinear Problems in the Physical Sciences and Biology. Edited by I. Stakgold, D. D. Joseph and D. H. Sattinger. VIII, 357 pages. 1973.

Vol. 323: J. L. Lions, Perturbations Singulières dans les Problèmes aux Limites et en Contrôle Optimal. XII, 645 pages. 1973.

Vol. 324: K. Kreith, Oscillation Theory. VI, 109 pages. 1973.

Vol. 325: C.-C. Chou, La Transformation de Fourier Complexe et L'Equation de Convolution. IX, 137 pages. 1973.

Vol. 326: A. Robert, Elliptic Curves. VIII, 264 pages. 1973.

Vol. 327: E. Matlis, One-Dimensional Cohen-Macaulay Rings. XII, 157 pages. 1973.

Vol. 328: J. R. Büchi and D. Siefkes, The Monadic Second Order Theory of All Countable Ordinals. VI, 217 pages. 1973.

Vol. 329: W. Trebels, Multipliers for (C, α)-Bounded Fourier Expansions in Banach Spaces and Approximation Theory. VII, 103 pages. 1973.

Vol. 330: Proceedings of the Second Japan-USSR Symposium on Probability Theory. Edited by G. Maruyama and Yu. V. Prokhorov. VI, 550 pages. 1973.

Vol. 331: Summer School on Topological Vector Spaces. Edited by L. Waelbroeck. VI, 226 pages. 1973.

Vol. 332: Séminaire Pierre Lelong (Analyse) Année 1971–1972. V, 131 pages. 1973.

Vol. 333: Numerische, insbesondere approximationstheoretische Behandlung von Funktionalgleichungen. Herausgegeben von R. Ansorge und W. Törnig. VI, 296 Seiten. 1973.

Vol. 334: F. Schweiger, The Metrical Theory of Jacobi-Perron Algorithm. V, 111 pages. 1973.

Vol. 335: H. Huck, R. Roitzsch, U. Simon, W. Vortisch, R. Walden, B. Wegner und W. Wendland, Beweismethoden der Differentialgeometrie im Grossen. IX, 159 Seiten. 1973.

Vol. 336: L'Analyse Harmonique dans le Domaine Complexe. Edité par E. J. Akutowicz. VIII, 169 pages. 1973.

Vol. 337: Cambridge Summer School in Mathematical Logic. Edited by A. R. D. Mathias and H. Rogers. IX, 660 pages. 1973.

Vol. 338: J. Lindenstrauss and L. Tzafriri, Classical Banach Spaces. IX, 243 pages. 1973.

Vol. 339: G. Kempf, F. Knudsen, D. Mumford and B. Saint-Donat, Toroidal Embeddings I. VIII, 209 pages. 1973.

Vol. 340: Groupes de Monodromie en Géométrie Algébrique. (SGA 7 II). Par P. Deligne et N. Katz. X, 438 pages. 1973.

Vol. 341: Algebraic K-Theory I, Higher K-Theories. Edited by H. Bass. XV, 335 pages. 1973.

continuation on page 489

Lecture Notes in Mathematics

For information about Vols. 1–276, please contact your bookseller or Springer-Verlag.

Vol. 277: Séminaire Banach. Edité par C. Houzel. VII, 229 pages. 1972.

Vol. 278: H. Jacquet, Automorphic Forms on GL(2) Part II. XIII, 142 pages. 1972.

Vol. 279: R. Bott, S. Gitler and I. M. James, Lectures on Algebraic and Differential Topology. V, 174 pages. 1972

Vol. 280: Conference on the Theory of Ordinary and Partial Differential Equations. Edited by W. N. Everitt and B. D. Sleeman. XV, 367 pages. 1972.

Vol. 281: Coherence in Categories. Edited by S. Mac Lane. VII, 235 pages. 1972.

Vol. 282: P. Flaschel und W. Klingenberg, Riemannsche-Hilbertmannigfaltigkeiten. Periodische Geodätische. VII, 211 Seiten. 1972.

Vol. 283: L. Illusie, Complexe Cotangent et Déformations II. VII, 304 pages. 1972.

Vol. 284: P. A. Meyer, Martingales and Stochastic Integrals I. VI, 89 pages. 1972.

Vol. 285: P. de la Harpe, Classical Banach-Lie Algebras and Banach-Lie Groups of Operators in Hilbert Space. III, 160 pages. 1972.

Vol. 286: S. Murakami, On Automorphisms of Siegel Domains. V, 95 pages. 1972.

Vol. 287: Hyperfunctions and Pseudo-Differential Equations. Edited by H. Komatsu. VII, 529 pages. 1973.

Vol. 288: Groupes de Monodromie en Géométrie Algébrique. (SGA 7 I). Dirigé par A. Grothendieck. IX, 523 pages. 1972.

Vol. 289: B. Fuglede, Finely Harmonic Functions. III, 188. 1972.

Vol. 290: D. B. Zagier, Equivariant Pontrjagin Classes and Applications to Orbit Spaces. IX, 130 pages. 1972.

Vol. 291: P. Orlik, Seifert Manifolds. VIII, 155 pages. 1972.

Vol. 292: W. D. Wallis, A. P. Street and J. S. Wallis, Combinatorics: Room Squares, Sum-Free Sets, Hadamard Matrices. V, 508 pages. 1972.

Vol. 293: R. A. De Vore, The Approximation of Continuous Functions by Positive Linear Operators. VIII, 289 pages. 1972.

Vol. 294: Stability of Stochastic Dynamical Systems. Edited by R. F. Curtain. IX, 332 pages. 1972.

Vol. 295: C. Dellacherie, Ensembles Analytiques Capacités. Mesures de Hausdorff. XII, 123 pages. 1972.

Vol. 296: Probability and Information Theory II. Edited by M. Behara, K. Krickeberg and J. Wolfowitz. V, 223 pages. 1973.

Vol. 297: J. Garnett, Analytic Capacity and Measure. IV, 138 pages. 1972.

Vol. 298: Proceedings of the Second Conference on Compact Transformation Groups. Part 1. XIII, 453 pages. 1972.

Vol. 299: Proceedings of the Second Conference on Compact Transformation Groups. Part 2. XIV, 327 pages. 1972.

Vol. 300: P. Eymard, Moyennes Invariantes et Représentations Unitaires. II, 113 pages. 1972.

Vol. 301: F. Pittnauer, Vorlesungen über asymptotische Reihen. VI, 186 Seiten. 1972.

Vol. 302: M. Demazure, Lectures on p-Divisible Groups. V, 98 pages. 1972.

Vol. 303: Graph Theory and Applications. Edited by Y. Alavi, D. R. Lick and A. T. White. IX, 329 pages. 1972.

Vol. 304: A. K. Bousfield and D. M. Kan, Homotopy Limits, Completions and Localizations. V, 348 pages. 1972.

Vol. 305: Théorie des Topos et Cohomologie Etale des Schémas. Tome 3. (SGA 4). Dirigé par M. Artin, A. Grothendieck et J. L. Verdier. VI, 640 pages. 1973.

Vol. 306: H. Luckhardt, Extensional Gödel Functional Interpretation. VI, 161 pages. 1973.

Vol. 307: J. L. Bretagnolle, S. D. Chatterji et P.-A. Meyer, Ecole d'Eté de Probabilités: Processus Stochastiques. VI, 198 pages. 1973.

Vol. 308: D. Knutson, λ-Rings and the Representation Theory of the Symmetric Group. IV, 203 pages. 1973.

Vol. 309: D. H. Sattinger, Topics in Stability and Bifurcation Theory. VI, 190 pages. 1973.

Vol. 310: B. Iversen, Generic Local Structure of the Morphisms in Commutative Algebra. IV, 108 pages. 1973.

Vol. 311: Conference on Commutative Algebra. Edited by J. W. Brewer and E. A. Rutter. VII, 251 pages. 1973.

Vol. 312: Symposium on Ordinary Differential Equations. Edited by W. A. Harris, Jr. and Y. Sibuya. VIII, 204 pages. 1973.

Vol. 313: K. Jörgens and J. Weidmann, Spectral Properties of Hamiltonian Operators. III, 140 pages. 1973.

Vol. 314: M. Deuring, Lectures on the Theory of Algebraic Functions of One Variable. VI, 151 pages. 1973.

Vol. 315: K. Bichteler, Integration Theory (with Special Attention to Vector Measures). VI, 357 pages. 1973.

Vol. 316: Symposium on Non-Well-Posed Problems and Logarithmic Convexity. Edited by R. J. Knops. V, 176 pages. 1973.

Vol. 317: Séminaire Bourbaki – vol. 1971/72. Exposés 400–417. IV, 361 pages. 1973.

Vol. 318: Recent Advances in Topological Dynamics. Edited by A. Beck. VIII, 285 pages. 1973.

Vol. 319: Conference on Group Theory. Edited by R. W. Gatterdam and K. W. Weston. V, 188 pages. 1973.

Vol. 320: Modular Functions of One Variable I. Edited by W. Kuyk. V, 195 pages. 1973.

Vol. 321: Séminaire de Probabilités VII. Edité par P. A. Meyer. VI, 322 pages. 1973.

Vol. 322: Nonlinear Problems in the Physical Sciences and Biology. Edited by I. Stakgold, D. D. Joseph and D. H. Sattinger. VIII, 357 pages. 1973.

Vol. 323: J. L. Lions, Perturbations Singulières dans les Problèmes aux Limites et en Contrôle Optimal. XII, 645 pages. 1973.

Vol. 324: K. Kreith, Oscillation Theory. VI, 109 pages. 1973.

Vol. 325: C.-C. Chou, La Transformation de Fourier Complexe et L'Equation de Convolution. IX, 137 pages. 1973.

Vol. 326: A. Robert, Elliptic Curves. VIII, 264 pages. 1973.

Vol. 327: E. Matlis, One-Dimensional Cohen-Macaulay Rings. XII, 157 pages. 1973.

Vol. 328: J. R. Büchi and D. Siefkes, The Monadic Second Order Theory of All Countable Ordinals. VI, 217 pages. 1973.

Vol. 329: W. Trebels, Multipliers for (C, α)-Bounded Fourier Expansions in Banach Spaces and Approximation Theory. VII, 103 pages. 1973.

Vol. 330: Proceedings of the Second Japan-USSR Symposium on Probability Theory. Edited by G. Maruyama and Yu. V. Prokhorov. VI, 550 pages. 1973.

Vol. 331: Summer School on Topological Vector Spaces. Edited by L. Waelbroeck. VI, 226 pages. 1973.

Vol. 332: Séminaire Pierre Lelong (Analyse) Année 1971-1972. V, 131 pages. 1973.

Vol. 333: Numerische, insbesondere approximationstheoretische Behandlung von Funktionalgleichungen. Herausgegeben von R. Ansorge und W. Törnig. VI, 296 Seiten. 1973.

Vol. 334: F. Schweiger, The Metrical Theory of Jacobi-Perron Algorithm. V, 111 pages. 1973.

Vol. 335: H. Huck, R. Roitzsch, U. Simon, W. Vortisch, R. Walden, B. Wegner und W. Wendland, Beweismethoden der Differentialgeometrie im Großen. II, 166 Seiten. 1973.

Vol. 336: L'Analyse Harmonique dans le Domaine Complexe. Edité par E. J. Akutowicz. VIII, 169 pages. 1973.

Vol. 337: Cambridge Summer School in Mathematical Logic. Edited by A. R. D. Mathias and H. Rogers. IX, 660 pages. 1973.

Vol: 338: J. Lindenstrauss and L. Tzafriri, Classical Banach Spaces. IX, 243 pages. 1973.

Vol. 339: G. Kempf, F. Knudsen, D. Mumford and B. Saint-Donat, Toroidal Embeddings I. VIII, 209 pages. 1973.

Vol. 340: Groupes de Monodromie en Géométrie Algébrique. (SGA 7 II). Par P. Deligne et N. Katz. X, 438 pages. 1973.

Vol. 341: Algebraic K-Theory I, Higher K-Theories. Edited by H. Bass. XV, 335 pages. 1973.

continuation on page 469

Lecture Notes in Mathematics

Edited by A. Dold and B. Eckmann

498

Model Theory and Algebra

A Memorial Tribute to Abraham Robinson

Edited by D. H. Saracino and V. B. Weispfenning

Springer-Verlag
Berlin · Heidelberg · New York 1975

Editors

Dr. Daniel H. Saracino
Department of Mathematics
Colgate University
Hamilton, New York 13346
USA

Dr. Volker B. Weispfenning
Mathematisches Institut
der Universität Heidelberg
Im Neuenheimer Feld 288
69 Heidelberg/BRD

Library of Congress Cataloging in Publication Data

Main entry under title:

Model theory and algebra.

(Lecture notes in mathematics ; 498)
"Bibliography of Robinson's works": p.
Includes index.
CONTENTS: Biography of Abraham Robinson.--Robinson, A. Algorithms in algebra.--Barwise, J. and Schlipf, J. On recursively saturated models of arithmetic. [etc.]
1. Model theory--Addresses, essays, lectures. 2. Algebra--Addresses, essays, lectures. 3. Robinson, Abraham, 1918-1974. I. Robinson, Abraham, 1918-1974. II. Weispfenning, V., 1944- III. Saracino, D., 1947- IV. Series: Lecture notes in mathematics (Berlin) ; 498.
QA3.L28 no. 498 [QA9.7] 510'.8s [511'.8] 75-40483

AMS Subject Classifications (1970): 01A70, 02B25, 02E10, 02F50, 02H05, 02H13, 02H15, 02H20, 02H25, 10N15, 12A20, 12D15, 12E05, 12E05, 12J15, 12L10, 12L15, 13A15, 13B20, 13B25, 13L05, 14H99, 16A40, 18A25, 20A10, 20E05, 20K10

ISBN 978-3-540-07538-7 Springer-Verlag New York · Heidelberg · Berlin

Abraham Robinson was "the one mathematical logician who accomplished incomparably more than anybody else in making this science fruitful for mathematics. I am sure his name will be remembered by mathematicians for centuries."

-- Kurt Gödel

ABRAHAM ROBINSON

October 6, 1918 - April 11, 1974

Foreword

The sudden fatal illness of Abraham Robinson came as a great shock to many people around the world. For Robinson was more than an excellent mathematician. He was also a person whom one came very quickly to like very much.

It was a wonderful thing to find in one person the combination of Abraham Robinson - cofounder of model theory and inventor of nonstandard analysis - on the one hand, and "Abby" - warm and humane human being - on the other. What a pleasure it was to have him stop by one's office in the morning and ask if one could spare the time for a walk to Naples Pizza for a cup of coffee. And on the way back one would be almost hesitantly asked if one could spare the time for a detour to the newsstand so he could pick up his New York Times.

Those swift sad months of November 1973 - April 1974 were for those at Yale tinged with a sense of unreality. He was gone before anyone could come to grips with what was happening. We sought a way of expressing our respect and our sense of personal loss. This volume was the best way we knew.

Perhaps a word is in order about the deliberately limited scope of the book. Surely many more people than those represented here would want to contribute to a collection in Robinson's honor. To keep a volume of reasonable size we restricted the contents to papers in "model theory and algebra", a subject with which he was deeply involved for most of his career. Furthermore, in attempting to create a personal tribute we sought papers primarily from young people who had worked with him in this area. Particularly noticeable is the omission of papers in nonstandard analysis. This omission has already been partly compensated for by some of the papers presented at the Robinson memorial conference held at Yale in May, 1975. The proceedings of this conference will appear separately as a special issue of the Israel Journal of Mathematics.

We would like to express our gratitude to Mrs. Renée Robinson for providing us with the photograph at the beginning of the book and for giving us permission to publish a version of Robinson's last paper. We also wish to thank Professor Kurt Gödel for allowing us to include the quotation on page v.

While we were in the very early stages of planning this volume, Professor G.H. Müller of Heidelberg suggested that the Springer Lecture Notes series might provide an appropriate format. We wish to thank him for arranging the publication of the book with Springer-Verlag, and to thank Springer for providing us with secretarial assistance. We are also grateful for the characteristic swiftness with which the manuscript was published.

Heidelberg, August 1975 D.S. and V.W.

TABLE OF CONTENTS

BIOGRAPHY

Abraham Robinson was born on October 6, 1918, in Waldenburg, Germany. He spent his boyhood in Germany and Palestine (now Israel) and graduated from the Jerusalem Gymnasium (Grammar School) in 1936. He was a student at the Hebrew University, Jerusalem from 1936 to 1943, including a term at the Sorbonne, Paris. He obtained the degree of M.Sc. from the Hebrew University in 1946 and the degrees of Ph.D. and D.Sc. from the University of London in 1949 and 1957 respectively. During the second world war he served in the Free French Air Force and later became a Scientific Officer at the Royal Aircraft Establishment, Farmborough, England. From 1946 to 1951 he was Senior Lecturer in Mathematics, later Deputy Head of the Department of Aerodynamics at the College of Aeronautics, Cranfield, England. Subsequently he was Associate Professor, then Professor, of Applied Mathematics at the University of Toronto, Canada (1951-1957), Professor of Mathematics and sometime chairman of the department of Mathematics, at the Hebrew University, Jerusalem, Israel (1957-1962), Professor of Mathematics and Philosophy at the University of California, Los Angeles (1962-1967), and Professor of Mathematics at Yale University 1967 - 1974 (Sterling Professor of Mathematics since 1971). He was at various times a Visiting Professor at the Universities of Princeton, Paris, Rome, Tübingen, Heidelberg, at the California Institute of Technology, and at the Weizmann Institute, Rehovoth, and a Visiting Fellow at St. Catherine's College, Oxford. His activities also included membership of the Fluid Notion Committee of the Aeronautical Research Council of Great Britain. In 1972 he was elected a Fellow of the American Academy of Arts and Sciences and in 1973 he received the Brouwer Medal from the Dutch Mathematical Society. He published nine books and over one hundred papers in Pure and Applied Mathematics. In 1944 Robinson married Renée Kopel of Vienna, Austria.

Robinson worked in widely separated areas of science. However, the common denominator to much of his research was his interest in applications. He was always fascinated by the problem of fashioning or refashioning a formal framework in order to fit a given problem, whether in Physics or in Pure Mathematics. Within classical Applied Mathematics, he was concerned chiefly with Fluid Mechanics, more particularly with the determination of the pressures and forces that act on a body in flight, under steady or unsteady conditions, from subsonic to supersonic speeds (ref. 3)

Some of his better known contributions in this area were concerned with delta wings and related shapes, while other papers dealt with the motion of small bodies in a viscous fluid and with the propagation of disturbances in fluids and solids. One of these led to an early example of a precise theory for a mixed boundary value problem for hyperbolic differential equations.

However, Robinson's major èffort went into the study of the relations between Logic and Mathematics proper. In his Ph.D. dissertation "On the Metamathematics of Algebraic Systems" 1949 (published in 1951, ref. 1) he helped to lay the foundations of the branch of Logic now known as Model Theory. He discussed, generally and in special cases, the mutual relationship between sets of axioms and the classes of structures (models) which satisfy them. The dissertation also contains a number of effective applications to Algebra. Among them is the theorem that an assertion X of the first order predicate calculus which is true in all commutative fields of characteristic zero is true also in all fields of characteristic $p > p_o$ where the natural number p_o depends on X. Among the basic tools which were introduced in the same work is the "method of diagrams".

In 1954 Robinson produced a widely applicable test (the model completeness test) for proving the completeness of various algebraic theories (a theory is complete if for any sentence in its vocabulary it contains either that sentence or its negation). An outgrowth of the line of thought which led to this test was the introduction of concepts which provide far-reaching generalizations of the notion of an algebraically closed field relative to the class of commutative fields and which embrace both previously known concepts, such as real closed fields and new concepts such as differentially closed fields. Beginning in 1969, Robinson introduced further generalizations of these notions by using the forcing methods introduced originally by Paul Cohen in Set Theory. By these means, Robinson was able to establish that even in Arithmetic one can introduce structures which are analogous to algebraically closed fields, and that the theory of these structures is complete (or, which ic the same, such that any two of these structures are elementarily equivalent). In another direction, Robinson showed that the compactness principle of the lower predicate calculus implies the existence of certain numerical bounds, in particular for the representability of positive definite functions as sums of squares (1955) where it had not been known previously, as well as in other cases, some known, some unknown.

Perhaps Robinson's best known contribution is Nonstandard Analysis (ref. 7). This area, which was introduced by him from 196o on, makes use of model theoretic notions and contributions in order to provide for the first time a satisfactory solution to the ancient problem of developing the Differential and Integral Calculus by means of infinitesimals. It turned out that the ideas which led to Nonstandard Analysis can be generalized so as to apply also to topological spaces and many other areas of Mathematics. The method has been used successfully e.g. for the solution of problems in Functional Analysis and in Complex Variable Theory and, more recently, in Mathematical Economics. It is, in many cases, an alternative to familiar classical methods, but it is too early to say how many mathematicians will choose to use it in their field.

As a logician, Robinson was also keenly interested in the Philosophy of Mathematics, although he published only a few papers in this area. He was opposed to the so-called "Platonic realism" which holds that mathematical objects and structures, even infinite ones, lead an independent existence which defines their properties uniquely in all cases.

-- Adapted from the official biography
published by the Yale News Bureau.

A Provisional Bibliography of Robinson's Works

The following is not an attempt at a definitive cataloguing of Robinson's publications, but rather a listing of the writings we were able to find records of in his collection of reprints and preprints. In some cases we have been unable to ascertain where (or even whether) the paper in question was ever published.

Books

1. On the Metamathematics of Algebra, North Holland Publ. Co., Amsterdam 1951
2. Théorie metamathématique des idéaux, Gauthier Villars, Paris, 1955
3. Wing Theory (with J.A.Lanzmann), Cambridge University Press, 1956
4. Complete Theories, North Holland Publ. Co., Amsterdam 1956
5. Introduction to Model Theory and to the Metamathematics of Algebra, North Holland Publ. Co., Amsterdam, 1963. Translated into Russian 1967; into Italian 1974. New Edition 1975
6. Numbers and Ideals, Holden-Day, 1965
7. Nonstandard Analysis, North Holland Publ. Co., Amsterdam, 1966 New Edition 1975
8. Contributions to Nonstandard Analysis (co-editor with W.A.J.Luxembourg), North Holland Publ. Co., Amsterdam 1972
9. Nonarchimedean Fields and Asymptotic Expansions (with A.H.Lightstone), North Holland Publ. Co., Amsterdam 1975

Papers

1. On the independence of the axiom of definiteness, Journal of Symbolic Logic 4 (1939), 69-72.
2. On nil ideals in general rings, 1939
3. On a certain variation of the distributive law for a commutative algebraic field, Proc. Royal Society of Edinburgh (A) 61 (1941), 92-1o1

4. Note on the interpretation of V-g recorders, (with S.V.Fagg and P.E.Montagnon), Reports and Memoranda of the Aeronautical Research Council of Great Britain No. 2o97, 1945.

5. The aerodynamic loading of wings with endplates, Reports and Memoranda of the Aeronautical Research Council of Great Britain No. 2342, 1945/195o.

6. Shock transmission in beams, Reports and Memoranda of the Aeronautical Research Council of Great Britain No. 2265, 1945/195o.

7. The wave drag of diamond-shaped aerofoils at zero incidence, Reports and Memoranda of the Aeronautical Research Council of Great Britian No. 2394, 1946/195o.

8. Aerofoil theory of a flat delta wing at supersonic speeds, Reports and Memoranda of the Aeronautical Research Council of Great Britain No. 2548, 1946/1952.

9. Flutter derivatives of a wing-tailplane combination (date uncertain

1o. The characterization of algebraic plane curves, (with T.S.Motzkin), Duke Math. Journal 14 (1947), 837-853.

11. Note on the application of the linearised theory for compressible flow to transonic speeds, (with A.D.Young), Reports and Memoranda of the Aeronautical Research Council of Great Britian No. 2399, 1947/1951.

12. The effect of the sweepback of delta wings on the performance of an aircraft at supersonic speeds, (with F.T.Davies), Reports and Memoranda of the Aeronautical Research Council of Great Britain No. 2476, 1947/1951.

13. Interference on a wing due to a body at supersonic speeds, (with S.Kirkby), Reports and Memoranda of the Aeronautical Research Council No. 25oo, 1947/1952.

14. Bound and trailing vortices in the theory of supersonic flow and the downwash in the wake of a delta wing, (with J.H.Hunter-Tod), Reports and Memoranda of the Aeronautical Research Council of Great Britain No. 24o9, 1948/1952.

15. On some problems of unsteady supersonic aerofoil theory, Proc. International Congress of Applied Mechanics 2 (1948), 5oo-514.

16. On source and vortex distribution in the linearized theory of steady supersonic flow, Quarterly Journal of Mechanics and Applied Mathematics 1 (1948), 4o8-432.

17. Rotary derivatives of delta wings at supersonic speeds, Journal of the Royal Aeronautical Society 52 (1948), 735-752.

18. The aerodynamic derivatives with respect to sideslip for a delta wing with small dihedral at zero incidence at supersonic speeds, (with J.H.Hunter-Tod), Reports and Memoranda of the Aeronautical Research Council of Great Britain No. 241o, 1948/1952.

19. On the metamathematics of algebraic systems, 1948

2o. On non-associative systems, Proc. Edinburgh Math. Soc. (2) 8 (1949), 111-118.

21. Numerical solution of integral equations (with S.Kirkby), 1949

22. On the integration of hyperbolic differential equations, Journal London Math. Soc. 25 (195o), 2o9-217.

23. On functional transformations and summability, Proc. London Math. Soc. (2) 52 (195o), 132-16o.

24. Les rapports entre le calcul déductif et l'interprétation sémantique, d'un système axiomatique, Colloque International de CNRS, Paris, 36 (195o), 35-52.

25. On the application of symbolic logic to algebra, Proc. International Congress of Mathematicians, Cambridge, Mass. 1 (195o/1952), 686-694.

26. Wave reflection near a wall, Proc. Cambridge Phil. Soc. 47 (1951), 528-544.

27. On axiomatic systems which possess finite models, Methodos (1951), 14o-149.

28. Aerofoil theory for swallowtail wings of small aspect ratio, Aero. Quarterly 4 (1952), 69-82.

29. L'application de la logique formelle aux mathématiques, Colloque international de logique mathématique, Paris, (1952), 51-64.

3o. Non-uniform supersonic flow, Quarterly Appl. Math. 1o (1953), 3o9-319.

31. Flow around compound lifting units, Symposium on High Speed Aerodynamics, Ottawa, Canada, (1953), 26-29.

32. Core-consistency and total inclusion for methods of summability, (with G.G.Lorentz), Canadian Journal of Mathematics 6 (1954), 27-34.

33. On some problems of unsteady wing theory, Second Canadian Symposium on Aerodynamics, Toronto, Canada, (1954), 1o6-122.

34. On predicates in algebraically closed fields, Journal of Symbolic Logic 19 (1954), 1o3-114.

35. Note on an embedding theorem for algebraic systems, J.London Mathematical Soc. 3o (1955), 249-252.

36. Mixed problems for hyperbolic partial differential equations, (with L.L.Campbell), Proc. London Math. Soc. 5 (1955), 129-147.

37. Metamathematical considerations on the relative irreducibility of polynomials, (with P.Gilmore), Canadian Journal of Mathematics 7 (1955), 483-489.

38. On ordered fields and definite functions, Mathematische Annalen 13o (1955), 257-271.

39. Further remarks on ordered fields and definite functions, Mathematische Annalen 13o (1956), 4o5-4o9.

4o. A result on consistency and its application to the theory of definition, Proc. Royal Dutch Academy of Sciences, Amsterdam, (A) 59 (1956), 47-58.

41. Note on a problem of L.Henkin, Journal of Symbolic Logic 21 (1956), 33-35.

42. Ordered structures and related concepts, Mathematical Interpretation of Formal Systems, Amsterdam, (1955), 51-56.

43. Completeness and persistence in the theory of models, Zeitschr. f. math. Logik und Grundlagen d. Math. 2 (1956), 15-26.

44. Solution of a problem by Erdös-Gillman-Henriksen, Proc. Amer. Math. Soc. 7 (1956), 9o8-9o9.

45. On the motion of small particles in a potential field of flow, Comm. Pure and Appl. Math. 9 (1956), 69-84.

46. Ordered structures and related concepts, in Mathematical Interpretation of Formal Systems, North Holland Publ. Co., 1955.

47. Wave propagation in a heterogeneous elastic medium, Journal of Math. and Physics 36 (1957), 21o-222.

48. Transient stresses in beams of variable characteristics,Quarterly Journal of Mech. and Appl. Math. 1o (1957), 148-159.

49. Some problems of definability in the lower predicate calculus, Fundamenta Mathematicae 44 (1957), 3o9-329.

5o. Syntactical transforms, (with A.H.Lightstone), Trans.Amer.Math. Soc. 86 (1957), 22o-245.

51. On the representation of Herbrand functions in algebraically closed fields, (with A.H.Lightstone), J. Symbolic Logic 22 (1957), 187-2o4.

52. Aperçu metamathématique sur les nombres réels, 1957.

53. Relative model-completeness and the elimination of quantifiers, Dialectica 12 (1958), 394-4o7.

54. Relative model-completeness and the elimination of quantifiers, Proceedings of the Cornell meeting, 1958.

55. Outline of an introduction to mathematical logic, Can. Math. Bull. 1 (1958), 41-54, 113-127, 193-2o8.

56. Proving a theorem (as done by man, logician, or machine), Proceedings of the Cornell meeting, 1958.

57. Applications to field theory, Proceedings of the Cornell meeting, 1958.

58. Outline of an introduction to mathematical logic, Can. Math. Bull. 2 (1959), 33-42.

59. Solution of a problem of Tarski, Fundamenta Mathematicae 47 (1959), 179-2o4.

6o. On the concept of a differentially closed field, Bull. Research Council of Israel (F) 8 (1959), 113-128.

61. Obstructions to arithmetical extension and the theorem of Los and Suszko, Proc. Royal Dutch Acadamy Sci., Amsterdam, (A) 62 (1959), 489-495.

62. Algèbre différentielle à valeurs locales, Dagli Atti del VI Congresso dell 'Unione Matématica Italiana, Napoli, 1959.

63. Model theory and non-standard arithmetic, Proc. IMU Symposium on Foundations of Math. Warsaw 1959, (1961), 265-3o2.

64. Local differential algebra, Trans. Amer. Math. Soc. 97 (196o), 427-456.

65. On the mechanization of the theory of equations, Bull. Research Council of Israel (F) 9 (196o), 47-7o.

66. Elementary properties of ordered Abelian groups, (with E.Zakon), Trans. Amer. Math. Soc. 96 (196o), 222-236

67. Recent developments in model theory, Logic, Methodology and Philosophy of Science, Proc. 196o International Congress, Stanford Univ. Press (196o/1962).

68. Local differential algebra - the analytic case, 196o.

69. On the construction of models, Essays on the Foundations of Math. (Fraenkel anniversary volume), Hebrew Univ. Press, Jerusalem (1961), 2o7-217.

7o. On the D-calculus for linear differential equations with constant coefficients, Math. Gazette 45 (1961), 2o2-2o6.

71. Non-standard analysis, Proc. Royal Dutch Academy of Sci. (A) 64 (1961), 432-44o.

72. A note on embedding problems, Fundamenta Mathematicae 5o (1962), 455-461.

73. Aerofoil theory, Chapter 72 in McGraw-Hill Handbook of Engineering Mechanics, ed. W.Flugge, New York, 1962.

74. Modern mathematics and the secondary schools, International Review of Education 8 (1963), 34-4o.

75. A basis for the mechanization of the theory of equations, Computer Programming and Formal Systems, Amsterdam (1962), 95-99.

76. Complex function theory over nonarchimedean fields, Technical-Scientific note No. 3o, USAF contract No. 61 (o52)-187, Jerusalem.

77. On the mechanization of the theory of numbers (with M.Machover), U.S. Office of Naval Research, Information Systems Brench, Jerusalem, Israel 1962.

78. Local partial differential algebra, Transactions of the American Mathematical Society, (with S.Halfin) 1o9, (1963), 165-18o.

79. On languages which are based on non-standard arithmetic, Nagoya Mathematical Journal, 22 (1963), 83-117.

8o. On symmetric bimatrix games (with Griesmer and Hoffman), IBM research paper, 1963.

81. Some remarks on threshold functions, Watson Research Center, 1963.

82. On generalized limits and linear functionals, Pacific Journal of Mathematics, 14 (1964), 263-283.

83. Random-Access Stored-Program Machines, an Approach to Programming Languages (with C.C.Elgot) Journal of the Association for Computing Machinery, 11 (1964), 365-399.

84. Between Logic and Mathematics, I.C.S.U. Review, 6 (1964), 218-226.

85. Formalism 64, Proceedings of the 1964 International Congress for Logic, Methodology and Philosophy of Science, 228-246.

86. On the theory of normal families, Acta Philosophica Fennica fasc. 18 (1965) 159-184 (Nevanlinna anniversary volume).

87. Mathematical logic and mechanical mathematics, Interdisciplinary Colloquium on mathematics in the behavioral sciences, Colloquium Documents, Western Management Science Institute, U.C.L.A., 1966.

88. Solution of an invariant subspace problem of K.T.Smith and P.R. Halmos (with A.R.Bernstein) Pacific Journal of Mathematics, 16 (1966), 421-431.

89. Topics in non-archimedean mathematics, The Theory of Models, North Holland Publ. Co., Amsterdam, (1965), 285-298.

9o. A new approach to the theory of algebraic numbers, Rendiconti della Academia Nazionali dei Lincei, ser. 8, vol. XI (1966), 222-225.

91. A new approach to the theory of algebraic numbers, II. Rendiconti della Academia Nazionali dei Lincei, ser. 8, vol. XI, 1966, pp 77o-774.

92. Non-standard theory of Dedekind rings. Proc. of the Koninklyke Nederlandse Akad. van Wetenschappen, Series A., 7o (1967), 444-453. Appeared also in Indagationes Mathematicae 29, 444-453.

93. Nonstandard arithmetic. Invited address. Bull. Amer. Math. Soc. 73 (1967), 818-843.

94. On some applications of model theory to algebra and analysis. Rendiconti di Matematica 25 (1967), 1-31.

95. (with C.C.Elgot and J.D.Rutledge) Multiple control computer models. Systems and Computer Science, Univ. of Toronto Press (1967), 6o-76. (Also was IBM research paper RC-1622, 1966).

96. On the metaphysics of calculus, Problems in the Philosophy of Mathematics, D.Lakatos, ed., North Holland, Amsterdam, 1967, 28-4o.

97. Model Theory in Contemporary Philosophy; a Survey. Vol. I: Logic and the Foundations of Mathematics, ed. Klibansky, Firenze, La Nuova Italia Editrice (1968), 61-73.

98. Some thoughts on the history of mathematics. Compositio Mathematica 2o (1968), 188-193. Appeared also in book form as Logic and Foundations of Mathematics, dedicated to A.Heyting on his 7oth birthday. Groningen, Walters-Noordhoff 1968.

99. (with A.E.Hurd) On flexural wave propagation in nonhomogeneous elastic plates. SIAM Journal of Applied Mathematics, 16 (1968), 1o81-1o89.

1oo. Topics in Nonstandard Algebraic Number Theory. Applications of Model Theory to Algebra, Analysis and Probability. Holt, Rinehart and Winston, New York (1969), 1-17.

1o1. (With Elias Zakon) A set-theoretical characterization of enlargements. Applications of Model Theory to Algebra, Analyis and Probability. Holt, Rinehart and Winston, New York (1969), 1o9-122.

1o2. Germs. Applications of Model Theory to Algebra, Analysis and Probability. Holt, Rinehart and Winston, New York (1969), 138-149.

1o3. Problems and methods of model theory. Centro Internazionale Matematico Estivo (C.I.M.E.)(1968), 183-266.

1o4. Elementary embeddings of fields of power series. Journal of Number Theory 2 (197o), 237-247.

1o5. (with Jon Barwise) Completing theories by forcing. Annals of Math. Logic 2 (197o), 119-142.

1o6. From a formalist's point of view, Dialectica, vol. 23, 197o, 45-49.

1o7. Compactification of groups and rings and nonstandard analysis. J. Symbolic Logic 34 (1969), 576-588.

1o8. Forcing in model theory. Proceedings of Symposia Mathematica, Instituto Nazionale di Alta Matematica, Vol. 5 (1971), 69-8o.

1o9. Non Standard Analyis. A filmed lecture presented by the Mathematical Association of America 197o. About one hour.

11o. Infinite forcing in model theory. Proceedings of the Second Scandinavian Logic Symposium. Studies in Logic and the Foundations of Mathematics Vol. 63 (1971), North-Holland Publishing Co. 317-34o.

111. Forcing in model theory. Proceedings of the International Congress of Mathematicians. Nice 197o (1971), 245-25o.

112. On the notion of algebraic closedness for non-commutative groups and fields. Journal of Symbolic Logic 36 (1971), 441-444.

113. Applications of logic to pure mathematics, 1971 (?).

114. Algebraic function fields and nonstandard arithmetic. Contributions to Nonstandard Analysis, Studies in Logic, North-Holland Publishing Company 69 (1972), 1-14.

115. (with E.Fisher) Inductive theories and their forcing companions. Israel J. Math. 12 (1972), 95-1o7.

116. (with Peter Kelemen) The nonstandard $\lambda:\emptyset_2^4(x)$ model. I,II.J.Math. Physics 13 (1972), 187o-1878.

117. Function theory on some non-archimedean fields. Amer. Math. Monthly 8o (1972), 87-1o9.

118. (with Donald J. Brown) A limit theorem in the cores of large standard exchange economies. Proc. of the National Academy of Sciences 69 (1972), 1258-126o.

119. On the real closure of a Hardy field. Theory of Sets in Topology. Hausdorff Memorial volume, Berlin (1972), 427-433.

12o. On bounds in the theory of polynomial ideals. Selected Questions of Algebra and Logic. Mal'cev Memorial Volume. Novosibirsk 1973, 245-252.

121. Ordered differential fields. J.Comb. Theory 14 (1973), 324-333.

122. Nonstandard points on algebraic curves. J. Number Theory 5 (1973), 3o1-327.

123. Metamathematical problems. J. Symbolic Logic 38 (1973), 5oo-516.

124. Numbers - What are they and what are they good for? Yale Scientific, May 1973, 14-16.

125. Standard and nonstandard number systems, Nieuw Archief voor Wiskunde, (3), XXI, 1973, 115-133.

126. Nonstandard arithmetic and generic arithmetic, Logic, Methodology, and Philosophy of Science IV, North Holland Publ. Co., 1973, 137-154.

127. A note on topological model theory. (Dedicated to A.Mostowski on his 6oth birthday.)Fund.Math. LXXXI (1974), 159-171.

128. Enlarged Sheaves, Victoria Symposium on Nonstandard Analysis (1972), Springer Lecture Notes number 369 (1974), 249-26o.

129. Generic Categories. (Presented at Logic Symposium in Orleans, France 9/1972).

13o. The Cores of Large Standard Exchange Economies (with D.J.Brown), to appear in Journal of Economic Theory.

131. Nonstandard Exchange Economies (with D.J.Brown), Econometrica

132. On Constrained Denotation, Russell Symposium, Duke University

133. A Decision method for elementary algebra and geometry, revisited; to appear

134. Concerning progress in the philosophy of mathematics, Proceedings of the Logic Colloquium at Bristol, 1973, H.E.Rose and J.C.Shepherdson, eds., North Holland Publ. Co.

135. Algorithms in Algebra (edited), this volume.

ALGORITHMS IN ALGEBRA

Abraham Robinson

Editors' Preface:

Robinson planned to deliver a lecture on the topic of "algorithms in algebra" at the meeting on algebra and logic held at Monash University (Australia) in January - February 1974. However, his illness made it impossible for him to attend the meeting; it also prevented him from ever finishing the paper. What he left was a preliminary manuscript.

It seemed clear that the paper was worth publishing in some form, and we attempted to produce a revised version adhering as closely as possible to Robinson's own plan and ideas. The most substantial changes were made in sections 5 and 8; in the former case we benefitted from a set of notes from a lecture Robinson gave on the subject at Yale.

It goes without saying that in publishing this manuscript we accept responsibility for any errors.

1. Introduction.

The notion of a computable function or relation in the domain of natural numbers is by now standard, and the fact that it is explicated correctly by the notion of recursivity (Church's thesis) is no longer open to doubt. Even so it is an intriguing philosophical problem to what category exactly this notion belongs (e.g., depending on one's school of thought, analytic, synthetic a priori, theoretical, empirical). Let me begin this talk by drawing your attention to the fact that the notion of computability in algebra is less clear, and that here it is even not obvious whether we are aiming at the explication of an objectively given notion, or at the description of the various activities of a number of individuals which they considered to be "effective" or "realizable in a finite nymber of steps".

A major figure in the history of effective methods in algebra was Kronecker. Among other things, he proposed [7] a method by which the reducibility of a polynomial of one variable with rational coefficients can be tested, and another by which the reducibility of polynomials of several variables is reduced to the reducibility of polynomials in a single variable. In a more advanced area he showed how to determine, effectively, the irreducible components of an algebraic variety (see below). However, the formal tools available in Kronecker's time precluded a precise determination of the notion of effectiveness, even if he had been disposed philosophically to embark on such an enterprise.

Kronecker was not the first mathematician to employ effective methods. He was the first to do so consciously, because until about 185o mathematicians were not even aware of a possible distinction between abstract and effective mathematics. Thus, the determination of the number of real roots of a polynomial in a given interval, which antedates that period (Sturm) is a beautiful example of an effective method, and, as you know, it is the basis of Tarski's decision method for the algebraic theory of real numbers. But even earlier (as in the first book which contains the term in its title) "Algebra" was regarded as a practical method of computation, and the name of the author of that book (al-Khowarizmi) was immortalized in the very word "algorithm".

2. Previous work.

The first paper that discusses the notion of effectiveness in algebra is that of Shepherdson and Frölich [15]. They reduce the notion of effectiveness in algebra to the corresponding notion in arithmetic by assuming that the structure within which a certain problem is to be solved effectively is given recursively, i.e. that it is, or is represented by, recursive functions. Among the questions treated by Shepherdson-Frölich is that of the existence of an effective method to test the reducibility of an equation. They conclude, following an earlier argument of van der Waerden, that there is no general decision procedure for this problem (uniformly applicable to all fields). More precisely, they "construct" (in the indicated sense) a particular field which has no decision procedure for this problem.

It may be argued that an effective procedure in an algebraic structure should be independent of the recursive nature of the structure, more generally it should be equally applicable to an algebraic structure (of a given type) which is not countable. In this vein, there exist several papers (Fraissé [3], Peter [9], Lambert [8]). Lambert introduces a kind of mixed recursive schemes which involve both elements of a given algebraic structure and natural numbers. The motivation for this is that even in an arbitrary algebraic structure, e.g. a group, which does not involve natural numbers a priori, they may intervene as soon as we try to introduce the powers of an element, by definition, as a function of two variables, a in the structure and n in the natural numbers. A theory of inductive definitions which has some similarities with Lambert's has been developed more recently by Moschovakis and others.

A theory of algebraic algorithms which is closer in spirit to contemporary model theory on the one hand and to computer programming on the other, has been developed in recent years by Erwin Engeler [2]. Let M be a model-theoretic structure. Engeler considers programs consisting of commands of the following kind. (i) operational instructions: do ψ then go to j (where ψ is an operation and j is the label of another instruction), and (ii) : if ϕ then go to j, else go to k, where ϕ is a statement concerning the structure whose truth or falsehood is revealed by oracle (i.e., is supposed known, for the purpose of carrying out the program). The operations are of the form $x_j := g_k(x_{j_1},\dots,x_{j_k})$, e.g. $x_j := x_{j+1}$, which implies that we may delete the content of cell x_j and replace it by the content of cell x_{j+1}. Engeler associates a formula of an infinitary language with each program, so that it holds in a structure iff the program is effective for it (terminates).

Finally, it is appropriate to mention here a notion introduced by Paul J. Cohen in connection with his work on decision procedures [1]. The papers listed above have stimulated the present work in varying degrees. However, they do not address themselves to the main problem considered here (see below).

3. Purpose of Present Investigation.

In the present paper, we shall study the connections between (i) the availability of an algebraic algorithm, (ii) definability in a first-order theory, and (iii) the existence of certain bounds in relation to an algebraic property.

The fact that (iii) can be a stepping stone to (i) is a matter of common experience. For example, suppose that we are given a field F and polynomials $f_1(x_1,\dots,x_n),\dots, f_k(x_1,\dots,x_n)$ and $g(x_1,\dots,x_n)$ in $F[x_1,\dots,x_n]$, for specified n, all of degrees less than positive integer d. We are asked to decide whether or not g belongs to the ideal generated by $f_1,\dots,f_k$, in other words, whether or not

(*) $$g = \Sigma h_j f_j$$

where $h_1,\dots,h_k$ are again polynomials in the given $F[x_1,\dots,x_n]$. It is not obvious how to determine this. However, once we are told that there is a bound $b = b(n,d)$ such that if (*) is satisfiable at all then it is satisfiable by polynomials h_j of degrees less than or equal to b, then we may substitute a polynomial of degree b with unknown coefficients in (*) and we then obtain a system of linear equations whose solvability settles the problem. It is in fact the actual determination of such a bound which seems involved historically, either directly, or through the proof of its existence by model-theoretic means. Moreover, as long as no bound is known, the solvability of (*) is represented by a predicate which is an infinite disjunction of existential predicates of the coefficients of f_j, g, but this is reduced to a finite disjunction and hence to a predicate of the lower predicate calculus, once a bound is known. Conversely, if we know that the infinite disjunction is equivalent to a predicate of the lower predicate calculus then it must already be equivalent to a finite subdisjunction, by the compactness theorem. While in the case under consideration this is not the way things went in the first place, there are other cases, in particular in connection with Hilbert's seventeenth problem, where the existence of such a bound was so determined in the first place.

In the present paper, we consider the converse question. Does the existence of an algebraic algorithm always imply the definability of the predicate in question in the lower predicate calculus, and hence, in cases where the predicate is known to be equivalent to an infinite disjunction, a reduction to a finite subdisjunction? We shall show that this is indeed the case, for an appropriate definition of the notion of an algebraic algorithm. Whether this definition is the right one is, as in the arithmetical case, not a matter of a purely mathematical argument. In the present context the argument is evidently stronger the weaker the definition. Accordingly, we shall base our argument on an algorithm whose effectiveness is in fact relative to a particular oracle, and discuss various possibilities within that definition subsequently.

The main application of our result will be a clarification of the situation in Differential Algebra where a number of important decision problems still await solution.

4. Auxiliary results from model theory.

Let K be a consistent set of axioms in the lower PC including equality, relations and functions, and let Σ be the class of models of K. We suppose that Σ is closed under unions of ascending chains. By the theorem of Chang-Los-Suszko this is the same as to assume that K is logically equivalent to a set of $\forall\exists$ sentences. Accordingly, we shall assume that K is a set of $\forall\exists$ sentences.

A set K^* in the vocabulary of K with class of models Σ^* is called the model completion of K if the following conditions are satisfied. (i) $\Sigma \supset \Sigma^*$, (ii) every $M \in \Sigma$ can be embedded in an $M^* \in \Sigma^*$, (iii) if $M \in \Sigma$ and $M_1, M_2 \in \Sigma^*$ are such that $M \subset M_1$, $M \subset M_2$, then for any sentence X in the vocabulary of M (i.e. with constants for any of the individuals of M), $M_1 \models X$ if and only if $M_2 \models X$.

It is known that for any given K as specified there can be up to logical equivalence not more than one K^* and to this extent we are justified in talking of the model completion [12]. The model completion of the theory of commutative fields (and also of course integral domains) is the theory of algebraically closed fields and the model completion of the theory of ordered fields is the theory of real closed ordered fields. The theory of groups and the theory of formally real fields have no model completions (although they have substitutes which do not concern us here [14]). K^* , if it exists, is also inductive and hence may be supposed to consist of $\forall\exists$ sentences.

Let $Q^*(x_1,\dots,x_n)$, $n \geq 1$, be any predicate in the language of K and let K^* be the model completion of K. Then there exists an existential predicate $Q(x_1,\dots,x_n)$ such that the following condition is satisfied.

Let $M \in \Sigma$, $M^* \in \Sigma^*$, $M^* \supset M$, and let $a_1,\ldots,a_n$ denote any elements of M. Then $M \models Q(a_1,\ldots,a_n)$ if and only if $M \models Q\ (a_1,\ldots,a_n)$. Q is called a <u>resultant</u> or <u>test</u> for Q. Moreover, if the theory of Σ is universal, i.e. if K may be taken to consist of universal axioms only, then for $n \geq 1$, $Q(x_1,\ldots,x_n)$ may be chosen so as to be free of quantifiers.

5. Algorithmic Instructions.

The language in which we shall formulate our algorithmic operations is the same as before, but in addition we use IF, PUT, and a symbol:= (We use PUT where in a computer language we might use DO, because DO is not appropriate to instantiations (see below).) We also use variables called <u>computational</u> <u>variables:</u> $\alpha, \beta, \ldots$

Let V be a fixed vocabulary, R a k-ary relation symbol not in V, and g a k-ary function symbol not in V. Q, Q', will always denote well-formed formulae in the vocabulary V.

We distinguish the following kinds of instructions.

5.1. A <u>standard</u> <u>instruction</u> I is of the form

$$\text{IF}\quad Q(\beta_1,\ldots,\beta_m)\quad \text{PUT}\quad \gamma := f(\xi_1,\ldots,\xi_n),$$

where γ is not one of the $\beta_1,\ldots,\beta_m,\xi_1,\ldots,\xi_n$ but where some of these may coincide. If $m = 0$ and $Q(x_1,\ldots,x_m)$ is a tautology then we call the instruction unconditional. Here, f is a composite function of our language. We say γ <u>is introduced by</u> I.

5.2. An <u>instantiation</u> I is of the form

$$\text{IF}\quad (\exists z_1)\ldots(\exists z_n)\quad Q(\beta_1,\ldots,\beta_m, z_1,\ldots,z_n)$$

$$\text{PUT}\quad \xi_1,\ldots,\xi_n\quad : Q(\beta_1,\ldots,\beta_m, \xi_1,\ldots,\xi_n)\ ,$$

where $\xi_1,\ldots,\xi_n$ are not among $\beta_1,\ldots,\beta_m$. We say $\xi_1,\ldots\ \xi_n$ <u>are introduced by</u> I .

5.3. <u>Final Instructions</u>

These are of one of the following forms:

(i)	IF	$Q(\beta_1,\ldots,\beta_m)$	PUT $R(\alpha_1,\ldots,\alpha_k)$
(ii)	IF	$Q(\beta_1,\ldots,\beta_m)$	PUT $\neg R(\alpha_1,\ldots,\alpha_k)$
(iii)	IF	$Q(\beta_1,\ldots,\beta_m)$	PUT $\alpha_k = g(\alpha_1,\ldots,\alpha_{k-1})$, k>0 .

We refer to instructions of form (i) and (ii) as positive and negative final instructions, respectively, and to $\alpha_1,\ldots,\alpha_k$ as final computational variables.

Fix a k-tuple $\alpha_1,\ldots,\alpha_k$ of computational variables. By a deduction d for R or g, respectively, with initial variables $\alpha_1,\ldots,\alpha_k$ we mean a finite sequence of instructions as above, such that the following conditions are satisfied. The last and only the last instruction in d is a final instruction I of form 5.3 (i), (ii) or 5.3(iii), respectively. In particular the final variables in I coincide with the initial variables of d. Every computational variable occuring in an instruction I in d is either an initial variable of d or has been introduced by a standard instruction or an instantiation I′ in d preceding I. A computational variable introduced by a standard instruction or an instantiation I in d does not occur in any instruction I' in d preceding I.

We now regard a deduction as a rule for interpreting the computational variables as elements of a specific structure, except for the final instruction which we interpret as defining an instance of a relation or function. In fact, the procedure is obvious. Suppose we are given a structure M which includes $\mathbf{V}$ in its vocabulary, $\alpha_1\ldots\alpha_\kappa \in M$, and a deduction d with initial variables $\alpha_1,\ldots,\alpha_k$. We interpret $\alpha_1,\ldots,\alpha_k$ by $\alpha_1,\ldots,\alpha_k$, respectively. If γ is introduced by a standard instruction of form 5.1 and $\beta_1,\ldots,\beta_m, \xi_1,\ldots,\xi_n$ have already been interpreted by $b_1,\ldots,b_m, c_1,\ldots, c_n \in M$, then we interpret γ by $f(c_1,\ldots,c_n) \in M$, if $M \models Q(b_1,\ldots, b_m)$. Otherwise we stop. If $\xi_1,\ldots,\xi_n$ are introduced by an instantiation of form 5.2 and $\beta_1,\ldots,\beta_m$ have already been interpreted by $b_1,\ldots,b_m \in M$, then we interpret $\xi_1,\ldots,\xi_n$ - to some extent arbitrarily - by any n-tuple $c_1,\ldots, c_n$ of elements of M which makes $Q(b_1,\ldots,b_m, c_1,\ldots,c_n)$ true in M in case $M \models (\exists z_1)\ldots(\exists z_n)\ Q(b_1,\ldots,b_m, z_1,\ldots,z_n)$. Otherwise we stop. We say d is effective at $a_1,\ldots, a_k \in M$, if some such assignment of elements of M to computational variables mapping $\alpha_1,\ldots,\alpha_k$ onto $\alpha_1,\ldots, \alpha_k$ can be carried out for all instructions in d (i.e.does not stop before the final instruction of d).

Now let Σ be an arithmetical class of models, with vocabulary V, and let π be a set of deductions for R with common initial variables $\alpha_1,\ldots\alpha_k$. Then we say π is a <u>program</u> <u>for</u> R in Σ , if

5.4. (i) (completeness condition) for all $M \in \Sigma$, $a_1,\ldots a_k \in M$, there exists $d \in \pi$ such that d is effective at $a_1,\ldots,a_k \in M$,

and (ii) (consistency condition) for all $d,d' \in \pi$, all $M \in \Sigma$ and all $a_1,\ldots,a_k \in M$, if d and d' are effective at $a_1,\ldots,a_k$ in M , then the final instructions of d and d' are both positive or both negative.

Similarly, we define a set π of deductions for g with common initial variables $\alpha_1,\ldots,\alpha_k$, $k > 0$, to be a <u>program</u> <u>for</u> g <u>in</u> Σ if

5.5 (i) (completeness condition) for all $M \in \Sigma$, and all $a_1,\ldots, a_{k-1} \in M$ there exists $a_k \in M$ and $d \in \pi$ such that d is effective at $a_1,\ldots,a_k$,

and (ii) (consistency condition) for all $M \in \Sigma$, and all $a_1,\ldots,a_{k-1} \in M$ there is at most one $a_k \in M$ such that a $d \in \pi$ is effective at $a_1,\ldots,a_k$.

Thus a program π for R (g) in Σ defines on every $M \in \Sigma$ a relation $R \subset M^k$ (a function $g: M^{k-1} \to M$) .
We will also consider the case that the relation R (or the function g) is defined in advance on every $M \in \Sigma$. Then we say the program π is <u>correct</u> if the relation (or function) defined by π coincides with R (with g) .

Next, we associate with every deduction d with initial variables $\alpha_1,\ldots, \alpha_k$ a formula $X_d(x_1,\ldots,x_k)$ in the vocabulary V . Choose a set of ordinary variables $x,y,\ldots$ in one-to-one correspondence with the computational variables. Denote the variables corresponding to $\alpha_1,\ldots,\alpha_k$ by $x_1,\ldots,x_k$. Let Y_d be the conjunction of all the following formulas:

(i) For every standard instruction of form 5.1 in d the formula $Q(y_1,\dots,y_m) \wedge z = f(z_1,\dots,z_n)$, where $y_1,\dots,y_m,z,z_1,\dots,z_n$ correspond to $\beta_1,\dots,\beta_m,\gamma,\xi_1,\dots,\xi_n$.

(ii) For every instantiation of form 5.2 in d the formula $Q(y_1,\dots,y_m, z_1,\dots,z_n)$, where $y_1,\dots,y_m,z_1,\dots,z_n$ correspond to $\beta_1,\dots,\beta_m,\xi_1,\dots,\xi_n$.

(iii) If d has final instruction of form 5.3, the formula $Q(y_1,\dots,y_m)$, where $y_1,\dots,y_m$ correspond to $\beta_1,\dots,\beta_m$.

Let $X_d(x_1,\dots,x_k)$ be the formula resulting from Y_d by existential quantification of all the free variables in Y_d except $x_1,\dots,x_k$.

It is now apparent from the definition of X_d that d is effective at $a_1,\dots,a_k \in M$ if and only if $M \models X_d(a_1,\dots,a_k)$. As a consequence conditions 5.4 and 5.5 can be expressed in terms of the formulas X_d: Let K be a set of sentences in the vocabulary V and let Σ be the class of models of K. Then 5.4 (i) and (ii) are equivalent to

5.6 (i) $$K \vdash (\forall x_1)\dots(\forall x_k) \bigvee_{d\in\pi} X_d(x_1,\dots,x_k)$$

and

(ii) $$K \vdash \bigwedge_{d\in\pi^+, d'\in\pi^-} (\forall x_1)\dots(\forall x_k) \neg (X_d(x_1,\dots,x_k) \wedge X_{d'}(x_1,\dots,x_k)),$$

where π^+ (π^-) is the set of deductions in π with positive (negative) final instruction.

Similarly, 5.5 (i) and (ii) are equivalent to

5.7 (i) $$K \vdash (\forall x_1)\dots(\forall x_{k-1}) \bigvee_{d\in\pi} (\exists x_k)\ X_d(x_1,\dots,x_k)$$

and

(ii) $$K \vdash \bigwedge_{d,d' \in \pi} (\forall x_1)\ldots(\forall x_{k-1})(\forall x_k)(\forall y)(X_d(x_1,\ldots,x_k) \wedge X_{d'}(x_1,\ldots,x_{k-1},y) \rightarrow x_k = y) .$$

Two programs are said to be <u>equivalent</u> in Σ, if they determine the same relation or function on every $M \in \Sigma$. Notice that every subset π' of a program π which is itself a program is equivalent in Σ to π.

We now see without difficulty:

5.8 <u>Basic Principle.</u> Every program π contains an equivalent sub-program π' which is finite.

<u>Proof.</u> Suppose π is a program for R in Σ and $\Sigma = \text{Mod}(K)$. Then by 5.6 (i)

$$K \vdash \bigvee_{d \in \pi} X_d(c_1,\ldots,c_k)$$, where $c_1,\ldots,c_k$ are new constants not in V. By the compactness theorem there exists a finite subset π' of π such that

$$K \vdash \bigvee_{d \in \pi'} X_d(c_1,\ldots,c_k) ,$$ and so

$$K \vdash (\forall x_1)\ldots(\forall x_k) \bigvee_{d \in \pi'} X_d(x_1,\ldots,x_k) .$$

Since 5.6 (ii) is trivially satisfied for π', π' is also a program for R in Σ and hence equivalent to π.

If π is a program for a function g in Σ the argument is similar using 5.7 instead of 5.6. This proves our assertion.

As a consequence we have now the following.

<u>Theorem 5.9</u> Let Σ be an arithmetical class and let π be a program for a relation R or a function g in Σ. Then R or g, respectively, is definable in Σ by a formula in the vocabulary V.

<u>Proof.</u> Suppose first that we are dealing with the case of a program π for a relation R in $\Sigma = (\text{Mod}(K)$. We may assume by the basic principle that π is finite.

Let $X^+(x_1,\ldots,x_k) = \bigvee_{d\varepsilon\pi^+} X_d(x_1,\ldots,x_k)$,

and let

$$X^-(x_1,\ldots,x_k) = \bigvee_{d\varepsilon\pi^-} X_d(x_1,\ldots,x_k) .$$

Then by 5.6

$$K \vdash (\forall x_1)\ldots(\forall x_k)(X^+(x_1,\ldots,x_k) \vee X^-(x_1,\ldots,x_k))$$

and

$$K \vdash (\forall x_1)\ldots(\forall x_k)\neg(X^+(x_1,\ldots,x_k)\wedge X^-(x_1,\ldots,x_k)) ,$$

in other words the exclusive ` or ´. Thus we may conclude that

$$(\forall x_1)\ldots(\forall x_k)(X^+(x_1,\ldots,x_k) \equiv R(x_1,\ldots,x_k))$$

and

$$(\forall x_1)\ldots(\forall x_k)(X^-(x_1,\ldots,x_k) \equiv \neg R(x_1,\ldots,x_k))$$

holds in Σ .

If π defines a function g in Σ , we may assume as above that π is finite. Then by 5.7 the formula

$$X(x_1,\ldots,x_k) = \bigvee_{d\varepsilon\pi} X_d(x_1,\ldots,x_k)$$

defines

$$g(x_1,\ldots,x_{k-1}) = x_k \text{ in } \Sigma .$$

<u>Remark.</u> Let π be a program for R in Σ . Suppose in particular that for all $d \varepsilon \pi$ the conditions Q occuring on the left hand side of the instructions in d are all <u>existential</u>. In that case X^+ and X^- are also existential, so that $R(x_1,\ldots,x_k)$ and its negation are both existential.

6. <u>Discussion.</u>

We now have to consider the question to what extent our computations may be said to be <u>algorithmic.</u> First of all, can we really carry out each individual step? This must be supposed to be the case if the conditions are all quantifier-free, since the ability to carry out an actual basic operation must be presumed. Equally, we cannot really be said to decide an arbitrary well-formed formula except by "oracle". However, if a predicate is D , i.e. both existential and universal, the question does not have a clear answer.

For suppose

$$Q(x_1,\dots,x_n) \equiv (\exists y_1)\dots(\exists y_m)Q_1(x_1,\dots,x_n, y_1,\dots,y_m)$$

$$\equiv (\forall z_1)\dots(\forall z_k)Q_2(x_1,\dots,x_n, z_1,\dots,z_k),$$

where Q_1 and Q_2 are free of quantifiers. Then

$$\neg Q(x_1,\dots,x_n) \equiv (\exists z_1)\dots(\exists z_k)\neg Q_2(x_1,\dots,x_n, z_1,\dots,z_k).$$

It follows that if the structure M has an effective enumeration as would be the case if M is recursive in any of the senses mentioned above (Shepherdson-Frölich-Rabin) then we can actually check in each particular case whether or not $Q(x_1,\dots,x_n)$ is verified. In particular, if in this case a program π has the property that all conditions occuring in instructions in π are D , then the relation or function determined by π is calculable. Also, in this case, we may find the x_j which are introduced by instantiation.

If we have elimination of quantifiers, then all predicates are equivalent to quantifier-free predicates. And in this case also, the predicate as given may be existential, so if it has first been verified, by elimination of quantifiers, we may then, in the case of a recursive structure, as before, again find the examples of the instantiation by enumeration.

Notice that our deductions are not programs, in the sense that they have no go-to instructions. This is irrelevant to the main conclusion in which we are interested here.

However, once we have reduced the given program to one, π' , consisting of a finite number of deductions, it is not difficult to turn this into a practical program, e.g. in the sense of Engeler. For this purpose, we need only a finite number of cells. We number the elements of π , $1,\dots,k$, and in each of these we number the instructions $a_{j\ell}, 1 \leq \ell \leq \lambda_j$. We now interpret the variables in one-to-one correspondence with cells. No erasing is necessary. The processing unit is supposed to carry out the individual step, i.e. enter data as by $x_j := a$ in the first available cell, verify conditions (by appealing to an oracle), and generally enter the name of an element of the given structure in the appropriate cell. In the case of instantiation this is somewhat indeterminate as indicated above. Here in fact, only if the structure is countable, and we are given an effective enumeration, is there any hope for success. In that case, we "play" the various deductions simultaneously, knowing that sooner or later one will arrive at a conclusion.

Let us now consider the converse problem. Suppose that we are given a program in the sense of Engeler. Can we transform it into a deduction of our kind?

The last step in Engeler's program is that a relation is to hold if the program terminates. Suppose that we also have another program which terminates if and only if the first program does not. We take our Engeler program, and apply to each variable x_j a second subscript which is raised by one whenever the variable occurs in an equation of the form x_j := an expression involving x_j.
Thus $x_j := x_j + 1$ becomes $x_{j,\ell+1} = x_{j,\ell} + 1$. And in any subsequent situation we also use the highest subscript that appeared previously. It is not difficult to see how to produce from finite pieces of the Engeler programs for R and $\neg R$ a set of deduction which constitute a program for R in our sense.

By contrast, our computability cannot be compared to the computability in the sense of Shepherdson-Frölich-Rabin. Thus, let us take the relative reducibility of polynomials in fields. This is expressed by an existential sentence. As such it is already computable by an existential condition in our language, although we know that it is not a computable problem in the sense of S-F-R. (Note here, that absolute reducibility is computable by all standards, since it permits elimination of quantifiers.)However, if our conditions are all quantifier-free, then our computations can be carried out in any recursive model.

Finally, we notice that in our set-up we do not have any reference to any universal bounds which may occur in a computation, for example the degree of a polynomial (see below). This is due to the fact that we are considering separately each given set of n data, e.g. the coefficients of a given polynomial. Since we have a total bound on the length of our computation it then follows that the number of coefficients of any other polynomial or even power series which may intervene must be subject to this bound also. We cannot go beyond this statement without further formalization.

7. Introduction of functions.

The introduction of functions in place of existential quantifiers on one hand simplifies formulae. Also, it makes our computations subject to the supplement to the ε - theorems [6], as follows.

Suppose, we are given a universal set of axioms K . Suppose also that we have as a conclusion from it a sentence of the form

$$(\forall x_1)\dots(\forall x_n)(\exists y_1)\dots(\exists y_m)\ Q(x_1,\dots,x_n,y_1,\dots,y_m) \ ,$$

Q free of quantifiers. Then [6] asserts that there exist $t_{kj}(x_1,\dots, x_n)$ such that

$$Q(x_1,\dots, x_n, t_{11}(x_1,\dots,x_n),\dots, t_{1m}(x_1,\dots,x_n)) \vee\dots\vee$$

$$Q(x_1,\dots, x_n, t_{\ell 1}(x_1,\dots,x_n),\dots, t_{\ell m}(x_1,\dots,x_n))$$

also is deducible from K . For $n = 0$, the terms in question are constant terms.

Now suppose that we have a set of $\forall\exists$ -axioms for a model-completion. We "Skolemize" the formulae by replacing each existential quantifier by a corresponding Skolem function symbol (e.g. in the case of an algebraically closed field). Then a certain measure of arbitraryness is introduced. For example, if ϕ_1 corresponds to $(\exists x)(x^2 - 2 = 0)$ and ϕ_2 corresponds to $(\exists x)(x^4 - 2 = 0)$ which are instances of the assertions that monic quadratic and biquadratic polynomials have a root then we cannot decide whether $\phi_2{}^2 - \phi_1 = 0$ or $\phi_2{}^2 + \phi_1 = 0$. In fact, given any field M one of these conditions may be satisfied in one algebraically closed extension of M and the other in another. Nevertheless in some cases we may still assume that the resulting set of axioms is model-complete.

To see this, consider the theory of real-closed ordered fields. Let M be a real-closed ordered field, and suppose that M' is an extension in which a particular existential sentence

$$X = (\exists x_1)\dots(\exists x_n)Q(x_1,\dots,x_n)$$

holds. (Note that, because of the indeterminacy mentioned above it is not true that the algebraic closure with respect to the Skolemized language of any field is uniquely determined by that field (i.e. we do not know if $(\sqrt[4]{t})^2 = \sqrt[2]{t}$ or $(\sqrt[4]{t})^2 = -\sqrt[2]{t}$).)

We define the square root function of a polynomial as its positive square root and the real root function of a polynomial of odd degree as its smallest. This can be represented by universal axioms, say

$$(\forall x_1)\dots(\forall x_{2k+1})[(\phi(x_1,\dots,x_{2k+1}))^{2k+1} + x_1\,\phi^{2k} + \dots + x_{2k+1} = 0]$$

$$(\forall x_1)\dots(\forall x_{2k+1})(\forall z)[z^{2k+1} + x_1 z^{2k} + \dots + x_{2k+1} = 0 \supset \phi(x_1,\dots,x_{2k+1}) \leq z]$$

With these definitions, it is not difficult to see that any ordered field has a unique extension which is prime for the situation. For this purpose, we only have to take the real closure and to define the ϕ as positive (in the case of a square root) or as the smallest root (in the case of a polynomial of odd degree).
Relative model-completeness can now be proved exactly as for real-closed ordered fields in the usual language. Accordingly, we have elimination of quantifiers, because we started out with a universal theory. Moreover, we even have completeness, because the real algebraic numbers prove to be a prime model.

The elimination of quantifiers makes instantiation more concrete in this case. Thus, suppose that we have a condition which is an existential statement $(\exists x_1)\dots(\exists x_n)Q(x_1,\dots,x_n,\alpha_1,\dots,\alpha_m)$.
In the first place we may suppose here that Q is free of quantifiers. In the second place, the entire predicate is equivalent to some $Q_1(\alpha_1,\dots,\alpha_m)$ which is quantifier-free. Also, if the given set of axioms is recursive then we may compute Q_1 from Q by proving $(\exists x_1)\dots(\exists x_n)Q \equiv Q_1$.

In particular we now have

$$Q_1(y_1,\dots,y_m) \supset (\exists x_1)\dots(\exists x_n)Q(x_1,\dots,x_n,y_1,\dots,y_m).$$

Hence by the second ε-theorem, we have terms

$$t_{11}(y_1,\dots,y_n),\dots,\ t_{jr}(y_1,\dots,y_n)$$

which instantiate the x, and we may actually find them by trial and error a finite number of times.

I can see no similar way to complement the corresponding set of axioms for algebraically closed fields.
[Editors' note: See the paper by Winkler in this volume.]

8. On a theorem of Polya.

We mentioned at the beginning of this paper one way to establish the existence of bounds for certain polynomial solutions: In the case of real numbers the representability of a positive definite polynomial by sums of squares of rational functions which is realized for <u>all</u> real closed fields, implies the existence of a bound on the number of squares required and on the degrees of the numerators and denominators (for a given bound on the degree of the polynomial and the number of variables). This yields a result even for the classical case of the real numbers. It is trivial that generally speaking the validity of the argument ceases if we have the equivalence of an infinite disjunction to a LPC condition only in one model of the arithmetical class, e.g. the real numbers alone. Thus, $x=x$ is certainly equivalent in that case to

$$y=0 \vee y^2>x \vee y^2+y^2 > x \vee y^2+y^2+y^2 > x \vee \ldots$$

yet this ceases to be true if we replace the disjunction by a finite subdisjunction, even in that model alone. A less trivial example is revealed by a study of a theorem of Polya which is given in [4] as a (supposedly simpler) companion of Artin's theorem. Let $F(x_1,\ldots,x_n)$ be a form (homogeneous polynomial of degree $k > 0$), such that $F(x_1,\ldots,x_n)$ is strictly positive for $x_j \geq 0,\ldots, \Sigma x_j > 0$. We confine ourselves to the domain of reals. Polya's theorem states that $F = G/H$ where G and H are forms with positive coefficients only. (More particularly we may choose $H = (x_1+\ldots x_n)^m$ for some m).

I am going to show that even if we allow general forms (there is clearly no point in permitting arbitrary nonhomogeneous polynomials) we can in this case not impose a bound on the degrees of the G and H in question. But even here, it is useful to consider in the first place an arbitrary real closed field R' such that $R \subset R'$, rather than the real numbers R.

An element $a \in R'$ is <u>infinitesimal</u> or infinitely close to zero (opposite: infinitely large)

if $|a| < r$ for all positive $r \in R$ and <u>finite</u> if $|a| < r$ for some positive $r \in R$. Let $F(x_1,\ldots,x_n)$ be a form of degree $k \geq 1$ with coefficients in R', $F = \Sigma\, \alpha_{i_1\ldots i_n} x_1^{i_1}\ldots x_n^{i_n}$, $\Sigma\, i_m=k$.
A point $x_1,\ldots,x_n$ is <u>finite</u> if all its coordinates are finite.
Then we have the following theorem.

Let Q_R, denote the set of all points $(x_1,\ldots,x_n)$ in R'^n such that $x_i \geq 0$ for $1 \leq i \leq n$ and $\Sigma x_i \neq 0$.

Theorem 8.1.

Let $F(x_1,\ldots,x_n)$ be a form with coefficients in R'. Suppose that all the coefficients of F are finite and that the non-zero ones are not infinitesimal. In order that there exist forms G and H with positive coefficients in R' such that $F = G/H$ and such that the coefficients of H are all finite and non-infinitesimal, it is necessary and sufficient that for all finite points $(x_1,\ldots,x_n)$ such that $x_i \geq 0$ for $1 \leq i \leq n$ and $\Sigma x_i \neq 0$, $F(x_1,\ldots,x_n)$ be positive non-infinitesimal.

Remark.

When we say that all the coefficients of G and H are positive, we mean that every possible monomial of appropriate degree must actually occur nontrivially.

Proof.

The condition is necessary. Clearly, if $F(x_1,\ldots,x_n) \leq 0$ at some finite point in $Q_{R'}$, then F cannot be written in the assumed form. To see that F does not take infinitesimal values on finite points in $Q_{R'}$, we argue as follows. Let $(\xi_1,\ldots,\xi_n)$ be such a point. Say ξ_j is not infinitesimal and consider the point $(0,\ldots,0,\xi_j, 0,\ldots,0)$ in $Q_{R'}$.
Let a be the coefficient of x_j^d in F (where d is the degree of F). Considering $F(0,\ldots,0,\xi_j, 0,\ldots,0)$ shows that $a > 0$ by the strict positivity of F in $Q_{R'}$. Furthermore a is not infinitesimal by assumption.. The equation $H(0,\ldots,0, \xi_j,0,\ldots,0) \cdot F(0,\ldots,0,\xi_j, 0,\ldots,0) = G(0,\ldots,0, \xi_j,0,\ldots,0)$ implies that the coefficient b of x_j^r in G is not infinitesimal (where $r = \deg G$). Therefore $G(\xi_1,\ldots,\xi_n)$ is not infinitesimal. Thus the equation $H(\xi_1,\ldots, \xi_n) F(\xi_1,\ldots,\xi_n) = G(\xi_1,\ldots,\xi_n)$ implies that $F(\xi_1,\ldots,\xi_n)$ is not infinitesimal.

Now we prove sufficiency. Although we are dealing with an arbitrary real closed field and not necessarily with a model of nonstandard analysis, we may use some of the notions and techniques of that sub-

ject [19] . Thus, every finite number $r \in R'$ is infinitely close to a unique standard real number r called the <u>standard part</u> of r, ${}^{\circ}r - r \simeq 0$, and the standard parts of sums and products of finite numbers are the sums and products of the standard parts of these numbers, respectively. The proof of sufficiency follows the outline of the "standard" proof and the reader is referred to it. The first step (for the example of three variables) involves taking the minimum of $F(x,y,z)$ for $x \geqslant 0,\ y \geqslant 0,\ z \geqslant 0,\ x+y+z = 1$.
By Tarski's theorem on real closed fields [16] , this minimum ν exists also for a form in R' , and there are (x_o,y_o,z_o) such that $F(x_o,y_o,z_o) = \nu$. Moreover, since $(x_o,y_o,z_o) \in Q_{R'}$, ν is not infinitesimal by the assumption on F . Accordingly we have a positive $\mu \in R$ such that ${}^{o}\nu > \mu$.
Next Polya introduces a function

$$\phi(x,y,z,t) = t^n \Sigma_n\, a_{\alpha\beta\gamma} \binom{xt^{-1}}{\alpha}\binom{yt^{-1}}{\beta}\binom{zt^{-1}}{\gamma} ,$$

where

$$F(x,y,z) = \Sigma_n\, a_{\alpha\beta\gamma} \frac{x^\alpha y^\beta z^\gamma}{\alpha!\beta!\gamma!} ,$$

the summation in both cases being over all triples (α,β,γ) of integers with $\alpha \geqslant o,\ \beta \geqslant o,\ \gamma \geqslant o,\ \alpha+\beta+\gamma = n$, and proves the identity, for every $k \geqslant n$,

$$(x+y+z)^{k-n}\, F(x,y,z) = (k-n)!k^n\, \Sigma_k\, \phi\left(\frac{a}{k},\frac{b}{k},\frac{c}{k},\frac{1}{k}\right) \frac{x^a y^b z^c}{a!b!c!}$$

which holds in any field of characteristic zero. Here again Σ_k denotes the sum over all triples (a,b,c) such that $a \geqslant o$, $b \geqslant o$, $c \geqslant o$, $a+b+c = k$.

Now, $\phi(x,y,z,t) \to F(x,y,z)$ as $t \to o$ which, since the coefficients are finite, may be interpreted in the standard sense. There is a positive $\varepsilon \in R$ such that

$$\phi(x,y,z,t) > F(x,y,z) - \frac{1}{2}\mu > \frac{1}{2}\mu > o \quad \text{for} \quad o < t < \varepsilon ,$$

in particular for $t = \frac{1}{k}$, where k is sufficiently large. This proves the theorem.

Remark 8.2.

The proof shows that H can in fact be chosen to be just $(x_1+\ldots+x_n)^m$ for some m. Also G can be chosen to have all its coefficients non-infinitesimal.

For the special case $R' = R$, the sufficiency content of our theorem does not add anything to Polya's. From the necessity part, however, we may derive the following "standard" result. Note that for $R' = R$ the statement $(x_1,\ldots,x_n) \in Q_R$ reduces to the "standard" statement that each $x_i \geqslant 0$ and $\Sigma x_i > 0$.

Theorem 8.3.

Suppose that we are given a form $F_\tau = \Sigma a_{i_1\ldots i_n}(\tau) x_1^{i_1}\ldots x_n^{i_n}$ where the a_i depend on a parameter τ which ranges over a set T in m-dimensional real space. Assume moreover that $|a_{i_1\ldots i_n}(\tau)|$ is bounded away from zero for all $i_1,\ldots,i_n$ and that F_τ is positive definite on Q_R for all $\tau \in T$. Suppose that there is a point $P \in R^m$ and a point $(\xi_1,\ldots,\xi_n) \in Q_R$ such that $F_\tau(\xi_1,\ldots,\xi_n) \to 0$ as $\tau \to P$ through values in T. Let k be any natural number. Then there is a neighborhood of P in T such that for every τ in this neighborhood F_τ cannot be represented by positive forms of degrees $< k$.

For the proof we use an enlargement $^*R$ of R. The previous theorem applies to $^*R$. It follows that if n is any infinite positive integer then within a radius $\frac{1}{n}$ around P in $^*T$, F cannot be represented by positive forms of finite degree with non-infinitesimal coefficients, in particular not by forms G,H such that $\deg G,H < k$ and such that all the coefficients of H are greater than some positive $r \in R$ (although it can be represented by forms of infinite degrees). This assertion can be represented by a sentence $X(n,r)$ in the full language of R. But since the sentence is true for all infinite n it will also be true for sufficiently large finite n, by a well known principle of nonstandard analysis. This proves the theorem.

For example, consider the form $x^2 - \tau xy + y^2$ where $1 < \tau < 2$. We choose the point P as $\tau = 2$ and $\xi_1 = \xi_2 = 1/2$. Then the conditions of the theorem are satisfied. It follows that although for each τ in the range in question the form is strictly positive, the degrees of forms G_τ, H_τ representing $F_\tau = x^2 - \tau xy + y^2$ tend to ∞ as τ approaches 2 .

Remark.

We note that we may associate with Polya's theorem a diophantine problem with side conditions. The question is whether one can solve $HF = G$ for given F and unknown forms H and G subject to the conditions that their coefficients are positive. The answer is that this problem has no algorithm in the sense of the present paper. By contrast, we have that the condition that F be strictly positive can be formulated in the LPC and hence can even be expressed in quantifier-free form in terms of addition, multiplication, subtraction and equality. But these are just not the same conditions for all real-closed fields. We may mention here that [4] contains an erroneous statement that Polya's theorem provides a decision procedure for deciding whether or not a form is strictly positive."We multiply repeatedly by Σx , and if the form is positive, we shall sooner or later obtain a form with positive coefficients"[4] . However, this is not a decision procedure, for if the form is not strictly positive then the procedure will never terminate. Of course, the elimination of quantifiers provides a decision procedure, as stated.

9. Algebraic and differential field theory.

Consider as a simple example the assertion which says that a system of linear equations has a solution.

Let

$$a_1^1 x_1 + \dots + a_n^1 x_n = b_1$$
$$a_1^2 x_1 + \dots + a_n^2 x_n = b_2$$
$$a_1^n x_1 + \dots + a_n^n x_n = b_n$$

be the system in question. The test is that the rank of the augmented matrix is equal to the rank of the (a_j^i) . It is not difficult to formulate the statement that the ranks are equal by a quantifier-free formula.

Next, let us consider the following problem. We are given polynomials $f_j(x_1,\dots,x_n)$, $g(x_1,\dots,x_n)$ of bounded degrees with variable coefficients. <u>To decide whether</u> $g \in (f_1,\dots,f_k)$ (with coefficients in the given field) . This can be carried out because there is a known bound on the degrees of $h_1,\dots,h_k$ for which we might have $g = \sum_i h_i f_i$ [5] . Thus the problem is reduced to one of solvability in -ms of the coefficients.

<u>Second problem.</u> To find a basis for $(f_1,\dots,f_k) : g$ where again bounds are given. (The symbol $(f_1,\dots,f_k) : g$ stands for division of ideals; $h \in (f_1,\dots,f_k) : g$ it and only if $hg \in (f_1,\dots,f_k)$). Notice that this problem is not a priori among the kind considered in our general part, since we do not know the degrees of the basis polynomials and their number from the outset. If we did, then we could regard the coefficients as functions to be calculated in the sense of the general theory. And indeed, the theory of Greta Hermann shows that such bounds exist, and that the coefficients in question can be calculated by rational operations.

<u>Third problem.</u> To find a basis for

$$(f_1,\dots,f_k) : g^\infty = \bigcup_{n=1}^{\infty} (f_1,\dots,f_k) : g^n .$$

To do this, we shall show that there is a uniform ν depending only on the degrees of $f_1,\dots,f_k, g$ such that $(f_1,\dots,f_k) : g^\nu = (f_1,\dots,f_k) : g^\infty$. First of all, notice that $(f_1,\dots,f_k) : g^n \subset (f_1,\dots,f_k) : g^{n+1} = ((f_1,\dots,f_k) : g^n) : g$. Also, if we have here equality then also $(f_1,\dots,f_k) : g^{n+1} = (f_1,\dots,f_k) : g^{n+2}$. For suppose $h \in (f_1,\dots,f_k) : g^{n+2}$. Then $hg^{n+2} = \sum h_j f_j$. We have to show that we can replace the left hand side by hg^{n+1} . But at any rate $hg \in (f_1,\dots,f_k) : g^{n+1}$ and so by assumption $hg \in (f_1,\dots,f_k) : g^n$, so $h \in (f_1,\dots,f_k) : g^{n+1}$. Moreover, by Hilbert's basis theorem, for given $(f_1,\dots,f_k)$ and g , the chain breaks off so there is a first ν for which we have equality, $(f_1,\dots,f_k) : g^\nu = (f_1,\dots,f_k) : g^{\nu+1} =$

$(f_1,\ldots,f_k) : g^\infty$.

Now to show that this ν may be chosen uniformly, we proceed as follows. We write down, for each n, the predicate $(f_1,\ldots,f_k) : g^n = (f_1,\ldots,f_k) : g^{n+1}$. This only requires writing down that all elements of the basis of $(f_1,\ldots,f_k) : g^{n+1}$ (which have been computed, as described), $k_1,\ldots,k_\ell$ say, belong to $(f_1,\ldots,f_k) : g^n$, i.e. they satisfy $k_j g^n \in (f_1,\ldots,f_k)$. All this requires only rational operations. Since they are completed after a finite number of steps in each case, all this is equivalent to some predicate Q_n of $\{f_1,\ldots,f_k,g\}$. Now consider the axioms of field theory K, together with $\neg Q_1, \neg Q_2,\ldots$ where we have replaced the variable coefficients of the f_j, g by new distinct constants. Then if $K, \neg Q_1, \neg Q_2,\ldots$ is consistent, it has a model in which Hilbert's basis theorem is not satisfied. This is impossible, so $K \vdash Q_1 \vee \ldots \vee Q_\nu$ for some ν. But since $K \vdash Q_\mu \supset Q_\nu$ for all $\mu < \nu$, we have $K \vdash Q_\nu$, proving the assertion.

Next we shall be concerned with prime ideals. We refer the reader to Ritt [11] for the notion of a chain and of a characteristic set. We have the following theorem. Let I_i be the initials of a chain $A_1, A_2,\ldots,A_r$, where A_1 is of positive class, i.e. is not a constant. Let G be any polynomial. Then there exist nonnegative integers t_i, $i = 1,\ldots, r$, and a polynomial R such that

$$I_1^{t_1}\ldots I_r^{t_r}\, G \equiv R \bmod (A_1,\ldots,A_r) ,$$

where R is <u>reduced</u> with respect to $A_1, A_2,\ldots, A_r$, that is to say, the degree of R in the last variable which occurs in A_j is lower than the degree of A_j in it.

Put $H = I_1^{t_1}\ldots I_r^{t_r}$. We are going to show :

<u>Theorem 9.1.</u> In order that G belong to the prime ideal J whose characteristic set is $(A_1,\ldots,A_r)$ (provided it is a characteristic set of any prime ideal) it is necessary and sufficient that $R = 0$.

Proof. Notice that every I_j must be reduced with respect to $(A_1, \ldots, A_r)$. For it is lower than A_j and by the definition of a chain must be lower than all the other A_i also. If $G \in J$ then the remainder is in J since the congruence is modulo $(A_1, \ldots, A_r)$. Since the remainder is reduced with respect to the $A_1, \ldots, A_r$, $R = 0$. Conversely since the I_j's are reduced with respect to $A_1, \ldots, A_r$, none of them is in J . So if $R = 0$, then $G \in J$, since J is prime. This proves the assertion.

We now come to one of Ritt's major problems in the constructive theory of algebraic equations. It is to determine whether a given chain is a characteristic set of a prime ideal, where the chain is of the form $A_1, A_2, \ldots, A_p, A_i$ containing the "parameters" $U_1, \ldots, U_q$ and the variables $Y_1, \ldots, Y_{i-1}$ and introducing the variable Y_i .

We shall show that for each such case, with the A_i having indeterminate coefficients and having given degrees, there is an $\forall\exists$ - predicate which determines whether the chain in question is the characteristic set of a prime ideal.

For $p = 1$, Ritt [11,p.88] shows that the condition is that the polynomial $A_1(Y_1)$ be irreducible regarded as a polynomial with co - efficients in $F(U_1, \ldots, U_q)$, where F is a given field whose diagram forms part of the axiomatic system K . The condition of irreducibility can be represented by a universal predicate.

For $p > 1$, Ritt proves that the following condition is necessary and sufficient.

(i) $A_1, \ldots, A_{p-1}$ is a characteristic set of a prime polynomial ideal.

(ii) If $u_1, \ldots, u_q$, $y_1, \ldots, y_{p-1}$ is a generic zero of $A_1, \ldots, A_{p-1}$ then when we substitute these for $U_1, \ldots, U_q$, $Y_1, \ldots, Y_{p-1}$ in A_p , we obtain a polynomiall $A_p(Y_p)$ which is irreducible in its field of coefficients.

To interpret these conditions, let us suppose that we have tested

- by an $\forall\exists$-predicate - that (i) is satisfied and let us consider (ii) . To substitute the generic zero really amounts to calculating with polynomials in $U_1,\dots, U_q, Y_1,\dots, Y_{p-1}$ modulo J_{p-1} , where J_{p-1} is the ideal determined by the generic point. Now, given $A_1,\dots, A_{p-1}$ we have, from Theorem 9.1 and the third problem,
$J_{p-1} = (A_1,\dots, A_{p-1}) : (I_1\dots I_{p-1})^\infty = (A_1,\dots, A_{p-1}) : (I_1\dots I_{p-1})^m$
where m depends only on the given bound for $A_1,\dots, A_{p-1}$. Accordingly, as mentioned, we can compute a basis for J_{p-1} .

Now, having found J_{p-1} , we know that we obtain a generic point of J_{p-1} simply by taking the residue class of $U_1,\dots,U_q,Y_1,\dots,Y_{p-1}$ in $F[U_1,\dots, U_q; Y_1,\dots,Y_{p-1}]$ modulo J_{p-1} . In other words the point $(u_1,\dots, u_q, y_1,\dots y_{p-1})$ is a given point in the field F' which is the field of quotients of $F[U_1,\dots, U_q, Y_1,\dots, Y_{p-1}] / J_{p-1}$. Any representation $A_p = H\cdot K$ in that field where neither H nor K is constant (as a polynomial in Y_p) is equivalent to a representation

$$(1) \qquad G\,A_p \equiv HK \mod J_{p-1} ,$$

where we have obtained G by clearing away denominators, $G \notin J_{p-1}$, and where H and K are not independent of Y_p . Now we may make sure that G is reduced with respect to $A_1,\dots, A_{p-1}$ by multiplying by appropriate powers of $I_1,\dots, I_{p-1}$, the result, R , mod J_{p-1} being reduced with respect to $A_1,\dots, A_{p-1}$ and hence of bounded degree. Hence, we may assume that G is reduced with respect to $A_1,\dots, A_{p-1}$. Similarly, it is enough to consider H's and K's which are reduced with respect to $A_1,\dots, A_{p-1}$, and reducibility is now expressed by the $\exists\forall$-assertion that there exist certain coefficients of G and of H and K with not all positive powers of Y_p in H or K having coefficients belonging to J_{p-1} such that $G\,A_p - HK$ belongs to J_{p-1} .

We have thus obtained an $\forall\exists$- test whether or not $A_1,\dots, A_p$ is the characteristic set of a prime ideal.

* * *

Next suppose we are given a system of polynomials $Q_1, \dots, Q_m$. We wish to develop a set of characteristic sets of prime ideals whose manifolds make up the manifolds of $Q_1, \dots, Q_m$. If we get these characteristic sets, we can also get their prime ideals, as above, i.e. their bases ...

References

1: P.J.Cohen, Decision procedures for real and p-adic fields, Communications in Pure and Applied Mathematics, Vol. XXII (1969), 131-152

2. E.Engeler, Formal Languages: Automata and Structures, Lectures in Advanced Mathematics, Markham Publ.Co Chicago 1968

3. R.Fraissé, Une notion de recursivité relative, Infinitistic Methods, Proc. Symp. on Foundations of Math., Warsaw 1961, 323-328

4. Hardy-Littlewood-Polya , Inequalities, Cambridge University Press, 1959

5. G.Hermann Die Frage der endlichen vielen Schritte in der Theorie der Polynomideale, Math. Ann. 95 (1926), 736-788

6. D.Hilbert - P.Bernays, Grundlagen der Mathematik, Vol. II, Springer-Verlag, Berlin 1939

7. L.Kronecker, Die Zerlegung der ganzen Größen eines natürlichen Rationalitätsbereichs in ihre irreductiblen Factoren, Kroneckers Werke, Teubner Verlag, Leipzig 1895, Vol. 2, 4o9-416

8. W.Lambert, A Notion of Effectiveness in Arbitrary Structures, J.S.L., Vol. 33, 1968, 577-6o2

9. R.Peter, Über die Verallgemeinerung der Theorie der rekursiven Funktionen für abstrakte Mengen geeigneter Struktur als Definitionsbereiche, Acta Math.Acad.Sci. Hung. 12 (3-4) (1961)

1o. M.Rabin, Computable Algebra: General Theory and Theory of Computable Fields, AMS Transactions 95 (196o) 341-36o

11. J.F.Ritt, Differential Algebra, AMS Colloquium Publications Vol. XXXIII, New York 195o

12. A.Robinson, Introduction to Model Theory and to the Metamathematics of Algebra, North Holland Publ. Co., Amsterdam 1963

13. ———, Nonstandard Analysis, North Holland Publ. Co., Amsterdam 1966

14. ———, Infinite Forcing in Model Theory, Proc. Second Scandinavian Logic Symposium, North Holland 1971, 317-34o

15. J.C.Shepherdson-A.Frölich, Effective Procedures in Field Theory, Trans. Royal Soc. London, ser. A, 248 (1956), 4o7-432

16. A.Tarski-J.C.C.McKinsey, A Decision Method for Elementary Algebra and Geometry, 2nd edition, Berkeley and Los Angeles, 1948/1951

CONTRIBUTED PAPERS

ON RECURSIVELY SATURATED MODELS OF ARITHMETIC[1]

Jon Barwise and John Schlipf
The University of Wisconsin-Madison

§1. Introduction. In his retiring presidential address to the ASL Abraham Robinson pointed out that one of the legitimate functions of the logician is "to use his own characteristic tools ... to gain a better understanding of the various and varigated kinds of structures, methods, theories and theorems that are to be found in mathematics" ([6], p. 500). In this note we use our characteristic tools, admissible sets with urelements from Barwise [1] and recursively saturated models from Schlipf [7], to shed a glimmer of light on the models that arise in non-standard analysis and some of the known theorems about them.

1.1 Definition. Let $\mathfrak{M} = \langle M, R_1, \ldots, R_k \rangle$ be a structure for a finite language L. We say that $\mathfrak{M}$ is recursively saturated if for every recursive set $\Phi(x, y_1, \ldots, y_n)$ of finitary formulas of L, the following infinite sentence is true in $\mathfrak{M}$:

$$\forall y_1 \ldots y_n [\bigwedge_{\Phi_0 \in S_\omega(\Phi)} \exists x \bigwedge \Phi_0(x, \vec{y}) \Rightarrow \exists x \bigwedge \Phi(x, \vec{y})]$$

where $S_\omega(\Phi)$ is the set of finite subsets of Φ.

It is not too hard to see that any model of Peano arithmetic (PA) which occurs as the integers in some model of non-standard analysis (or in some non ω-model of ZF) is recursively saturated. The principle goal of this paper is to:

(a) isolate a weak subsystem of analysis, called Δ^1_1-PA

(b) prove that the recursively saturated models of PA are exactly those models that can be expanded to models of Δ^1_1-PA

(c) derive certain corollaries from (b).

---

[1]The preparation of this paper was supported by Grant NSF GP-43882X. The first author is an Alfred P. Sloan Fellow. Part of the research for this paper was done while the second author held a NSF Graduate Fellowship at the Univ. of Wis.

The importance of recursively saturated structures first became apparent in connection with admissible sets with urelements when Schlipf proved Theorem 2.1(i) below. Since then, a number of other characterizations and uses of recursive saturation have been discovered. Useful facts about recursively saturated models can be found in [1] and [7]. Some of these will be pulled out of the hat as needed.

We now turn from general structures to models of Peano arithmetic. We assume PA is formulated in a first order language L with symbols 0, + and $\times$ and variables $x_1, x_2, \ldots$. We use $\mathfrak{n} = \langle N, 0, +, \times \rangle$ to range over arbitrary non-standard models of PA.

We identify analysis with second order number theory as usual. Thus we expand our language to a language L^* by adding a membership symbol ϵ and new variables $X_1, X_2, \ldots$. We allow atomic expressions of the form $x_i \epsilon X_j$ in formulas. A structure for L^* consists of a pair $\langle \mathfrak{n}, \mathfrak{X} \rangle$ where $\mathfrak{n} = \langle N, 0, +, \times \rangle$ is a structure for L, $\mathfrak{X}$ is a non-empty set of subsets of N, ϵ is interpreted by membership, and the second order quantifiers range over $\mathfrak{X}$.

By the axiom of induction we mean the following single second-order axiom:

$$\forall X[0 \epsilon X \wedge \forall x(x \epsilon X \Rightarrow (x+1) \epsilon X) \Rightarrow \forall x(x \epsilon X)] .$$

The reader should be warned that, in the theories we consider here, this axiom does not imply all instances of the axiom scheme

$$\varphi(0) \wedge \forall x(\varphi(x) \Rightarrow \varphi(x+1)) \Rightarrow \forall x\, \varphi(x) .$$

Before one can apply induction, one first has to form the set $X = \{x \mid \varphi(x)\}$. This we assume we can always do if φ is in $L_{\omega\omega}$, and generally even if φ is first order in $L^*_{\omega\omega}$, but we do not assume it for all φ. (The situation is analogous to the difference between Gödel-Bernays set theory and Morse-Kelley set theory. In GB one asserts the axiom of foundation, not the infinite scheme.)

By a first order formula of L^* we mean a formula with (possibly) set variables but no set quantifiers. A Σ^1_1 formula is one of the form $\exists X\varphi(X, \ldots)$ where φ is first order. A formula is essentially Σ^1_1 if one could transform it to a Σ^1_1 formula by the manipulations

$$\exists x\, \exists X\, \varphi(x,X) \Rightarrow \exists X\, \exists x\, \varphi(x,X)$$

$$(*) \qquad \forall x\, \exists X\, \varphi(x,X) \Rightarrow \exists Y\, \forall x\, \varphi(x,Y_x)$$

familiar from recursion theory (where $Y_x = \{y \mid 2^x 3^y \in Y\}$). More formally, the <u>essentially Σ^1_1 formulas</u> (resp. <u>essentially Π^1_1 formulas</u>) form the smallest class containing all first order formulas and closed under $\wedge, \vee, \forall x_1, \exists x_1, \exists X_1$ (resp. $\forall X_1$).

The axiom scheme of Δ^1_1-<u>Comprehension</u> (Δ^1_1-Comp) asserts for all essentially Π^1_1 formulas $\Phi(x)$ and all essentially Σ^1_1 formulas $\Psi(x)$, the universal closure of the following:

$$\forall x(\Phi \Leftrightarrow \Psi) \Rightarrow \exists X\, \forall x(x \in X \Leftrightarrow \Phi(x)) .$$

(Both individual and set variables may occur free.)

The <u>axiom of Σ^1_1-choice</u>, Σ^1_1-AC, asserts (*) above for all essentially Σ^1_1 formulas $\varphi(x,X)$. (Thus Σ^1_1-AC proves that every essentially Σ^1_1 formula is equivalent to a Σ^1_1 formula.)

We use Δ^1_1-PA to denote the theory PA with the axiom of induction and the scheme Δ^1_1-Comp.

Our main result is the following:

1.2 <u>Theorem</u>. Let $\mathfrak{n}$ be a non-standard model of PA. Then the following are equivalent:

(i) $\mathfrak{n}$ is recursively saturated.

(ii) For some $\mathfrak{X} \subseteq \mathcal{P}(\mathfrak{n})$, $\langle \mathfrak{n}, \mathfrak{X} \rangle \models \Delta^1_1$-PA.

(iii) For some $\mathfrak{X} \subseteq \mathcal{P}(\mathfrak{n})$, $\langle \mathfrak{n}, \mathfrak{X} \rangle \models \Delta^1_1$-PA + Σ^1_1-AC.

The implication (iii) $\Rightarrow$ (ii) is trivial. We prove (i) $\Rightarrow$ (iii) and some corollaries in §2; and (ii) $\Rightarrow$ (i), plus some more corollaries, in §3.

§2. Definable Sets. For any $\mathfrak{M}$ let $Df(\mathfrak{M})$ be the collection of all $X \subseteq M$ of the form

$$X = \{x : \mathfrak{M} \models \varphi(x, \vec{y})\}$$

for some first order formula $\varphi(x_0, x_1, \ldots, x_n)$ and some $y_1 \ldots y_n \in M$. We need to use the following result.

2.1 Theorem. Let $\mathfrak{M} = \langle M; R_1, \ldots, R_k \rangle$ be a structure for L.

(i) $\mathfrak{M}$ is recursively saturated iff $\mathbb{HYP}_{\mathfrak{M}}$, the smallest admissible set with $\mathfrak{M}$ as a set of urelements, has ordinal ω—written $o(\mathbb{HYP}_{\mathfrak{M}}) = \omega$.

(ii) If $o(\mathbb{HYP}_{\mathfrak{M}}) = \omega$, then the subsets of $\mathfrak{M}$ in $\mathbb{HYP}_{\mathfrak{M}}$ are just the elements of $Df(\mathfrak{M})$.

(iii) Every structure $\mathfrak{M}$ has an elementary extension $\mathfrak{M}'$ of the same cardinality which is recursively saturated.

The proofs of these results can be found in [1]. The implication (i) ⇒ (iii) of Theorem 1.3 follows from the next result.

2.2 Theorem. Let $\mathfrak{N} \models PA$ be recursively saturated. Then $\langle \mathfrak{N}, Df(\mathfrak{N})\rangle$ is a model of Δ^1_1-PA plus Σ^1_1-AC.

Proof. It is clear that $\langle \mathfrak{N}, Df(\mathfrak{N})\rangle$ satisfies the axiom of induction since PA asserts induction for definable sets. This part does not require the recursive saturation of $\mathfrak{N}$, but the others do—we use Theorem 2.1. We begin by making some simple systematic remarks on the relationship between $\mathfrak{N}$ and $\mathbb{HYP}_{\mathfrak{N}}$:

(i) Since $\mathfrak{N} \in \mathbb{HYP}_{\mathfrak{N}}$, first order quantifiers over $\mathfrak{N}$ count as bounded quantifiers (bounded by the universe N of $\mathfrak{N}$).

(ii) Since $\mathfrak{N}$ is recursively saturated, a second order quantifier over $Df(\mathfrak{N})$ counts as a first order quantifier over $\mathbb{HYP}_{\mathfrak{N}}$ by Theorem 2.1 (ii). More specifically,

$$\forall X(\ldots X\ldots) \text{ becomes } \forall a(a \subseteq N \Rightarrow (\ldots a\ldots))$$

$$\exists X(\ldots X\ldots) \text{ becomes } \exists a(a \subseteq N \wedge (\ldots a\ldots)) .$$

So an essentially Π^1_1 formula Φ becomes a first order Π formula φ of set theory; and an essentially Σ^1_1 formula, a first order Σ formula. Given these observations, we get almost immediately that $\langle \mathfrak{n}, Df(\mathfrak{n})\rangle \models \Delta^1_1$-Comp. For suppose $\langle \mathfrak{n}, Df(\mathfrak{n})\rangle \models \forall x[\Phi(x) \Longleftrightarrow \Psi(x)]$, where Φ, Ψ are essentially Π^1_1, Σ^1_1 respectively. We may safely ignore the parameters in these formulas. Let φ, ψ be the corresponding first order formulas of set theory (which are Π and Σ respectively). For all $x \in N$,

$$\langle \mathfrak{n}, Df(\mathfrak{n})\rangle \models \Phi(x) \text{ iff } \mathbb{HYP}_{\mathfrak{n}} \models \varphi(x)$$

$$\langle \mathfrak{n}, Df(\mathfrak{n})\rangle \models \Psi(x) \text{ iff } \mathbb{HYP}_{\mathfrak{n}} \models \psi(x) .$$

So

$$\mathbb{HYP}_{\mathfrak{n}} \models \forall x \in N\, [\varphi(x) \Longleftrightarrow \psi(x)] .$$

We now apply Δ-separation in the admissible set $\mathbb{HYP}_{\mathfrak{n}}$ to conclude that

$$X = \{x \in N \mid \mathbb{HYP}_{\mathfrak{n}} \models \varphi(x)\}$$

is in $\mathbb{HYP}_{\mathfrak{n}}$. But then $X \in Df(\mathfrak{n})$ by 2.1(ii), and $\langle \mathfrak{n}, Df(\mathfrak{n})\rangle$ satisfies $\forall x[x \in X \Longleftrightarrow \Phi(x)]$, as desired.

The proof that $\langle \mathfrak{n}, Df(\mathfrak{n})\rangle \models \Sigma^1_1$-AC is almost as easy once we have the following lemma about admissible sets:

2.3 Lemma. For any $\mathfrak{m} = \langle M; R_1, \ldots, R_k\rangle$, $\mathbb{HYP}_{\mathfrak{m}}$ is a model of the sentence expresing: If $\mathfrak{m}$ is well ordered, so is every element. (By being wellordered we mean the obvious: that every subset of $\mathfrak{m}$ has a least element, not that every definable subclass of $\mathfrak{m}$ has a least element.)

We shall return to the proof of 2.3 after finishing the proof of 2.1. Let $\Phi(x,X)$ be an essentially Σ^1_1 formula and suppose $\langle \mathfrak{n}, Df(\mathfrak{n})\rangle$ is a model of $\forall x\, \exists X\, \varphi(x,X)$. Let φ be the rewritten version of Φ in the language of set theory, so that φ is a Σ formula and

$$\forall x \in N\, \exists a[a \subseteq N \wedge \varphi(x,a)]$$

is true in $\mathbb{HYP}_{\mathfrak{n}}$. By Σ-reflection, there is a transitive $b \in \mathbb{HYP}_{\mathfrak{n}}$ such that

$$\forall x \in N\, \exists a \in b[a \subseteq N \wedge \varphi(x,a)^{(b)}].$$

Since $\mathbb{HYP}_{\mathfrak{n}}$ contains only definable subsets of $\mathfrak{n}$, $\mathbb{HYP}_{\mathfrak{n}}$ is a model of the sentence expressing "$\mathfrak{n}$ is wellordered". So by the lemma, there is a linear ordering $<$ of b in $\mathbb{HYP}_{\mathfrak{n}}$ such that

$$\text{"}\langle b,<\rangle \text{ is wellordered"}$$

is true in $\mathbb{HYP}_{\mathfrak{n}}$. So every nonempty subset c of b which is in $\mathbb{HYP}_{\mathfrak{n}}$ has a $<$-least element. Define a $\mathbb{HYP}_{\mathfrak{n}}$-recursive function f with domain N by

$$f(x) = \text{the } <\text{-least member } a \text{ of } b \text{ such that } \varphi(x,a)^{(b)}.$$

This is $\mathbb{HYP}_{\mathfrak{n}}$-recursive since $\varphi(x,a)^{(b)}$ is a Δ_0-formula. Let $Y = \{2^x 3^y \mid y \in f(x)\}$, where multiplication and exponentiation take place in $\mathfrak{n}$. By Σ-replacement, $Y \in \mathbb{HYP}_{\mathfrak{n}}$, so $Y \in \mathrm{Df}(\mathfrak{n})$, and $\langle \mathfrak{n}, \mathrm{Df}(\mathfrak{n})\rangle \models \forall x\, \Phi(x, Y_x)$. □

<u>Remark.</u> We could have skipped the proof of Δ^1_1-Comp since arithmetic comprehension plus Σ^1_1-AC implies Δ^1_1-Comp. (On the other hand, Steel [8] has announced that Δ^1_1-Comp does not imply Σ^1_1-AC, even the presence of the full scheme of induction.)

<u>Remark</u> (for the reader familiar with the Moschovakis [5] theory of inductive definitions). One might be puzzled by Theorem 2.2 in view of the fact that Moschovakis proves that for "almost acceptable" structures $\mathfrak{m} = \langle M, R_1 \ldots R_\ell\rangle$, one has (a) the hyperelementary sets form the smallest model of Δ^1_1-Comp but (b) there is a hyperelementary set not in $\mathrm{Df}(\mathfrak{m})$. The conclusion, of course, is that recursively saturated models of PA are <u>not</u> almost acceptable. In fact, for $\mathfrak{n} \models$ PA the following are equivalent:

(i) $\mathfrak{n}$ is recursively saturated;

(ii) $\mathfrak{n}$ is not almost acceptable;

(iii) Every hyperelementary subset of $\mathfrak{n}$ is in Df($\mathfrak{n}$).

Proof of Lemma 2.3. We content ourselves with proving the lemma in the case $o(\mathbb{HYP}_{\mathfrak{m}}) = \omega$, since that is what we used above. Recall from [1] that, in this case,

$$\mathbb{HYP}_{\mathfrak{m}} = \bigcup_n L(M,n)$$

where L(M, n) is defined by recursion as follows: We adjoin to Gödel's basic operations $\mathfrak{F}_1 \ldots \mathfrak{F}_8$ a few extra ($\mathfrak{F}_9 \ldots \mathfrak{F}_N$) to take care of the atomic relations on urelements (and to make each L(M, n) transitive). For any b we let

$$\mathcal{D}(b) = b \cup \{\mathfrak{F}_i(x,y) \mid 1 \le i \le n;\ x, y \in b\} .$$

Then

$$L(M,0) = M$$

$$L(M,n+1) = \mathcal{D}(L(M,n) \cup \{L(M,n)\}) .$$

Now if $\mathbb{HYP}_{\mathfrak{m}}$ contains a relation $<$ which it believes to wellorder M, that relation can be lifted up to a relation $<_n$ $\mathbb{HYP}_{\mathfrak{m}}$ such that $<_n$ appears to wellorder L(M, n), just as for the ordinary constructible sets. But every $b \in \mathbb{HYP}_{\mathfrak{m}}$ is a subset of some L(M, n). □

The following corollary of 2.2 is of some interest for the program, undertaken by Takeuti and others, of developing, to the extent possible, real mathematical analysis (calculus, theory of series, etc.) in a conservative extension of PA. To state the general result we let T denote some consistent first order theory in the language of arithmetic, PA $\subseteq$ T, and we let T^* be T plus the axioms of induction, Δ^1_1-Comp, and Σ^1_1-AC.

2.4 <u>Corollary</u>. T^* is a conservative extension of T. That is, if φ is a first order sentence provable from T^*, then φ is provable from T.

This corollary is immediate from 2.2 and 2.1(iii). We can improve it to Σ^1_1 sentences as follows:

2.5 Corollary. Suppose $\exists X\varphi(X)$ is a Σ^1_1 sentence provable from T^*. Then there exists a finite number of formulas $\psi_1(x),\dots,\psi_n(x)$ of $L_{\omega\omega}$ such that

$$\varphi(\{x|\psi_1(x)\})\vee\dots\vee\varphi(\{x|\psi_n(x)\})$$

is a theorem of T.

Proof. If not, by compactness there is a model of $T\cup\{\neg\varphi(\{x|\psi(x)\})\,\psi\in L_{\omega\omega}\}$. By Theorem 2.1(iii), this theory has a recursively saturated model $\mathfrak{n}$. By Theorem 2.2, $\langle\mathfrak{n},\mathrm{Df}(\mathfrak{n})\rangle$ is a model of T^* and hence, by the hypothesis of 2.5, is a model of $\exists X\varphi(X)$. Thus there is a definable X such that $\varphi(X)$ is true in $\mathfrak{n}$--so for some $\theta(x,y_1,\dots,y_k)$ of $L_{\omega\omega}$, and for some $a_1\dots a_k$ in N, $\mathfrak{n}$ is a model of

$$\varphi(\{x|\theta(x,a_1,\dots,a_k)\})$$

and hence is a model of

$$\exists y_1\dots y_k\ \varphi(\{x|\theta(x,y_1,\dots,y_k)\}).$$

The problem now reduces to eliminating the $y_1\dots y_k$. Suppose $k=1$. Let $\psi(x)$ be the formula expressing:

$$\theta(x,y_1)\ \text{for the least}\ y_1\ \text{such that}\ \varphi(\{x|\theta(x,y_1)\}).$$

Then $\mathfrak{n}$ is a model of $\varphi(\{x|\psi(x)\})$, a contradiction. Now suppose $k=2$. Let $\psi(x)$ be the formula expressing:

> $\theta(x,y_1,y_2)$ holds for the least y_1 such that
> there is a z, $\varphi(\{x|\theta(x,y_1,z)\})$, and
> where y_2 is the least z such that $\varphi(\{x|\theta(x,y_1,z)\})$.

Then again $\mathfrak{n}\models(\varphi(\{x|\psi(x)\})$, a contradiction. The general case is similar. □

Since T^* contains Σ_1^1-AC, Corollary 2.5 also yields results for essentially Σ_1^1 formulas. For example if $T^* \vdash \forall z \exists X \varphi(z,X)$, where φ is first order, then there are formulas $\psi_1(x,z), \ldots, \psi_n(x,z)$ such that T proves:

$$\forall z \varphi(z, \{x|\psi_1(x,z)\}) \vee \ldots \vee \forall z \varphi(z, \{x|\psi_n(x,z)\}) .$$

The reader can easily verify this from 2.5. From this one can go on to verify that any essentially Σ_1^1 sentence provable in T^* is provable in TA = T + Arithmetic Comprehension. But TA is weaker than T^*. First, the axioms of induction and Δ_1^1-Comp are not provable in TA. And second, any model of T can be trivially expanded to a model of TA.

We do not dwell at great length on these matters since our principal concern is with recursively saturated models. But before we turn to Section 3, and the other half of the main theorem, we want to clear up one point. We have looked only at models of Δ_1^1-PA of the form $\langle \mathfrak{n}, \mathrm{Df}(\mathfrak{n})\rangle$, but there are models of other forms. For example, if $\mathfrak{n}$ is countable and recursively saturated then the set

$$\bigcup \{\mathfrak{X} \mid \langle \mathfrak{n}, \mathfrak{X}\rangle \models \Delta_1^1\text{-PA} + \Sigma_1^1\text{-AC}\}$$

has power $2^{\aleph_0}$. This is any easy consequence of the version of Makkai's Theorem in section IV4 of [1]. It is not clear, however, how many different $\mathfrak{X}$ there are, or exactly which $\mathfrak{X}$ can occur (or which X, for that matter).

§3. Models of Δ^1_1-PA. The results of §2 show that Δ^1_1-PA is a very weak theory. One might get the impression that it is so weak as to be totally useless. The proof of the following theorem shows this impression to be wrong. The remaining implication (ii) ⟹ (i) of Theorem 1.2 is contained in the following result.

3.1 Theorem. If $\mathfrak{n}$ is nonstandard and $\langle \mathfrak{n}, \mathfrak{X} \rangle$ is a model of Δ^1_1-PA then $\mathfrak{n}$ is recursively saturated.

Proof. Identify the standard elements of $\mathfrak{n}$ with the natural numbers. Our strategy is to show that if $\mathfrak{n}$ is not recursively saturated then there is an infinite set Y of standard integers which is Δ^1_1 definable in $\langle \mathfrak{n}, \mathfrak{X} \rangle$ so that, by Δ^1_1-Comp, $Y \in \mathfrak{X}$. But then $\omega = \{y \mid \exists x(y < x \wedge x \in Y)]\}$ is also in $\mathfrak{X}$, by Δ^1_1-Comp, and this violates the axiom of induction. We identify first order formulas of L with their Gödel numbers. We assume that the Gödel numbering has been chosen so that (the number of) φ is greater than (a) the number n such that the free variables of φ are among $x_0, \ldots, x_n$ and (b) (the Gödel numbers of) all proper subformulas of φ. Assume that $\mathfrak{n}$ is not recursively saturated and let $\Phi(x_0, x_1, \ldots, x_n)$ be a recursive set of formulas such that, for some $y_1 \ldots y_n$, $\Phi(x_0, y_1 \ldots y_n)$ is finitely satisfiable but not satisfiable. Let $\hat{\Phi}$ be the set of (standard and nonstandard) formulas of $\mathfrak{n}$ defined by the defining formula of Φ. Since Φ is recursive, we can assume that $\Phi = \{\varphi \in \hat{\Phi} \mid \varphi$ is standard$\}$. Since $\Phi(x_0, \vec{y})$ is not satisfiable, we have

$$\forall x \ \exists \text{ least } \varphi_x \in \hat{\Phi}[\mathfrak{n} \models \neg \varphi_x[x, \vec{y}]]$$

and this $\varphi_x \in \Phi$. On the other hand, the set of all such φ_x is infinite since $\Phi(x_0, y)$ is finitely satisfiable. The set $Y = \{\varphi_x \mid x \in \mathfrak{n}\}$ is thus an infinite set of standard integers. Our task is to show that Y is Δ^1_1 definable over $\langle \mathfrak{n}, \mathfrak{X} \rangle$.

Let Sat(x,X) be the first-order formula in the expanded language which asserts that X is the satisfaction relation for formulas $\leq x$. In particular, it has the following properties:

i) if k is standard and $X \subseteq \mathfrak{n}$ then $(\mathfrak{n}, X) \models \mathrm{Sat}(k, X)$ iff $X = \{$codes for pairs $\langle \varphi, s \rangle \mid \varphi \leq k$, s is a sequence of length $\leq k$ and $\mathfrak{n} \models \varphi[s]\}$

ii) if k is standard then $\langle \mathfrak{n}, \mathfrak{X} \rangle \models \exists ! X\, \mathrm{Sat}(k, X)$ (since satisfaction for a finite number of formulas is first order definable)

iii) if k is infinite and $(\mathfrak{n}, X) \models \mathrm{Sat}(k, X)$ then for all standard φ, s, $\langle \varphi, s \rangle \in X$ iff $\mathfrak{n} \models \varphi[s]$.

(Warning: We can assert that for each standard k, $\langle \mathfrak{n}, \mathfrak{X} \rangle \models \exists X\, \mathrm{Sat}(k, X)$, but we cannot assert that there is such an infinite k. This would require the scheme of foundation, not just the axiom.)

Using these properties of Sat it is easy to give a Δ^1_1 definition of Y over $\langle \mathfrak{n}, \mathfrak{X} \rangle$. The following gives the essentially Σ^1_1 definition of $\varphi \in Y$:

$$\begin{aligned}&\varphi \in \hat{\Phi} \wedge \exists z[z = \neg \varphi \wedge \exists x \exists X \text{ such that} \\ &\mathrm{Sat}(z, X), \text{ and} \\ &\langle \neg\varphi, \langle x, y_1 \ldots y_n \rangle\rangle \in X \\ &\forall \psi < \varphi[\psi \in \hat{\Phi} \rightarrow \langle \psi, \langle x, y_1 \ldots y_n \rangle\rangle \in X]\end{aligned}$$

The essentially Π^1_1 definition of $\varphi \in Y$ is given by:

$$\begin{aligned}&\varphi \in \hat{\Phi} \wedge \exists y[y = \neg \varphi \wedge \exists x\, \forall X[\text{If } \mathrm{Sat}(y, X) \text{ then} \\ &\langle \neg\varphi, \langle x, y_1 \ldots y_n \rangle\rangle \in X \text{ and} \\ &\forall \psi < \varphi[\psi \in \hat{\Phi} \rightarrow \langle \psi, \langle x, y_1 \ldots y_n \rangle\rangle \in X]]] \quad .\end{aligned}$$

This concludes the proof. □

Remark. We only used set parameters in Δ^1_1-comp in one place in the above proof —to go from Y to ω. This could have been avoided by defining ω directly as Δ^1_1, so the proof shows that if $\langle \mathfrak{n}, \mathfrak{X} \rangle$ is a model of Δ^1_1-comp$^-$ (where set parameters are not permitted) then $\mathfrak{n}$ is recursively saturated.

In [2], Ehrenfeucht and Kreisel showed that not every model $\mathfrak{N}$ of the complete theory of arithmetic can be expanded to a model $\langle \mathfrak{N}, \mathfrak{X} \rangle$ of analysis, where by analysis they meant the scheme of full comprehension (which implies the scheme of full induction). In particular, they showed that models of the form $\mathfrak{N}[x]$ cannot be so expanded. The following corollary extends this by showing that these models cannot even be expanded to a model of Δ^1_1-PA.

By using the notation $\mathfrak{N}[x]$, we mean that $\mathfrak{N}$ is generated by x, i.e., if $\mathfrak{N}' \prec \mathfrak{N}[x]$ and $x \in \mathfrak{N}'$ then $\mathfrak{N}' = \mathfrak{N}[x]$. Similarly we use $\mathfrak{N}[\leq x]$ to indicate that $\mathfrak{N}$ is generated by $\{y \mid y \leq x\}$.

3.2 Corollary. If $\mathfrak{N}$ is a nonstandard model of PA of the form $\mathfrak{N}[x]$ or $\mathfrak{N}[\leq x]$ then $\mathfrak{N}$ cannot be expanded to a model of Δ^1_1-PA.

Proof. We show that neither $\mathfrak{N}[x]$ nor $\mathfrak{N}[\leq x]$ can be recursively saturated. In the first case the recursive type $\Phi(z, x)$ omitted is the one asserting that z is not definable from x; in the second, the one asserting that z is not definable from y's $\leq$ x. It is clear that these types are recursive, and that they are omitted. So it only remains to show they are finitely satisfiable. In the $\mathfrak{N}[x]$ case this is trivial--a finite number of formulas can define only finitely many elements. So suppose in $\mathfrak{N}[\leq x]$ that every $z \in \mathfrak{N}[\leq x]$ is definable from a fixed finite set of formulas and parameters $\leq$ x. This violates the axiom of replacement, which is provable in PA. □

3.3 Definition. Let T be a theory whose models are of the form $\langle \mathfrak{N}, \mathfrak{X} \rangle$ where $\mathfrak{X} \subseteq P(\mathfrak{N})$. By the Hard Core of $\mathfrak{N}$ over T we mean $\bigcap \{\mathfrak{X} : \langle \mathfrak{N}, \mathfrak{X} \rangle \models T\}$.

The well-known Gandy-Kreisel-Tait Theorem (see [3]) states that if T is any consistent Π^1_1-theory of second order arithmetic, containing Δ^1_1-Comp, then the Hard Core of T over the standard model of PA is the set of hyperarithmetic sets.

As an immediate corollary to 3.1 we have:

3.4 <u>Corollary</u>. If T is a second order theory of arithmetic, Δ^1_1-PA $\subseteq$ T, and if $\mathfrak{n}$ is a nonstandard, non-recursively-satursated model of PA, then the Hard Core of T over $\mathfrak{n}$ is $\emptyset$.

3.5 <u>Corollary</u>. Let $\mathfrak{n}$ be a nonstandard model of PA. Then the Hard Core of Δ^1_1-PA over $\mathfrak{n}$ (or of Σ^1_1-AC over $\mathfrak{n}$) is just $\emptyset$ or Df($\mathfrak{n}$).

Proof. We prove the Δ^1_1-PA case the other is identical. Suppose $\langle \mathfrak{n}, \mathfrak{X} \rangle \models \Delta^1_1$-PA. Then $\mathfrak{n}$ is recursively saturated. So $\langle \mathfrak{n}, \mathrm{Df}(\mathfrak{n}) \rangle \models \Delta^1_1$-PA, so the Hard Core $\subseteq$ Df($\mathfrak{n}$). But if $\langle \mathfrak{n}, \mathfrak{X} \rangle \models \Delta^1_1$-PA, we easily see that Df($\mathfrak{n}$) $\subseteq \mathfrak{X}$. So Df($\mathfrak{n}$) is the Hard Core of Δ^1_1-PA over $\mathfrak{n}$. □

What is more interesting is that the same result holds for every stronger axiomatizable theory T^*, at least for $\mathfrak{n}$ countable. The following was noticed independently by H. Friedman. In view of Theorem 3.1, it is a special case of Theorem IV 1.1 of [1].

3.6 <u>Theorem</u>. Let T^* be an r.e. second order theory of L^* with Δ^1_1-PA $\subseteq T^*$. Let $\mathfrak{n}$ be a countable nonstandard model such that, for some $\mathfrak{X}$, $\langle \mathfrak{n}, \mathfrak{X} \rangle \models T^*$. The Hard Core of T^* over $\mathfrak{n}$ is Df($\mathfrak{n}$).

Proof. By Theorem IV 1.1 of [1], the Hard Core of T^* is contained in $\mathbb{HYP}_{\mathfrak{n}}$. By 3.1, $\mathfrak{n}$ is recursively saturated, so $o(\mathbb{HYP}_{\mathfrak{n}}) = \omega$, and by 2.1 (iii), the Hard Core is contained in Df($\mathfrak{n}$). But clearly Df($\mathfrak{n}$) $\subseteq \mathfrak{X}$ for any $\mathfrak{X}$ which makes $\langle \mathfrak{n}, \mathfrak{X} \rangle$ a model of T^*. □

4. Epilogue. We conclude by making some (perhaps controversial) remarks on subsystems of analysis.

Let Δ^1_1-CA be the theory Δ^1_1-PA plus the full second-order scheme of induction. Δ^1_1-CA is not a conservative extension of PA since, for example, Con(PA) is provable in Δ^1_1-CA but not in PA. The theory Δ^1_1-CA, and stronger theories like π^1_1-CA, have been studied extensively by proof theoretic methods, but there does not seem to be a good model theory of such subsystems. Our Theorem 1.2, on the other hand, shows that Δ^1_1-PA does have an interesting model theory. So it seems to suggest that the study of other subsystems of analysis, and their associated model theory, might proceed more fruitfully with the axiom of induction, rather than scheme.

Moral: make your induction match your comprehension.

References

1. Barwise, K. J., Admissible Sets and Structures, to appear in Springer Verlag series "Perspectives in Mathematical Logic".

2. Ehrenfeucht, A., and G. Kreisel, Strong Models of Arithmetic, Bull.de l'Acad. Polonaise des Sciences, XV (1966) pp. 107-110.

3. Gandy, R. O., G. Kreisel, and Tait, W., Set existence, Bull. Acad. Polon. Ser. Sci. Math. Astron. Phys., 8 (1960) pp. 577-582.

4. Moschovakis, Y., Predicative classes, Axiomatic Set Theory, Part I, Amer. Math. Soc., (1971) pp. 247-264.

5. Moschovakis, Y., Elementary Induction on Abstract Structures, North Holland Publ. Amsterdam (1974).

6. Robinson, A., Metamathematical problems, J. Symbolic Logic, 38 (1973) pp. 500-516.

7. Schlipf, J. S., Some Hyperelementary Aspects of Model Theory, Doctoral Dissertation, The University of Wisconsin, in preparation.

8. Steal, J., Forcing with tagged trees (abstract), Notices of Amer. Math. Soc. 21 (1974) pp. A-627-8.

A NOTE ON EXISTENTIALLY COMPLETE DIVISION RINGS

Maurice Boffa

A few years ago, Abraham Robinson introduced and developed the forcing notion in the context of model theory. He pointed out [6] the possibility of applications in the domain of algebra, especially for (non commutative) groups and division rings. Applications to division rings became possible as early as P.M.Cohn [3] proved the amalgamation property for these structures. A study of existentially complete division rings was done by several people [1,5,7]. For the basic notions and results about this matter we refer the reader to these papers as well as to the forthcoming lecture notes [4].

Throughout this note, all the division rings will have the same characteristic (0 or any prime). A field will be a commutative division ring. $\langle a_1,\dots,a_n\rangle$ will denote the division subring generated by $a_1,\dots,a_n$ and $C_D(E)$ will be the centralizer of E in D.

Lemma 1. Assume that $a_1,\dots,a_n,a,b,c$ are elements of some existentially complete division ring D such that $a,b,c\notin\langle a_1,\dots,a_n\rangle$. Then there is an inner automorphism t of D such that $t(a_i)=a_i$ $(i=1,\dots,n)$, $b\notin\langle a_1,\dots,a_n,t(a)\rangle$ and $t(a).c\neq c.t(a)$.

Proof. Let $D'=\{x' \mid x\in D\}$ be a disjoint copy of D and let L be an existentially complete division ring including the free product of D and D' amalgamating a_i and a_i' for $i=1,\dots,n$. L has an inner automorphism v such that $v(a_i)=a_i$ $(i=1,\dots,n)$ and $v(a)=a'$.

Since $b\notin\langle a_1,\dots,a_n,a'\rangle$ and $a'c\neq ca'$, L satisfies

$$(\exists u)\left(\bigwedge_{i=1}^{n} u^{-1}a_iu=a_i \text{ and } b\notin\langle a_1,\dots,a_n,u^{-1}au\rangle \text{ and } u^{-1}auc\neq cu^{-1}au\right)$$

which can be put in an existential form, also satisfied by D.

Lemma 2. Assume that D is a countable existentially complete division ring, that D_1 is a division subring of D which is not included in any finitely generated division subring of D, and that b is some non central element of D. Then there is an isomorphism $D \to D'$ such that $b \notin D' \subset D$ and $C_D(D_1') = C_{D'}(D_1')$, where D_1' is the image of D_1 under the isomorphism.

Proof. Put on D a well-ordering of type ω. Define inductively a sequence $(a_n)_{n<\omega}$ as follows: a_{2n} is the first element of $D - \langle a_0, a_1, \ldots, a_{2n-1} \rangle$ and a_{2n+1} is the first element of $D_1 - \langle a_0, a_1, \ldots, a_{2n} \rangle$. Using the lemma 1, we can define inductively a sequence $(a_n')_{n<\omega}$ of elements of D such that $\langle a_0, a_1, \ldots, a_n \rangle \cong \langle a_0', a_1', \ldots, a_n' \rangle$ (by $a_i \mapsto a_i'$), $b \notin \langle a_0', a_1', \ldots, a_n' \rangle$ and a_{2n+1}' does not commute with the first element of $D - \langle a_0', a_1', \ldots, a_{2n}' \rangle$ which commutes with $a_1', a_3', \ldots, a_{2n-1}'$. Our lemma is then satisfied by the division subring D' of D generated by $(a_n')_{n<\omega}$.

The following proposition improves the result of W.H.Wheeler [4, 7] that each countable existentially complete division ring D has an extension E of power $\aleph_1$ such that $D \prec_{\infty\omega} E$ (which is equivalent to saying that E and D have the same finitely generated division subrings).

Proposition. Assume that D is a countable existentially complete division ring and that D_1 is a division subring of D such that $C_D(D_1) \subseteq D_1$. Then D has an extension E of power $\aleph_1$ such that $D \prec_{\infty\omega} E$ and $C_E(D_1) = C_D(D_1)$.

Proof. By a transfinite induction using the lemma 2, there is an ascending chain $(E_\alpha)_{\alpha<\aleph_1}$ such that $E_0=D$, $E_{\alpha+1}$ is a proper isomorphic extension of E_α such that $C_{E_{\alpha+1}}(D_1)=C_{E_\alpha}(D_1)=C_D(D_1)$, and $E_\lambda=\bigcup_{\alpha<\lambda} E_\alpha$ for λ limit. The proposition is then satisfied by $E=\bigcup_{\alpha<\aleph_1} E_\alpha$.

Corollary 1. If D is a countable existentially complete division ring with prime subfield k, then D has an extension E of power $\aleph_1$ such that $D \prec_{\infty\omega} E$ and $C_E(D)=k$.

Proof. Apply the proposition with $D_1=D$.

Corollary 2. Assume that M is a maximal subfield of a countable existentially complete division ring D. Then D has an extension E of power $\aleph_1$ such that $D \prec_{\infty\omega} E$ and M is a maximal subfield of E.

Proof. Apply the proposition with $D_1=M$.

Improving [2], W.H.Wheeler [4] proved that each countable field which is not finitely generated can be embedded as a maximal subfield in each countable existentially complete division ring. Thus, by corollary 2, each countable field which is not finitely generated can be embedded as a maximal subfield in some existentially complete division ring of power $\aleph_1$. But it is not true that each countable field which is not finitely generated can be embedded as a maximal subfield in each existentially complete division ring of power $\aleph_1$. In fact it is easy to prove (using the result of [4] that each countable existentially complete division ring D has a proper isomorphic extension containing a transcendental element centralizing D) that each countable existentially complete division ring D has an extension E of power $\aleph_1$ such that

$D \prec_{\infty\omega} E$ and E has no countable maximal subfield. Many problems can be asked about maximal subfields of existentially complete division rings of power $\aleph_1$, for example the following one: is there an existentially complete division ring of power $\aleph_1$ in which each countable field which is not finitely generated can be embedded as a maximal subfield?

REFERENCES

[1] M.Boffa and P.Van Praag. Sur les corps génériques. C.R.Acad.Sc. Paris 274(1972), p.1325-1327.

[2] ------. Sur les sous-champs maximaux des corps génériques dénombrables. Ibid.275(1972), p.945-947.

[3] P.M.Cohn. The embedding of firs in skew fields. Proc.London Math. Soc. 23(1971), p.193-213.

[4] J.Hirschfeld and W.H.Wheeler. Forcing,arithmetic,and division rings.(to appear)

[5] A.Macintyre. On algebraically closed division rings.(preprint)

[6] A.Robinson. On the notion of algebraic closedness for non commutative groups and fields. J.Symb.Logic 36(1971), p.441-444.

[7] W.H.Wheeler. Algebraically closed division rings,forcing,and the analytical hierarchy.(Dissertation,Yale University,1972)

UNIVERSITE DE L'ETAT A MONS
UNIVERSITE LIBRE DE BRUXELLES

IDEALS OF INTEGERS IN NONSTANDARD NUMBER FIELDS

Gregory L. Cherlin

Mathematisches Institut, Universität Heidelberg

Research supported by NSF Grant P43706

Contents: Pages

Index of special notations:

§2.	A	a nonstandard ring of algebraic integers
	$\mathcal{P}$	the set of definable prime ideals of A
	$\mathcal{D}$	the Boolean algebra of A-definable subsets of $\mathcal{P}$
	$S_{a,b}$	$\{P \in \mathcal{P} : P \text{ divides } (a,b)\}$
	$D(I)$	the filter generated by $\{S_a : a \in I\}$
	$M(D)$	the ideal $\{a : S_a \in D\}$
§3.	A_D	the definable ultrapower $\mathrm{Def}(A^{\mathcal{P}})/D$
	$\mathbb{1}_D$	the prime of A_D represented by $\mathbb{1} : \mathcal{P} \xrightarrow{=} \mathcal{P}$

Special notations:

§4.	s_I	$\{n \in \mathbb{N}^*: I \subseteq M^n\}$ (if M is definable)
	M^s	$\bigcap_{n \leq s} M^n$ (if M is definable)
	$\exp(a,M)$	$\exp(a,P)_D$ where $a = \prod P^{\exp(a,P)}$
	M^s	$\{a: \exp(a,M) \geq s\}$
§5.	$\mathfrak{M}$	the maximal ideal space of A; the bounded ultrafilters on D
	TL	the lattice of ideals of the lattice L
	X	a topological space with base M and fibers $T\mathbb{N}^*_D$
	s or s^X	a compactly supported semicontinuous section of X
	$\prod P^{s_I(P)}$	I
	$\Gamma(S,\mathbb{N}^*)$	the lattice of definable functions from S to $\mathbb{N}^*$
	s^T	a compactly supported continuous section of the sheaf $T\Gamma(\cdot,\mathbb{N}^*)$ over $\mathfrak{M}$
§6.	B	the integral closure of A in a finite dimensional extension of the quotient field of A
	$\mathfrak{p}, \mathfrak{m}, P, M$	prime and maximal ideals in A (resp. B)

§1. Introduction

We will investigate the ring of nonstandard integers in a number field. We have in the first place the standard number fields K, which are just the finite dimensional extensions of the rational field $\mathbb{Q}$, and within a number field K we consider the ring A of algebraic integers. By the phrase "nonstandard number field" we might mean one of the following:

1. An elementary extension K^* of a number field K.
2. A finite dimensional extension of an elementary extension Q^* of Q.
3. A model $\mathcal{K}$ of the theory T of all number fields.

These classes of models are successively larger; for example if p is an infinite prime number in Q^* then $Q^*[\sqrt{p}]$ does not fall under type 1, as an infinite prime number cannot be ramified in an elementary extension of a number field K.

If we wish to study number theory over $\mathbb{Z}^*$ (an elementary extension of the ring of integers) then the natural class of structures to consider is the collection of models of type 2. It would of course be desirable to extend our results to all the models of type 3, but our brush is not that broad (nor our palate that extensive).

In the consideration of the models of type 2, which will be referred to henceforth as "nonstandard number fields" (caveat lector), a great simplification is effected by the following definability theorem of J. Robinson:

THEOREM [1] Let K be a number field of dimension n, let A be the ring of algebraic integers in K, and let $\mathbb{Z}$ be the ring of rational integers. Then A and $\mathbb{Z}$ are both definable in K, by formulas depending only on n.

(The basic idea in the proof is to try to define the set of integers by the formula:

$$a \text{ is an integer iff } "\forall t((\varphi(0,t) \,\&\, \forall x\, (\varphi(x,t) \longrightarrow \varphi(x+1,t) \vee x = a) \longrightarrow \varphi(a,t))"$$

where φ is chosen artfully; the main algebraic ingredient consists of theorems concerning the representation of numbers by quadratic forms.)

It follows also that $\mathbb{Z}$ is definable over A.

For each integer n the previous theorem may be viewed as a statement concerning Q. When applied to Q^* this produces:

Corollary 1.1 Let K be an extension of Q^* of dimension n, and let A be the integral closure of $\mathbb{Z}^*$ in K. Then A and Z^* are definable over K. ⊣

In particular if we let T^n be the theory of n-dimensional extensions of Q and let T_o^n be the theory of the rings of algebraic integers in n-dimensional extensions of Q then:

Corollary 1.2 The models of T^n are the n-dimensional extensions of nonstandard rational fields Q^*.

The models of T_o^n are the rings of integers over $\mathbb{Z}^*$ in models of T^n. ⊣

We know of no similar result or counterexample concerning the definability of nonstandard models of arithmetic within models of the general type 3 above.

So much by way of introduction. We will deal here primarily with the classification of prime and maximal ideals in the ring of integers of a nonstandard number field and the usual associated notions from commutative algebra. The basic facts can be rather easily summarized in a couple of pictures. The space of prime ideals in $\mathbb{Z}$ looks more or less like this:

Fig. 1

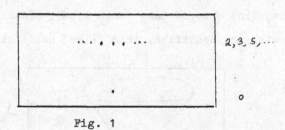

Over any nonstandard model the space of definable prime ideals looks much the same. However in general:

$$\{P \subseteq A : P \text{ is definable}\} \subsetneq \{P \subseteq A : P \text{ is maximal}\} \subset$$
$$\subsetneq \{P \subseteq A : P \text{ is prime}\};$$

thus in the first place there will be additional undefinable maximal ideals as "limit points" of the definable ones (more precise statements can of course be found in the body of the paper). We may represent the space $\mathcal{M}$ of maximal ideals as follows:

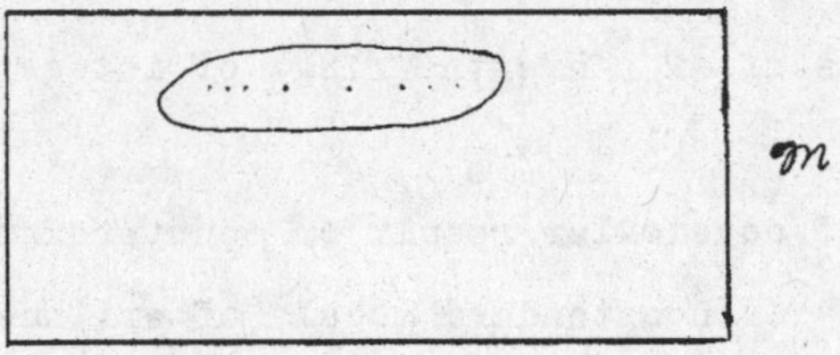

Fig. 2

$\mathcal{M}$ is a locally compact totally disconnected Hausdorff space (in an appropriate topology). Throw in all the prime ideals and we get the "cone" over $\mathcal{M}$ depicted below, bearing in mind the following two facts:

1. Every prime ideal P is contained in a <u>unique</u> maximal M.
2. The "line" $[0,M] = \{P: (0) \subseteq P \subseteq M\}$ consists of all Dedekind cuts in some dense linear ordering. As no cuts are identified, this "line" contains many gaps.

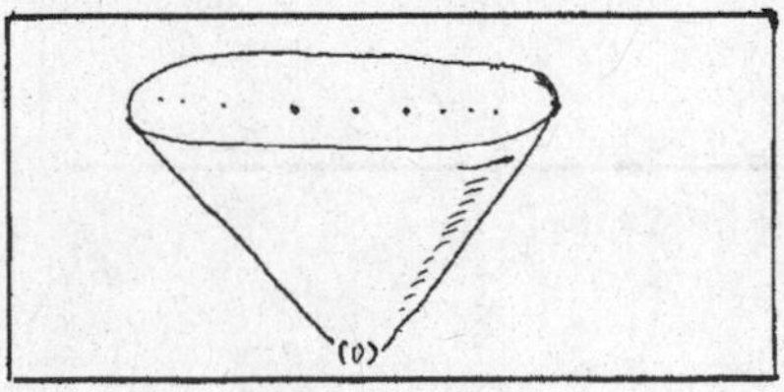

Fig. 3

§2. Maximal ideals

We consider an elementary extension Q^* of Q and an n-dimensional extension K of Q^* (n is finite). Let A be the integral closure of $\mathbb{Z}^*$ in K. Any such ring A will be called a nonstandard ring of integers.

Since Z^* is an elementary extension of Z, J. Robinson's definability theorem tells us that the definable ideals of A behave in every respect like ideals in a standard ring of algebraic integers. For example: if I is definable in A we may write $I = \prod P^{s(P)}$ where P varies over a definable set of definable primes and s is a definable function. As we would like to take such assertions for granted in the future, we will give a more thorough explanation of the present instance.

Let $\mathcal{P}$ be the set of definable primes of A. By transfer from Q each P in $\mathcal{P}$ is at worst doubly generated. We may think of $\mathcal{P}$ as a definable subset of A in several equivalent ways. For example, we may first construe $\mathcal{P}$ as a definable subset of $A \times A$ on which equality is represented by a definable equivalence relation. Then we may use a definable bijection $f: N^* \longleftrightarrow A$ to identify $A \times A$ with A or with N^* and we can select a definable subset of $\mathcal{P}$ (or of the current incarnation of $\mathcal{P}$) containing one representative of each equivalence class. Then with a slight abuse of language we may refer to "definable functions" from $\mathcal{P}$ to N^* and the like.

Now consider the assertion $I = \prod P^{s(P)}$. Notice that the expression $P^{s(P)}$ is to be taken with a grain of salt if $s(P)$ is infinite; any reasonable interpretation will do. If in addition $s(P) > 0$ for infinitely many P (as is frequently the case) then the product $\prod P^{s(P)}$ must again be suitably interpreted.

We now begin the study of the ideals of A by classifying all the maximal ideals. This will involve the Boolean algebra of all definable subsets of $\mathcal{P}$, which we will call $\mathcal{D}$, or more explicitly $\mathcal{D}(A)$. (A subset $S \subseteq A$ will be considered definable iff it is first order definable over A using parameters from A.) Any definable 1-1 enumeration of $\mathcal{P}$ with domain $\mathbb{N}^*$ induces an isomorphism of $\mathcal{D}$ with the algebra of definable subsets of $\mathbb{N}^*$, which is of no great importance apart from the fact that it provides us with a more concrete conception of $\mathcal{D}$.

Most of the definable sets S in which we will be interested are <u>bounded</u> in the sense that A satisfies "S is finite". This can be taken to mean for example that S is the range of a definable function whose domain is an initial segment $[0,n]$ of $\mathbb{N}^*$, or equivalently that the set of norms of ideals in S is bounded in $\mathbb{N}^*$ (hence our choice of terminology). The important observation is that a definable set $S \subseteq \mathcal{P}$ is bounded iff S consists of all the prime divisors of some definable nonzero ideal $(a,b) \subseteq A$, in which case we will write $S = S_{a,b}$.

<u>Notation</u>: $S_{a,b} = \{P \in \mathcal{P} : P \text{ divides } (a,b)\}$

We can now describe the classification of maximal ideals:

<u>Definition</u> 2.1

1. If I is an ideal of A let D(I) be the filter on $\mathcal{D}(A)$ generated by $\{S_a : a \in I\}$.
2. If D is a filter on $\mathcal{D}(A)$ let $M(D) = \{a \in A : S_a \in D\}$.

Notice that if $I \neq (0)$ then D(I) contains a bounded set. Such a filter will be called a <u>bounded</u> filter.

<u>Proposition</u> 2.2

1. If P is a nonzero prime ideal of A then D(P) is a bounded ultrafilter on $\mathcal{D}$.
2. If D is a bounded ultrafilter on $\mathcal{D}$ then M(D) is a

maximal ideal of A.

3. The correspondence $D \longleftrightarrow M(D)$ is a 1-1 correspondence between bounded ultrafilters on $\mathcal{D}$ and maximal ideals of A.

4. If P is a nonzero prime ideal of A then $M(D(P))$ is the unique maximal ideal containing P.

Proof: We note first that $D(I)$ is just the closure under $\supseteq$ of the set $\{S_{a,b}: a,b \in I\}$ because if $a,b,a',b' \in I$ are arbitrary then for suitable c,d we have $(a,b,a',b') = (c,d)$, so that $S_{a,b} \cap S_{a',b'} = S_{c,d}$.

1. Suppose that P is prime. To see that $D(P)$ is an ultrafilter it suffices to consider a partition (X,Y) of any bounded set $S \in D(P)$; we must then show that X or Y is in $D(P)$. We may assume in fact that $S = S_{a,b}$ with $a,b \in P$, and write: $(a,b) = \prod_S Q^{s(Q)}$ for suitable $s(Q)$ in $\mathbb{N}^*$.

Let $I = \prod_X Q^{s(Q)}$ and $J = \prod_Y Q^{s(Q)}$.

Then P contains $IJ = (a,b)$, hence P contains I or J. If for instance $P \supseteq I = (c,d)$ then $X = S_{c,d} \in D(P)$.

2. If D is any filter on $\mathcal{D}$ it is clear that $M(D)$ is an ideal, since $S_{a+b} \supseteq S_a \cap S_b$ and $S_{ac} \supseteq S_a$. Suppose then that D is a bounded ultrafilter, $M = M(D)$, and $a \notin M(D)$. Let $S = S_a$. Then S is not in D, so S is disjoint from some bounded set $T \in D$. Choose b,c to be generators for $\prod_{P \in T} P$. Then $b,c \in M(D)$ and $(a,b,c) = A$ (since (a,b,c) has no definable prime divisors), and this proves that $M(D)$ is maximal.

3. We will show that the maps $D \longrightarrow M(D)$, $M \longrightarrow D(M)$ are inverses of each other.

a). Fix a bounded ultrafilter D. Clearly $D \subseteq D(M(D))$ and since D is maximal therefore $D = D(M(D))$.

b). Fix a maximal ideal M. Clearly $M \subseteq M(D(M))$ and hence $M = M(D(M))$.

4. In the first place $M(D(P))$ is a maximal ideal containing P. If we consider any ideal I containing P then $D(I) \supseteq D(P)$ and hence $D(I) = D(P)$. Therefore $I \subseteq M(D(I)) = M(D(P))$ and thus $M(D(P))$ is the only maximal ideal containing P. ⊣

The previous proposition furnishes an adequate description of the space of maximal ideals, and tells us in addition that each prime ideal lies below a unique maximal ideal. We now attend to the classification of prime ideals P lying below a given maximal ideal M. We will call any ideal I with the property that it is contained in no maximal ideal other than M an M-ideal. The classification of such ideals depends on the definable ultrapower construction.

§3. Definable ultrapowers

Let us recall the definable ultrapower construction. If A is a structure with a definable subset $I \subseteq A$, and if D is an ultrafilter on the Boolean algebra of A-definable subsets of I, we let $\mathrm{Def}(A^I)$ be the space of definable functions from I to A, and we obtain the definable ultrapower $\mathrm{Def}(A^I)/D$ by factoring out the equivalence relation:

$$f \sim g \quad \text{iff} \quad \{i: f(i) = g(i)\} \in D.$$

Thus to any function f is associated its equivalence class f_D in the definable ultrapower. To each a in A we assign the equivalence class a_D of the constant function $a: I \longrightarrow \{a\}$ and this induces the diagonal embedding $\Delta: A \longrightarrow \mathrm{Def}(A^I)/D$, which is an elementary embedding if A possesses definable

Skolem functions. If D is principal then Δ is an isomorphism, and if D is nonprincipal then Δ cannot be an isomorphism as the identity function $\mathbb{1}: I \xrightarrow{=} I$ represents a new element.

When $\Delta: A \longrightarrow Def(A^I)/D$ is an elementary embedding, then each definable function $f: I \longrightarrow A$ represents a function $f^*: I^* \longrightarrow A^* = Def(A^I)/D$ by transfer from A (if "$f(x) = y$" is defined by $\varphi(x,y)$ in A, use $\varphi(x,y)$ to define "$f^*(x) = y$" in A^* as well). We have $f^*(\mathbb{1}_D) = f_D$.

We are about to apply this construction to nonstandard rings of integers, so that in particular the existence of definable Skolem functions is assured. We will take $I = \mathcal{P}$ (viewing $\mathcal{P}$ as a subset of A) and all ultrafilters will be over $\mathcal{D}$. We will denote $Def(A^{\mathcal{P}})/D$ more simply by A_D. However the notation A_M is reserved for the usual operation of localization at M; there will be little opportunity for confusion in the present context because we make no serious use of localization.

Now fix a maximal ideal M, let $D = D(M)$, and consider A_D. In A_D the identity function $\mathbb{1}: \mathcal{P} \xrightarrow{=} \mathcal{P}$ represents a doubly generated maximal ideal $\mathbb{1}_D = (\alpha_D, \beta_D)$, where α, β are suitable definable functions. We will exploit the connection between the possibly undefinable ideal M and the definable ideal $\mathbb{1}_D$ in order to carry out the classification of M-ideals in §4. The rest of this section is devoted to a preliminary analysis of this connection.

The following theorem serves as an illustration of the general principle that the analysis of undefinable ideals may be reduced to the analysis of definable ideals. We have a fairly good understanding of the rings A/P, with P definable and maximal, simply by transfer from the standard case (see also [2]). The situation is <u>a priori</u> less clear if P is not definable, but in fact we have:

Theorem 3.1 If M is a maximal ideal of A and $D = D(M)$ then $\mathcal{M}_D \cap A = M$ and the induced homomorphism $h\colon A/M \longrightarrow A_D/\mathcal{M}_D$ is an isomorphism.

Proof: For the first assertion, fix $a \in A$. Then

a is in $\mathcal{M}_D$ iff

A_D satisfies "$a_D \in \mathcal{M}_D$" iff

$\{P \in \mathcal{P}\colon a \in P\} \in D$ iff

$S_a \in D$ iff

$a \in M(D) = M$, as desired.

Thus h is a monomorphism. To see that h is surjective choose $f_D \in A_D$ and fix a bounded set $S \in D$. We will show that $f_D/\mathcal{M}_D$ is in range(h).

By the Chinese remainder theorem (suitably interpreted in A) we can find an a in A satisfying:

(*) $a \equiv f(P) \pmod P$ for all $P \in S$.

Then A_D satisfies:

$a_D \equiv f_D \pmod{\mathcal{M}_D}$,

so $h(a/M) = f_D/M_D$. ⊣

The reduction of the classification of M-ideals to the classification of $\mathcal{M}_D$-ideals is carried out similarly. First however a point concerning the transfer of second order structure from A to A_D deserves clarification.

Given a definable subset X of A defined by the formula $\chi(x)$, we set $X_D = \{f_D\colon A_D \models \chi(f_D)\}$. Equivalently $X_D = \{f_D\colon \{P\colon f(P) \in X\} \in D\}$, and we can therefore extend this notation to undefinable sets X as well. This is inappropriate from our present point of view because it does not produce the following desirable equation:

$M_D = \mathcal{M}_D$.

In fact if I is any ideal of A and we define I_D by the suggested procedure, then we will simply get $I_D = IA_D$. (The argument is as follows. Fix f_D in I_D and choose a,b in M so that $S_{a,b} \subseteq \{P: f(P) \in I\}$. Let I' be the ideal $(f(P): P \in S_{a,b}) \subseteq I$. Then $I' = (c,d)$ for suitable c,d and thus for any $P \in S_{a,b}$ we can write $f(P) = cf_1(P) + df_2(P)$ with some definable choice of f_1, f_2. Then $f_D = cf_{1D} + df_{2D}$ is in IA_D.) On the other hand IA_D cannot be finitely generated unless I is, as is easily verified.

We can show in fact that when M is undefinable MA_D not only fails to be maximal, but is not even prime. MA_D will be contained in at least 3 different maximal ideals, exactly one of which is definable- namely $\mathcal{M}_D$. The proof of these facts, including a description of those ideals M of A lying below exactly 3 maximal ideals of A_D, occupies the rest of this section.

We will first consider definable ideals over MA_D.

<u>Theorem</u> 3.2 If M is contained in the definable proper ideal I of A_D then $I = \mathcal{M}_D$.

<u>Proof</u>: A definable proper ideal I of A_D corresponds to two definable functions $f_1, f_2: \mathcal{P} \longrightarrow A$ such that $I = (f_{1D}, f_{2D})$. We may assume that for each $Q \in \mathcal{P}$ $(f_1(Q), f_2(Q)) \neq$ $\neq A$ and then fix a definable function $g: \mathcal{P} \longrightarrow \mathcal{P}$ such that for each $Q \in \mathcal{P}$: $g(Q) \supseteq (f_1(Q), f_2(Q))$. Then g_D is a prime ideal containing M. We claim:

(*) for $S \in D$, we have $g[S] \in D$.

Suppose on the contrary that (*) fails, so that we may choose $S \in D$, $T \in D$ with $g[S] \cap T = \emptyset$. By the Chinese remainder theorem choose $a \in A$ so that:

$a \equiv 0 \pmod{Q}$ for $Q \in T$ and

$a \equiv 1 \pmod{Q}$ for $Q \in g[S]$.

Then $a \in M \subseteq g_D$, so $\{Q \in \mathcal{P} : a \in g_D(Q)\} \in D$. Hence $S \cap \{Q \in \mathcal{P} : a \notin g_D(Q)\} \in D$, but this last set is empty, which is a contradiction.

Thus (*) holds. Since D is an ultrafilter it follows, as is well known (cf. [3]) that $\{Q : g(Q) = Q\} \in D$, i.e. $g_D = \mathbb{1}_D$.

So far we have proved that $I \subseteq \mathbb{1}_D$. Writing $I = \mathbb{1}_D I'$, we claim that $I' = A_D$. Otherwise the argument we have given shows that $I' \subseteq \mathbb{1}_D$, so that $M \subseteq I \subseteq \mathbb{1}_D^2$. This possibility is easily eliminated:

Fix a bounded $S \in D$ and let $J = \prod_S Q \subseteq M$. If $M \subseteq \mathbb{1}_D^2$ then $\mathbb{1}_D^2$ divides JA_D, hence $\{Q \in \mathcal{P} : Q^2 \text{ divides } J\} \in D$. As this last set is empty we have a contradiction. ⊣

Similarly we can prove:

<u>Theorem</u> 3.3 Let M, N be maximal ideals of A, $D = D(M)$. The following are equivalent:

1. $N = A \cap I$ for some definable ideal I of A_D.
2. $N = A \cap \overline{N}$ for some definable maximal ideal $\overline{N}$ of A_D.
3. $D(N) = f(D(M))$ for some definable $f : \mathcal{P} \longrightarrow \mathcal{P}$. ⊣

In general the choice of $\overline{N}$ in 2 is not unique (an example is the definable version of Hirschfeld's example, p. 6 of [3]).

It is easy to construct nonstandard rings of integers A such that any undefinable maximal ideal becomes definable (in the sense of Theorem 3.3) in the nonstandard extension corresponding to any other undefinable ideal. In general these and various other more complicated structures may be produced by combinatorial arguments involving the explicit construction of definable ultrafilters, which are then used to form definable ultrapowers of N.

Now we will come back to the question: how many maximal

ideals of A_D can contain MA_D? We will prove that the number of such ideals is at least 3 and can equal 3. If A is countable it is also clear that this number cannot lie strictly between ω and 2^ω, because it is the cardinality of a compact Hausdorff space with a countable basis of open sets. We conjecture that the other values ($[4,\omega] \cup \{2^\omega\}$) can be attained.

Theorem 3.4. Let M be an undefinable ideal of A, D = D(M). Then the number of maximal ideals of A_D containing MA_D is at least 3, and may equal 3.

Proof: Fix a definable linear ordering < of $\mathcal{P}$ (and hence also of $\mathcal{P}_D$). Let $D^* = \{S_D : S \in D\}$. We claim that either of the sets

$$X = \{P_D \in \mathcal{P}_D : P_D <_D \mathbb{1}_D\},\ Y = \{P_D \in \mathcal{P}_D : P_D\ {}_D\!> \mathbb{1}_D\}$$

may be adjoined to D*; that is we claim that $D^* \cup \{X\}$ and $D^* \cup \{Y\}$ generate proper filters. Taking into account the principal filter $D(\mathbb{1}_D)$ this will yield at least 3 ultrafilters, and hence at least 3 maximal ideals.

It suffices to check that $D^* \cup \{Y\}$ generates a proper filter. Suppose on the contrary that for some set S of D we have $S_D \cap Y = \emptyset$, i.e. there is no definable $f: \mathcal{P} \longrightarrow S$ such that $f_D\ {}_D\!> \mathbb{1}_D$. This is ridiculous, for if we define a function $f: S \longrightarrow S$ in such a way that $f(P) > P$ for all $P \in S$ (with the exception of the largest P in S, if there is one) then clearly $f_D\ {}_D\!> \mathbb{1}_D$.

Now we give an example in which only 3 maximal ideals occur above M. Let D be an ultrafilter with the property that for any definable binary relation $\varphi(x,y)$ there is a set $S \in D$ of order indiscernibles for φ (i.e. for x,y,x_1,y_1 in S if $x < y$ and $x_1 < y_1$ then $[\varphi(x,y) \Leftrightarrow \varphi(x_1,y_1)]$). Such filters are dense in the space of nonprincipal filters if the nonstandard model is countable (this is just Ramsey's theorem using definable sets of order indiscernibles). Let X, Y be as above, and let $D_1 = \langle D^* \cup \{X\}\rangle$, $D_2 = \langle D^* \cup \{Y\}\rangle$, $M_1 = M(D_1)$, $M_2 = M(D_2)$. Our claim is of course that D_1, D_2 are the only nonprincipal ultrafilters of $\mathcal{D}(A_D)$ containing D^*. Notice that $MA_D = \mathbb{1}_D M_1 M_2$, so this gives us the prime factorization of MA_D in A_D.

Fix a definable set Z in $\mathcal{P}_D$. Assume that $\mathbb{1}_D \notin Z$. We must show that $Z \in D_1$ or D_2. Let $\varphi(y)$ be a definition of Z. Since every element of A_D is definable from $\mathbb{1}_D$ using parameters from A, we may assume that φ is of the form $\varphi(\mathbb{1}_D, y)$ where all parameters not exhibited lie in A.

Now fix a set $S \in D$ of order indiscernibles for the formulas $\varphi(x,y)$ and $\varphi(y,x)$. Then on S we have either:

1. $P,Q \in S$ and $P < Q \Rightarrow \varphi(P,Q)$

or 2. $P,Q \in S$ and $P < Q \Rightarrow \neg\varphi(P,Q)$

and similarly for $\varphi(Q,P)$.

We will establish in a moment that in case 1 $Z \in D_2$. Similarly in case 2 the complement Z' of Z is in D_1, with analogous statements in the cases 1',2' corresponding to $\varphi(Q,P)$. Let us just verify first that this completes the proof.

Clearly if case 1 or 1' applies then Z is in D_1 or D_2, as desired. On the other hand if cases 2, 2' both apply then the complement Z' of Z is in D_1 and D_2, and since $\mathbb{1}_D \notin Z$ it follows that Z' is already in D, in which case the filter $\langle D^* \cup \{Z\}\rangle$ is already improper.

So now let us assume that for $P < Q$ in S, $\varphi(P,Q)$ holds. We claim that $S_D \cap Y \subseteq Z$, i.e. that if $f: \mathcal{P} \longrightarrow \mathcal{P}$ is definable then: $f_D \in S_D \cap Y \Rightarrow \varphi(\mathbb{1}_D, f_D)$.

Fix $f_D \in S_D \cap Y$. We may assume that $f: S \longrightarrow S$ and that

$P < f(P)$ for $P \in S$ (with perhaps one exception). In this case we have $\varphi(P,f(P))$ for (almost) all $P \in S$, and hence $\varphi(1_D,f_D)$, as desired. ⊣

§4. M-ideals

Let us consider initially a definable maximal ideal M in a nonstandard ring of algebraic integers A. We will classify the M-ideals, that is the ideals I of A contained in M and in no other maximal ideal. If I is itself definable then of course $I = M^n$ for a suitable nonstandard natural number n in $\mathbb{N}^*$ (we take $\mathbb{N}^* = \{n \in Z^*: n \geq 1\}$; 0 will be omitted). In general let $s_I = \{n: I \subseteq M^n\}$. Then s_I is a Dedekind cut, in the sense that s_I is a proper initial segment of $\mathbb{N}^*$. We will refer to the set of all such cuts as the Dedekind completion of $\mathbb{N}^*$, and we write $m \leq s$, $m > s$, $m = s$ according as $m \in s$, $m \notin s$, $s = [1,m]$.

Remark: $\bigcap_{m\leq s} M^m = \bigcup_{n\geq s} M^n$ for any Dedekind cut s.

Proof: Clearly $\bigcap_{m\leq s} M^m \supseteq \bigcup_{n\geq s} M^n$, and conversely if $a \in \bigcap_{m\leq s} M^m$ then the exponent n of M in the factorization of (a) into definable ideals must be at least s, so $a \in \bigcup_{n\geq s} M^n$. ⊣

Notation: $M^s = \bigcap_{m\leq s} M^m$.

It will be convenient to have several characterizations of M-ideals, valid also in the undefinable case.

Lemma 4.1 Let M be a maximal ideal of A, $I \subseteq M$. Then the following are equivalent:

1. I is an M-ideal
2. $D(I) = D(M)$
3. For some $n \in \mathbb{N}^*$ the set $M^{(n)} = \{a^n: a \in M\}$ is contained in I.

Proof: Clearly 1 <=> 2, and 3 => 2 since for each a

$S_a = S_{a^n}$.

2 => 3: Fix $a \in I$ and factor $(a) = \prod_{\mathcal{P}} P^{s(P)}$. Let $n = \sup_{\mathcal{P}}(s(P))$. We claim $M^{(n)} \subseteq I$. Indeed fix $b \in M^{(n)}$, write $(b) = \prod_{\mathcal{P}} P^{t(P)}$, and let $S = \{P \in \mathcal{P} : t(P) > 0\}$. Then $t \geq s$ on S. Since $S \in D(M) = D(I)$, fix $(c,d) \subseteq I$ such that $S_{c,d} \subseteq S$. Then $b \in (a,c,d) \subseteq I$, as desired. ⊣

Notice also that if M is definable then 3 is equivalent to:

3'. For some n $M^n \subseteq I$.

<u>Proposition</u> 4.2 Suppose M is a definable maximal ideal. Then each M-ideal may be written uniquely as $I = M^{s_I}$, where s_I is a Dedekind cut in $\mathbb{N}^*$.

<u>Proof</u>: Notice first that each ideal I of the form M^s is an M-ideal, for if we choose $n > s$ then $M^n \subseteq I$.

Suppose then that I is an M-ideal. If a is any element of I and (a) has the factorization $(a) = M^k J$ with $(J,M) = A$ then $M^k \subseteq I$, because for some n $M^n \subseteq I$ and therefore $M^k \subseteq (M^n, M^k J) \subseteq I$.

Clearly $I \subseteq M^{s_I}$. If $n \geq s_I$ we must show that $M^n \subseteq I$; it then follows that $I \supseteq \bigcup_{n \geq s_I} M^n = M^{s_I}$ and the proof will be complete.

Fix $n > s_I$. Then $I \not\subseteq M^n$, so we may choose $a \in I - M^n$ and write $(a) = M^k J$ with $(M,J) = A$, $k \leq n-1$. Thus $M^{n-1} \subseteq M^k \subseteq I$. This completes the proof. ⊣

Now we can use the technique of §3 to reduce the undefinable case to the definable case. In the present instance this seems to require slightly more care than is really desirable.

<u>Proposition</u> 4.3 The map $I \xrightarrow{\pi} I \cap A$ from $\mathcal{M}_D$-ideals of A_D to M-ideals of A is a 1-1 correspondence.

<u>Proof</u>: We should verify first that π takes $\mathcal{M}_D$-ideals to

M-ideals. Assume therefore that I is a $\mathbb{1}_D$-ideal, and that $\mathbb{1}_D^{n_D} \subseteq I$. Fix a bounded set S in D and let $n = \sup_S n_D(Q)$. Fix a in M. We will prove that a^n is in I, so that $I \cap A$ is an M-ideal. This is in fact obvious: a^{n_D} is in I, and a^{n_D} divides a^n.

We will prove the proposition by defining an inverse for π, an extension map e from M-ideals J of A to $\mathbb{1}_D$-ideals $I = e(J)$ in A_D. As we have noted, JA_D is in general not an M-ideal. If we localize A_D at $\mathbb{1}_D$ and then intersect the localization $(JA_D)_{\mathbb{1}_D}$ with A_D we will in fact get a suitable ideal $e(J)$. We will take a different approach, equivalent to the foregoing.

Let J be an M-ideal of A and define a Dedekind cut s_J in $\mathbb{N}_D^*$ by:

$n_D \in s_J$ iff $J \subseteq \mathbb{1}_D^{n_D}$.

Set $e(J) = \mathbb{1}_D^{s_J}$. Evidently $e(J)$ is a $\mathbb{1}_D$-ideal. Thus we need only verify that $\pi \circ e$, $e \circ \pi$ are both identity mappings and the proof will be complete.

<u>Computation of $\pi \circ e$:</u>

$\pi \circ e(J) = \mathbb{1}_D^{s_J} \cap A$, which clearly contains J. If on the other hand $a \notin J$ let $(a) = \mathbb{1}_D^n I$ in A_D with $n \in \mathbb{N}_D^*$, $(I, \mathbb{1}_D) = A_D$. Then a is not in $\mathbb{1}_D^{n+1}$, and if we prove that $J \subseteq \mathbb{1}_D^{n+1}$ then we have $a \notin \pi \circ e(J)$ which shows that $\pi \circ e(J) = J$.

We proceed by contradiction. Suppose $b \in J$, $b \notin \mathbb{1}_D^{n+1}$. Then $(b) = \mathbb{1}_D^k I'$ with $k \leq n$, $(I', \mathbb{1}_D) = A_D$. Now $a \in \mathbb{1}_D^k$, so for some set $S \in D(M)$ $a \in \prod_S Q^{k(Q)}$. Fix $u, v \in J$ so that $S_{u,v} \subseteq S$. Then $(b,u,v) = \prod Q^{k(Q)}$, so $a \in J$, a contradiction.

<u>Computation of $e \circ \pi$:</u>

Let I be a $\mathbb{1}_D$-ideal of A_D. Then $e \circ \pi(I) = \mathbb{1}_D^{s_{I \cap A}}$. We must show that $s_{I \cap A} = s_I$. By definition:

1. $n_D \in s_{I\cap A}$ iff $\mathbb{1}_D^{n_D} \supsetneq I \cap A$
2. $n_D \in s_I$ iff $\mathbb{1}_D^{n_D} \supsetneq I$.

Clearly 2 => 1. We will show that ¬2 => ¬1 to complete the proof.

Fix $f_D \in I$, $f_D \notin \mathbb{1}_D^{n_D}$. Use the Chinese remainder theorem to choose a in A,

$a \equiv f(Q) \pmod{Q^{k(Q)}}$ for all Q in some $S \in D(M)$,

where k_D is chosen so large that $\mathbb{1}_D^{k_D} \subseteq I$. Then a is in $I \cap A$ and $a \equiv f_D \pmod{\mathbb{1}_D^{n_D}}$, so $a \notin \mathbb{1}_D^{n_D}$.

This proves ¬1, as desired. ⊣

Notation: If I is an M-ideal of A let s_I be the following Dedekind cut of $\mathbb{N}_D^*$:

$\{n_D : \mathbb{1}_D^{n_D} \supsetneq I\}$.

Combining Propositions 4.2, 4.3 we have:

Corollary 4.4 For any maximal ideal M of A, any M-ideal I of A may be written uniquely as $I = \mathbb{1}_D^{s_I} \cap A$. ⊣

For any maximal ideal M of A and Dedekind cut s of $\mathbb{N}_D^*$ (D = D(M)) let us agree momentarily to write $M^s = \mathbb{1}_D^s \cap A$. Then we restate the preceding corollary and proposition as:

Theorem 4.5 Any M-ideal I of A may be expressed uniquely as $I = M^s$ where the exponent s is a Dedekind cut in $\mathbb{N}_D^*$. There is a 1-1 correspondence between ideals M^s of A and ideals $\mathbb{1}_D^s$ of A_D. ⊣

This result is more satisfactory if we adopt the following point of view. For each element a of A write the factorization of a into primes in the form:

$(a) = \prod Q^{\exp(a,Q)}$

where $\exp(a,\cdot)$ is a definable function from $\mathcal{P}$ to $\mathbb{N}^*$. For any maximal ideal M of A let D = D(M) and set $\exp(a,M) = \exp(a,\cdot)_D$, which lies in $\mathbb{N}_D^*$. An easy computation shows that for any Dedekind cut s in N_D^*:

$$M^s = \{a \in A: \exp(a,M) \geq s\}.$$

Hence this formula should be taken as the definition of M^s. It is then easy to prove the first part of Theorem 4.5 without looking very closely at A_D.

Let us specialize this result in order to classify the prime ideals P of A. Since each prime ideal P is an M-ideal for a suitable M, and since the M-ideals are classified by cuts in A_D, the prime ideals will be classified by certain "prime cuts". (We will use a less culinary terminology.)

Definition 4.6 If s is a Dedekind cut in $\mathbb{N}^*$ call s:

1. a nonprincipal cut iff s is closed under addition of 1.
2. an additive cut iff s is closed under addition.

We may introduce two equivalence relations on $\mathbb{N}^*$ by:

$m \approx n$ iff $|m-n|$ is finite

$m \sim n$ iff $|\log(m/n)|$ is finite.

Nonprincipal and additive cuts correspond to cuts in the quotient orderings $\mathbb{N}^*/\approx$ and $\mathbb{N}^*/\sim$. There is a natural bijection between nonprincipal and additive cuts; namely the injection $i: \mathbb{N}^* \longrightarrow \mathbb{N}^*$ defined by $i(n) = 2^n$ induces an order isomorphism between the two quotient orderings $\mathbb{N}^*/\approx$, $\mathbb{N}^*/\sim$ and this induces a bijection between the cuts in the quotient orderings. Thus if s is nonprincipal we may define the additive cut 2^s, and conversely we associate to any additive cut the nonprincipal cut $\log s$.

In order to reduce the classification of prime M-ideals to the classification of prime 1_D-ideals we supplement Theorem 4.5 as follows:

Proposition 4.7 $M^s M^t = M^{s+t}$ (addition of Dedekind cuts). Equivalently, the correspondence of Proposition 4.3 preserves multiplication.

Proof: A little care must be exercised in the definition of addition for Dedekind cuts. We define s+t by:

$k \geq s+t$ iff for some $i \geq s$, $j \geq t$ $(k = i+j)$.

In general $s+t \neq \{i+j: i \leq s, j \leq t\}$ (examples abound).

Now since $M^s = \bigcup_{i \geq s} M^i$, $M^t = \bigcup_{j \geq t} M^j$, and $M^{s+t} = \bigcup_{k \geq s+t} M^k = \bigcup_{i \geq s, j \geq t} M^{i+j}$, therefore it suffices to prove the proposition when s,t,s+t are integers i,j,i+j in $\mathbb{N}^*_D$.

We use the formula $M^i = \{a \in A: \exp(a,M) \geq i\}$ mentioned above. Then clearly $M^iM^j \subseteq M^{i+j}$. For the converse inclusion fix $a \in M^{i+j}$ and write $(a) = \prod Q^{k(Q)}$ where $k_D \geq i+j$. We may set $k = k_1 + k_2$ with $k_{1D} \geq i$, $k_{2D} \geq j$. Let

$$I = \prod Q^{k_1(Q)}, \quad J = \prod Q^{k_2(Q)}.$$

Then $a \in IJ \subseteq M^iM^j$, as desired. ⊣

Theorem 4.8 The M-ideal M^s is prime iff s is additive or $s = 1$. Hence in particular in the correspondence between M-ideals and $\mathbb{1}_D$-ideals prime ideals on one side correspond to prime ideals on the other side.

Proof: The second statement follows immediately from the first, since corresponding ideals are associated with the same cut. This correspondence between prime M-ideals and prime $\mathbb{1}_D$-ideals, which might seem a priori evident, appears in fact to require some proof.

Let us now fix an M-ideal M^s and assume that s is additive. Then M^s is prime, for if $ab \in M^s$ then $\exp(a,M) + \exp(b,M) = \exp(ab,M) \geq s$, and thus by the additivity of s $\exp(a,M) \geq s$ or $\exp(b,M) \geq s$, so that a or b is in M^s.

If conversely M^s is prime and $m_D, n_D < s$ then $M^s \not\supseteq M^{m_D}, M^{n_D}$, so $M^s \not\supseteq M^{m_D}M^{n_D} = M^{m_D+n_D}$ which proves that $m_D + n_D < s$. It follows at once that if $s > 1$ then s is additive. ⊣

<u>Corollary</u> 4.9 An ideal I is prime iff:

1. I is contained in a unique maximal ideal

and 2. for all a, $a^2 \in I \Rightarrow a \in I$. ⊣

<u>Corollary</u> 4.10 Let A be a countable nonstandard ring of algebraic integers. Then A has exactly $2^{\aleph_0}$ maximal ideals, and contained in each maximal ideal is a linearly ordered set of $2^{\aleph_0}$ prime ideals, order isomorphic with the space of Dedekind cuts in the rational interval $[0,1]$ (making no identifications between cuts).

<u>Proof</u>: The last statement is simply a description of the order type of the additive Dedekind cuts. We introduced the nonprincipal cuts to make this structure clearer; as is well known the order type of $N^*/\approx$ for N^* countable is the rational interval $[0,1)$--the missing endpoint corresponds to the prime ideal (0) lying under all M. ⊣

This justifies Figure 3, §1.

§5. <u>General ideals</u>

We will now give a general description of the lattice of all ideals of A. If L is any lattice we will let TL be the lattice of all proper ideals of L. When L is an ordered set we continue to refer to the ideals of L as "Dedekind cuts". Our intention is to express each ideal I of A in the form:

$$(5^*)\quad I = \prod_{M \in \mathfrak{M}} M^{s_I(M)} .$$

Here $\mathfrak{M}$ is the space of maximal ideals and s_I takes values in the set of Dedekind cuts $T\mathbb{N}^*_D$ $(D = D(M);\ M \in \mathfrak{M})$. One way to make this intelligible is as follows.

Begin by defining the set

$X = \{(M,c) : M \in \mathfrak{M},\ c \in T(\mathbb{N}^*_D),\ D = D(M)\}.$

The space $\mathfrak{M}$ of maximal ideals is to be viewed as an open dense subset of the Stone space of $\mathcal{D}(A)$. X is "fibered" over $\mathfrak{M}$ by the sets $T\mathbb{N}^*_D$. Topologize X by associating to any definable $S \subseteq \mathfrak{M}$ and any definable $f,g: S \to \mathbb{N}^*$ the basic open set

$O_{S,f,g} = \{(M,c) \in X: M \in S,\ g_D < c < f_D \quad (D = D(M))\}.$

Here f_D, $g_D \in \mathbb{N}^*_D$ are as in §3.

Any function $s: \mathfrak{M} \to X$ satisfying $\pi \circ s = \mathbb{1}$ (with π the projection of X to $\mathfrak{M}$ and $\mathbb{1}$ the identity on $\mathfrak{M}$) is called a <u>section</u> of X. We will not use the continuous sections of X; rather we will look at the <u>semicontinuous</u> sections of X, which are defined to be those sections of X which are continuous with respect to the following rather coarse topology:

$O_{S,f} = \{(M,c) \in X: M \in S,\ c < f_D \quad (D = D(M))\}.$

<u>Definition</u> We correlate ideals of A and semicontinuous sections of X as follows:

If I is an ideal of A let $s_I: \mathfrak{M} \to X$ be defined by: $s_I(M) = (M,c)$ where c is determined by

$I_M = M^c$ (here $I_M = IA_M$ - that is, I localized at M; this is an M-ideal).

<u>Theorem</u> 5.1 The correspondence $I \to s_I$ is an isomorphism between the lattice of ideals of A and the lattice of compactly supported semicontinuous sections of X. Under this correspondence multiplication of ideals corresponds to pointwise addition of sections, the addition being addition of cuts.

<u>Proof</u>: Concerning the addition of cuts compare Proposition

4.7.

It is clear that each s_I is compactly supported and that the map $I \mapsto s_I$ is 1-1 since two ideals which have the same localization at each maximal ideal must coincide. We therefore must verify that the range of $I \mapsto s_I$ is precisely the set of compactly supported semicontinuous sections.

Observe in the first place that $s_I = \inf_{a \in I} s_{(a)}$. Thus the sections s_I are exactly the infima of "principal" sections. Since for any a,b we have $s_{(a,b)} = \inf(s_{(a)}, s_{(b)})$ we are reduced to showing that any semicontinuous section is an infimum of the rather special continuous sections of the form $s_{(a,b)}$ (we deal throughout only with compactly supported sections). This is a straightforward bit of point set topology (using the Chinese remainder theorem) and we leave it to the reader. ⊣

In particular we now have a reasonable interpretation of (5*).

<u>Corollary</u> 5.2 If $\underline{M} = \{M_1, \ldots, M_k\}$ is a finite set of maximal ideals then there is a 1-1 correspondence between the set of $\underline{M}$-ideals — i.e. ideals contained in no maximal ideals other than $M_1, \ldots, M_k$ -- and the set $T(N_1^*) x \ldots x T(N_k^*)$ where $N_i^* = N_{D(M_i)}^*$. ⊣

We will devote the rest of this section to exposing the defects of a more natural classification of the ideals of A. We begin by associating with each definable set S (= basic open neighborhood in $\mathfrak{M}$) the lattice $\Gamma(S, N^*)$ of all definable functions from S to N*. Thinking of the function $\Gamma(\cdot, N^*)$ as a lattice-valued presheaf over $\mathfrak{M}$, the stalks will be given by $\Gamma_M = N_{D(M)}^*$. Now consider the covariant functor $T: DL \to DL$ on the category of Distributive Lattices which takes a lattice L to its lattice of ideals. Then $T\Gamma(\cdot, N^*)$ is a presheaf of distributive lattices. Now every ideal corresponds naturally to a compactly supported continuous section of $T\Gamma$ -- in other words to a suitable filter on $\Gamma(\mathcal{P}, \mathbb{N}^*)$ -- via the correspondence $I \longrightarrow \langle s_a : a \in I\rangle$ (here $(a) = \prod_{P \in \mathcal{P}} P^{s_a(P)}$). Thus we can classify the ideals by compactly supported sections of a sheaf, and if the functor T happened to commute with sufficiently many direct limits we would then compute $(T\Gamma)_M = T(\Gamma_M)$ = Dedekind cuts in $\mathbb{N}^*_M$, and we would have recovered X, equipped however with a sheaf topology. However none of this occurs and $T\Gamma$ appears to be a rather unpleasant object. The following observations may clarify the difference between $T\Gamma$ and X:

1. If M is a nonprincipal maximal ideal then the corresponding section s^X_M of X is given by:
$$s^X_M(N) = \begin{cases} 0 & M \neq N \\ 1 & M = N \end{cases}$$
whereas $s^{T\Gamma}_M$ is continuous.
2. If furthermore $a \in M - M^2$ and $I = (a)$ then $\{N : s^{T\Gamma}_I(N) = s^{T\Gamma}_M(N)\}$ intersected with S_a is an open subset of $\{M\}$, hence empty, whereas $s^X_I(M) = s^X_M(M)$.
3. The natural maps $\varphi_M : (T\Gamma)_M \longrightarrow T(\Gamma_M)$ induce a map $\varphi : T\Gamma \longrightarrow X$. By observation 2 this map is not 1-1, so the difference between $T\Gamma$ and X is not just a matter of the topology.

§6. A little number theory

We fix a nonstandard ring of integers A with quotient field K and an n-dimensional extension L of K with ring of integers B. Also fix a prime ideal $\mathfrak{p}$ of A associated with the ultrafilter D. We will use the method of §3 to reduce questions concerning the primes of B over $\mathfrak{p}$ to the definable case (the definable case is trivial because n is finite).

Begin with the commutative diagram:

$$\text{(I)}\quad \begin{array}{ccc} B & \longrightarrow & B_D \\ \uparrow & & \uparrow \\ A & \longrightarrow & A_D \end{array} \qquad (B_D = \mathrm{Def}(B^{\mathcal{P}})/D \text{ where } \mathcal{P} \text{ is the set of primes of } A.)$$

We will be concerned with residue fields and prime factorization in the context of diagram I. Let $\mathfrak{m}$ be the unique maximal ideal containing $\mathfrak{p}$ in A and let $\mathfrak{m}_D$ be the corresponding definable ideal of A_D ($\mathcal{I}_D$ in the notation of §4). We have $\mathfrak{p} = \mathfrak{m}^s$ with s an additive cut in $\mathbb{N}_D^*$; write also $\mathfrak{p}_D = \mathfrak{m}_D^s$. The general version of Theorem 3.1 is:

<u>Theorem</u> 6.1 $\mathfrak{p}_D \cap A = \mathfrak{p}$ and the induced homomorphism $h: A/\mathfrak{p} \longrightarrow A_D/\mathfrak{p}_D$ is an isomorphism.

<u>Proof</u>: The first statement is discussed in §4, before and after Theorem 4.5. For the second statement use the proof of Theorem 3.1, replacing the equation (*) used there by:

(**) $a \equiv f(P) \pmod{P^k}$ for each P in S

where $k \in \mathbb{N}^*$ is chosen so that $k > s$. ⊣

This works equally well for arbitrary M-ideals, naturally.

As a complement to Theorem 6.1 we should supply some information concerning the quotients $A/\mathfrak{p}$ when $\mathfrak{m}$ is already definable. We have:

<u>Theorem</u> 6.2 Let $\mathfrak{p}$ be a nonmaximal prime of the nonstandard ring of integers A, and suppose that $\mathfrak{p}$ is contained in the definable maximal ideal $\mathfrak{m}$. Then $A/\mathfrak{p}$ is elementarily equivalent to the ring of $\mathfrak{m}$-adic integers in the Henselization of K with respect to the $\mathfrak{m}$-adic valuation (the value group here is $\mathbb{Z}^*$). (A simple description of such theories is found in the work of Ax-Kochen-Ershov, described in [4]; in our proof of Theorem 6.2 we rely heavily on these results.)

<u>Proof</u>: Let K' be the Henselization of K. Then the theory of K' is determined by the following information [4]:

1. K' is a valued field satisfying Hensel's lemma, with a value group elementarily equivalent to $\mathbf{Z}$.
2. In the unramified case it is enough to know in addition the theory of the residue field $\overline{K}'$; in the ramified case (which only occurs when the residue field is actually finite, in the present context) one must know the theory of $\overline{K}'$ with respect to certain distinguished elements (one starts with an element π of order 1 at $\mathfrak{m}$, considers the coefficients $a_0, \ldots, a_{e-1}$ of the corresponding Eisenstein polynomial, and expands each a_i modulo finite powers p^j of the residual characteristic using Teichmüller representatives a_{ij}; the residues $\bar{a}_{ij}$ must then be distinguished in $\overline{K}'$).

We must show that with respect to these two types of information, the quotient field of $A/\mathfrak{p}$ has the same properties as K'. Happily, point 2 creates no problems since the residue fields (with distinguished elements) are canonically identified: $A/\mathfrak{p}\ /\ \mathfrak{m}/\mathfrak{p} \simeq A/\mathfrak{m}$.

Thus it suffices to prove that $A/\mathfrak{p}$ satisfies Hensel's lemma; namely we assert:

(HL) Fix $f(x) = a_0 + a_1x + \ldots + a_Nx^N$ with coefficients in $A/\mathfrak{p}$ and suppose that f has a root α modulo $\mathfrak{m}$ such that $f'(\alpha) \notin \mathfrak{m}$. Then f has a root modulo $\mathfrak{p}$.

The proof is trivial: write $\mathfrak{p} = \mathfrak{m}^s$, fix $k > s$ in $\mathbb{N}^*$, and by transfer from standard models conclude that there is a root modulo $\mathfrak{m}^k$, hence certainly modulo $\mathfrak{p}$. ⊣

We proceed to the question of prime factorization in B. First consider the situation: $I \subseteq \mathfrak{m} \subseteq A \subseteq B$ with $\mathfrak{m}$ a definable maximal ideal and I an $\mathfrak{m}$ -ideal. The definable ideal $\mathfrak{m}B$ factors in B as:

$$\mathfrak{m}B = \prod_{1\leq i\leq k} Q_i^{e_i}$$

Here the Q_i are finitely many definable maximal ideals of B and the e_i are (finite) ramification indices. Writing $I = \mathfrak{m}^s$ with s a Dedekind cut in $\mathbb{N}^*$ we see that

$$IB = \prod_{1\leq i\leq k} Q_i^{e_i s}.$$

If $I = \mathfrak{p}$ is a nonmaximal prime then the cut s is additive, so $e_i s = s$ and we have:

$$\mathfrak{p}B = \prod Q_i^{e_i} \qquad \text{if } \mathfrak{p} = \mathfrak{m}.$$
$$\mathfrak{p}B = \prod Q_i^{s} \qquad \text{if } \mathfrak{p} \subsetneq \mathfrak{m}.$$

It follows that the prime ideal $\mathfrak{p}$ is contained in the maximal ideals $Q_1,\ldots,Q_k$ and no others, and that the primes P of B lying over $\mathfrak{p}$ (by which we mean $P \cap A = \mathfrak{p}$) are exactly:

$$\{Q_1,\ldots,Q_k\} \quad \text{if } \mathfrak{p} = \mathfrak{m}.$$
$$\{Q_1^s,\ldots,Q_k^s\} \quad \text{if } \mathfrak{p} \subsetneq \mathfrak{m}.$$

The last statement requires further verification. It suffices to show that if $Q = Q_i$, $e = e_i$ then for any Dedekind cut s in N^*:

$$Q^{es} \cap A = \mathfrak{m}^s.$$

Clearly $\mathfrak{m}^s \subseteq Q^{es} \cap A$. The other inclusion is also straightforward since if $a \in Q^{es} \cap A$ then for some $n \geq s$ we have $a \in Q^{en} \cap A = \mathfrak{m}^n$. Thus $a \in \mathfrak{m}^s$.

Now we return to the context of Diagram (I). We have a prime ideal $\mathfrak{p}$ of A contained in a maximal ideal $\mathfrak{m}$ of A, and in A_D we have corresponding ideals $\mathfrak{p}_D$, $\mathfrak{m}_D$. We assume $\mathfrak{p} = \mathfrak{m}^s$, $s > 1$, so by the above we may write:

$$\mathfrak{m}_D B_D = \prod Q_i^{e_i}; \quad \mathfrak{p}_D B_D = \prod Q_i^{s}.$$

We set $M_i = Q_i \cap B$. We claim that:

6.3.1. $\mathfrak{m} B = \prod M_i^{e_i}$

6.3.2. $\mathfrak{p} B = \prod M_i^{s}$

6.3.3. The prime ideals lying over $\mathfrak{m}$ are $M_1, \ldots, M_k$

6.3.4. The prime ideals lying over $\mathfrak{p}$ are $M_1^s, \ldots, M_k^s$.

This will follow from an extension of the commutative diagram (I). Fix any maximal ideal $M \supseteq \mathfrak{m}$ with associated ultrafilter $E = D(M)$ and consider the diagram:

(II)

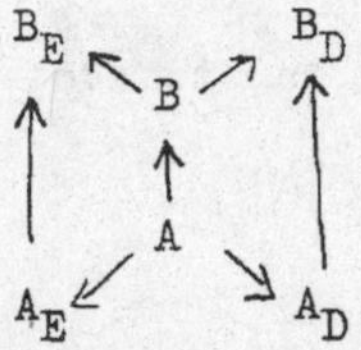

$D = D(\mathfrak{m})$; $E = D(M)$.

We have left some room in (II) for arrows between B_E and B_D, A_E and A_D, because there are <u>canonical</u> <u>isomorphisms</u>

$$B_E \simeq B_D$$

$$A_E \simeq A_D$$

which leave the diagram commutative.

Indeed, let $\mathcal{P}$ be the set of primes in A, and let $\mathcal{Q}$ be the set of primes in B. Each element of $\mathcal{P}$ has at most n factors in $\mathcal{Q}$. Thus there is an integer k such that almost every element of $\mathcal{P}$ (mod D) has exactly k factors in $\mathcal{Q}$. Let the factors of P in $\mathcal{Q}$ be $\{Q_1, \ldots, Q_k\}$ and choose the notation so that the maps $h_i: P \longmapsto Q_i$ are definable. We may construe D as a filter on $\mathcal{D}(B)$ if we replace each $P \in \mathcal{P}$ by the corresponding $Q_1, \ldots, Q_k \in \mathcal{Q}$. Then D admits precisely k extensions to ultrafilters E on (B), namely

$$E_i = \langle D \vee \mathrm{range}(h_i) \rangle.$$

The function h_i induces an isomorphism between D and E_i, which induces the isomorphisms $B_E \simeq B_D$, $A_E \simeq A_D$.

Notice that in B_{E_i} the identity function $\mathbb{1}:\mathfrak{Q} = \mathfrak{Q}$ represents a definable ideal sitting over $M_i = M_{E_i}$. Of course if we consider $B_{E_i} \simeq B_D \simeq B_{E_j}$ then the identity functions in B_{E_i}, B_{E_j} do not correspond under the isomorphism, as is clear upon inspection of the definitions.

Assertion 6.3.3 is immediate in view of the correspondence between maximal ideals and ultrafilters. Since $\mathfrak{m}B$ is an $\{M_1,\ldots,M_k\}$-ideal it follows (by Corollary 5.2) that

$$\mathfrak{m}B = \prod M_i^{s(M_i)} \quad \text{for a suitable section } s.$$

Since $Q_i = \mathbb{1}_{E_i}$ the exponent $s(M_i)$ may be defined as the set of m such that $\mathfrak{m}B \subseteq Q_i^m$. Clearly $\mathfrak{m}B \subseteq Q_i^{e_i}$ for each i. We claim $\mathfrak{m}B \not\subseteq Q_i^{e_i+1}$, which will prove 6.3.1. Indeed, write $\mathfrak{m}_D = \mathbb{1}_D = (\alpha_D, \beta_D)$ (in A_D) and choose $f_D = \alpha_D x_D + \beta_D y_D \in \mathfrak{m}_D B_D$ with $f_D \notin Q_i^{e_i+1}$. Fix $S \in D$ and choose $a, b \in A$, $x, y \in B$ with

$$a \equiv \alpha_D,\ b \equiv \beta_D,\ x \equiv x_D,\ y \equiv y_D \pmod{P^k} \text{ for all } P \in S.$$

Here $k \geq 2$.

Then $ax + by \in \mathfrak{m}B$ and $ax + by - \alpha_D x_D - \beta_D y_D \equiv 0 \pmod{\mathfrak{m}_D^k}$ and hence also modulo $Q_i^{e_i+1}$. Thus $ax + by \notin Q_i^{e_i+1}$, which completes the proof of 6.3.1.

6.3.2 is proved similarly, with a somewhat larger value of k. As for 6.3.4, this follows from the commutativity of (I), intersecting Q_i^s around both ways. ⊣

This completes our discussion of residue class field degrees and ramification indices. Notice that to compute an invariant like a ramification index it suffices to view a maximal ideal $\mathfrak{m}$ as the average over an ultrafilter of definable maximal ideals. The index for $\mathfrak{m}$ can then be computed as the average value of the index over the various definable approximations to $\mathfrak{m}$.

References:

1. J. Robinson, "The undecidability of algebraic rings and fields," PAMS 10, pp. 950-956 (1959).
2. J. Ax, "The elementary theory of finite fields," Ann. Math. 88, pp. 239-271 (1968).
3. G. Cherlin and J. Hirschfeld, "Ultrafilters and ultraproducts in non-standard analysis," in Contributions to Nonstandard Analysis, Luxemburg and Robinson, eds., North Holland (1972).
4. S. Kochen, "The model theory of local fields," lecture notes from Kiel (1974), to appear in the series Lecture Notes in Mathematics, Springer Verlag.

Note. (added July 1975).

The problem immediately preceding Theorem 3.4 has been largely solved by Alex Wilkie: the number of maximal ideals in question must be odd finite, $\aleph_0$, or $2^{\aleph_0}$. Furthermore any odd number or $\aleph_0$ may in fact be obtained. Whether $2^{\aleph_0}$ is actually possible remains probable but unsettled.

CATEGORIES OF LOCAL FUNCTORS

Paul C. Eklof[1]
University of California, Irvine

Introduction. The subtitle of this paper is "An application of the study of $L_{\infty\kappa}$-equivalence-preserving functors to the problem of the existence of complete systems of invariants for certain classes of abelian groups." The author first became interested in the goal of using model-theoretic tools to provide negative solutions to the above-mentioned problem when he was an instructor at Yale. The theorems of section 5 represent a first attempt in this direction. Theorem 5.3 and Corollary 5.6 assert that certain classes of groups do not have complete systems of invariants which satisfy certain conditions. One of these conditions is that the functors involved preserve $L_{\infty\omega_1}$-equivalence. It thus becomes necessary to study such functors to determine how restrictive this condition is,and to establish algebraic criteria for this condition to hold. The first four sections of the paper are devoted to this study, and it is hoped that the results are of independent interest.

The starting point for this study is the notion, introduced by S. Feferman, of κ-local functor, which gives sufficient algebraic conditions for a functor to preserve $L_{\infty\kappa}$-equivalence. However, these conditions are not satisfied in some important examples of functors which preserve $L_{\infty\kappa}$-equivalence. G. Sabbagh provided a generalization of Feferman's definition covering these examples but the functors he defined do not satisfy nice closure properties e.g. they are not closed under composition. In section 4 we introduce a further generalization of the notion of κ-local in order to define categories of functors

[1]Preparation of this paper partially supported by NSF Grant GP-43910.

preserving $L_{\infty\kappa}$-equivalence which have nice closure properties.

In section 1 we introduce the categorical setting for our work: roughly, we consider categories of models of universal-Horn theories of algebras. In section 2 we review the notion of $L_{\infty\kappa}$-equivalence; an interesting sidelight here is that in our setting the notion can be expressed in category-theoretic terms using one auxiliary notion: that of a substructure generated by fewer than κ elements. In section 3 we discuss the notion of κ-local functor and a generalization of the notion.

1. Universal-Horn categories. Throughout this paper L will denote a countable first-order predicate calculus with function symbols (including constant symbols) but no relation symbols. Structures for L will be called L-algebras. Let $\underset{\sim}{Alg}_L$ denote the category of L-algebras and L-homomorphisms. Because we have no relation symbols, an embedding of L-algebras is simply a one-one L-homomorphism. As a consequence of the following we see that embedding is a category-theoretic notion in this setting.

1.1 PROPOSITION. A morphism f in $\underset{\sim}{Alg}_L$ is a monomorphism if and only if f is one-one.

Proof. It is clear that if f is one-one then f is a monomorphism in $\underset{\sim}{Alg}_L$. Conversely, suppose that f: $A \longrightarrow B$ is a morphism such that there exist $a_1, a_2 \in A$ with $a_1 \neq a_2$ and $f(a_1)=f(a_2)$. Let C be the subalgebra of $A \times A$ generated by (a_1, a_2) and let $g_i: C \longrightarrow A$ be the restriction to C of projection of $A \times A$ on the ith factor, for i=1, 2. Then $fg_1 = fg_2$ but $g_1 \neq g_2$. Thus f is not a monomorphism. ⊣

Recall that a sentence φ of L is called a universal Horn sentence if φ is logically equivalent to a sentence of the form

$$(*) \qquad \forall x_1 \ldots \forall x_n [\theta_1 \vee \ldots \vee \theta_m]$$

where $\theta_1, \ldots, \theta_{m-1}$ are negated atomic formulas and θ_m is either atomic or negated atomic. An elementary class is closed under substructures and products it and only if it is the class of models of a set of universal Horn sentences. (See for example [C-K; Thm. 5.2.4] and [Co;p. 235]).

In general we follow the terminology and notation of Mitchell [Mi]; for example a direct limit is a colimit with respect to a directed set [Mi;p. 47]; also, a subobject of A is a monomorphism into A.

We will be interested in certain nice subcategories of $\underset{\sim}{\mathrm{Alg}}_L$ which are characterized in the next proposition.

1.2 PROPOSITION. Let $\underset{\sim}{A}$ be a full subcategory of $\underset{\sim}{\mathrm{Alg}}_L$ and suppose:

(i) $\underset{\sim}{A}$ is a complete subcategory of $\underset{\sim}{\mathrm{Alg}}_L$ closed under isomorphism and subobjects in $\underset{\sim}{\mathrm{Alg}}_L$.

Then $\underset{\sim}{A}$ is cocomplete. If $\underset{\sim}{A}$ also satisfies

(ii) the inclusion functor: $\underset{\sim}{A} \to \underset{\sim}{\mathrm{Alg}}_L$ preserves direct limits,

then it follows that

(iii) $\mathrm{Ob}\underset{\sim}{A}$ is the class of models of a set of universal-Horn sentences of L.

Conversely, if $\underset{\sim}{A}$ satisfies (iii) and also

(iv) $\underset{\sim}{A}$ has a final object,

then $\underset{\sim}{A}$ satisfies (i) and (ii).

Proof. That $\underset{\sim}{A}$ is cocomplete may be proved by an application of the adjoint functor theorem, but for future reference we shall give an explicit construction of colimits (obtained from the proof of the adjoint functor theorem). Let D be a diagram in $\underset{\sim}{A}$ over a diagram scheme $\Sigma=(I, M, d)$. Let S denote the disjoint union of the D_i, $i \in I$, and let $\lambda=\text{Card}(S)+\aleph_0$. Let P be the set of all (set) mappings, $f:S \to A_f$ such that $A_f \in \text{Ob}\underset{\sim}{A}$ and $\text{Card}(A_f) \leq \lambda$, and $\{f \upharpoonright D_i \mid i \in I\}$ is a compatible family. (Notice that P is non-empty since $\underset{\sim}{A}$ has a final object, the empty product in $\underset{\sim}{A}$). Let $B=\prod_{f \in P} A_f$ and let $\beta:S\to B$ be the (set) map induced by the maps $f:S\to A_f$. Let C be the subalgebra of B generated by $\beta(S)$ and let η_i be β restricted to D_i for each $i \in I$. Then $C \in \text{Ob}\underset{\sim}{A}$ since $\underset{\sim}{A}$ satisfies (i), and $\eta_i \in \underset{\sim}{A}(D_i, C)$ since $\underset{\sim}{A}$ is a full subcategory of $\underset{\sim}{Alg}_L$. One may check that $\{D_i \xrightarrow{\eta_i} C \mid i \in I\}$ is a colimit of D, since every compatible family $\{D_i \longrightarrow X \mid i \in I\}$ "factors through" some f in P.

To prove that $\text{Ob}\underset{\sim}{A}$ is elementary, it suffices to prove that $\text{Ob}\underset{\sim}{A}$ is closed under ultraproducts (since $\text{Ob}\,\underset{\sim}{A}$ is closed under substructures). But the ultraproduct $\prod_{i \in I} A_i/U$ is the direct limit (in $\underset{\sim}{Alg}_L$) of the family of morphisms $\prod_{i \in X} A_i \to \prod_{i \in Y} A_i$ where X, Y range over elements of U with $X \supseteq Y$. It follows then from (ii) that $\underset{\sim}{A}$ is closed under ultraproducts. By the remarks preceding the proposition, $\text{Ob}\underset{\sim}{A}$ is the class of models of a set of universal - Horn sentences of L.

Now suppose that $\text{Ob}\underset{\sim}{A} = \text{Mod}(T)$, where T is a set of universal - Horn sentences of L. Suppose also that T has a final object i.e. the empty product. Then $\text{Ob}\underset{\sim}{A}$ is closed under products and under subobjects in $\underset{\sim}{Alg}_L$. If D is a diagram in $\underset{\sim}{A}$ over $\Sigma = (I, M, d)$, let L be the subalgebra of $\prod_{i \in I} D_i$ consisting of all $\alpha \in \prod_{i \in I} D_i$ such that $D(m)(\alpha(i)) = \alpha(j)$ whenever $d(m) = (i, j)$, $m \in M$. Let p_j be the restriction to L of the projection: $\prod_{i \in I} D_i \to D_j$.

Then $L \in \text{Ob}\underset{\sim}{A}$ and $\{p_j : L \longrightarrow D_j \mid j \in I\}$ is the limit in $\underset{\sim}{Alg}_L$ and in $\underset{\sim}{A}$ of D. This proves (i).

For (ii), we observe that the usual construction of a direct limit of a diagram D as the disjoint union of the D_i, $i \in I$, modulo the equivalence relation induced by the $D(m)$, $m \in M$, does not lead out of $\text{Ob}\underset{\sim}{A}$ since the sentences in T are universal - Horn. ⊣

1.3 REMARKS. (1) An example which shows that assumption (ii) is necessary to conclude (iii) is the category of reduced torsion-free abelian groups. (An abelian group A is reduced if it does not contain any non-trivial divisible subgroup).

(2) In the terminology of Mal'cev [Mal; p. 59], a full subcategory of $\underset{\sim}{Alg}_L$ satisfying (i) is called quasifree in $\underset{\sim}{Alg}_L$, and a class of algebras satisfying (iii) + (iv) is called a quasivariety. The equivalence of (i) + (ii) and (iii) + (iv) is a theorem of Mal'cev ([Mal; Thm. 3, p. 419] or [Ma2; Cor. 3, p. 214]). It is not hard to see that if $\underset{\sim}{A}$ satisfies the hypotheses of Proposition 1.2, then the final object in $\underset{\sim}{A}$ consists of a single element which satisfies every atomic formula. Hence $\underset{\sim}{A}$ is the class of models of a set of sentences of the form (*) where θ_m is atomic.

We define a uHf category (or quasivariety) to be a full subcategory $\underset{\sim}{A}$ of $\underset{\sim}{Alg}_L$ satisfying the conditions of Proposition 1.2. Since $\underset{\sim}{A}$ is closed under subobjects and products in $\underset{\sim}{Alg}_L$ the same argument as in 1.1 proves the following.

1.4 PROPOSITION. If $\underset{\sim}{A}$ is a uHf category and f is a morphism in $\underset{\sim}{A}$, f is a monomorphism if and only if f is one-one. ⊣

From now on, all categories referred to will be assumed to be uHf categories.

The uHf categories include many of the most important categories arising in algebra, as we see in the following examples.

1.5 EXAMPLES. (1) The following are uHf categories: $\underset{\sim}{\mathrm{Alg}}_L$; $\underset{\sim}{\mathrm{Semi}}$, the category of semi-groups; $\underset{\sim}{\mathrm{Semi}}$*, the category of cancellation semi-groups; $\underset{\sim}{\mathrm{Set}}$, the category of sets; $\underset{\sim}{\mathrm{Grp}}$, the category of groups; $\underset{\sim}{\mathrm{Ab}}$, the category of abelian groups; $\underset{\sim}{\mathrm{Ab}}$*, the category of torsion-free abelian groups; $\underset{\sim}{\mathrm{Rng}}$, the category of rings (with 1); $\underset{\sim}{\mathrm{Rng}}$*, the category of rings without nilpotent elements; $\underset{\sim}{\mathrm{CRng}}$, the category of commutative rings; $\underset{\sim}{\mathrm{R\text{-}Mod}}$, the category of left modules over a fixed ring R; $\underset{\sim}{\mathrm{R\text{-}Lie}}$, the category of Lie algebras over R. The asterisked categories are quasi-varieties which are not varieties.

(2) If $\underset{\sim}{A}$ is a uHf category and Σ is a diagram scheme, let $\underset{\sim}{A}_\Sigma$ denote the category of diagrams in $\underset{\sim}{A}$ over Σ. It is easily seen that $\underset{\sim}{A}_\Sigma$ is (equivalent to) a uHf category. (The objects of $\underset{\sim}{A}_\Sigma$ are best thought of as many-sorted structures, where there are as many sorts as vertices of Σ).

If $\underset{\sim}{A}$ and Σ are as above, let Lim (resp. Colim) denote the functor: $\underset{\sim}{A}_\Sigma \to \underset{\sim}{A}$ which takes a diagram D in $\underset{\sim}{A}_\Sigma$ to the limit object of D (resp. colimit object of D) and which takes a morphism $f: D_1 \to D_2$ in $\underset{\sim}{A}_\Sigma$ to the naturally induced morphism: $\mathrm{Lim}\,(D_1) \to \mathrm{Lim}\,(D_2)$ (resp. $\mathrm{Colim}\,(D_1) \to \mathrm{Colim}\,(D_2)$). Let $\mathcal{U}$ denote the category of all uHf - categories. Note that $\mathcal{U}$ is a <u>two-dimensional</u> <u>category</u>: it consists of <u>0-cells</u> (uHf - categories), <u>1 - cells</u> (functors between uHf - categories) and <u>2-cells</u> (natural transformations of functors), and it has two laws of composition (for functors and for natural transformations).

For any two-dimensional category $\mathcal{C}$ and two 0-cells $\underset{\sim}{A}$ and $\underset{\sim}{B}$ in $\mathcal{C}$, the class of all 1-cells from $\underset{\sim}{A}$ to $\underset{\sim}{B}$ together with all 2-cells between 1-cells from $\underset{\sim}{A}$ to $\underset{\sim}{B}$ constitutes a one-dimensional (ordinary) category which we will denote $\mathcal{C}(\underset{\sim}{A}, \underset{\sim}{B})$. The two-dimensional category $\mathcal{C}$ is called <u>locally complete</u>

(resp. <u>locally cocomplete</u>) if for every pair of 0-cells $\underset{\sim}{A}$ and $\underset{\sim}{B}$ in $\mathcal{C}$, $\mathcal{C}(\underset{\sim}{A},\underset{\sim}{B})$ is complete (resp. cocomplete).

1.6 PROPOSITION. $\mathcal{U}$ <u>is locally complete and locally cocomplete.</u>

<u>Proof</u>: Suppose D is a diagram in $\mathcal{U}(\underset{\sim}{A},\underset{\sim}{B})_\Sigma$. Let $\hat{D}: \underset{\sim}{A} \to \underset{\sim}{B}_\Sigma$ be the functor which takes A to D "evaluated at A", in the obvious sense. Then Lim o $\hat{D}$ is the limit of D, and Colim o $\hat{D}$ is the colimit of D. ⊣

2. <u>$L_{\infty\kappa}$ - equivalence.</u> Let κ be an infinite cardinal and let $\underset{\sim}{A}$ be a uHf category. We wish to express in category-theoretic terms the equivalence of two objects of $\underset{\sim}{A}$ with respect to the infinitary language $L_{\infty\kappa}$, making use of one auxiliary notion: that of a κ - generated subobject. (For readers not familiar with the definition of $L_{\infty\kappa}$ - equivalence of structures A_0 and A_1 - denoted $A_0 \equiv_{\infty\kappa} A_1$ - Theorem 2.1 may be taken as a definition).

For any object A of $\underset{\sim}{A}$, let Sub(A) denote the class of subobjects of A and let $\mathrm{Sub}_\kappa(A)$ denote the class of <u>κ - generated</u> subobjects $e \in \mathrm{Sub}(A)$ i.e dom(e) is generated by $<\kappa$ elements.

If $e: A' \to A$ is a subobject of A and $f: B \to A$ is an arbitrary morphism, write $f \leq e$ if there is a morphism γ (necessarily unique) such that $f = e\gamma$. If $g: C \to A$ is a morphism write $g \leq f$ if for every subobject e of A, $f \leq e$ implies $g \leq e$. It is easy to see that $g \leq f$ if and only if range (g) $\subseteq$ range (f).

If A_0 and A_1 are objects of $\underset{\sim}{A}$, a <u>partial isomorphism from A_0 to A_1</u>, is defined to be a triple $\tilde{f} = (f, e_0, e_1)$ where $e_i \in \mathrm{Sub}(A_i)$ and f is an isomorphism from dom(e_0) to dom(e_1). If $\tilde{f}' = (f', e_0', e_1')$ is another partial isomorphism from A_0 to A_1 write $\tilde{f}' \geq \tilde{f}$ if $e_0' \geq e_0$, $e_1' \geq e_1$, and

$$\begin{array}{ccc} \mathrm{dom}(e_0') & \xrightarrow{f'} & \mathrm{dom}(e_1') \\ \gamma_0 \uparrow & & \uparrow \gamma_1 \\ \mathrm{dom}(e_0) & \xrightarrow{f} & \mathrm{dom}(e_1) \end{array}$$

commutes, where γ_i is the unique morphism such that $e_i = e'_i \gamma_i$.

(If the partial isomorphism $\tilde{f} = (f, e_0, e_1)$ is such that e_i is an inclusion map of a substructure of A_i into A_i for i=0,1 we will sometimes identify $\tilde{f}$ with f. However in the next section we will need to make explicit reference to the morphisms e_0 and e_1; thus it is convenient to define a partial isomorphism in the way we have done).

For each infinite cardinal κ, we define a descending sequence of sets of partial isomorphisms from A_0 to A_1 by transfinite induction. Let $PI^0_\kappa(A_0, A_1)$ = the set of all partial isomorphisms (f, e_0, e_1) from A_0 to A_1, such that $e_i \in Sub_\kappa(A_i)$ for i=0, 1. If μ is a limit ordinal, define

$$PI^\mu_\kappa(A_0, A_1) = \bigcap_{\nu<\mu} PI^\nu_\kappa(A_0, A_1).$$

If μ is a successor ordinal - say $\mu = \nu + 1$ - define

$PI^\mu_\kappa(A_0, A_1)$ = the set of all $\tilde{f} = (f, e_0, e_1) \in PI^\nu_\kappa(A_0, A_1)$ such that given <u>either</u> $u_0 \in Sub_\kappa(A_0)$ <u>or</u> $u_1 \in Sub_\kappa(A_1)$, there exists $\tilde{f}' = (f', e_0', e_1') \in PI^\nu_\kappa(A_0, A_1)$ such that $\tilde{f}' \geq \tilde{f}$ and $e_0' \geq u_0$ or $e_1' \geq u_1$, respectively.

Finally, define

$$PI^\infty_\kappa(A_0, A_1) = \bigcap_\mu PI^\mu_\kappa(A_0, A_1)$$

where the intersection is over all ordinals μ.

The following theorem is a restatement of the back-and-forth characterization of $L_{\infty\kappa}$ - equivalence (cf. [B]).

2.1 THEOREM. $A_0 \equiv_{\infty\kappa} A_1$ <u>if and only if</u> $PI^\infty_\kappa(A_0, A_1) \neq \emptyset$. ⊣

Where there is no ambiguity we will sometimes write PI^ν_κ for $PI^\nu_\kappa(A_0, A_1)$. A functor $F: \underset{\sim}{A} \to \underset{\sim}{B}$ is said to <u>preserve $L_{\infty\kappa}$ - equivalence</u> if whenever $A_0, A_1 \in Ob\underset{\sim}{A}$ such that $A_0 \equiv_{\infty\kappa} A_1$ then $F(A_0) \equiv_{\infty\kappa} F(A_1)$. By Theorem 2.1, F preserves $L_{\infty\kappa}$ - equivalence if and only for all

$A_0, A_1 \in \mathrm{Ob}A$, $PI_\kappa^\infty(A_0, A_1) \neq \emptyset$ implies $PI_\kappa^\infty(F(A_0), F(A_1)) \neq \emptyset$.

Let $\mathcal{U}_\kappa$ denote the subcategory of $\mathcal{U}$ defined as follows: the 0-cells of $\mathcal{U}_\kappa$ are all the uHf categories; the 1-cells of $\mathcal{U}_\kappa$ are all functors between uHf categories which preserve $L_{\infty\kappa}$ equivalence; the 2-cells of $\mathcal{U}_\kappa$ are all natural transformations between functors in $\mathcal{U}_\kappa$. For the next two sections we shall investigate the category $\mathcal{U}_\kappa$.

3. <u>Local Functors</u>. We will consider some conditions on a functor F in $\mathcal{U}$ which imply that F preserves $L_{\infty\kappa}$ - equivalence, conditions which naturally arise in algebra and which are (relatively) easy to verify.

Following Feferman[Fe] we define a functor $F: \underset{\sim}{A} \to \underset{\sim}{B}$ to be <u>κ - local</u> if:

(i) for every $A \in \mathrm{Ob}\underset{\sim}{A}$ and every $u \in \mathrm{Sub}_\kappa(F(A))$ there exists $e \in \mathrm{Sub}_\kappa(A)$ such that $u \leq F(e)$; and

(ii) F preserves monomorphisms.

Feferman proved that κ - local functors preserve $L_{\infty\kappa}$ - equivalence. But as we shall see in the examples which follow, this class of functors does not include some natural examples of functors in $\mathcal{U}_\kappa$. Inspired by a suggestion of G. Sabbagh (private communication) we make the following definition. A functor $F: \underset{\sim}{A} \to \underset{\sim}{B}$ is <u>generalized κ - local</u> (gen. κ-local) if F satisfies (i) above; and also

(ii')for every object A and every $e \in \mathrm{Sub}_\kappa(A)$, if $x, y \in \mathrm{dom}\ F(e)$ such that $F(e)(x) = F(e)(y)$ then there exists $e' \in \mathrm{Sub}_\kappa(A)$ such that $e \leq e'$ and $F(\gamma)(x) = F(\gamma)(y)$, where γ is the unique morphism such that $e = e'\gamma$.

(Sabbagh's original suggestion was closer to the characterization contained in Proposition 3. 3).

Very roughly, (ii') says that if F fails to preserve monomorphisms it is precisely because it fails to preserve monomorphisms of κ - generated structures. It is clear that (ii) implies (ii′) and hence every κ - local functor is generalized κ - local.

3.1 EXAMPLES. (1) The functor from $\underset{\sim}{\text{Rng}}$ to $\underset{\sim}{\text{Rng}}$ taking R to R[X] (resp. R to R[[X]]) is ω - local (resp. ω_1 - local). [Fe; p. 77].

(2) For any right R - module M, the functor $M\otimes_R$: $\underset{\sim}{\text{R-Mod}} \to \underset{\sim}{\text{Ab}}$ is gen. ω - local but not ω - local.

(3) Let $\Sigma = (I, M, d)$ where $I = \{0,1\}$, $M = \{m_0\}$, $d(m_0) = (0,1)$. The functor from $\underset{\sim}{\text{Ab}}_\Sigma$ to $\underset{\sim}{\text{Ab}}$ taking D to the cokernel of $D(m_0)$ is gen. ω-local but not ω - local. (This is a special case of Corollary 3.5).

(4) In this example and the next we refer to section 6 of Fuchs [Fu 1]. The functors of examples 1, 2, 3, 4 (p. 25) are ω - local. (Here we regard them as functors from $\underset{\sim}{\text{Ab}}$ to $\underset{\sim}{\text{Ab}}$). The functors of examples 5 and 6 (p. 26) are gen. ω - local but not ω - local. If F is any subfunctor of the identity which is κ - local, then the quotient functor of the identity F* (defined as in [Fu 1; Thm. 6.1]) is gen. κ - local.

(5) Let χ be a class of groups and define the functors V_χ and W_χ as in [Fu 1; p. 27]. If κ is such that $\overset{=}{X} < \kappa$ for all $X \in \chi$ then W_χ is κ - local. If χ is an elementary class then V_χ is ω - local. (To see this, given $a \in V_\chi(A)$ apply the Compactness Theorem to $\text{Diag}_+(A) \cup T \cup \{c_a \neq 0\}$, where χ is the class of models of T).

(6) In this example we refer to section 85 of [Fu 2]. For any type t, the functor from $\underset{\sim}{\text{Ab}}$*, the category of torsion-free abelian groups, to $\underset{\sim}{\text{Ab}}$* taking A to A(t) (resp. A*(t)) is ω_1- local. It is ω - local if and only if $t = (0, 0, \ldots 0, \ldots)$.

(7) Let h be a height in the sense of Warfield [W] i.e. a formal product $\prod_p p^{v(p)}$ where v(p) is an ordinal. (We exclude $v(p) = \infty$). The functor from $\underset{\sim}{\text{Ab}}$ to $\underset{\sim}{\text{Ab}}$ taking A to $hA \overset{\text{def}}{=} \bigcap_p p^{v(p)} A$ is κ - local if $\sum_p |v(p)| < \kappa$.

The key result about generalized κ-local functors is the following.

3.2 PROPOSITION. If F is generalized κ-local, then F preserves $L_{\infty\kappa}$-equivalence.

Proof. For any $\tilde{f}=(f, e_0, e_1) \in PI_\kappa^1(A_0, A_1)$ we shall define $\tilde{f}^*=(f^*, e_0^*, e_1^*) \in PI_\kappa^0(F(A_0), F(A_1))$ such that if $\tilde{f} \in PI_\kappa^{1+\mu}(A_0, A_1)$ then $\tilde{f}^* \in PI_\kappa^\mu(F(A_0), F(A_1))$. This will clearly suffice to prove the lemma.

Let $\mathrm{dom}(e_i^*)=\{F(e_i)(x) \mid x \in \mathrm{dom}\ F(e_i)\}$ and let e_i^* be the inclusion of $\mathrm{dom}(e_i^*)$ in $F(A_i)$. Define f^* by $f^*(F(e_0)(x))=F(e_1)(F(f)(x))$. We must check that f^* is well-defined. Suppose that $F(e_0)(x)=F(e_0)(y)$. By hypothesis there exists $u_0' \in \mathrm{Sub}_\kappa(A_0)$ such that $u_0' \geq e_0$ and $F(\gamma_0)(x)=F(\gamma_0)(y)$, where $e_0=u_0'\gamma_0$. Since $\tilde{f} \in PI_\kappa^1(A_0, A_1)$ there exists $\tilde{f}'=(f', e_0', e_1') \in PI_\kappa^0(A_0, A_1)$ such that $\tilde{f}' \geq \tilde{f}$ and $u_0' \geq e_0'$. We may assume $u_0'=e_0'$. If γ_1 is such that $e_1=e_1'\gamma_1$, then by definition of $\tilde{f}' \geq \tilde{f}$, we have $f'\gamma_0=\gamma_1 f$ and hence $F(f')F(\gamma_0)=F(\gamma_1)F(f)$. It follows that $F(\gamma_1)F(f)(x)=F(\gamma_1)F(f)(y)$. Since $e_1=e_1'\gamma_1$, $F(e_1)(F(f)(x))=F(e_1)(F(f)(y))$. Hence f^* is well-defined.

A similar argument shows that f^* has a well-defined inverse. It follows that f^* is an isomorphism and therefore $(f^*, e_0^*, e_1^*) \in PI_\kappa^0(F(A_0), F(A_1))$.

We prove by induction on μ that if $\tilde{f} \in PI_\kappa^{1+\mu}$, then $\tilde{f}^* \in PI_\kappa^\mu$. Suppose $\mu=\upsilon+1$ and $\tilde{f} \in PI_\kappa^{1+\mu}$. By induction $\tilde{f}^* \in PI_\kappa^\upsilon$. To prove $\tilde{f}^* \in PI_\kappa^\mu$, consider $u_0 \in \mathrm{Sub}_\kappa(F(A_0))$. By (i) of the definition of κ-local and by the definition of $PI_\kappa^{1+\mu}(A_0, A_1)$, there exists $\tilde{g}=(g, v_0, v_1) \in PI_\kappa^{1+\upsilon}(A_0, A_1)$ such that $\tilde{g} \geq \tilde{f}$ and $u_0 \leq F(v_0)$. By induction, $\tilde{g}^*=(g^*, v_0^*, v_1^*) \in PI_\kappa^\upsilon(F(A_0), F(A_1))$. Now $u_0 \leq F(v_0)$ implies $\{u_0(x) \mid x \in \mathrm{dom}(u_0)\} \subseteq \{F(v_i)(x) \mid x \in \mathrm{dom}\ F(v_i)\}$ and therefore by definition of v_0^*, $u_0 \leq v_0^*$. A similar result holds if we start with

$u_1 \in Sub_\kappa(F(A_1))$. Hence $f^* \in PI^\mu_\kappa$. ⊣

The definition of a generalized κ-local functor is easy to verify in practice but not very elegant. In the next result we give a neat equivalent formulation for regular cardinals κ. We say that a diagram scheme $\Sigma=(I, M, d)$ is <u>κ-directed</u> if I is a partially ordered set which is κ-directed (i.e. every subset of cardinality $< \kappa$ has an upper bound) and if there is precisely one arrow from i to j if $i \leq j$ and no arrows from i to j otherwise. A diagram D in $\underset{\sim}{A}_\Sigma$ is <u>monomorphic</u> if for every $m \in M$, D(m) is a monomorphism in $\underset{\sim}{A}$.

3.3 PROPOSITION. <u>Let F be a functor from $\underset{\sim}{A}$ to $\underset{\sim}{B}$. A necessary condition for F to be generalized κ-local is that for every monomorphic κ-directed diagram D in $\underset{\sim}{A}_\Sigma$, colimit $(F \circ D)$ equals F(colimit D). If κ is regular, the condition is also sufficient.</u>

<u>Proof.</u>. Suppose that F is generalized κ-local and that $\{D_i \xrightarrow{e_i} C \mid i \in I\}$ is the colimit of D. Note that each e_i is a subobject. To prove that colimit $(F \circ D)$ = F (colimit D) it suffices to prove that if $\{F(D_i) \xrightarrow{f_i} X \mid i \in I\}$ is a compatible family (w.r.t. $F \circ D$) then there exists $\psi : F(C) \to X$ such that $f_i = \psi F(e_i)$ for all $i \in I$. By condition (i) of the definition of κ-local and the fact that Σ is κ-directed, for every $x \in F(C)$ there exists $i \in I$ such that $x = F(e_i)(y)$ for some $y \in F(D_i)$. Define $\psi(x) = f_i(y)$. We leave as an exercise for the reader the proof that ψ is well-defined; the proof uses (ii').

For the converse, suppose that F preserves colimits over monomorphic κ-directed diagrams. Since κ is regular, the diagram D in $\underset{\sim}{A}$ of all κ-generated substructures of A with inclusion maps between substructures is κ-directed. The colimit of D is A together with the inclusions of κ-generated substructures into A. Conditions (i) and (ii') of the definition of generalized κ-local now follow from the hypothesis and from (ii) of Proposition 1.2.⊣

3.4 COROLLARY. If $F: \underset{\sim}{A} \longrightarrow \underset{\sim}{B}$ is a left adjoint then F is generalized ω-local.

Proof. Left adjoints preserve colimits [Mac, pp 114 f].⊣

3.5 COROLLARY. The functor Colim: $\underset{\sim}{A}_{\Sigma} \longrightarrow \underset{\sim}{A}$ is generalized ω-local.

Proof. Colim is left adjoint to the diagonal functor $\Delta: \underset{\sim}{A} \longrightarrow \underset{\sim}{A}_{\Sigma}$ [Mac;p. 86].⊣

In general, Colim is not ω-local since it does not preserve monomorphisms. For the functor Lim we have the following result whose proof follows easily from the construction of limits given in the proof of Proposition 1.2.

3.6 PROPOSITION. For any $\Sigma = (I, M, d)$, the functor Lim: $\underset{\sim}{A}_{\Sigma} \longrightarrow \underset{\sim}{A}$ is κ-local if κ is greater than Card(I)+$\aleph_0$ ⊣

We leave to the reader the verification of the following result.

3.7 PROPOSITION. If $\lambda > \kappa$ are infinite cardinals and F is κ-local (resp. generalized κ-local), then F is λ-local (resp. generalized λ-local).⊣

3.8 COROLLARY. The functor Colim is generalized κ-local for every infinite κ⊣

4. Categories of local functors. The generalized κ-local functors as we have seen include a large class of important examples. But the definition of generalized κ-local is deficient in that these functors do not have nice closure properties viz. it is not clear that the composition, limit, or colimit of gen. κ-local functors is gen. κ-local. In this section we remedy this defect by producing a class of functors including the generalized κ-local functors which has nice closure properties.

Let $\mathcal{C}$ be a two-dimensional category. A subcategory $\mathcal{C}'$ of $\mathcal{C}$ is called a locally full subcategory if for every pair of 0-cells $\underset{\sim}{A}$ and $\underset{\sim}{B}$ of $\mathcal{C}'$, $\mathcal{C}'(\underset{\sim}{A}, \underset{\sim}{B})$ is a full subcategory of $\mathcal{C}(\underset{\sim}{A}, \underset{\sim}{B})$. For example $\mathcal{U}_\kappa$ is a locally full subcategory of $\mathcal{U}$.

A diagram D over a scheme $\Sigma = (I, M, d)$ is called a κ-diagram if card $(I) < \kappa$. A category C is κ-complete (resp. κ-cocomplete) if every κ-diagram in C has a limit (resp. colimit) in C. A subcategory C' of C is a κ-complete subcategory if C' is κ-complete and the inclusion functor of C' into C preserves limits over κ-diagrams.

A subcategory $\mathcal{C}'$ of the two-dimensional category $\mathcal{C}$ is a locally complete (resp. locally cocomplete; locally κ-complete; locally κ-cocomplete) subcategory of $\mathcal{C}$ if for every pair $\underset{\sim}{A}, \underset{\sim}{B}$ of 0-cells of $\mathcal{C}'$, $\mathcal{C}'(\underset{\sim}{A}, \underset{\sim}{B})$ is a complete (resp. cocomplete; κ-complete; κ-cocomplete) subcategory of $\mathcal{C}(\underset{\sim}{A}, \underset{\sim}{B})$.

It is not clear whether $\mathcal{U}_\kappa$ is locally complete or locally cocomplete in $\mathcal{U}$. But the following theorem asserts that there is a subcategory of $\mathcal{U}_\kappa$ which has nice closure properties and which is "large" in the sense that it contains the generalized κ-local functors.

4.1 THEOREM. For any regular infinite cardinal κ, there is a locally full subcategory $\mathcal{V}^*_\kappa$ of $\mathcal{U}_\kappa$ such that $\mathcal{V}^*_\kappa$ contains all generalized κ-local functors and $\mathcal{U}^*_\kappa$ is a locally κ-complete and locally cocomplete subcategory of $\mathcal{U}$.

Proof. Notice that since $\mathcal{U}^*_\kappa$ is to contain all generalized κ-local functors it must have the same 0-cells (categories) as $\mathcal{U}$. Since $\mathcal{U}^*_\kappa$ is to be locally full in $\mathcal{U}_\kappa$, to define $\mathcal{U}^*_\kappa$ it suffices to define the 1-cells (functors) in $\mathcal{U}^*_\kappa$.

For any ordinal δ, call a functor $F: \underset{\sim}{A} \to \underset{\sim}{B}$ of $\mathcal{U}$ (κ, δ)-local if

(i) for every $A \in \mathrm{Ob}\, \underset{\sim}{A}$ and every $u \in \mathrm{Sub}_\kappa(F(A))$ there exists $e \in \mathrm{Sub}_\kappa(A)$ such that $u \le F(e)$; and

(ii)$_\delta$ for every $\tilde{f} = (f, e_0, e_1)$ in $\mathrm{PI}^\delta_\kappa(A_0, A_1)$ there exists $f^* = (f, e_0^*, e_1^*)$ in $\mathrm{PI}^0_\kappa(F(A_0), F(A_1))$ and morphisms γ_0, γ_1 such that

(*) 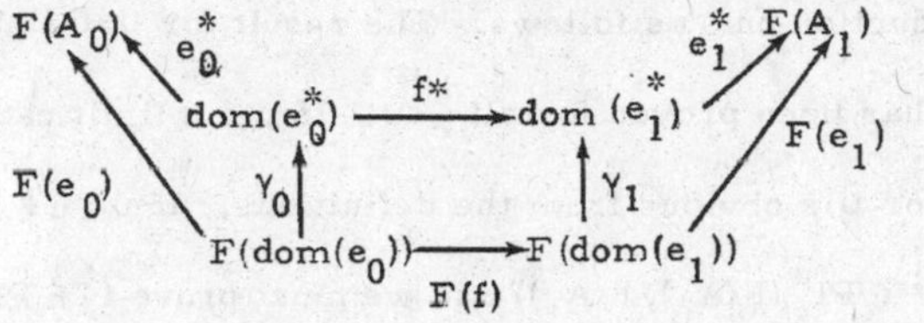

commutes.

Let us say that $F(\tilde{f})$ lifts to $\tilde{f}^*$ if e_0^* and e_1^* are inclusion maps and there is a commutative diagram (*) in which γ_0 and γ_1 are surjective. Then if F is (κ, δ)-local, for every $\tilde{f} \in \mathrm{PI}^\delta_\kappa$, there is one and only one $\tilde{f}^*$ such that $F(\tilde{f})$ lifts to $\tilde{f}^*$. Observe the following facts.

4.2 (1) If F is (κ, δ)-local and $\mu \ge \delta$, then F is (κ, μ)-local;

(2) F is κ-local if and only if F is $(\kappa, 0)$-local;

(3) if F is generalized κ-local, then F is $(\kappa, 1)$-local;

(4) if F is (κ, δ)-local, $\tilde{f} \in \mathrm{PI}^{\delta+\upsilon}_\kappa(A_0, A_1)$ and $F(\tilde{f})$ lifts to $\tilde{f}^*$, then $\tilde{f}^* \in \mathrm{PI}^\upsilon_\kappa(F(A_0), F(A_1))$;

(5) if F is (κ, δ)-local, then F preserves $L_{\infty\kappa}$-equivalence.

Proof. Part (1) follows immediately from the definitions. As for (2), if F is κ-local then F is $(\kappa, 0)$-local since for any $\tilde{f} \in PI^0_\kappa(A_0, A_1)$, we may take $\tilde{f}*$ to be $(F(f), F(e_0), F(e_1))$. Conversely if F is $(\kappa, 0)$-local and $e_0 : S_0 \longrightarrow A_0$ is any monomorphism where $Card(S_0) < \kappa$ consider $\tilde{f} = (1_{S_0}, e_0, e_1) \in PI^0_\kappa(A_0, A_1)$ where $A_1 = e_0(S_0)$ and $e_1 : S_0 \longrightarrow A_1$ is the map e_0 regarded as an isomorphism of S_0 onto A_1. Since $F(\tilde{f})$ lifts to an $\tilde{f}* \in PI^0_\kappa(F(A_0), F(A_1))$ and $F(1_{S_0})$ and $F(e_1)$ are isomorphisms, it follows easily that $F(e_0)$ is one-one i.e. a monomorphism. Then using (i) of the definition of $(\kappa, 0)$-local, one may prove that F preserves monomorphisms with arbitrary domains.

Part (3) follows from the proof of Proposition 3.2.

Part (4) is proved by induction on υ as follows. The result for $\upsilon = 0$ is by definition. Suppose (4) has been proved for all $\mu < \upsilon$. If υ is a limit cardinal the desired conclusion for υ is obvious from the definitions. If $\upsilon = \mu + 1$, we know by induction that $\tilde{f}* \in PI^\mu_\kappa(F(A_0), F(A_1))$ and we must prove $\tilde{f}* \in PI^{\mu+1}_\kappa$. Let $u_0 \in Sub_\kappa(F(A_0))$. By (i), there exists $e'_0 \geq e_0$ such that $u_0 \leq F(e'_0)$. Since $\tilde{f} \in PI^{\mu+1}_\kappa$ there exists $\tilde{g} = (g, v_0, v_1)$ in $PI^\mu_\kappa(A_0, A_1)$ such that $\tilde{g} \geq \tilde{f}$ and $v_0 \geq e'_0$. If $F(\tilde{g})$ lifts to $\tilde{g}* = (g*, v^*_0, v^*_1)$ in PI^μ_κ then it is not hard to see that $\tilde{g}* \geq \tilde{f}*$ and $v^*_0 \geq u_0$. Since an analogous conclusion holds when we start with $u_1 \in Sub_\kappa(F(A_1))$, we conclude that $\tilde{f}* \in PI^{\mu+1}_\kappa(F(A_0), F(A_1))$ and (4) is proved. Finally, (5) follows immediately from (4) since if $\tilde{f} \in PI^\infty_\kappa(A_0, A_1)$ and $F(\tilde{f})$ lifts to $\tilde{f}*$ then $\tilde{f}* \in PI^\infty_\kappa(F(A_0), F(A_1))$. This completes the proof of 4.2.

Let us say that F is (κ, ∞)-local if F is (κ, δ)-local for some ordinal δ. For a fixed κ, if F is (κ, ∞)-local, write $Loc(F) = \delta$ if δ is minimal such that F is (κ, δ)-local. We are going to define the 1-cells in $\mathcal{U}^*_\kappa$ to be the (κ, ∞)-local functors. First we prove a lemma which will insure that the (κ, ∞)-local functors are closed under composition.

4.3 LEMMA. Let $F: \underset{\sim}{A} \to \underset{\sim}{B}$ and $G: \underset{\sim}{B} \to \underset{\sim}{C}$ be (κ, ∞)-local functors. Then $G \circ F$ is (κ, ∞)-local; moreover $\mathrm{Loc}(G \circ F) \le \mathrm{Loc}(F) + \mathrm{Loc}(G)$.

Proof. Suppose $\mathrm{Loc}(F) = \mu$ and $\mathrm{Loc}(G) = \upsilon$. Let $\tilde{f} = (f, e_0, e_1) \in PI_\kappa^{\mu+\upsilon}(A_0, A_1)$. We must prove that there is a diagram of the form (*) with bottom row = GF(f). By 4.2.4 there exists $\tilde{f}^* \in PI_\kappa^\upsilon(F(A_0), F(A_1))$ such that $F(\tilde{f})$ lifts to $\tilde{f}^*$. Moreover since $\mathrm{Loc}(G) = \upsilon$, there exists $\tilde{g} = (g, u_0, u_1)$ such that $G(\tilde{f}^*)$ lifts to $\tilde{g}$. Thus we obtain the following diagrams.

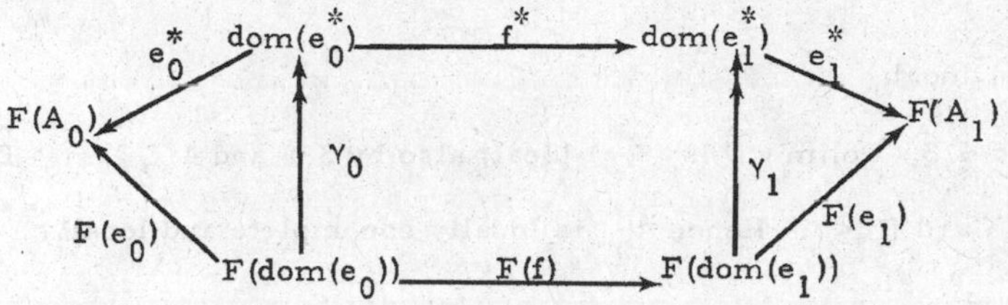

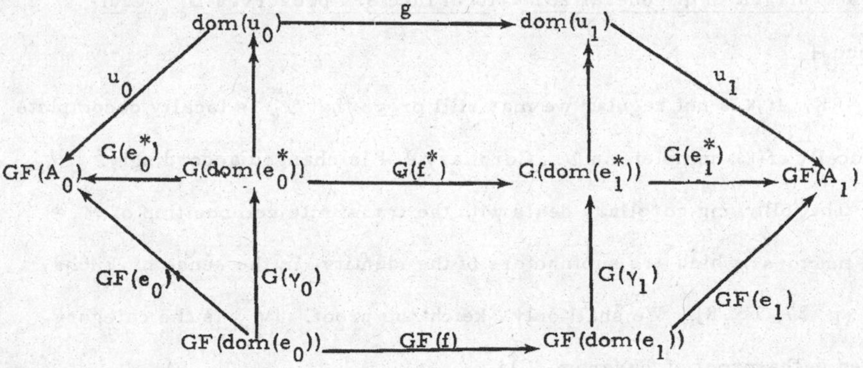

The outer portion of the last diagram provides the desired diagram. (Note that GF(f) lifts to a restriction of g, since $G(\gamma_0)$ and $G(\gamma_1)$ are not necessarily surjective). This completes the proof of 4.3.

In view of 4.3 we can define $\mathcal{U}_\kappa^*$ to be the locally full subcategory of $\mathcal{U}$ whose 1-cells consist of all $(\kappa\text{-}\infty)$-local functors. By 4.2.5 $\mathcal{U}_\kappa^*$ is a subcategory of $\mathcal{V}_\kappa$. It remains to prove that $\mathcal{V}_\kappa^*$ is locally κ-complete and locally cocomplete in $\mathcal{U}$. Let D be a diagram in $\mathcal{U}_\kappa^*(\underset{\sim}{A}, \underset{\sim}{B})$ over $\Sigma = (I, M, d)$. Let $\hat{D}: \underset{\sim}{A} \longrightarrow \underset{\sim}{B}_\Sigma$ be defined as in Proposition 1.6. Recall that $\lim \hat{D} = \mathrm{Lim} \circ \hat{D}$ and $\mathrm{colim}\, \hat{D} = \mathrm{Colim} \circ \hat{D}$. Using the fact that κ is regular we can see that $\hat{D}$ satisfies (i) of the definition of (κ, δ)-local. Moreover it is tedious but not difficult to check that D satisfies $(\mathrm{ii})_\delta$ of the definition for $\delta = \sup\{\mathrm{Loc}(D_i) : i \in I\}$.

Hence $\hat{D}$ is (κ, ∞)-local.

By 3.8 and 4.3, $\mathrm{Colim} \circ \hat{D}$ is (κ, ∞)-local; also by 3.8 and 4.3, $\mathrm{Lim} \circ \hat{D}$ is (κ, ∞)-local if Card $(I) < \kappa$. Hence $\mathcal{U}_\kappa^*$ is locally cocomplete and locally κ-complete in $\mathcal{U}$.⊣

4.4 COROLLARY. Let κ be regular. The composition (resp. colimit; limit over a κ-diagram) of generalized κ-local functors preserves $L_{\infty\kappa}$-equivalence.⊣

REMARK. If κ is not regular we may still prove that $\mathcal{V}_\kappa^*$ is locally cocomplete and locally $\mathrm{cf}(\kappa)$-complete in $\mathcal{V}$. Corollary 4.4 is changed accordingly.

The following corollary deals with the transfinite composition of (κ, ∞)-local functors which are subfunctors of the identity, in the sense of Fuchs [Fu 1; p. 27, Ex. 8]. We shall only sketch the proof. ($\mathcal{U}_\kappa^*$ is the category defined in the proof of Theorem 4.1).

4.5 COROLLARY. If $\{F_\sigma \mid \sigma<\tau\} \subseteq \mathcal{U}^*_\kappa(\underset{\sim}{Ab}, \underset{\sim}{Ab})$ is a sequence of subfunctors of the identity which are (κ, ∞)-local, then $\prod_{\sigma<\tau} F_\sigma$ preserves $L_{\infty\kappa}$ - equivalence.

Proof. Since subfunctors of the identity preserve inclusions, we shall confine our attention to monomorphisms which are inclusions. Thus we identify sub-objects with substructures. Also, we identify a partial isomorphism $f = (f, e_0, e_1)$ with f when e_0 and e_1 are inclusions. Hence $f \in PI^0_\kappa(A_0, A_1)$ is an isomorphism of substructures of A_0 and A_1 respectively. Let $G_\tau = \prod_{\sigma<\tau} F_\sigma$ and let $f \in PI^\infty_\kappa(A_0, A_1)$. Define $g_\tau = f \mid (\text{dom } f) \cap G_\tau(A_0)$. Then one may prove by induction on τ that g_τ is an isomorphism onto $(\text{codom } f) \cap G_\tau(A_1)$ and that g_τ is in $PI^\infty_\kappa(G_\tau(A_0), G_\tau(A_1))$. We shall do one of the key steps in the inductive proof. Suppose $\tau = \mu + 1$ and the result has been proved for μ. Let $x \in (\text{dom } f) \cap G_\tau(A_0)$. Note that $G_\tau(A_0) = F_\mu(G_\mu(A_0))$. Since F_μ is $(\kappa, 0)$-local there exists $B_0 \in \text{Sub}_\kappa(G_\mu(A_0))$ such that $x \in F_\mu(B_0) \subseteq G_\tau(A_0)$. Since $g_\mu \in PI^\infty_\kappa(G_\mu(A_0), G_\mu(A_1))$, there exists $g'_\mu \in PI^\infty_\kappa$ extending g_μ such that $B_0 \subseteq \text{dom}(g'_\mu)$. By definition of a subfunctor of the identity, $F_\mu(g'_\mu) = g'_\mu \mid F_\mu(\text{dom } g'_\mu)$. Hence $g_\tau(x) = g_\mu(x) = F_\mu(g'_\mu)(x) \in F_\mu(\text{codom } g'_\mu) \subseteq G_\tau(A_1)$. Thus g_τ maps into $(\text{codom } f) \cap G_\tau(A_1)$. A similar argument applied to f^{-1} shows that g_τ maps onto $(\text{codom } f) \cap G_\tau(A_1)$. Finally another inductive argument shows that if $f \in PI^\upsilon_\kappa(A_0, A_1)$, then $g_\tau \in PI^\upsilon_\kappa(G_\tau(A_0), G_\tau(A_1))$. Thus we see that $G_\tau \in \mathcal{U}_\kappa$. ⊣

4.6 EXAMPLE. The functor $F: \underset{\sim}{Ab} \to \underset{\sim}{Ab}$ defined by $F(A) = pA$ is ω- local (cf. Example 3.1.4). Hence for every ordinal τ, the functor: $A \rightsquigarrow p^\tau A$ preserves $L_{\infty\kappa}$ -equivalence for any $\kappa \geq \omega$.

5. Complete systems of invariants. Let $\mathfrak{S}$ be a subclass of the class of objects of a uHf category $\underset{\sim}{A}$. Let us define a complete system of functorial invariants for $\mathfrak{S}$ to be a class of functors $F_i : \underset{\sim}{A} \to \underset{\sim}{B}_i$ in $\mathcal{U}$, $i \in I$, such that for every pair of objects A_0, A_1 in $\mathfrak{S}$, A_0 is isomorphic to A_1 if and only if $F_i(A_0)$ is isomorphic to $F_i(A_1)$ for every $i \in I$.

Of course, by this definition the set consisting of the identity functor on A is a complete system of functorial invariants for $\mathfrak{S} = \mathrm{Ob}\underset{\sim}{A}$. For a complete system of functorial invariants to be useful it is necessary that the isomorphism problem for objects in the image of the F_i's should be more tractable than the original isomorphism problem for $\mathfrak{S}$. To make that idea more precise we make the following definition.

Let $\underset{\sim}{A}$ be a uHf category of L-algebras.

Denote the coproduct of objects $\{A_i \mid i \in I\}$ in $\underset{\sim}{A}$ by $\coprod_{i \in I} A_i$. For any ordinal μ denote by $A^{(\mu)}$ the coproduct of a sequence of copies of A indexed by μ. A subclass $\mathfrak{Y}$ of $\mathrm{Ob}\underset{\sim}{A}$ is said to have a κ- structure theory if card L $< \kappa$, $\mathfrak{Y}$ is closed under coproducts, and there is a subset $\mathfrak{C} = \{C_\nu \mid \nu < \sigma\}$ of $\mathfrak{Y}$ satisfying:

(i) each C_ν is generated by $< \kappa$ elements;

(ii) every object Y of $\mathfrak{Y}$ is isomorphic to a coproduct $\coprod_{\nu<\sigma} C_\nu^{(\lambda_\nu)}$, where each λ_ν is a cardinal ≥ 0 and the λ_ν's are uniquely determined by Y; and

(iii) whenever $\mu_\nu \leq \mu'_\nu$ for all $\nu < \sigma$, then the canonical morphism:

$$\coprod_{\nu<\sigma} C_\nu^{(\mu_\nu)} \to \coprod_{\nu<\sigma} C_\nu^{(\mu'_\nu)}$$

is an embedding.

Obviously, the isomorphism problem for a class $\mathfrak{Y}$ with a κ - structure theory is tractable (relative to knowledge of $\mathfrak{C}$); indeed the cardinals λ_ν, $\nu < \sigma$, form a "complete set of cardinal invariants" (in the usual informal sense).

5.1 EXAMPLES. (1) If F is a field and F-Mod is the category of vector spaces over F, then Ob(F-Mod) has an ω - structure theory. (Here $\mathfrak{C} = \{F\}$).

(2) The class Ob(Set) has an ω - structure theory. (Here $\mathfrak{C} = \{1\}$).

(3) If $\mathfrak{D} \subseteq$ Ob(Ab) is the class of divisible abelian groups, $\mathfrak{D}$ has an ω_1 - structure theory. (Here $\mathfrak{C} = \{Q\} \cup \{Z(p^\infty) : p \text{ prime}\}$).

We shall be considering complete systems of functorial invariants which take value in classes with structure theories. Let us look at some examples.

5.2 EXAMPLES. (1) Let p be a prime and let $\mathfrak{T}_p$ be the class of all totally projective p-primary groups [Fu 2; §82]. For each $A \in \mathfrak{T}_p$ let $\mathrm{Div}(A)$ denote the divisible part of A and let $R(A)$ be $A/\mathrm{Div}(A)$. For each ordinal ν let $F_\nu(A)$ be $p^\nu R(A)[p] / p^{\nu+1} R(A)[p]$. Then $\{\mathrm{Div}\} \cup \{F_\nu : \nu \in \mathrm{Ord}\}$ is a complete system of invariants for $\mathfrak{T}_p$ [Fu 2; Thms 83.3, 83.5]. Moreover, by the results of sections 3 and 4, $\mathrm{Div} \in \mathcal{U}^*_{\omega_1}$ and $F_\nu \in \mathcal{U}_\omega$. (For the latter we use Example 4.5. Note that F_ν is not necessarily in $\mathcal{U}^*_\omega$ if $\nu \geq \omega$). Also Div takes values in a class with an ω_1 - structure theory. (Example 5.1.3) and F_ν takes values in a class with an ω - structure theory. (Example 5.1.1).

(2) Let $\mathfrak{D}$ be the class of all completely decomposable torsion-free abelian groups [Fu 2; §86]. Let Div and R be as above. For each type $t \neq (\infty, \infty, \ldots)$ choose a prime p_t such that the component of t corresponding to p_t is not ∞ [Fu 2; §85]. For each A in $\mathfrak{D}$, let $G_t'(A) = A(t)/A^*(t)$ and let $G_t(A) = G_t'(A)/p_t G_t'(A)$. Then $\{\mathrm{Div}\} \cup \{G_t \mid t \text{ a type} \neq (\infty, \infty, \ldots)\}$ is a complete system of functorial invariants for $\mathfrak{D}$. Moreover, $G_t \in \mathcal{U}^*_{\omega_1}$ (cf. Example 3.1.6) and takes values in a class with an ω - structure theory (Example 5.1.1). (Of course $\mathfrak{D}$ itself has an ω_1 - structure theory; the proof of this fact makes use of the functors G_t').

The two examples considered above constitute perhaps the most important examples of complete systems of invariants in abelian group theory. The classes $\mathfrak{T}_p$ and $\mathfrak{D}$ are closed under direct summands and under direct sums. We shall prove that if we close these classes under some kinds of direct limits then we no longer have a complete system of functorial invariants with the nice properties of those in the examples.

By a direct limit of $\aleph_1$-pure embeddings we mean a direct limit of a diagram in which all maps are $\aleph_1$-pure embeddings. A direct limit is well-ordered if it is a direct limit over a directed set which is well-ordered. If $\mathfrak{S}$ is a class of structures, $\mathfrak{S}_\kappa$ denotes the elements of $\mathfrak{S}$ of cardinality $< \kappa$.

5.3 THEOREM. Let $\mathfrak{S}$ be the closure of $\mathfrak{T}_p$ (or $\mathfrak{D}$) under well-ordered direct limits of $\aleph_1$-pure embeddings. There is no complete system of functorial invariants for $\mathfrak{S}_{\omega_2}$ satisfying the following properties:

(a) every functor in the system is in $\mathcal{U}_{\omega_1}$;

(b) every functor in the system takes values in a class with an ω_1-structure theory; and

(c) for each functor F in the system and each $A \in \mathfrak{S}_{\omega_2}$, $F(A)$ has cardinality $\leq \omega_1$.

REMARKS. (i) Note that the countable members of $\mathfrak{S}$ are in $\mathfrak{T}_p$ or $\mathfrak{D}$, as the case may be, and that every countable subgroup of a member of $\mathfrak{S}$ is contained in a countable subgroup which is in $\mathfrak{T}_p$ or $\mathfrak{D}$.

(ii) A functor satisfies condition (c) if it is in $\mathcal{U}^*_\omega$ and takes finitely-generated structures to finitely-generated structures.

Proof. The proof of the theorem consists of two steps: (I) there exist groups A_0 and A_1 in $\mathfrak{S}$ of cardinality ω_1 such that $A_0 \equiv_{\infty\omega_1} A_1$ but $A_0 \not\cong A_1$; and (II) if B_0 and B_1 are members of a class with an ω_1-structure theory, and $B_0 \equiv_{\infty\omega_1} B_1$, and $\max\{\text{card}(B_0), \text{card}(B_1)\} \leq \omega_1$, then $B_0 \cong B_1$.

Suppose that (I) and (II) have been proved and that there is a complete system $\{F_i : i \in I\}$ of functorial invariants for $\mathfrak{S}$ satisfying properties (a), (b), (c). Let A_0 and A_1 be as in (I). For every $i \in I$, F_i is in $\mathcal{U}_{\omega_1}$ by (a), so $F_i(A_0) \equiv_{\infty\omega_1} F_i(A_1)$ by definition of $\mathcal{U}_{\omega_1}$. Therefore by (b), (c) and (II), $F_i(A_0) \cong F_i(A_1)$. But this contradicts the definition of a complete system of functorial invariants and the fact that $A_0 \not\cong A_1$.

For the proof of step (I) we refer to [E]. Consider first the case when $\mathfrak{S}$ is the closure of $\mathfrak{D}$. Let A_0 be the free abelian group of cardinality ω_1. By Corollary 3.2 of [E], there is a non-free group A_1 of cardinality ω_1 such that $A_0 \equiv_{\infty\omega_1} A_1$. An examination of the construction in [E] shows that A_1 is the union of a chain $\{G_\nu : \nu < \omega_1\}$ of countable free abelian groups such that for each $\nu < \mu < \omega_1$ G_ν is a direct summand of G_μ. (Referring to the proof of Theorem 2.2 of [E], take G_ν to be the group $A_{\nu+1}$ of that proof). Hence A_1 belongs to $\mathfrak{S}$.

More generally, we may choose A_0 to be any completely decomposable homogeneous group of type $t \neq (\infty, \infty, \ldots)$. Then there is an A_1 which is not completely decomposable such that $A_0 \equiv_{\infty\omega_1} A_1$ (The proof makes use of Theorem 86.6 and Lemma 86.8 of [Fu 2]). In fact, using an idea of Shelah[S;§1] we can prove that there are $2^{\aleph_1}$ different such A_1's for each A_0.

A similar argument applies in the case when $\mathfrak{S}$ is the closure of $\mathfrak{T}_p$. In this case we may take A_0 to be any direct sum of cyclic p-groups such that for each n, $p^n A$ is not countable. Again we can prove by an argument like that in [S] that for each such A_0 there are $2^{\aleph_1}$ groups A_1 such that $A_0 \equiv_{\infty\omega_1} A_1$.

Step (II) follows from the next proposition.

5.4 PROPOSITION. Let $\kappa \leq \rho$ be cardinals with ρ regular. Let $\mathfrak{N}$ be a subclass of a uHf - category $\underset{\sim}{A}$ such that $\mathfrak{N}$ has a κ - structure theory. If B and B′ are elements of $\mathfrak{N}$ such that $B \equiv_{\infty\rho} B'$ and $\max\{\mathrm{card}(B), \mathrm{card}(B')\} \leq \rho$, then B is isomorphic to B′.

First we need some information about cardinalities of coproducts. Define card (L) = the cardinality of the set of formulas of L.

5.5. (1) For any coproduct $\coprod_{i\in I} A_i$, card $(\coprod_{i\in I} A_i) \le$ card(L)+ $\sum_{i\in I}$ card A_i. Also for any subset Y of $\coprod_{i\in I} A_i$ there exists a subset $J \subseteq I$ of cardinality $\le$ card(Y) + $\aleph_0$ such that $Y \subseteq \coprod_{i\in J} A_i$;

(2) If $\mathfrak{C}$ is as in the definition of a κ - structure theory and $C \in \mathfrak{C}$, then for any cardinal λ, card $(C^{(\lambda)}) \ge \lambda$.

Proof. The first part follows from the construction of colimits given in Proposition 1.2. In the situation of part (iii) of the definition of a κ- structure theory, let us call the image of $\coprod_{\nu<\sigma} C_\nu^{(\mu'_\nu)}$ under the canonical embedding a block in $\coprod_{\nu<\sigma} C_\nu^{(\mu_\nu)}$. We shall identify $\coprod_{\nu<\sigma} C_\nu^{(\mu'_\nu)}$ with its image in $\coprod_{\nu<\sigma} C_\nu^{(\mu_\nu)}$. Now for part (2), it suffices to prove that the ascending chain of blocks $\{C^{(\mu)} \mid \mu<\lambda\}$ in $C^{(\lambda)}$ is strictly increasing. But this follows since by the definition of a κ - structure theory, $C \amalg C \ncong C$. This proves 5.5.

Proof of 5.4. Suppose $B = \coprod_{\nu<\sigma} C_\nu^{(\lambda_\nu)}$ and $B' = \coprod_{\nu<\sigma} C_\nu^{(\lambda'_\nu)}$ where $\lambda_\nu, \lambda'_\nu$ are cardinals ≥ 0 (using the notation of the definition of a κ - structure theory). We must prove that $\lambda_\nu = \lambda'_\nu$ for all $\nu<\sigma$. Notice that by Lemma 5.5 (2), $\lambda_\nu, \lambda'_\nu \le \rho$. Suppose that for some $\mu<\sigma$, $\lambda_\mu < \lambda'_\mu$. Then $\lambda_\mu < \rho$. By 5.5 (1), card $(C_\mu^{(\lambda_\mu)}) < \rho$. Hence there exists $f_0 : W_0 \longrightarrow W'_0$ in $PI_\rho^\infty(B, B')$ such that W_0 is the block $C_\mu^{(\lambda_\mu)}$. We consider two cases. Suppose first that $\lambda'_\mu < \rho$. Then by Lemma 5.5(1) and the definition of PI_ρ^∞, there exists $f_1 : W_1 \longrightarrow W'_1$ in $PI_\rho^\infty(B, B')$ extending f_0 and such that W'_1 is a block in B' which contains the block $C_\mu^{(\lambda'_\mu)}$. By induction we obtain $f_n : W_n \longrightarrow W'_n$ in $PI_\rho^\infty(B, B')$ such that f_n extends f_{n-1} and W_n (resp. W'_n) is a block if n is even (resp. odd). The union $g = \bigcup_n f_n$ is an isomorphism of a block $\bigcup_n W_n$ in B with a block $\bigcup_n W'_n$ in B'. But this contradicts the unique representation of coproducts of elements of $\mathfrak{C}$, since $\bigcup_n W_n$ contains the block $C_\mu^{(\lambda_\mu)}$ and $\bigcup_n W'_n$ contains the block $C_\mu^{(\lambda'_\mu)}$.

In the second case, suppose $\lambda'_\mu = \rho$. Let f_0 be as above. Since card $(W'_0) < \rho$ there is by 5.5, a block W''_0 in B' of cardinality $< \rho$ containing W'_0. Obviously W''_0 does not contain the block $C_\mu^{(\lambda'_\mu)}$ since card $C_\mu^{(\lambda'_\mu)} = \rho$. Hence there exists $f_1 : W_1 \to W'_1$ in $PI_\rho^\infty(B, B')$ such that W'_1 is a block of the form $W''_0 \amalg C_\mu$. Define $f_n : W_n \to W'_n$ for $n \geq 2$ as above so that W_n (resp. W'_n) is a block if n is even (resp. odd). The union $g: W = \bigcup_n W_n \to W' = \bigcup_n W'_n$ is an isomorphism of blocks which induces an isomorphism: $W/W_0 \to W'/W'_0$. (Here W/W_0 denotes the cokernel of $W_0 \subseteq W$). But $W/W_0 = \coprod_{\nu<\sigma} C_\nu^{(\gamma_\nu)}$ where $\gamma_\mu = 0$ because $W_0 = C_\mu^{(\lambda_\mu)}$, while $W'/W'_0 \cong \coprod_{\nu<\sigma} C_\nu^{(\gamma'_\nu)}$ where $\gamma_\mu > 0$ since W'_1/W'_0 is a block in W' and $W'_1 = W''_0 \amalg C_\mu$. This contradicts the uniqueness of coproducts of elements of $\mathfrak{C}$. Therefore the proposition and the theorem are proved. ⊣

The proof of the theorem also yields the following Corollary which is relevant to problems 51 (p. 55) and 77 (p. 84) of [Fu 2].

5.6 COROLLARY. If $\mathfrak{S}$ is either the class of separable p-groups or the class of separable torsion-free groups then the conclusion of Theorem 5.3 holds for $\mathfrak{S}_{\omega_2}$.

Proof. It is only necessary to observe that step (I) holds for the classes in question. For the class of separable p-groups, if $A_0 \equiv_{\infty\omega_1} A_1$ and A_0 is a direct sum of cyclic groups then A_1 is separable. For the class of separable torsion-free groups see Mekler [Me] where an $\aleph_1$- separable non-free group A_1 of cardinality ω_1 is constructed which is $L_{\infty\omega_1}$ - equivalent to a free group. ⊣

Added in proof: W. Hodges in A Normal Form for Algebraic Constructions II (to appear in the proceedings of the Louvain model theory conference) has given a model-theoretic definition of a class of functors preserving $L_{\infty\kappa}$-equivalence with the closure properties of Theorem 4.1.

REFERENCES

[B] M. Benda, Reduced products and non-standard logics, J. Symbolic Logic 34 (1969), pp. 424-436.

[C-K] C.C. Chang and H.J. Keisler, Model Theory, North-Holland, Amsterdam, 1973.

[Co] P. Cohn, Universal Algebra, Harper and Row, London, 1965.

[E] P. Eklof, On the existence of κ- free abelian groups, Proceedings of the Amer. Math. Soc. 47 (1975), pp. 65-72.

[Fe] S. Feferman, Infinitary properties, local functors, and systems of ordinal functions, Conference in Mathematical Logic (London, 1970), Lecture notes in Math., no. 255, Springer-Verlag, Berlin, 1972, pp. 63-97.

[Fu 1] L. Fuchs, Infinite Abelian Groups, vol I, Academic Press, New York, 1970.

[Fu 2] L. Fuchs, Infinite Abelian Groups, vol II, Academic Press, New York, 1973.

[Ma 1] A Mal'cev, The Metamathematics of Algebraic Systems, North-Holland, 1971.

[Ma 2] A. Mal'cev, Algebraic Systems, Springer-Verlag, Berlin 1973.

[Mac] S. MacLane, Categories for the Working Mathematician, Springer-Verlag, Berlin, 1972.

[Me] A. Mekler, Ph.D dissertation, Stanford University, in preparation.

[Mi] B. Mitchell, Theory of Categories, Academic Press, New York, 1965.

[S] S. Shelah, Infinite abelian groups-Whitehead problem and some constructions, Is. J. Math. 18 (1974), 243-256.

[W] R. Warfield, Simply-presented groups, preprint, 1974.

IMPREDICATIVITY OF THE EXISTENCE OF THE LARGEST DIVISIBLE SUBGROUP OF AN ABELIAN p-GROUP

Solomon Feferman[1)]
Stanford University

§1. The structure theory of Abelian p-groups G (cf. [4]) makes essential use of the largest divisible subgroup of G,

(1) $$\mathrm{Div}(G) = \cup H[H \subseteq G\ ,\ H \text{ divisible}]\ .$$

$\mathrm{Div}(G)$ may also be described as $p^{\lambda}G$ where λ is the length of G, i.e., the least ordinal for which $p^{\lambda}G = p^{\lambda+1}G$. The definition of $\mathrm{Div}(G)$ has a typically impredicative character, which may be established by non-definability results with respect to the class HYP of hyperarithmetic sets. For positive definability results it is obvious that if G has hyperarithmetic structure, then $\mathrm{Div}(G) \in \Sigma^1_1$. By general considerations on inductive definitions one sees under the same conditions that $\mathrm{length}(G) \leq \omega_1^{rec}$ (the least non-recursive or, equivalently, least non-HYP ordinal). $\mathrm{Div}(G)$ and $\mathrm{length}(G)$ are classified similarly for arbitrary countable G in §2.

Barwise proved the existence of a recursive G for which $\mathrm{Div}(G) \notin \mathrm{HYP}$.[2)] As will be explained at the conclusion, this does not establish the impredicativity of the existence of the largest divisible subgroup. Instead one must consider

(2) $$\mathrm{Div}^*(G) = \cup H[H \subseteq G\ ,\ H \text{ divisible}\ ,\ H \in \mathrm{HYP}]\ .$$

It happens that in the example given by Barwise, $\mathrm{Div}^*(G) = 0$. The main result here is as follows:

[1)] The results of this paper were obtained during the author's visit in the U.E.R. de Mathématiques, Université Paris VII in 1973. Preparation supported by NSF grant GP 43901 X.
[2)] Personal communication.

1.1. <u>Theorem</u>. <u>There is a recursive 2-group</u> G <u>such that</u>

(i) $\mathrm{Div}^*(G) \notin \mathrm{HYP}$ <u>and</u> $\mathrm{Div}(G) \notin \mathrm{HYP}$, <u>and</u>

(ii) <u>length</u>$(G) = \omega_1^{rec}$.

The method of proof of 1.1 is to reduce it to related questions of definability for trees. With each tree T in the natural numbers is associated a group G_T with generators $\{g_t \mid t \in T\}$ given with divisibility relations exactly following T . This induces a divisibility picture in G_T ; divisible subgroups of G_T correspond to unfounded subsets of T . This correspondence is not 1-1 but it can be shown that enough information about T can be recaptured in G_T by use of suitable normal forms. There may be independent interest in this kind of relationship between trees and groups (and possibly other algebraic structures). For 1.1 one uses the existence of a recursive tree whose unfounded part (even in a * sense) is not in HYP and whose well-founded part has ordinal ω_1^{rec} .[3)]

§2. We consider throughout countable infinite Abelian groups $(G, +_G)$ given with an enumeration $\varphi : N \xrightarrow[\text{onto}]{1-1} G$, where N is the set of natural numbers. All definability notions in N are transferred to G via φ . For example, G is said to be recursive in A if the set $\{(x,y,z) \mid \varphi(x) +_G \varphi(y) = \varphi(z)\} \leq_{rec} A$. For simplicity, one can restrict attention to $G = N$, $\varphi = id_N$. We shall also write + for $+_G$ when there is no ambiguity.

2.1. <u>Theorem</u>. <u>Suppose</u> G <u>is recursive in</u> A . <u>Then</u>

(i) $\mathrm{Div}(G) \in \Sigma_1^{1(A)}$ <u>and</u> (ii) $\mathrm{Div}^*(G) \in \Pi_1^{1(A)}$.

<u>Proof</u>. (i) is immediate by definition. (ii) is by relativization to A of Kleene's theorem in [5], according to which if R is recursive and

1) $$m \in B \Longleftrightarrow \exists \alpha_{HYP} \forall x\, R(m, \bar{\alpha}(x))$$

then

2) $$B \in \Pi_1^1 .$$

---

3) It is of interest to compare this with Barwise's example, which is extracted from a non-well-founded model (of a part of ZF) whose ordinal standard part is exactly ω_1^{rec} .

2.2. Theorem. Suppose G is a p-group recursive in A. Then $\text{length}(G) \leq \omega_1^A$.

Proof. By relativization of the following argument to A. Let $\lambda = \text{length}(G)$ and let $X_\sigma = G - p^\sigma G$ for each ordinal σ. Thus

1) $X_0 = 0$

2) $X_{\sigma+1} = \{x \mid \forall y(x = py \Longrightarrow y \in X_\sigma)\}$

3) $X_\sigma = \bigcup_{\tau<\sigma} X_\tau$ for limit σ, and

4) λ is the least σ with $X_\sigma = X_{\sigma+1}$.

X_λ is thus the smallest set satisfying the closure condition $\forall x[\,\mathfrak{a}(x,X) \Longrightarrow x \in X]$ where $\mathfrak{a}(x,X) \Longleftrightarrow \forall y(x = py \Longrightarrow y \in X)$. By Spector [8], every set inductively defined by means of $\mathfrak{a}$ in Π_1^0 has closure ordinal $\leq \omega_1^{rec}$.

Note that it also follows by Spector's work that for each $\sigma < \omega_1^A$, X_σ and hence $p^\sigma G$ belong to $HYP^{(A)}$. Theorem 2.2 is due to Barwise.

§3. In this section we collect the information which will be needed about trees in N; these are identified with sets of sequences as follows. Given $s = \langle s_0,\ldots,s_n\rangle$ we write $\ell h(s) = n+1$ and $s \subseteq t$ if t is an extension of s. If $\ell h(s) > 1$ we put $s^+ = \langle s_0,\ldots,s_{n-1}\rangle$, so $s^+ \subseteq s$, $\ell h(s) = \ell h(s^+) + 1$. A set X of sequences is said to be closed if

(1) $$s \in X \text{ and } \ell h(s) > 1 \text{ implies } s^+ \in X.$$

The closure of X under $^+$ is denoted by $(X)^+$. T is called a tree if it is a closed set of non-empty sequences. Given T, we write $t \preceq_T s$ if $t \in T$ and $s \subseteq t$; $t \prec_T s$ is written for $t \preceq_T s$ and $t \neq s$. (When T is fixed, the subscript is dropped.) For definability purposes, sequences of numbers are identified with numbers in a standard primitive recursive way.

Given T, the well-founded part of T is defined by

(2) $$Wf(T) = \{s \mid \neg \exists \alpha[\alpha(0) = s \wedge \forall n(\alpha(n+1) \prec \alpha(n))]\}.$$

For $s \in Wf(T)$, the subtree $T_s = \{t \mid (s*t) \in T\}$ is well-founded; we put $|s|$ = the ordinal length of T_s. Thus $|s| = 0$ for s minimal and

$|s| = \sup\{|t| + 1 \mid t \prec s\}$ otherwise. We put

(3) $$|Wf(T)| = \sup\{|s| \mid s \in Wf(T)\}\ .$$

The notion of well-foundedness relativized to HYP leads to the following definition :

(4) $$Wf^*(T) = \{s \mid \neg \exists \alpha_{HYP}[\alpha(0) = s \wedge \forall n(\alpha(n+1) \prec \alpha(n))]\}\ .$$

This generalizes the set 0^* introduced in [3].[4)]

Given any $X \subseteq T$, we call X unfounded if

(5) $$\forall s \in X\ \exists t \in X\ (t \prec s)\ .$$

Define

(6) $$Unf(T) = \bigcup X[X \subseteq T \text{ and } X \text{ is (closed and) unfounded}]\ ,$$

and

(7) $$Unf^*(T) = \bigcup X[X \subseteq T\ ,\ X \text{ is (closed and) unfounded and } X \in HYP]\ .$$

It is easily seen that the same sets are defined if we omit the condition that X is closed, since X is unfounded iff $(X)^+$ is unfounded and $X \in HYP \Longrightarrow (X)^+ \in HYP$. The following is also quite direct.

3.1. Lemma. (i) $Unf(T) = T - Wf(T)$.

(ii) $Unf^*(T) \supseteq T - Wf^*(T)$.

(iii) $Wf(T) \subseteq Wf^*(T)$ and $Unf^*(T) \subseteq Unf(T)$.

(iv) $Unf(T)$ is the largest (closed) unfounded subset of T .

4) To be precise, given any partial ordering $\leq$, we may associate the tree $T_{\leq}$ of all sequences $s = \langle s_0,\dots,s_n\rangle$ which are strictly descending in it, i.e., for which $s_0 > \dots > s_n$. For a certain r.e. relation $\leq$ we have $\langle x\rangle \in Wf(T_{\leq})$ iff $x \in 0$, and $\langle x\rangle \in Wf^*(T_{\leq})$ iff $x \in 0^*$.

3.2. Lemma. Suppose $T \in HYP$. Then

(i) $Wf(T) \in \Pi^1_1$ and $Unf(T) \in \Sigma^1_1$

(ii) $|Wf(T)| \le \omega_1^{rec}$

(iii) $Wf^*(T) \in \Sigma^1_1$ and $Unf^*(T) \in \Pi^1_1$

(iv) $Unf^*(T) = T - Wf^*(T)$.

Proof. For (ii), we can embed $\preceq_T$ in a HYP well-ordering by proceeding lexicographically. Thus, for each $s \in Wf(T)$, $|s| < \omega_1^{rec}$.

(iii) uses Kleene's [5] as in 2.1.

(iv) holds because if $X \in HYP$ and is unfounded, then there exists a HYP function γ which chooses for each $s \in X$ some $t \in X$ with $\gamma(s) = t \prec s$. Given $s_0 \in X$, put $\alpha(0) = s_0$ and $\alpha(n+1) = \gamma(\alpha(n))$; then $\alpha \in HYP$ and descends from s_0 , i.e., $s_0 \notin Wf^*(T)$.

3.3. Theorem. There exists a primitive recursive tree T for which:

(i) $Unf^*(T) \notin HYP$ and $Unf(T) \notin HYP$, and

(ii) $|Wf(T)| = \omega_1^{rec}$.

Proof. We use here Spector's converse [7] to Kleene's theorem, from which there exists primitive recursive R such that for all m :

1) $$m \in O \Longleftrightarrow \exists \beta_{HYP} \forall x\, R(m,\bar{\beta}(x)) .$$

Let T be the set of all sequences s of the form $s = \langle m, b_0, \ldots, b_{n-1} \rangle$ where $R(m, \langle b_0, \ldots, b_{n-1} \rangle)$.

By 3.2 (iv),

2) $$Unf^*(T) = \{s \mid s \in T \text{ and } \exists \alpha_{HYP}[\alpha(0) = s \wedge \forall n(\alpha(n+1) \prec \alpha(n))]\} .$$

Hence

3) $$m \in O \Longleftrightarrow \langle m \rangle \in Unf^*(T) .$$

It follows that $Unf^*(T) \notin \Sigma^1_1$ so not in HYP .

To prove $\mathrm{Unf}(T) \notin \mathrm{HYP}$ it suffices to establish (ii), since otherwise $\mathrm{Wf}(T) \in \mathrm{HYP}$ and then $|\mathrm{Wf}(T)| < \omega_1^{rec}$.

Now for (ii) suppose $|\mathrm{Wf}(T)| < \omega_1^{rec}$. Let $a \in O$ with $|\mathrm{Wf}(T)| = |a|$. Then $s \in \mathrm{Wf}(T)$ iff there exists an order-preserving mapping of $\{t | t \preceq s\}$ into $\{b | b \leq_O a\}$. Hence $\mathrm{Wf}(T) \in \Sigma_1^1$ so $\mathrm{Wf}(T) \in \mathrm{HYP}$. We claim that in this case $\mathrm{Wf}(T) = \mathrm{Wf}^*(T)$. For suppose there exists $s \in \mathrm{Wf}^*(T) - \mathrm{Wf}(T)$. Let $X = \{t | t \preceq s\} - \mathrm{Wf}(T)$; thus X is HYP and unfounded. But then $s \in \mathrm{Wf}^*(t)$ implies $s \notin X$ which is a contradiction. To complete the argument, since $\mathrm{Wf}(T) \in \mathrm{HYP}$ and $\mathrm{Wf}(T) = \mathrm{Wf}^*(T)$ we have $\mathrm{Wf}^*(T) \in \mathrm{HYP}$, contradicting $\mathrm{Unf}^*(T) \notin \mathrm{HYP}$. Thus $|\mathrm{Wf}(T)| = \omega_1^{rec}$, as was to be established.[5)]

§4. We assume throughout this section that T is any tree in N. An Abelian group G_T is associated with T by means of the following recursive presentation. G_T has generators g_s for each $s \in T$, satisfying relations:

(1) (i) $g_s + g_s = g_{(s^+)}$ when $\ell h(s) > 1$, and
(ii) $g_s + g_s = 0$ when $\ell h(s) = 1$.

More precisely, G_T is F_T/R_T where F_T is the free Abelian group on the set of generators $\{g_s | s \in T\}$ and R_T is the subgroup generated by $\{g_s + g_s - g_{s^+} | \ell h(s) > 1\} \cup \{g_s + g_s | \ell h(s) = 1\}$. $\Sigma_{i=1}^n m_i g_{s_i}$ or $(m_1 g_{s_1} + \ldots + m_n g_{s_n})$ will be written for typical elements both of F_T and of G_T ; the m_i are integers and $n \geq 0$. We use $=$ for the equality relation in G_T and $\doteq$ for the equality relation in F_T. Thus

(2)
$$\Sigma_{i=1}^n g_{s_i} \doteq \Sigma_{j=1}^m g_{t_i}$$

if and only if $n = m$ and $\langle s_i \rangle_{i \leq n}$ is a permutation of $\langle t_i \rangle_{i \leq n}$. Obviously

(3)
$$2^{\ell h(s)} g_s = 0$$

for any $s \in T$. This proves the first part of the following.

---

5) The argument here is a variant of that given in [3] to show that if $a \in O^* - O$, then the set of predecessors of $\underline{a}$ in O has order-type ω_1^{rec}.

4.1. Theorem. (i) G_T is an Abelian 2-group.

(ii) G_T is recursive in T.

To prove (ii) we need to provide an assignment of normal forms, which is carried out as follows. By (3) it is sufficient to restrict attention to representation of elements of G_T in the form $\Sigma_{i=1}^{n} m_i g_{t_i}$ where $m_i > 0$, $n \geq 0$.

An element of F_T is said to be in standard form if it is given as

(4) $$g \doteq \Sigma_{i=1}^{n} m_i g_{t_i} \quad \text{where } n \geq 0, \text{ and } i < j \Longrightarrow m_i > 0, t_i \neq t_j,$$
$$\ell h(t_i) \leq \ell h(t_j) \text{ and if } \ell h(t_i) = \ell h(t_j), \text{ then } t_i < t_j.$$

Obviously, every element g of G_T is equal to an element $\sigma(g)$ in standard form, and σ is primitive recursive.

An element g of G_T is said to be in normal form if it is given as

(5) $$g \doteq \Sigma_{i=1}^{p} g_{s_i} \quad \text{where } p \geq 0, \text{ and } i < j \Longrightarrow s_i < s_j.$$

For $\sigma(g) \doteq \Sigma_{i=1}^{n} m_i g_{t_i}$ we put $\ell h^*(g) = \ell h(t_n)$ and $rk(g)$ = the number of i with $\ell h(t_i) = \ell h^*(g)$. The definition of $\nu(g)$ is by recursion on $\ell h^*(g)$ and for given $\ell h^*(g)$ by recursion on $rk(g)$.

When $\sigma(g) \doteq 0$ we put $\nu(g) \doteq 0$. When $\sigma(g) \doteq \Sigma_{i=1}^{n} m_i g_{t_i}$ with $n > 0$, we write $m = m_n$ and $t = t_n$ so

(6) $$\sigma(g) \doteq (\Sigma_{i=1}^{n-1} m_i g_{t_i}) + m g_t .$$

Then we take

(7) $$\nu(g) = \begin{cases} \nu(\Sigma_{i=1}^{n-1} m_i g_{t_i}) & \text{if } \ell h(t) = 1 \text{ and } m \text{ is even} \\ \nu(\Sigma_{i=1}^{n-1} m_i g_{t_i}) + g_t & \text{if } \ell h(t) = 1 \text{ and } m \text{ is odd} \\ \nu(\Sigma_{i=1}^{n-1} m_i g_{t_i} + k g_{(t^+)}) & \text{if } \ell h(t) > 1 \text{ and } m = 2k \\ \nu(\Sigma_{i=1}^{n-1} m_i g_{t_i} + k g_{(t^+)}) + g_t & \text{if } \ell h(t) > 1 \text{ and } m = 2k+1 . \end{cases}$$

This reduces the definition of $\nu(g)$ to $\nu(g')$ where (in $\sigma(g')$) either ℓh^* is lowered or ℓh^* is fixed and rk is lowered. ν is obviously primitive recursive.

4.2. Lemma. (i) For each $g \in F_T$, $\nu(g)$ is in normal form and $g \doteq \nu(g)$.

(ii) $g \doteq h \Longrightarrow \nu(g) \doteq \nu(h)$.

(iii) If $g \doteq \Sigma_{i=1}^{n} g_{t_i}$ and $\nu(g) \doteq \Sigma_{j=1}^{k} g_{s_j}$, then for each $j \leq k$ there exists $i \leq n$ with $t_i \preceq s_j$.

Proof. (i) is easily seen by induction.

To prove uniqueness (ii), we take the set G^o consisting just of the words in normal form. Given $g,h \in G^o$ define $g \oplus h = \nu(g+h)$. It may be verified that G^o forms an Abelian group under $\oplus$, generated by $\{g_s | s \in T\}$ and satisfying (1) (i), (ii). Hence the map $g_t \mapsto g_t$ induces a homomorphism γ of G_T onto G^o. γ is the identity on G^o, and thus for any g, $\gamma(g) = \gamma(\nu(g)) = \nu(g)$. This implies (ii). (iii) is proved using (7) above. In the passage from $\nu(g)$ to $\nu(g') = \nu(\sigma(g'))$, the only new g_s which enters is $s = t^+$.

4.2 (i), (ii) and the recursiveness of ν now establishes 4.1 (ii). We may read 4.2 (iii) as saying that every term in $\nu(g)$ has some ancestor in g. The next tells us that if $g = 2g'$, then every term in $\nu(g)$ has a proper ancestor in g'.

4.3. Lemma. If $g \doteq \Sigma_{j=1}^{k} g_{s_j}$ is in normal form and $g' = \Sigma_{i=1}^{n} g_{t_i}$ and $g = 2g'$, then for each $j \leq k$ there exists $i \leq n$ with $t_i \prec s_j$.

Proof. $2g' = \Sigma_{i=1}^{n} 2g_{t_i} = \Sigma_i \, g_{(t_i^+)}$, where this sum extends only over those i with $\ell h(t_i) > 1$. Since $\nu(\Sigma_i \, g_{(t_i^+)}) = \Sigma_{j=1}^{k} g_{s_j}$ by 4.2 (ii), we can then apply 4.2 (iii).

§5. In this section we turn to global relationships between T and G_T. For simplicity we assume throughout that T is recursive, though this is used only in questions of definability. At the end we will be able to establish 1.1 (i) using a particular T.

Given $X \subseteq T$, define

(1) $\mathrm{Gen}(X) =$ the subgroup of G_T generated by $\{g_s | s \in X\}$.

Obviously $\mathrm{Gen}(X) = \mathrm{Gen}(X^+)$ where X^+ is the closure of X in T .

5.1. Lemma. (i) If X is unfounded, then $\mathrm{Gen}(X)$ is divisible.
(ii) If $X \in HYP$, then $\mathrm{Gen}(X) \in HYP$.

Proof. (i) We may assume X closed. For each $s \in X$ there exists $t \in X$ with $t \prec s$, hence there exists $t \in X$ with $t^+ = s$; then $2g_t = g_s$. It follows that for every $g \doteq \Sigma_i g_{s_i}$ in $\mathrm{Gen}(X)$ there exists $g' = \Sigma_j g_{t_j}$ in $\mathrm{Gen}(X)$ with $g = 2g'$.
(ii) is obvious.

We may also associate with each subgroup H of G_T a subset of T by:

(2) $S(H) =$ closure $\{s | g_s$ appears in $\nu(h)$ for some $h \in H\}$.

5.2. Lemma. (i) $H_1 \subseteq H_2 \Longrightarrow S(H_1) \subseteq S(H_2)$.
(ii) H divisible $\Longrightarrow S(H)$ closed and unfounded.
(iii) $H \subseteq \mathrm{Gen}(S(H))$.
(iv) If $H \in HYP$, then $S(H) \in HYP$.

Proof. (i), (iii), (iv) are evident.
For (ii), suppose $s \in S(H)$. There exists a normal form $g \doteq \Sigma_{j=1}^{k} g_{s_j}$ of some $g \in H$ and some j_0 with $s_{j_0} \prec s$. By divisibility of H we can find some $h \in H$ with $g = 2h$, $h \doteq \Sigma_{i=1}^{n} g_{t_i}$ in normal form. Using 4.3, we see that there exists i_0 with $t_{i_0} \prec s_{j_0}$. Since $t_{i_0} \in S(H)$ by definition, we have produced for each $s \in H$ some $t \in H$ with $t \prec s$.

Obviously for any X we have $X \subseteq S(\mathrm{Gen}(X))$.

5.3. Lemma. *If* X *is closed, then* $S(\mathrm{Gen}(X)) = X$.

Proof. If $s \in S(\mathrm{Gen}(X))$, then $s_{j_0} \preceq s$ for some s_{j_0} where $g_{s_{j_0}}$ appears in a normal form $\Sigma_{j=1}^{k} g_{s_j}$ in $\mathrm{Gen}(X)$. This normal form is itself $\nu(\Sigma_{i=1}^{n} g_{t_i})$ for some $t_1, \ldots, t_n$ all of which lie in X . By 4.2 (iii) s_{j_0} has an ancestor t_{i_0} . Since X is closed, $s \in X$.

5.4. Theorem. (i) $\mathrm{Div}(G_T) = \mathrm{Gen}(\mathrm{Unf}(T))$.

(ii) $\mathrm{Div}^*(G_T) \subseteq \mathrm{Gen}(\mathrm{Unf}^*(T))$.

Proof. (i) $\mathrm{Gen}(\mathrm{Unf}(T))$ is divisible by 5.1 (i) so $\mathrm{Gen}(\mathrm{Unf}(T)) \subseteq \mathrm{Div}(G_T)$. Conversely, if H is divisible, then $S(H)$ is closed and unfounded by 5.2 (ii) so $S(H) \subseteq \mathrm{Unf}(T)$. Hence $\mathrm{Gen}(S(H)) \subseteq \mathrm{Gen}(\mathrm{Unf}(T))$. $\bigcup H[H \text{ divisible}] \subseteq \mathrm{Gen}(\mathrm{Unf}(T))$ follows by 5.2 (iii).

For (ii) we apply the same argument to $H \in \mathrm{HYP}$.

5.5. Lemma. *If* $\mathrm{Div}^*(G_T) \in \mathrm{HYP}$, *then* $\mathrm{Div}^*(G_T) = \mathrm{Gen}(\mathrm{Unf}^*(T))$ *and* $\mathrm{Unf}^*(T) \in \mathrm{HYP}$.

Proof. Let $H = \mathrm{Div}^*(G_T)$, which is a divisible HYP subgroup of G_T . (This is because if H_1 , H_2 are HYP divisible subgroups, the same holds for $H_1 + H_2$.) We know $S(H)$ is closed, unfounded, and in HYP . Consider any X which is closed and unfounded with $X \in \mathrm{HYP}$. Then $\mathrm{Gen}(X)$ is divisible and in HYP by 5.1, so $\mathrm{Gen}(X) \subseteq \mathrm{Div}^*(G_T) = H$. But $H \subseteq \mathrm{Gen}(S(H))$ so $S(\mathrm{Gen}(X)) \subseteq S(\mathrm{Gen}(S(H)))$, i.e., $X \subseteq S(H)$ by 5.3. It follows that $S(H) = \mathrm{Unf}^*(T)$, and $\mathrm{Unf}^*(T) \in \mathrm{HYP}$. But then $\mathrm{Gen}(\mathrm{Unf}^*(T)) \subseteq \mathrm{Div}^*(G_T)$ by 5.1 again.

To prove Theorem 1.1 (i), let T be any recursive tree satisfying 3.3, i.e., with $\mathrm{Unf}^*(T) \notin \mathrm{HYP}$ and $\mathrm{Unf}(T) \notin \mathrm{HYP}$. Then G_T is a recursive Abelian 2-group, and $\mathrm{Div}^*(G_T) \notin \mathrm{HYP}$ by 5.5. If $\mathrm{Div}(G_T) \in \mathrm{HYP}$, then $S(\mathrm{Div}(G_T)) \in \mathrm{HYP}$, which would give $\mathrm{Unf}(T) \in \mathrm{HYP}$ by 5.3 and 5.4 (i).

§6. We now establish a relationship between the length of G_T and the ordinal of $Wf(T)$ for any T. Put $|s| = \infty$ if $s \notin Wf(T)$.

6.1. Lemma. Suppose $g = \dot{\Sigma}_{j=1}^{k} g_{s_j}$ is in normal form where some $s_j \in Wf(T)$. Let $\sigma = \min_{1\leq j\leq k} |s_j|$. Then $g \in 2^{\sigma}G_T - 2^{\sigma+1}G_T$.

Proof. By induction on σ. We may rearrange terms so that $|s_1| = \sigma$.

(i) If $\sigma = 0$, then s_1 is minimal in T. It follows that $g \notin 2G_T$ for otherwise there would exist $g' = \dot{\Sigma}_{i=1}^{n} g_{t_i}$ in normal form with $g = 2g'$, leading to a contradiction by 4.3.

(ii) Suppose $\sigma = \tau+1$ and that the statement is true for all $\tau' \leq \tau$. Since $|s_1| = \tau+1$ there exists t_1 with $s_1 = |t_1{}^+|$ and $|t_1| = \tau$. Further for each $j \leq k$, $|s_j| \geq \tau+1$ so there exists t_j with $s_j = t_j{}^+$ and $|t_j| \geq \tau$. Let $g' = \Sigma_{j=1}^{k} g_{t_j}$; g' is in normal form (up to rearrangement) since $t_i = t_j$ implies $t_i{}^+ = t_j{}^+$. By induction hypothesis $g' \in 2^{\tau}G_T$; then $g = 2g' \in 2^{\tau+1}G_T = 2^{\sigma}G_T$. Suppose $g \in 2^{\sigma+1}G_T$. Then there would exist $g'' = \Sigma_{i=1}^{n} g_{r_i}$ in normal form with $g'' \in 2^{\sigma}G_T$, $g = 2g''$. There is some i with $r_i \prec s_1$, so $\rho = \min_{1\leq i\leq n} |r_i| \leq \tau$. By induction hypothesis $g'' \notin 2^{\rho+1}G_T$, contradicting $g'' \in 2^{\sigma}G_T$.

(iii) Suppose σ is a limit number and that the statement is true for each $\tau < \sigma$. With $|s_1| = \sigma$ there exists for each $\tau < \sigma$ a t_1 such that $t_1{}^+ = s_1$ and $|t_1| \geq \tau$. Again for each $j \leq k$, $|s_j| \geq \sigma$ so there exists t_j with $s_j = t_j{}^+$ and $|t_j| \geq \tau$. Let $g' = \Sigma_{j=1}^{k} g_{t_j}$, which is in normal form; thus $g' \in 2^{\tau}G_T$. $g = 2g' \in 2^{\tau+1}G_T \subseteq 2^{\tau}G_T$. Since this holds for each $\tau < \sigma$ we have $g \in 2^{\sigma}G_T$. To conclude, we show $g \notin 2^{\sigma+1}G$. Suppose $g = 2g''$ where $g'' \in 2^{\sigma}G_T$, $g'' = \Sigma_{i=1}^{n} g_{r_i}$ in normal form. Again there is some i with $r_i \prec s_1$ so $\rho = \min_{1\leq i\leq n} |r_i| < \sigma$. Hence $g'' \notin 2^{\rho+1}G_T$, contradicting $g'' \in 2^{\sigma}G_T$.

We may now prove 1.1 (ii), again using G_T for any recursive tree satisfying 3.3. Since $|Wf(T)| = \omega_1^{rec}$ we have an $s \in T$ with $|s| = \sigma$ for each $\sigma < \omega_1^{rec}$. Then $2^{\sigma}G_T \neq 2^{\sigma+1}G_T$ by the preceding lemma. Hence $\mathrm{length}(G_T) \geq \omega_1^{rec}$, while $\mathrm{length}(G_T) \leq \omega_1^{rec}$ by 2.2.

§7. This is not the place to discuss the idea of predicativity. That has been studied in terms of certain formal systems of analysis and systems of higher type and set theory which are reducible to them (cf. [1], [2]). The feature of these systems which concerns us here is that their second order part has an ω-model in HYP. Hence any statement formulated in second order terms which is false in HYP is independent of the given formal systems, and in that sense is established to be impredicative. The first example of such an independence result was given by Kreisel for the Cantor-Bendixson theorem [6].

When we restrict ourselves to enumerated groups G or equivalently to $G \subseteq N$, the statement ψ of the existence of a largest divisible subgroup of any Abelian p-group G is in second order form:

$$(1) \qquad \forall G\,(Ab_p(G) \Longrightarrow \exists D(D \subseteq G \wedge Div(D) \wedge \forall H\,(H \subseteq G \wedge Div(H) \Longrightarrow H \subseteq D))] .$$

Here $Ab_p(G)$ expresses that G is an Abelian p-group, and $Div(H)$ expresses that H is a divisible subgroup of G; both of these are first-order formulas in G, $+_G$, and H.

To show that (1) is not true in HYP we must provide an Abelian p-group in HYP such that

$$(2) \qquad \neg\, \exists D_{HYP}[D \subseteq G \wedge Div(D) \wedge \forall H_{HYP}[H \subseteq G \wedge Div(H) \Longrightarrow H \subseteq D)] .$$

This is equivalent to saying that

$$(3) \qquad Div^*(G) \notin HYP .$$

Thus the first part of Theorem 1.1 (i) establishes the impredicativity of (1).

On the other hand, knowing for some $G \in HYP$ that $Div(G) \notin HYP$ would not by itself provide sufficient grounds to prove the impredicativity of (1). In fact, the example $G^{(B)}$ found by Barwise which was mentioned in §1 shows the difference in a striking way. $G^{(B)}$ had $H \notin HYP$ for all divisible $H \subsetneq G$ with $H \neq 0$. Hence within the ω-model HYP, $G^{(B)}$ <u>does</u> have a largest divisible subgroup, namely 0 --since all non-HYP divisible subgroups of $G^{(B)}$ are excluded.

A phenomenon in a way akin to this already appears in [6]. Recall that the Cantor-Bendixson theorem (C-B) for $\mathbb{R}$ tells us that every closed set F is the union of a perfect set P_F and a countable (scattered) set $S_F = \{s_0, \ldots, s_n, \ldots\}$. For a second order formulation of C-B, one represents closed sets F (say) by the set of all pairs of end-points of rational open intervals disjoint from F. The perfect kernel P_F of F may be described as the largest perfect subset of F, or as the intersection of the sequence of derived sets $F^{(\sigma)}$. Kreisel gave an example of an arithmetical(Π^0_1) closed set F such that there is no HYP enumeration of S_F. In the "real world," the perfect kernel P_F of F is non-trivial. However, P_F contains no HYP reals. Hence in the ω-model HYP, F does have a perfect kernel--namely the empty set. This permits us to draw a slightly stronger conclusion than the independence of (C-B), namely the impredicativity of the following statement: every uncountable closed set contains a non-empty perfect subset.

In all these examples, one has also considered naturally associated ordinals. In the one just mentioned $\sigma = \omega_1^{rec}$ for the least σ with $F^{(\sigma)} = F^{(\sigma+1)}$. For both the group $G = G_T$ constructed in 1.1 and $G = G^{(B)}$ we have $\text{length}(G) = \omega_1^{rec}$. While the ordinal ω_1^{rec} is impredicative, these do not lead to further independence results for second order statements since the notion of well-ordering relativized to HYP is not absolute. However, the statement Θ that every Abelian p-group has a length makes sense as a statement of set theory.[6)] Θ is independent of the predicatively reducible system PS_1 of [2], since PS_1 has a model in which all ordinals are $< \omega_1^{rec}$. Of course, there are much stronger impredicative fragments of set theory with the same property (as there are strong impredicative subsystems of analysis with models in HYP). Thus such independence results always give more information than impredicativity of a classical statement.

---

6) In this connection, Barwise observed the following generalization of Theorem 2.2: if A is admissible and $G \in A$, then $\text{length}(G) \leq \text{ord}(A)$.

Bibliography

[1] S. Feferman, Systems of predicative analysis, J. Symbolic Logic 29 (1964) 1-30.

[2] S. Feferman, Predicatively reducible systems of set theory, to appear in Proc. Symp. Pure Math. 13, Part II.

[3] S. Feferman and C. Spector, Incompleteness along paths in progressions of theories, J. Symbolic Logic 27 (1962) 383-390.

[4] I. Kaplansky, *Infinite Abelian Groups* (Univ. of Michigan Press, 1969).

[5] S. C. Kleene, Quantification of number-theoretic functions, Compositio Mathematica 14 (1959) 23-40.

[6] G. Kreisel, Analysis of the Cantor-Bendixson theorem by means of the analytic hierarchy, Bull. Acad. Pol. Sci. 7 (1959) 621-626.

[7] C. Spector, Hyperarithmetical quantifiers, Fund. Math. 48 (1959) 313-320.

[8] C. Spector, Inductively defined sets of natural numbers, in *Infinitistic methods* (Pergamon, 1961) 97-102.

ELEMENTARY EQUIVALENCE CLASSES OF GENERIC STRUCTURES AND EXISTENTIALLY COMPLETE STRUCTURES

E. Fisher, H. Simmons, and W. Wheeler

The theory of forcing in model theory associates three classes of structures with each first order theory T: the class $\mathcal{E}_T$ of existentially complete structures, the class $\mathfrak{F}_T$ of finitely generic structures, and the class $\mathcal{G}_T$ of infinitely generic structures [8, 15]. The number of elementary equivalence classes of existentially complete structures, of finitely generic structures, and of infinitely generic structures will be denoted by $n_{\mathcal{E}}(T)$, $n_{\mathfrak{F}}(T)$, and $n_{\mathcal{G}}(T)$, respectively. The problem of interest is that of determining the ranges of these functions as T varies over all countable, first order theories. The model theory and topology in the solution are, however, equally interesting for their own sake.

The solution of the original problem is that, subject to certain natural restrictions, anything is possible. Since all finitely generic structures and all infinitely generic structures are existentially complete, $n_{\mathcal{E}}(T)$ is always at least as large as the maximum of $n_{\mathfrak{F}}(T)$ and $n_{\mathcal{G}}(T)$. These values are at least 1, since a countable theory always has both finitely generic structures and infinitely generic structures. The three classes coincide when T

has a model-companion, in which case $n_{\mathcal{E}}(T) = n_{\mathcal{F}}(T) = n_{\mathcal{G}}(T)$. In any case, $n_{\mathcal{F}}(T)$ is always less than or equal to $n_{\mathcal{G}}(T)$. These facts are more or less well-known. Thus, $1 \le n_{\mathcal{F}}(T) \le n_{\mathcal{G}}(T) \le n_{\mathcal{E}}(T) \le 2^{\aleph_0}$. In fact, $n_{\mathcal{F}}(T) = n_{\mathcal{G}}(T)$ except for one case in which $n_{\mathcal{F}}(T) = \aleph_0 < 2^{\aleph_0} = n_{\mathcal{G}}(T)$. As one would expect, the only infinite values of these functions are $\aleph_0$ and $2^{\aleph_0}$. Examples show that these are the only restrictions, except possibly for the case $n_{\mathcal{F}}(T) = n_{\mathcal{G}}(T) = 1 < n_{\mathcal{E}}(T) < \aleph_0$ where examples have been found only for $n_{\mathcal{E}}(T) = 2^r$, r a positive integer. (See remark 4.)

The methods of this paper use both model theory and topology. The fundamental fact is the existence of a one-to-one correspondence between the elementary equivalence classes of each class of structures and the points of a G_δ subset of an appropriate, compact, metrizable space. After the correspondence has been established, well-known facts about such subsets can be applied. The interesting work therefore is the model theory used in establishing the one-to-one correspondence. The correspondence for the class of infinitely generic structures is effected by using the components of the theory; this role for the components is to be expected in view of the results in [5]. The existence of a correspondence in the case of finitely generic structures is a far more delicate matter, because the components of the theory may not correspond to the elementary equivalence classes of finitely generic structures for the theory. Thus, the characterization in §2 of the finite forcing components and the results, analogous to those in [5], for finitely generic structures are interesting in themselves.

Finally, the correspondence for the class of existentially complete structures is based upon the characterization of their theories in terms of omitting certain types [8, Lemma 7.16, page 129].

This paper consists of five sections: Preliminaries, Components and the function $n_{\mathfrak{G}}(T)$, Finite forcing components, Finite forcing components and the function $n_{\mathfrak{F}}(T)$, and Existentially complete structures and $n_{\mathcal{E}}(T)$.

§ 0 Preliminaries

The formal languages used in this paper will be first order with logical symbols =, $\wedge$, $\vee$, $\sim$, $\exists$, and $\forall$, variables $v_0, v_1, \ldots$, and arbitrary constant, predicate, and function symbols. Formulas of a language will be denoted by lower case Greek letters φ, ψ, χ, etc. The notation $\varphi(v_0, \ldots, v_n)$ indicates that the free variables of φ are among $v_0, \ldots, v_n$. In this context, $\varphi(a_0, \ldots, a_n)$ is the sentence obtained by replacing each occurrence of v_i with an occurrence of the constant a_i for $i = 0, \ldots, n$. The notations $\bar{v}$ and $\bar{a}$ will denote the sequences $v_0, \ldots, v_n$ and $a_0, \ldots, a_n$, respectively. Accordingly, the formula $\varphi(v_0, \ldots, v_n)$ may be written $\varphi(\bar{v})$. The letter T will denote a (consistent) theory. The collection of universal sentences deducible from T will be denoted by $T_\forall$. The notation $T \vdash \varphi$ denotes that φ is deducible from T.

Structures will be denoted by $\mathfrak{M}$ and $\mathfrak{N}$ either with or without embellishments such as superscripts or subscripts. The class of models of T will be denoted by Mod(T). If $\varphi(v_0,\ldots,v_n)$ is a formula in the language of T and $a_0,\ldots, a_n$ are names for elements of a structure $\mathfrak{M}$ ($a_0,\ldots, a_n$ may not be in the language of T), then $\varphi(a_0,\ldots, a_n)$ is said to be <u>defined in</u> $\mathfrak{M}$. A structure is a model of $T_\forall$ if and only if it is a substructure of a model T. The theory T has the <u>joint embedding property</u> if any two models of T have a common extension which is a model of T. Equivalently, T has the joint embedding property if and only if either the sentence φ or the sentence ψ is in $T_\forall$ whenever the sentence $\varphi \vee \psi$ is in $T_\forall$. Two examples of theories with the joint embedding property are a complete theory and the universal theory of a structure.

The Lindenbaum algebra of a theory T will be denoted by $\mathcal{B}(T)$. The equivalence class of a sentence φ in this algebra will be denoted by $[\varphi]$. The Boolean subalgebra of $\mathcal{B}(T)$ which is generated by the equivalence classes of existential sentences will be denoted by $\mathcal{B}_1(T)$.

A type is a collection $\Delta = \{\varphi_i(\bar{v}) : i \in I\}$ of formulas of the language of T with free variables among $v_0,\ldots, v_n$. A type Δ is <u>locally omitted</u> by the theory T if, for each formula $\psi(\bar{v})$ such that $T \cup \{\exists\bar{v}\, \psi(\bar{v})\}$ is consistent, there is an index i such that $T \cup \{\exists\bar{v}(\psi(\bar{v}) \wedge \sim \varphi_i(\bar{v})\}$ is consistent. The Omitting Types Theorem [3, page 79, Theorem 2.2.9] asserts that if a countable theory locally omits each type in a countable collection of types, then it has a model which omits each type in the collection. If the theory is complete, then the converse is true also.

The reader is referred to references 1,8,10, 13 and 15 for

discussions of finite forcing, to references 8, 14, and 15 for discussions of infinite forcing, and to references 8, 15, and 18 for discussions of existential completeness. All results about forcing and existential completeness used in this paper may be found in reference 8.

The following theorem of elementary topology [17] will be applied in sections 1,3, and 4.

Theorem A. Let X be a complete metric space with a countable basis. Suppose S is a G_δ subset of X which satisfies one of the following conditions:

(i) the isolated points of S are not dense in S, or

(ii) S is uncountable.

Then S includes a perfect set, and the cardinality of S is $2^{\aleph_0}$.

This theorem can be proved, for instance, by using the fact that G_δ subset of a complete metric space can be assigned a complete metric which defines an equivalent topology on the subset. Also, this theorem holds for a compact, Hausdorff space with a countable basis, since such a space may be assigned a complete metric.

§ 1 Components and the function $n_{\mathcal{G}}(T)$

The components of a theory T are relevant for the determination of $n_{\mathcal{G}}(T)$ because they determine the elementary equivalence classes of infinitely generic structures. A deductively closed set of universal sentences which has the same vocabulary as T, includes $T_\forall$, and has the joint embedding property, is called an irreducible ideal of T. A component of T is a minimal irreducible ideal of T. A component is, from the point of view of model theory, just the universal theory of an existentially complete structure for T. To verify this, one proceeds as follows. The universal theory of a model of T is an irreducible ideal of T. Suppose J is the universal theory of an existentially complete structure $\mathfrak{M}$ for T. If $J' \subsetneq J$ were an irreducible ideal for T, then there would be a sentence ψ in J such that $J' \cup \{\sim\psi\}$ would be consistent. Then the joint embedding property for J' would imply that $\mathfrak{M}$ was included in a model of $\sim\psi$ contradicting that $\mathfrak{M}$ is existentially complete. Conversely, suppose that J is a component, and let $\mathfrak{M}$ be an existentially complete structure for J. Any model of T which extends $\mathfrak{M}$ must be a model of J also, since the universal theory of the extension would be an irreducible ideal included in J. Therefore, $\mathfrak{M}$ is existentially complete for T.

The following theorem summarizes the main result of E. Fisher and A. Robinson in [5].

Theorem B. (1) If J is a component of T, then $\mathcal{G}_T \cap \mathrm{Mod}(J) = \mathcal{G}_J$.

(2) The classes $\mathcal{G}_J$ as J varies over all the components of T partition $\mathcal{G}_T$ into nonempty subclasses. Moreover, these subclasses are precisely the elementary equivalence classes of infinitely generic structures for T.

Part (2) is a consequence of part (1) and the preceding remarks on components. An infinitely generic structure $\mathfrak{M}$ for T is existentially complete, so part (1) asserts that $\mathfrak{M}$ is infinitely generic for the component $J = Th(\mathfrak{M})_\forall$. Then $\mathfrak{M}$ is a model of J^F. But J^F is complete, since J has the joint embedding property; so, $J^F = Th(\mathfrak{M})$. Distinct components have distinct infinite forcing companions. Thus, two infinitely generic structures for T are elementarily equivalent if and only if they are models of the same component of T.

Several alternative characterizations of components will be used in this paper. If J is a set of universal sentences in the language of T, then $*J$ will denote the dual complement of J, i.e., $*J = \{\sim\psi : \psi$ is a universal sentence in the language of T and ψ is not in $J\}$. A set I of existential sentences in the language of T will be called a $T_\forall$-__maximal__ __existential__ __set__ if $T_\forall \cup I$ is consistent and each existential sentence in the language of T which is consistent with $T_\forall \cup I$ is actually in I itself.

__Lemma 1.1__. The following are equivalent for a theory T and a deductively closed set J of universal sentences in the language of T:

(1) J is a component of T;

(2) $*J$ is a $T_\forall$-maximal existential set;

(3) $T_\forall \cup *J$ is consistent and, for each sentence ψ in J, there is a sentence φ in $*J$ such that $T_\forall \vdash \varphi \rightarrow \psi$.

(4) The set $F = \{[\chi] \in \mathcal{B}_1(T_\forall) : \exists\, \varphi$ in $*J$ such that $T_\forall \vdash \varphi \rightarrow \chi\}$ is an ultrafilter in $\mathcal{B}_1(T_\forall)$.

Proof. (1) ⇒ (2). Let $\mathfrak{M}$ be an existentially complete structure for T such that $J = Th(\mathfrak{M})_\forall$. Then $*J = Th(\mathfrak{M})_\exists$. If φ is an existential sentence consistent with $T_\forall \cup *J$, then $\mathfrak{M}$ can be embedded in a model of $T_\forall \cup \{\varphi\}$ and so must satisfy φ. Hence, $*J$ is a $T_\forall$-maximal existential set.

(2) ⇒ (3). Since $T_\forall \cup *J$ is consistent, there is an existentially complete structure $\mathfrak{M}$ for T which satisfies $*J$. Since $*J$ is a $T_\forall$-maximal existential set, $*J = Th(\mathfrak{M})_\exists$. The latter part of (3) is a special case of a well-known property of existentially complete structures [8, Proposition 1.6, page 22; 18, Theorem 2.1, page 297].

(3) ⇒ (4). Since $T_\forall \cup *J$ is consistent, there is an existentially complete structure $\mathfrak{M}$ which satisfies $*J$. Since each sentence in J is a consequence of $T_\forall \cup *J$, $*J = Th(\mathfrak{M})_\exists$. Clearly, $[\chi]$ in F implies that χ is in $Th(\mathfrak{M})$. Since each universal sentence in $Th(\mathfrak{M})$ is implied by an existential sentence in $Th(\mathfrak{M})$, each Boolean combination of existential sentences in $Th(\mathfrak{M})$ is implied by an existential sentence in $Th(\mathfrak{M})$. Hence, $F = \{[\chi] \in \mathcal{B}_1(T_\forall) : \chi$ is in $Th(\mathfrak{M})\}$, which is an ultrafilter in $\mathcal{B}_1(T_\forall)$.

(4) ⇒ (1). Since F is an ultrafilter, $T_\forall \cup *J$ is consistent and therefore is satisfied by some existentially complete structure $\mathfrak{M}$ for T. Clearly, for each $[\chi]$ in $\mathcal{B}_1(T_\forall)$, $\mathfrak{M}$ satisfies χ if and only if $[\chi]$ is in F. Consequently, $Th(\mathfrak{M})_\exists = *J$, so $Th(\mathfrak{M})_\forall = J$ and J is a component of T .

The preceding lemma is similar to Theorem 1.7 of [5]. The differences, especially part (2) of Lemma 1.1 which has no analogue in Theorem 1.7 of [5], are due to the restriction to the Boolean algebra $\mathcal{B}_1(T_\forall)$.

Corollary 1.2. There is a one-to-one correspondence between components of T and ultrafilters in $\mathcal{B}_1(T_\forall)$ with the property that below each member of the ultrafilter is another member which is the equivalence class of an existential sentence.

Proof. Suppose that F is such an ultrafilter. If J is the set of universal sentences whose equivalence classes are in F, then $F = \{[x] \in \mathcal{B}_1(T_\forall) : \exists \varphi \in {}^*J$ such that $T \vdash \varphi \to \chi\}$, so J is a component of T.

Conversely, if J is a component of T, then $^*J$ generates such an ultrafilter F in $\mathcal{B}_1(T_\forall)$, and the universal sentences whose equivalence classes occur in F are precisely the elements of J.

The Stone space $\mathcal{S}(\mathcal{B})$ of a Boolean algebra $\mathcal{B}$ is the topological space whose points are the ultrafilters of $\mathcal{B}$ and whose topology of open sets has as a basis the sets $u(a) = \{F : F$ is an ultrafilter in $\mathcal{B}$ and a is an element of $F\}$ as a varies over all members of $\mathcal{B}$. The Stone space of a Boolean algebra is a compact, Hausdorff space.

Theorem 1.3. The ultrafilters corresponding to the components of the theory T form a G_δ subset of the Stone space $\mathcal{S}(\mathcal{B}_1(T_\forall))$.

Proof. Define an open set $U(\chi)$ for each equivalence class $[\chi]$ in $\mathcal{B}_1(T_\forall)$ by

$U(\chi) = u([\sim\chi]) \cup \bigcup \{u([\varphi]) : \varphi$ is an existential sentence and $T_\forall \vdash \varphi \to \chi$ }.

Let $\mathcal{C} = \bigcap\{U(\chi) : [\chi] \in \mathcal{B}_1(T_\forall)\}$. An ultrafilter F is in $\mathcal{C}$ if and only if, for each sentence χ, either $[\chi]$ is not in F or there is an existential sentence φ such that $T_\forall \vdash \varphi \to \chi$ and $[\varphi]$ is in F. Thus, an ultrafilter is in $\mathcal{C}$ if and only if it corresponds to a component.

A component will be called principal if there is a single existential sentence which generates the ultrafilter corresponding to that component. Clearly, a component is principal if and only if the corresponding ultrafilter is principal and hence isolated in $\mathcal{S}(\mathcal{B}_1(T_\forall))$.

Theorem 1.4. The following are true for a countable theory T.

(1) If the principal components of T are not dense in $\mathcal{C}$, then T has $2^{\aleph_0}$ many components. In particular, if T has no principal components, then T has $2^{\aleph_0}$ many components.

(2) If T has uncountably many components, then T has $2^{\aleph_0}$ many components.

Proof. This theorem is a consequence of Theorem A and Theorem 1.3.

Corollary 1.5. If $n_{\mathcal{C}}(T) > \aleph_0$, then $n_{\mathcal{C}}(T) = 2^{\aleph_0}$.

Corollary 1.6. (1) If T^F has no complete extension which is obtained by the addition of one existential sentence, then T^F has $2^{\aleph_0}$ many completions.

(2) If there is no existential sentence φ such that

$T \cup \{\varphi\}$ has the joint embedding property, then T^F has $2^{\aleph_0}$ many completions.

Proof. (1) If J were a principal component whose ultrafilter was generated by an existential sentence φ, then the deductive closure of $T^F \cup \{\varphi\}$ would be the complete theory J^F. Consequently, the hypothesis implies that T has no principal components.

Theorem 1.7. The range of the function $n_{\mathcal{G}}(T)$ as T varies over countable theories is $\{1,2, \ldots, \aleph_0, 2^{\aleph_0}\}$.

Proof. That the range is included in this set follows from Corollary 1.5 and the fact that $1 \leq n_{\mathcal{G}}(T) \leq 2^{\aleph_0}$. If T_n is the theory of fields of characteristic $p_1, p_2, \ldots,$ or p_n, where these are the first n primes, then $n_{\mathcal{G}}(T) = n$. If T is the theory of fields, then $n_{\mathcal{G}}(T) = \aleph_0$. Finally, if T is the theory of commutative rings without nilpotent elements, then $n_{\mathcal{G}}(T) = 2^{\aleph_0}$ [9]. Another example for which $n_{\mathcal{G}}(T) = 2^{\aleph_0}$ is the theory T of an equivalence relation R(x,y) with the axiom

$\forall x \forall y \forall z (R(x,x) \wedge (R(x,y) \rightarrow R(y,x)) \wedge ((R(x,y) \wedge R(y,z)) \rightarrow R(x,z)))$

in a language with infinitely many constant symbols $a_0, a_1, \ldots$.

§2 Finite forcing components

The determination of the elementary equivalence classes of finitely generic structures for T can be reduced to the consideration of components also. The finite forcing companion of a theory, like the infinite forcing companion, is uniquely determined by the theory's universal consequences and is complete if and only if the theory has the joint embedding property. Since each finitely generic structure for a theory is existentially complete, the structure's universal theory is a component. Thus, two finitely generic structures for a theory are elementarily equivalent if and only if they are models of the same component. However, in contrast to the case of infinite forcing, there is a theory with a component which does not have a model that is finitely generic for the theory [12, § 3]. A component of T will be called a <u>finite forcing component</u> of T if there is a finitely generic structure for T which is a model of the component. The determination of the range of $n_{\mathfrak{F}}(T)$ must await the characterization of the finite forcing components of T.

While seeking a characterization of finite forcing components, one is interested also in the questions of whether a finitely generic structure for T is finitely generic for the component determined by it and, conversely, of whether a finitely generic structure for a component of T is finitely generic for T also. These questions are meaningful only for finite forcing components of course. In fact, the finitely generic structures for a finite forcing component are precisely the finitely generic structures for T, which are models of the component. This yields a partitioning of the finitely generic structures for T, which is analogous to the partitioning of the infinitely generic structures described in Theorem B, parts (1) and (2).

Some definitions and conventions are necessary. As before, T is a countable theory. The language of T is expanded by the introduction of an infinite set $\{a_i : i < \omega\}$ of new constant symbols. All conditions will be formed from this expanded language. If P is a condition, then $P(a_0, \ldots, a_n)$ denotes that the constants occurring in P but not in T are among $a_0, \ldots, a_n$, $P(v_0, \ldots, v_n)$ is the set of formulas obtained by replacing each occurrence of a_i by an occurrence of v_i for $i = 0, \ldots, n$, and $\wedge P(v_0, \ldots, v_n)$ is the conjunction of all formulas in $P(v_0, \ldots, v_n)$. The notation $P \Vdash_T \varphi$ means that P is a condition relative to T, φ is a formula of the expanded language, and P finitely forces φ relative to T. The notation $P \Vdash_T^w \varphi$ is an abbreviation for $P \Vdash_T \sim\sim \varphi$.

A model of a theory T' is said to <u>complete</u> T' or to be a <u>completing model of</u> T' if it is an elementary substructure of each of its extensions in the class of models of T'. J. Barwise and A. Robinson [1] have shown (i) that a countable theory T' is forcing complete, i.e., is its own finite forcing companion, if and only if, for each sentence φ consistent with T', there is a model of $T' \cup \{\varphi\}$ which completes T', and (ii) that a structure is finitely generic for T' if and only if it completes T'^f.

Theorem 2.1. **The following are equivalent for a component J of the countable theory T.**

(1) J is a finite forcing component, i.e., $\mathfrak{F}_T \cap \mathrm{Mod}(J) \neq \emptyset$.

(2) $\mathfrak{F}_T \cap \mathrm{Mod}(*J) \neq \emptyset$.

(3) For each condition P consistent with $T_\forall \cup *J$ and each formula $\psi(v_0, \ldots, v_n)$ in the language of T, there is a condition Q extending P and consistent with $T_\forall \cup *J$ such that either $Q \Vdash_T \psi(a_0, \ldots, a_n)$ or $Q \Vdash_T \sim\psi(a_0, \ldots, a_n)$.

(4) $T^f \cup *J$ has a completing model.

(5) J^f is the deductive closure of $T^f \cup *J$.

(6) For each sentence φ in J^f, there is a condition P consistent with $T_\forall \cup *J$ such that $P \Vdash^w_T \varphi$.

(7) For each formula $\psi(v_0, \ldots, v_n)$ in the language of T for which $T^f \cup *J \cup \{\exists v_0 \ldots \exists v_n \psi(v_0,\ldots,v_n)\}$ is consistent, there is a quantifier-free formula $\varphi(v_0,\ldots, v_n, v_{n+1}, \ldots, v_m)$ in the language of T such that

$\exists v_0 \ldots \exists v_n \exists v_{n+1} \ldots \exists v_m \varphi(v_0,\ldots,v_n, v_{n+1},\ldots,v_m)$ is in $*J$ and

$$T^f \cup *J \vdash \forall v_0 \ldots \forall v_n \forall v_{n+1} \ldots \forall v_m (\varphi(v_0,\ldots,v_n, v_{n+1},\ldots,v_m) \rightarrow \psi(v_0,\ldots,v_n)).$$

(8) The finitely generic structures for J are precisely the finitely generic structures for T which are models of J, i.e., $\mathfrak{F}_T \cap \mathrm{Mod}(J) = \mathfrak{F}_J$.

Several comments on the significance of various parts of this theorem may be helpful for the reader. The equivalence of parts (1) and (3) provides a characterization of finite forcing components which suffices for the proof that the set of finite forcing components corresponds to a G_δ subset of $\mathcal{S}(\mathcal{B}_1(T_\forall))$. For this reason, the line of proof will include the sequence (1) ⇒ (2) ⇒ (3) ⇒ (1). Part (8) is the analogue of part (2) of Theorem B and shows that the classes $\mathfrak{F}_J$ as J ranges over the finite forcing components of T partition the class $\mathfrak{F}_T$. Parts (4) through (7) are concerned with completing models. Parts (4) and (5) are necessary to show that (1) implies (8). Parts (6) and (7) characterize the finite forcing components in terms of the existence of completing models and the omission of the corresponding collection of types. This characterization is not necessary to later developments, but is included because of the increasing tendency to think of finite forcing in terms of completing models rather than finite forcing. The proofs of (6) ⇒ (7) and (7) ⇒ (4) are necessarily more complicated than the proofs of the other implications, because the theory T^f axiomatizes the weak finite forcing relation rather than the finite forcing relation itself.

<u>Proof</u>. (1) ⇒ (2). Each member of $\mathfrak{F}_T \cap \mathrm{Mod}(J)$ is existentially complete for T, so its universal theory is a component of T. On the other hand, its universal theory includes the component J, so the two are equal. Thus, the existential theory of each member of $\mathfrak{F}_T \cap \mathrm{Mod}(J)$ is $*J$.

(2) ⇒ (3). The hypothesis that P is consistent with $T_\forall \cup *J$ implies that the sentence $\exists v_0 \ldots \exists v_n (\wedge P(v_0,\ldots,v_n))$ is in $*J$, because $*J$ is a $T_\forall$-maximal existential set. Let $\mathfrak{M}$ be a

member of $\mathcal{F}_T \cap \mathrm{Mod}(*J)$. Assign the constants $a_0, \ldots, a_n$ so that $\mathfrak{M}$ satisfies $\wedge P(a_0, \ldots, a_n)$. Since $\mathfrak{M}$ is finitely generic for T, there is a condition Q' in the diagram of $\mathfrak{M}$ such that $Q' \Vdash_T \psi(a_0, \ldots, a_n)$ or $Q' \Vdash_T \sim\psi(a_0, \ldots, a_n)$. The condition $Q = P \cup Q'$ satisfies the conclusion of part (4).

(3) ⇒ (1). Let $\{\varphi_n : n < \omega\}$ be an enumeration of all the sentences in the expanded language. Construct inductively a complete sequence of forcing conditions as follows. Let $P_0 = \emptyset$. Assume $P_0 \subseteq \ldots \subseteq P_n$ have been chosen such that each is consistent with $T_\forall \cup *J$ and $P_{i+1} \Vdash \varphi_i$ or $P_{i+1} \Vdash \sim \varphi_i$. There is, by hypothesis, a condition Q containing P_n and consistent with $T_\forall \cup *J$ such that $Q \Vdash \varphi_n$ or $Q \Vdash \sim \varphi_n$. Let $P_{n+1} = Q$.

The sequence thus constructed is a complete sequence of forcing conditions relative to T, so it determines a unique, finitely generic structure $\mathfrak{M}$ for T. Since each condition P_n is consistent with $T_\forall \cup *J$, it cannot force the negation of any sentence in $*J$. Consequently, $\mathfrak{M}$ is a model of $*J$, hence, of J, since J is deducible from $T_\forall \cup *J$.

(4) ⇒ (5). The deductive closure of $T^f \cup *J$ is forcing complete, because $T^f \cup *J$ has a completing model. Since $J = (T^f \cup *J)_\forall$, the deductive closure of $T^f \cup *J = (T^f \cup *J)^f = J^f$.

(5) ⇒ (6). The deductive closure of $T^f \cup *J$ is the set $\{\varphi : \varphi$ is a formula in the language of T and there is a condition P such that $P \Vdash^w_T \varphi$ and $T^f \cup *J \vdash \exists v_0 \ldots \exists v_n \wedge P(v_0, \ldots, v_n)\}$. Furthermore, the last requirement is true for a condition P if and only if P is consistent with $T_\forall \cup *J$, since $*J$ is a $T_\forall$-maximal existential set.

(6) ⇒ (7). Assume that $\psi(v_0,\ldots, v_n)$ is a formula in the language of T for which $T^f \cup *J \cup \{\exists v_0 \ldots \exists v_n\, \psi(v_0,\ldots, v_n)\}$ is consistent. Either $\exists v_0 \ldots \exists v_n\, \psi(v_0,\ldots, v_n)$ or its negation is in the complete theory J^f. Suppose that the negation is in J^f. This entails the existence of a condition P consistent with $T_\forall \cup *J$ such that $P \Vdash^w_T \sim\exists v_0 \ldots \exists v_n\, \psi(v_0,\ldots, v_n)$. But then T^f implies $\exists v_0 \ldots \exists v_m \wedge P(v_0,\ldots,v_m) \to \sim\exists v_0 \ldots \exists v_n\, \psi(v_0,\ldots,v_n)$ and $T^f \cup *J$ implies $\exists v_0 \ldots \exists v_m \wedge P(v_0,\ldots, v_m)$, so $T^f \cup *J$ implies $\sim\exists v_0 \ldots \exists v_n\, \psi(v_0,\ldots,v_n)$ contrary to the initial assumption. Therefore, $\exists v_0 \ldots \exists v_n\, \psi(v_0,\ldots,v_n)$ is in J^f, and there is a condition P relative to J such that $P \Vdash_J \psi(a_0,\ldots,a_n)$. Consequently, J^f contains the sentence

$\forall v_0 \ldots \forall v_n\, \forall v_{n+1} \ldots \forall v_m (\wedge P(v_0,\ldots,v_n, v_{n+1},\ldots, v_m) \to \psi(v_0,\ldots,v_n))$;

denote this sentence by χ. According to part (6), there is a condition Q consistent with $T_\forall \cup *J$ such that $Q \Vdash^w_T \chi$. Then T^f implies $\forall v_0 \ldots \forall v_r (\wedge Q(v_0,\ldots,v_r) \to \chi)$ and $T^f \cup *J$ implies $\exists v_0 \ldots \exists v_r \wedge Q(v_0,\ldots,v_r)$ so $T^f \cup *J$ implies χ.

(7) ⇒ (4). In order to simplify notation in this part, the symbols $\bar{w}$, $\bar{x}$, and $\bar{z}$ as well as $\bar{v}$ will be used to denote finite sequences of variables. All formulas in this part will be in the language of T.

For each formula $\exists \bar{v}\, \psi(\bar{v},\bar{w})$, let

$$\Delta_{\exists\bar{v}\, \psi(\bar{v},\bar{w})} = \{\exists v\, \psi(\bar{v},\bar{w})\} \cup \{\sim\exists \bar{v}\, \exists \bar{z}\, \varphi(\bar{v},\bar{w},\bar{z}) : \varphi \text{ is a quantifier-free formula and } T^f \cup *J \text{ implies } \forall \bar{v}\, \forall \bar{w}\, \forall \bar{z} (\varphi(\bar{v},\bar{w},\bar{z}) \to \psi(\bar{v},\bar{w}))\}.$$

One must show that each of these types is locally omitted by

$T^f \cup *J$.

Suppose that $\chi(\bar{w})$ is a formula for which $T^f \cup *J \cup \{\exists\bar{w}\ \chi(\bar{w})\}$ is consistent. By hypothesis, there is a quantifier-free formula $\sigma(\bar{w},\bar{x})$ for which $T^f \cup *J \vdash \forall\bar{w}\ \forall\bar{x}(\sigma(\bar{w},\bar{x}) \to \chi(\bar{w}))$ and $\exists\bar{w}\ \exists\bar{x}\ \sigma(\bar{w},\bar{x})$ is in $*J$. Suppose further that $\exists\bar{w}(\chi(\bar{w}) \wedge \sim\exists\bar{v}\ \psi(\bar{v},\bar{w}))$ is inconsistent with $T^f \cup *J$. This entails that

$T^f \cup *J \vdash \forall\bar{w}\ \forall\bar{x}(\sigma(\bar{w},\bar{x}) \to \exists\bar{v}\ \ \psi(\bar{v},\bar{w}))$, so that

$T^f \cup *J \vdash \exists\bar{w}\ \exists\bar{x}(\sigma(\bar{w},\bar{x}) \wedge \exists\bar{v}\ \ \psi(\bar{v},\bar{w}))$. By hypothesis again, there is a quantifier-free formula $\tau(\bar{v},\bar{w},\bar{x},\bar{z})$ such that $\exists\bar{v}\ \exists\bar{w}\ \exists\bar{x}\ \exists\bar{z}\ \tau(\bar{v},\bar{w},\bar{x},\bar{z})$ is in $*J$ and $T^f \cup *J$ implies $\forall\bar{v}\ \forall\bar{w}\ \forall\bar{x}\ \forall\bar{z}\ (\tau(\bar{v},\bar{w},\bar{x},\bar{z}) \to (\sigma(\bar{w},\bar{x}) \wedge \psi(\bar{v},\bar{w})))$. But then the formula $\sim\exists\bar{v}\ \exists\bar{x}\ \exists\bar{z}\ \tau(\bar{v},\bar{w},\bar{x},\bar{z})$ is in $\Delta_{\exists\bar{v}\ \psi(\bar{v},\bar{w})}$ and $T^f \cup *J \vdash \exists\bar{w}(\chi(\bar{w}) \wedge \exists\bar{v}\ \exists\bar{x}\ \exists\bar{z}\ \tau(\bar{v},\bar{w},\bar{x},\bar{z}))$. Hence, the type $\Delta_{\exists\bar{v}\ \psi(\bar{v},\bar{w})}$ is locally omitted by $T^f \cup *J$.

Let $\mathfrak{M}$ be a model of $T^f \cup *J$ which omits all types of the form $\Delta_{\exists\bar{v}\ \psi(\bar{v},\bar{w})}$. The structure $\mathfrak{M}$ completes $T^f \cup *J$.

(8) ⇒ (1). The class $\mathfrak{F}_J$ is nonempty, since J is countable.

(2) ⇒ (4). Clearly, every model of $*J$ which completes T^f also completes $T^f \cup *J$.

(5) ⇒(8). First, assume that $\mathfrak{M}$ is a member of $\mathfrak{F}_J$. The structure $\mathfrak{M}$ completes $T^f \cup *J$, since $\mathfrak{M}$ completes the deductive closure J^f of $T^f \cup *J$. Suppose that $\mathfrak{M}$ is a substructure of a model $\mathfrak{M}'$ of T^f. The model $\mathfrak{M}'$ must satisfy $*J$, since $*J$ consists of existential sentences; so, $\mathfrak{M}$ is an elementary substructure of $\mathfrak{M}'$. Thus $\mathfrak{M}$ completes T^f and is in $\mathfrak{F}_T$.

Conversely, assume that $\mathfrak{M}$ is a member of $\mathfrak{F}_T \cap \mathrm{Mod}(J)$.

Each member of $\mathfrak{F}_T \cap \mathbf{Mod}(J)$ is a model of $T^f \cup *J$ (see(1) ⇒ (2)) Accordingly, $\mathfrak{M}$ is a model of the deductive closure J^f of $T^f \cup *J$. If $\mathfrak{M}$ is a substructure of a model $\mathfrak{M}'$ of J^f, then $\mathfrak{M}$ is an elementary substructure of $\mathfrak{M}'$, since $\mathfrak{M}$ completes T^f. Hence, $\mathfrak{M}$ completes J^f and is in $\mathfrak{F}_J$.

A natural conjecture for a characterization of a finite forcing component would be that $T^f \cup J$ is a complete theory or, the possibly weaker hypothesis, that $T^f \cup *J$ is a complete theory (since J is deducible from $T^f \cup *J$). Part (5) of the preceding theorem implies that $T^f \cup *J$ is complete whenever J is a finite forcing component. However, there are examples in which $T^f \cup *J$ is complete but J is not a finite forcing component (i.e., $T^f \cup *J$ is complete but its deductive closure is not J^f). The following example was suggested by an example of M. Pouzet [12, §3].

Example 1. The vocabulary will consist of a binary predicate $R(x,y)$ and a unary predicate $P_n(x)$ for each positive, odd integer.

Let Σ be the class of finite structures for this language which have an odd number of elements and in which (i) $R(x,y)$ is an equivalence relation, (ii) each equivalence class except one contains exactly two elements and the exceptional equivalence class contains only one element, and (iii) $P_n(x)$ holds for an element of the structure if and only if the structure has exactly n elements (and so $P_n(x)$ holds either for all elements of the structure or for no elements of the structure). Finally, let $T = Th(\Sigma)$.

Each structure in Σ is a model of T and completes

T, for such a structure cannot be embedded in a strictly larger structure because it satisfies $\forall x\, P_n(x)$ for some n.

If a sentence φ is consistent with T, then φ is true in some member of Σ, which, as just noted, completes T, so $T = T^f$.

The theory T has infinite models, so the set of sentences $I = \{\exists v_1 \ldots \exists v_n (\bigwedge_{i \neq j} \sim(v_i = v_j)) : n = 1, 2, \ldots\}$ is consistent with T. Furthermore, $T \cup I$ is categorical in every infinite power, so $T \cup I$ is complete and has the joint embedding property. Consequently, $J = (T \cup I)_\forall$ is a component of T, *J includes I, and $T^f \cup {}^*J$ is complete. Yet no infinite model of T can complete T, for an infinite model $\mathfrak{M}$ of T can be embedded in a larger model of T in which the equivalence class of $\mathfrak{M}$ with one element has been enlarged to an equivalence class with two elements. In other words, T has no finitely generic models of infinite cardinality. Hence, J is not a finite forcing component even though $T^f \cup {}^*J$ is complete.

The preceding example is relevant also to the question raised by R. Cusin in [4] as to whether every "existential type" (or more precisely, existentially generated type) is a "sur-existential type". (In fact, part (7) of Theorem 2.1 was motivated by the definition of sur-existential types.) Let $\mathfrak{M}$ be an infinite model of the theory T of Example 1. Then the 0-type $Th(\mathfrak{M})$ is existential, since $Th(\mathfrak{M})$ is the deductive closure of $T^f \cup {}^*J$. But it is not sur-existential, because $Th(\mathfrak{M})$ includes the sentence $\exists x \forall y (x = y \vee \sim R(x,y))$ although there is no existential formula $\varphi(x)$ such that $\exists x\, \varphi(x)$ is in $Th(\mathfrak{M})$ and $T^f (=T) \vdash \forall x(\varphi(x) \rightarrow \forall y(x = y \vee \sim R(x,y)))$.

Moreover, if $a_0, \ldots, a_n$ are members of doubleton equivalence classes of $\mathfrak{M}$, then the complete type of $a_0, \ldots, a_n$ in $\mathfrak{M}$ is an existential type but not a sur-existential type.

§ 3 Finite forcing components and the function $n_{\mathfrak{F}}(T)$

That the elementary equivalence classes of finitely generic structures correspond to a G_δ subset of $\mathcal{S}(\mathcal{B}_1(T_\forall))$ can now be proven using parts (1) and (3) of Theorem 2.1. Then Theorem A can be applied, and, in addition, the relationship between $n_{\mathfrak{G}}(T)$ and $n_{\mathfrak{F}}(T)$ can be readily described.

Theorem 3.1. The points of the Stone space $\mathcal{S}(\mathcal{B}_1(T_\forall))$ which correspond to the finite forcing components of T form a G_δ subset of the space.

Proof. Define an open set $V(P,\psi)$ for each ordered pair consisting of a condition $P(a_0,\ldots,a_n)$ and a formula $\psi(v_0,\ldots,v_n)$ in the language of T by

$$V(P,\psi) = u([\ \sim\exists v_0\ldots\exists v_n \wedge P(v_0,\ldots, v_n)])$$

$$\cup (\bigcup \{u([\exists v_0\ldots\exists v_n\, \exists v_{n+1}\ldots\exists v_m \wedge Q(v_0,\ldots,v_n, v_{n+1},\ldots,v_m)]) :$$

$Q(a_0,\ldots, a_n, a_{n+1},\ldots, a_m)$ is a condition extending

P and either

$$Q \Vdash_T \psi(a_0,\ldots, a_n) \text{ or } Q \Vdash_T \sim\psi(a_0,\ldots, a_n)\}).$$

Define a subset $\mathfrak{F}C$ by

$\mathfrak{F}C = C \cap (\bigcap \{V(P,\psi):\ P$ is a condition and ψ is a formula$\})$,

where C is the G_δ subset of points corresponding to components as defined in Theorem 1.3. Clearly, $\mathfrak{FC}$ is a G_δ subset of $\mathcal{S}(\mathcal{B}_1(T_\forall))$. Part (3) of Theorem 2.1 implies that a component of T is a finite forcing component if and only if the corresponding point is in $\mathfrak{FC}$.

<u>Theorem 3.2</u>. If T has uncountably many finite forcing components, then T has $2^{\aleph_0}$ many finite forcing components.

<u>Proof</u>. This theorem is a consequence of Theorem A and Theorem 3.1.

<u>Corollary 3.3</u>. If $n_{\mathfrak{F}}(T) > \aleph_0$, then $n_{\mathfrak{F}}(T) = 2^{\aleph_0}$.

<u>Theorem 3.4.</u> The range of the function $n_{\mathfrak{F}}(T)$ as T varies over countable theories is $\{1,2, \ldots, \aleph_0, 2^{\aleph_0}\}$.

<u>Proof.</u> The proof is identical to the proof of Theorem 1.7., for all theories mentioned therein have model-companions so that $n_{\mathfrak{F}}(T) = n_{\mathcal{C}}(T)$.

The values of $n_{\mathfrak{F}}(T)$ and $n_{\mathcal{C}}(T)$ are well correlated because $\mathfrak{FC}$ is a dense subset of C.

<u>Lemma 3.5.</u> The collection $\{C \cap u([\varphi]) : \varphi$ is an existential sentence in the language of $T\}$ is a basis for the induced topology on C.

<u>Proof.</u> Suppose that an ultrafilter F is a member of $C \cap u([\chi])$ for some element $[\chi]$ of $\mathcal{B}_1(T_\forall)$. Since F corresponds to a component, there is an existential sentence φ such that $[\varphi]$ is in F and $[\varphi] \leq [\chi]$. This implies that F is a

member of $u([\varphi]) \subseteq u([\chi])$, so F is in $\mathcal{C} \cap u([\varphi]) \subseteq \mathcal{C} \cap u([\chi])$.

Theorem 3.6. $\mathfrak{FC}$ is a dense subset of $\mathcal{C}$.

Proof. Suppose that φ is an existential sentence consistent with T. There is a condition P such that $P \Vdash_T \varphi$. Since T is countable, there is a complete sequence $P = P_0 \subseteq P_1 \subseteq \ldots$ of forcing conditions. This sequence determines a unique, finitely generic structure $\mathfrak{M}$ for T. The structure $\mathfrak{M}$ satisfies φ. The universal theory of $\mathfrak{M}$ is a finite forcing component for T. Thus, the corresponding ultrafilter in $\mathfrak{FC}$ contains $[\varphi]$ and so is in $\mathcal{C} \cap u([\varphi])$.

Corollary 3.7. Each principal component of T is a finite forcing component.

Proof. A principal component corresponds to an isolated point of $\mathcal{C}$, hence, a point of $\mathfrak{FC}$.

Corollary 3.8. If T has only finitely many components, then each component is a finite forcing component.

Proof. The hypothesis implies that each component is principal.

Corollary 3.9. If T has infinitely many components, then it has infinitely many finite forcing components.

Proof. An infinite, Hausdorff space has no finite, dense subsets.

The preceding results and remarks yield the following theorem.

<u>Theorem 3.10</u>. The inequality $n_{\mathfrak{F}}(T) \le n_{\mathcal{G}}(T)$ holds for all countable theories T. If one of the values is finite, then both are finite and are equal. The range of each function is $\{1,2,\ldots,\aleph_0,2^{\aleph_0}\}$.

Although the theory T in Example 1 has one component which is not a finite forcing component, nevertheless $n_{\mathfrak{F}}(T) = \aleph_0 = n_{\mathcal{G}}(T)$. The next example, in which $n_{\mathfrak{F}}(T) = \aleph_0 < 2^{\aleph_0} = n_{\mathcal{G}}(T)$, shows that theorem 3.10 cannot be improved.

<u>Example 2.</u> This example is closely related to Example 1. The vocabulary will consist of a binary predicate $R(x,y)$, a unary predicate $P_n(x)$ for each positive integer n, and an infinite collection $a_0, a_1, \ldots$ of individual constant symbols.

Let Σ_n for each positive integer n consist of the finite structures for this language which have exactly n elements and in which (i) $R(x, y)$ is an equivalence relation and each equivalence class has at most two elements, (ii) $P_n(x)$ holds for all elements of the structure and $P_m(x)$ for $m \ne n$ holds for <u>no</u> elements of the structure, and (iii) $a_j = a_0$ for $j \ge n$. Let $\Sigma = \bigcup \{\Sigma_n : n = 1, 2, \ldots\}$. Finally, let $T = Th(\Sigma)$.

Each model in Σ completes T (since it cannot be embedded in any larger model) and so is finitely generic for T. Therefore, each sentence consistent with T is true in a model which completes T, so $T = T^f$. (In fact, $T = T^F$ also. First, each structure in Σ is infinitely generic for T, so $T \supseteq T^F$. Secondly, each sentence true in an infinite model of $T_\forall$ is true also in a member of Σ by an Ehrenfeucht and Fraïssé back and forth argument, so any sentence not in T^F is not in T, i.e., $T^F \supseteq T$.)

Moreover, the models in Σ are the only finitely generic odels for T. Suppose $\mathfrak{M}$ is not in Σ and is existentially comlete for T. Since $\mathfrak{M}$ is not in Σ, $\mathfrak{M}$ must satisfy $\forall x \sim P_n(x)$ or all n; so, $\mathfrak{M}$ is included in strictly larger models of T. onsequently, $\mathfrak{M}$ must be infinite, because it is existentially omplete. The existential completeness of $\mathfrak{M}$ implies also that satisfies the sentence $\forall x\, \exists y(R(x,y) \wedge \sim(x = y))$. However, $\mathfrak{M}$ annot complete T, for the set of sentences $T \cup$ diagram of $\mathfrak{M}$ $\cup \{\exists x\, \forall y(x = y \vee \sim R(x,y))\}$ is consistent (a standard compactness rgument so that $\mathfrak{M}$ has an extension which is a model of T and f $\exists x\, \forall y(x = y \vee \sim R(x,y))$. Thus, $\mathfrak{M}$ is not finitely generic for .

For each infinite set K of nonnegative integers, $K = \{n_1 < n_2 < \ldots\}$, let $I_K = \{R(a_{n_{2m+1}}, a_{n_{2m+2}}) : m = 0, 1, \ldots\}$ $\cup \{\sim R(a_i, a_j)$: there does not exist a nonnegative integer m such that

$$i = n_{2m+1},\ j = n_{2m+2}\} \cup \{\sim(a_i = a_j) : i \neq j\}.$$

Each I_K can be extended to a component J_K such that J_K and $J_{K'}$ are distinct whenever $K \neq K'$, since $K \neq K'$ implies that $I_K \cup I_{K'}$ is nconsistent. This yields $2^{\aleph_0}$ components none of which is a finite forcing component. Thus, $n_{\mathfrak{F}}(T) = \aleph_0$ and $n_{\mathfrak{G}}(T) = 2^{\aleph_0}$.

An example with a finite vocabulary, if preferred, may be obtained by implicity adding infinitely many constants to M. Pouzet's example through the method of R. Vaught [20, page 319] and then proceeding as above.

<u>Theorem 3.11</u>. If the principal components of a countable theory T are not dense in $\exists C$ (or in C, since $\exists C$ is dense in C), then T has $2^{\aleph_0}$ many finite forcing components. In particular, if a countable theory T has no principal components, then it has $2^{\aleph_0}$ many finite forcing components.

<u>Proof.</u> Both parts are consequences of Theorem A, Theorem 3.1, and Corollary 3.7.

<u>Corollary 3.12</u>. (1) If T^f has no complete extension which is obtained by the addition of one existential sentence, then T^f has $2^{\aleph_0}$ many completions.

(2) (H. Simmons, [18]) If no finite extension of T^f has the joint embedding property, then T^f has $2^{\aleph_0}$ many completions.

<u>Proof.</u> The hypothesis of each part implies that there are no principal components.

§4 Existentially complete structures for a component

The problem of counting elementary equivalence classes of existentially complete structures is at once both similar and dissimilar to those for infinitely generic structures and finitely generic structures. Since the latter structures are existentially complete, one has in fact been counting elementary equivalence classes of existentially complete structures already.

However, there is an important distinction between these problems. Whereas the elementary equivalence classes of infinitely generic structures or of finitely generic structures are in one-to-one correspondence with components or finite forcing components, a theory may have only one component but infinitely many elementary equivalence classes of existentially complete structures (D. Goldrei, A. Macintyre, and H. Simmons [6], J. Hirschfeld [7], J. Hirschfeld and W. Wheeler [8], A. Macintyre [11], D. Saracino [16], W. Wheeler [21]) . In other words, while each existentially complete structure is a model of a unique component, elementarily inequivalent, existentially complete structures may be models of the same component. Thus, the appropriate method of counting the elementary equivalence classes of existentially complete structures is first to determine the components and then to determine the elementary equivalence classes of existentially complete structures for each component.

The theories of existentially complete structures for a theory T form a G_δ subset of the Stone space of the Lindenbaum algebra of $T_\forall$. The crux of the proof is the fact that a complete theory T' which includes $T_\forall$ is the theory of an existentially complete structure for T if and only if, for each universal formula $\psi(v_0,\ldots,v_n)$ in the language of T, the type $\{\psi(v_0,\ldots v_n)\} \cup \{\sim\varphi(v_0,\ldots,v_n) : \varphi$ is an existential formula in the language of T and $T_\forall \vdash \forall v_0\ldots\forall v_n(\varphi(v_0,\ldots,v_n) \rightarrow \psi(v_0, \ldots,v_n))\}$ is locally omitted by T'. This assertion follows immediately from the fact that a structure $\mathfrak{M}$ is existentially complete for T if and only if it is a model of $T_\forall$ and, for each universal formula $\psi(a_0,\ldots,a_n)$ defined and true in $\mathfrak{M}$, there is an existential formula $\varphi(v_0,\ldots,v_n)$

in the language of T such that $\mathfrak{M}$ satisfies $\varphi(a_0,\ldots,a_n)$ and $T_\forall \vdash \forall v_0\ldots\forall v_n(\varphi(v_0,\ldots,v_n) \to \psi(v_0,\ldots,v_n))$.

Theorem 4.1. The points of the Stone space $\mathcal{S}(\mathcal{B}(T_\forall))$ which correspond to the theories of existentially complete structures for T form a G_δ subset of the space.

Proof. Define an open set $W(\psi,\chi)$ for each ordered pair consisting of a universal formula $\psi(v_0,\ldots,v_n)$ in the language of T and an arbitrary formula $\chi(v_0,\ldots,v_n)$ in the language of T by

$$W(\psi,\chi) = u([\sim\exists v_0\ldots\exists v_n\ \chi(v_0,\ldots,v_n)]) \cup u([\exists v_0\ldots\exists v_n(\chi(v_0,\ldots,v_n) \wedge \sim\psi(v_0,\ldots,v_n))]) \cup (\bigcup\{u([\exists v_0\ldots\exists v_n(\chi(v_0,\ldots,v_n) \wedge \varphi(v_0,\ldots,v_n))]) : \varphi \text{ is an existential formula in the language of } T \text{ and } T_\forall \vdash \forall v_0\ldots\forall v_n(\varphi(v_0,\ldots,v_n) \to \psi(v_0,\ldots,v_n))\}.$$

Define a set $\mathfrak{J}$ by

$$\mathfrak{J} = \bigcap\{W(\psi,\chi) : \psi \text{ is a universal formula and } \chi \text{ is an arbitrary formula}\}.$$

Clearly, the set $\mathfrak{J}$ is a G_δ subset of $\mathcal{S}(\mathcal{B}(T_\forall))$. As each point in this Stone space corresponds to a complete theory which includes $T_\forall$, it follows from earlier remarks that $\mathfrak{J}$ consists precisely of the points corresponding to the theories of existentially complete structures for T.

Theorem 4.2. If $n_{\mathcal{E}}(T) > \aleph_0$, then $n_{\mathcal{E}}(T) = 2^{\aleph_0}$.

Proof. This theorem is a consequence of Theorem A and Theorem 4.1.

<u>Theorem 4.3</u>. If the number of elementary equivalence classes of existentially complete structures for T which are models of a component J of T is uncountable, then it is $2^{\aleph_0}$.

<u>Proof.</u> This theorem follows from the preceding theorem and the fact that the existentially complete structures for T which are models of J are just the existentially complete structures of J.

Thus, the number of elementary equivalence classes of existentially complete models of a component of a countable theory must be either finite, $\aleph_0$, or $2^{\aleph_0}$. The example in [16] has one component and $\aleph_0$ many elementary equivalence classes of existentially complete structures. The theory of division rings of a specified characteristic has one component and $2^{\aleph_0}$ many elementary equivalence classes of existentially complete structures [8,11,21]. The same is true of arithmetic [6,7,8]. The theory of fields of a specified characteristic has exactly one elementary equivalence class of existentially complete structures. The following example, a variant of the example in [16], has one component and exactly two elementary equivalence classes of existentially complete structures.

<u>Example 3.</u> The vocabulary will consist of individual constant symbols a_n and b_n for each integer n, a unary predicate $I(x)$, a binary predicate $P(x,z)$, a binary predicate $G(x,y)$, and a ternary predicate $F(x, y, z)$.

Let T be the theory of indexed partitions $(P(x,z))$ in which the index set $(I(x))$ is disjoint from the union of the partitioning sets, $F(x, y, z)$ for each index x is an injective mapping of the index set into the set indexed by x which misses at most one point of the set indexed by x, and $G(x, y)$ is a bijection of the entire model which preserves the index set and the partition relation, has no finite cycles, and satisfies the conditions that, if $x \xrightarrow{G} \bar{x}$ and $z \xrightarrow{G} \bar{z}$, then $F(x, y, z)$ holds if and only if $F(\bar{x}, y, \bar{z})$ holds. Also, T must assert that each a_n is an index, each b_n is in the partitioning set indexed by a_n, there is no index y such that $F(a_n, y, b_n)$, and $G(a_n, a_{n+1})$ and $G(b_n, b_{n+1})$.

The minimal model for T is the structure with universe $\{(k,m) : k \text{ and } m \text{ are integers}\} \cup \{(k, 1/2) : k \text{ is an integer}\}$ where $a_n = (n,0)$, $b_n = (n, 1/2)$, and in which $I(x)$ holds if and only if $x = (k,0)$ for some integer k, $P(x,y)$ holds if and only if $x = (k,0)$ and $y = (k,r)$ for some integer k and some number r which is either $1/2$ or a nonzero integer, $F(x,y,z)$ holds if and only if $x = (k,0)$, $y = (m,0)$, and $z = \begin{cases} (k,m) & \text{if } m < 0 \\ (k,m+1) & \text{if } m \geq 0 \end{cases}$ for some integers k and m, and $G(x,y)$ holds if and only if $x = (k, r)$ and $y = (k+1, r)$ for some integer k and a number r which is either $1/2$ or an integer.

Explicit axioms for T are listed below.

$I(a_n) \wedge \sim I(b_n)$ for all integers n;

$\sim(a_i = a_j)$ for all integers i and j with $i \neq j$;

$\forall x \forall y (P(x,y) \rightarrow (I(x) \wedge \sim I(y)))$;

$\forall x_1 \forall x_2 \forall y((P(x_1,y) \wedge P(x_2,y)) \rightarrow x_1 = x_2)$;

$\forall x(I(x) \rightarrow \exists y\, P(x,y)) \wedge \forall y(\sim I(y) \rightarrow \exists x\, P(x,y))$;

$P(a_n, b_n)$ for all integers n;

$\forall x \forall y \forall z(F(x,y,z) \rightarrow (I(x) \wedge I(y) \wedge \sim I(z) \wedge P(x,z))$;

$\forall x \forall y_1 \forall y_2 \forall z_1 \forall z_2 (((F(x,y_1,z_1) \wedge F(x,y_1,z_2)) \rightarrow z_1 = z_2)$
$\wedge ((F(x,y_1,z_1) \wedge F(x,y_2,z_1)) \rightarrow y_1 = y_2))$;

$\forall x \forall z_1 \forall z_2((P(x,z_1) \wedge P(x,z_2) \wedge \sim (z_1 = z_2)) \rightarrow \exists y(F(x,y,z_1) \vee$
$F(x,y,z_2)))$;

$\forall x \forall y((I(x) \wedge I(y)) \rightarrow \exists z\, F(x,y,z))$;

$\forall y \sim F(a_n,y,b_n)$ for all integers n;

$\forall x_1 \forall x_2 \forall y_1 \forall y_2 (((G(x_1,y_1) \wedge G(x_1,y_2)) \rightarrow y_1 = y_2)$
$\wedge ((G(x_1,y_1) \wedge G(x_2,y_1)) \rightarrow x_1 = x_2))$;

$\forall x \exists y\, G(x,y) \wedge \forall y \exists x\, G(x,y)$;

$\forall x_1 \ldots \forall x_n \sim(G(x_1,x_2) \wedge \ldots \wedge G(x_{n-1}, x_n) \wedge G(x_n, x_1))$
for all positive integers n;

$\forall x \forall y(G(x,y) \rightarrow (I(x) \leftrightarrow I(y)))$;

$\forall x_1 \forall x_2 \forall y \forall z_1 \forall z_2 ((G(x_1,x_2) \wedge G(z_1,z_2)) \rightarrow ((P(x_1,z_1) \leftrightarrow P(x_2,z_2))$
$\wedge (F(x_1,y,z_1) \leftrightarrow F(x_2,y,z_2))))$;

$G(a_n,a_{n+1})$, $G(b_n, b_{n+1})$ for all integers n.

Each existentially complete structure for T is a model of T, because all axioms of T are $\forall\exists$ formulas. In fact, a structure is existentially complete for T if and only if it is a model of T and satisfies the infinitary sentence

$\forall x \forall z ((P(x,z) \wedge (\bigwedge_{n \in Z} \sim(x = a_n))) \rightarrow \exists y\, F(x,y,z))$ where Z is the set of integers.

Suppose that $\mathfrak{M}$ is an existentially complete model of T. Define two elements c and d of the set of members of $\mathfrak{M}$ which satisfy $I(x)$ to be G-equivalent if there are elements $c_1, \ldots, c_n$

such that $G(c_i, c_{i+1})$ holds for $i = 1, \ldots, n-1$ and $c_1 = c$, $c_n = d$ or $c_1 = d$, $c_n = c$. This is an equivalence relation. Each equivalence class will be called an orbit of $\mathfrak{M}$.

One existentially complete model can be embedded in a second existentially complete model if and only if the second has at least as many orbits as the first. Therefore, T has only one component.

An existentially complete model with only one orbit satisfies the sentence $\forall x(I(x) \rightarrow \exists z(P(x,z) \wedge \forall y \sim F(x,y,z)))$. An existentially complete model with more than one orbit satisfies the negation of this sentence. Two existentially complete models are isomorphic if and only if they have the same number of orbits. Moreover, an existentially complete model $\mathfrak{M}$ with at least two orbits can be elementarily embedded in an existentially complete model $\mathfrak{N}$ if and only if $\mathfrak{N}$ has at least as many orbits as $\mathfrak{M}$. Hence, $n_{\mathcal{E}}(T) = 2$.

The only essential difference between Example 3 and the example in [16] is the bijection $G(x,y)$. The function G ensures that index set either is precisely $\{a_n : n$ is an integer$\}$ or else contains infinitely many elements in addition to the a_n's. Therefore, the restriction to a model of the theory in [16] is either finitely generic or infinitely generic for that theory.

This example can be generalized easily to obtain theories with one component and exactly 2^r elementary equivalence classes of existentially complete structures for any positive integer r. We are not aware of any examples for positive integers which are not of the form 2^r. (See remark 4.)

Suppose that X is the range of the function $n_{\mathcal{E}}(T)$ as T varies over countable theories with the joint embedding property. Let $h : X \to \{1,2,\ldots, \aleph_0, 2^{\aleph_0}\}$ be an arbitrary function. One can construct, through a judicious use of unary predicates and constants, a theory T_h such that, for each x in X, the number of components of T with precisely x elementary equivalence classes of existentially complete structures is $h(x)$.

Remarks

1. R. Cusin in [4] uses topological methods to show that a countable theory with the joint embedding property but without a prime, finitely generic structure must have $2^{\aleph_0}$ many, nonisomorphic, countable, finitely generic structures. The proof is that the set of sur-existential types corresponds to a (dense) G_δ subset of a complete, metric space and that the absence of a prime, finitely generic structure implies that the isolated points of the subset are not dense in the subset. Hence, part (i) of Theorem A is applicable, and there must be $2^{\aleph_0}$ many sur-existential types.

2. That the only uncountable value of the functions $n_{\mathcal{E}}(T)$, $n_{\mathfrak{F}}(T)$, and $n_{\mathfrak{G}}(T)$ is $2^{\aleph_0}$ can be proven in other ways of course. One alternative, which does not use topology, is to construct binary splitting trees such that each branch corresponds to the theory of an existentially complete, finitely generic, or infinitely generic structure, respectively. The splitting depends upon the fact that each finite subbranch with the property (*) that it is satisfied

by uncountably many, non-elementarily equivalent structures of the appropriate class can be split by adjoining either φ or $\sim\varphi$ so that property (*) is true of each enlarged subbranch. The only subtlety is that, for $\mathcal{E}_T$ or $\mathcal{F}_T$, one must alternate splitting steps with steps which will ensure that all relevant types, for example, the finite forcing types, are locally omitted by each theory corresponding to a branch of the tree.

A second alternative is to show that certain subsets of appropriate Stone spaces are analytic sets. Part (ii) of Theorem A is true also for analytic sets and so can be applied still. An appropriate Stone space for existentially complete or finitely generic structures is the Stone space $\mathcal{S}(\mathcal{B}(L))$ of the Lindenbaum algebra of the language L. That the subset of points corresponding to the theories of existentially complete structures or of finitely generic structures is analytic follows from Theorem C and the fact that each class is axiomatized by a sentence of $L_{\omega_1\omega}$.

<u>Theorem C.</u> The set of complete, $L_{\omega\omega}$ theories of models of a sentence of $L_{\omega_1\omega}$ for a countable language L is an analytic subset of the Lindenbaum algebra of L.

The case of infinitely generic structures is treated by showing that the set $\hat{\mathcal{C}}$ of points of $\mathcal{S}(\mathcal{B}_1(\mathcal{L}))$ corresponding to components of T is an analytic subset. This follows from the fact that there is a continuous mapping of the Stone space $\mathcal{S}(\mathcal{B}(L))$ onto $\mathcal{S}(\mathcal{B}_1(\mathcal{L}))$ such that the set $\hat{\mathcal{C}}$ is the image of the set of points corresponding to theories of existentially complete structures. Also, the image of the set of points corresponding to theories of finitely generic structures is just the set of points corresponding to finite forcing components. This method, although gross in that it misses the G_δ nature of the various subsets, may be applicable to a wider class of problems.

3. The preceding remark raises the question of whether the choice of a Stone space is significant. The answer is negative with regard to analyticity and is negative also in regard to the G_δ nature of the set of theories of finitely generic structures or of existentially complete structures.

The various Boolean algebras are related as follows:

$$\begin{array}{ccc}
\mathcal{B}_1(\mathcal{L}) & \xrightarrow[\text{injective}]{f} & \mathcal{B}(\mathcal{L}) \\
\text{surjective}\ \Big\downarrow h & & k \Big\downarrow\ \text{surjective} \\
\mathcal{B}_1(T_\forall) & \xrightarrow[\text{injective}]{g} & \mathcal{B}(T_\forall)
\end{array}$$

Of course, only $\mathcal{B}(\mathcal{L})$ and $\mathcal{B}(T_\forall)$ are relevant for theories of existentially complete structures.

The dual diagram of Stone spaces and continuous mappings is

$$\begin{array}{ccc}
\mathcal{S}(\mathcal{B}_1(\mathcal{L})) & \xleftarrow[\text{surjective}]{\bar{f}} & \mathcal{S}(\mathcal{B}(\mathcal{L})) \\
\text{injective}\ \Big\uparrow \bar{h} & & \bar{k} \Big\uparrow\ \text{injective} \\
\mathcal{S}(\mathcal{B}_1(T_\forall)) & \xleftarrow[\text{surjective}]{\bar{g}} & \mathcal{S}(\mathcal{B}(T_\forall))
\end{array}$$

The images $\text{im}\,\bar{h}$ and $\text{im}\,\bar{k}$ are closed, because they are continuous images of compact spaces in Hausdorff spaces.

Consequently, the inverse image of an analytic subset under either $\bar{h}$ or $\bar{k}$ is analytic. Suppose for example, that A is an analytic subset of $\mathcal{S}(\mathcal{B}(\mathcal{L}))$. Then $A \cap \text{im}\ \bar{k}$ is an analytic subset of $\mathcal{S}(\mathcal{B}(\mathcal{L}))$ and of $\text{im}\,\bar{k}$. Since $\bar{k}$ is a homeomorphism onto its image, $\bar{k}^{-1}(A \cap \text{im } \bar{k})$ is an analytic subset of $\mathcal{S}(\mathcal{B}(T_\forall))$. Thus, the analyticity mentioned in the preceding remark may be pulled back to $\mathcal{S}(\mathcal{B}_1(T_\forall))$.

On the other hand, the image of a G_δ subset under either $\bar{h}$ or $\bar{k}$ is a G_δ subset of $\mathcal{S}(\mathcal{B}_1(\mathcal{L}))$ or $\mathcal{S}(\mathcal{B}(\mathcal{L}))$, respectively. Suppose, for example, that B is a G_δ subset of $\mathcal{S}(\mathcal{B}(\mathcal{L}))$. Then $\bar{k}(B)$ is a G_δ subset of im $\bar{k}$. Since im $\bar{k}$ is closed in a metric space, it is a G_δ subset of $\mathcal{S}(\mathcal{B}(\mathcal{L}))$. Hence $\bar{k}(B)$ is a G_δ subset of a G_δ subset of $\mathcal{S}(\mathcal{B}(\mathcal{L}))$ and so is G_δ itself.

It remains to show that the set of points of $\mathcal{S}(\mathcal{B}(T_\forall))$ which correspond to theories of finitely generic structures is a G_δ subset. A direct proof, similar to that of Theorem 4.1, can be given using the facts:

(1) a theory is the theory of a finitely generic structure for T if and only if it includes $T_\forall$ and locally omits each type in a certain, countable collection, and

(2) the result, implicit in Theorem 4.1, that the set of points corresponding to theories which contain a given theory and locally omit each type in a countable collection is a G_δ subset. On the other hand, one may reason as follows. The inverse image of $\mathcal{FC}$ under $\bar{g}$ is a G_δ set, so $\bar{g}^{-1} \cap (\bigcap \{ u([\varphi]) : \varphi \text{ is in } T^f \})$ is a G_δ subset of $\mathcal{S}(\mathcal{B}(T_\forall))$. A point is in this set if and only if it corresponds to a theory which includes $T^f \cup {}^*J$ for some finite forcing component and hence is the theory of a finitely generic structure for T.

4. J. Schmerl has found theories T_n for $n \geq 3$ such that each theory T_n has the joint embedding property and has exactly n elementary equivalence classes of existentially complete structures. The reader is referred to his paper in this volume [22] .

BIBLIOGRAPHY

1. Barwise, J., and Robinson, A. "Completing Theories by Forcing." Annals of Mathematical Logic, vol. 2 (1970), pp. 119-142.

2. Bell, J.L., and Slomson, A.B. Models and Ultraproducts. Amsterdam, North-Holland Publishing Company, 1969.

3. Chang, C.C., and Keisler, H.J. Model Theory. Amsterdam, North-Holland Publishing Company, 1973.

4. Cusin, R. "The number of countable, generic models for finite forcing." Fundamenta Mathematica, vol. 84 (1974), pp. 265-270.

5. Fisher, E., and Robinson, A. "Inductive Theories and Their Forcing Companions." Israel Journal of Mathematics, vol. 12 (1972), pp. 95-107.

6. Goldrei, D., Macintyre, A., and Simmons, H. "The Forcing Companions of Number Theories." Israel Journal of Mathematics, vol. 14 (1973), pp. 317-337.

7. Hirschfeld, J. Existentially Complete and Generic Structures in Arithmetic. Dissertation, Yale University, 1972.

8. Hirschfeld, J., and Wheeler, W. Forcing, Arithmetic, and Division Rings. To appear.

9. Lipshitz, L., and Saracino, D. "The model-companion of the theory of commutative rings without nilpotent elements." Proceedings of the American Mathematical Society, vol. 38 (1973), pp. 381-387.

10. Macintyre, A. "Omitting Quantifier-Free Types in Generic Structures." Journal of Symbolic Logic, vol. 37 (1972), pp. 512-520.

11. Macintyre, A. "On Algebraically Closed Division Rings." Preprint.

12. Pouzet, M. "Extensions completes d'une theorie forcing complete." Israel Journal of Mathematics, vol. 16 (1973), pp. 212-215.

13. Robinson, A. "Forcing in Model Theory." Symposia Mathematica, vol. 5 (1970), pp. 64-82.

14. Robinson, A. "Infinite Forcing in Model Theory." Proceedings of the Second Scandinavian Logic Symposium, Oslo, 1970. Amsterdam, North-Holland Publishing Co., 1971.

15. Robinson, A. "Forcing in Model Theory." Actes des Congres International des Mathematiciens, Nice, 1970.

16. Saracino, D. "A counterexample in the theory of model-companions." To appear.

17. Sierpinski, Waclaw. General Topology (trans. by C. Cecilia Krieger). Toronto, University of Toronto Press, 1952.

18. Simmons, H. "Existentially Closed Structures." Journal of Symbolic Logic, vol. 37 (1972), pp. 293-310.

19. ________. "Le nombre de structures générique d'une théorie." C. R. Acad. Sci. Paris, Ser A., vol. 277 (1973), pp. 487-489.

20. Vaught, R.L. "Denumerable models of complete theories." Infinitistic Methods. New York, Pergamon Press, 1961.

21. Wheeler, W. Algebraically Closed Division Rings, Forcing, and the Analytical Hierarchy. Dissertation, Yale University, 1972.

22. Schmerl, J. "The Number of Equivalence Classes of Existentially Complete Structures." This volume.

THE NUMBER OF EQUIVALENCE CLASSES OF EXISTENTIALLY COMPLETE STRUCTURES

James H. Schmerl

University of Connecticut
Storrs, Connecticut

In this note we prove the following

Theorem: For each $n < \omega$ there is a countable first-order theory T_n which has the joint-embedding property and which has exactly $n+1$ elementary equivalence classes of existentially complete structures.

Let us denote the number of elementary equivalence classes of existentially complete structures for T by $n_{\mathcal{E}}(T)$. A study of $n_{\mathcal{E}}(T)$, as well as the number of finitely generic and infinitely generic structures for T, is undertaken by Fisher, Simmons, and Wheeler in [1]. (We refer the reader to [1] and its bibliography for all relevant background material.) They give a nearly complete treatment in their study, the one exception being theories T which have only one elementary equivalence class of infinitely generic structures (i.e. have the joint-embedding property) and for which $n_{\mathcal{E}}(T) < \omega$. However, they did succeed in finding, for each $n < \omega$, a theory T with the joint-embedding property for which $n_{\mathcal{E}}(T) = 2^n$.

In constructing our examples, we were influenced by Ehrenfeucht's example [2] of a complete theory with exactly 3 non-isomorphic countable models.

Proof. Let $2 \leq n < \omega$. ($n < 2$ is handled in [1].) Let $\mathbb{N}$ denote the non-negative integers and $\mathbb{Q}$ the rationals. Let

$$A = \{x \in \mathbb{Q} : \exists y \in \mathbb{N}(2y \leq x \leq 2y + 1)\} .$$

Let $U_2, \ldots, U_n \subseteq A$ be such that

(1) $U_2 \cup \ldots \cup U_n = A$;

(2) $2 \leq i < j \leq n \Rightarrow U_i \cap U_j = \emptyset$;

(3) $\mathbb{N} \subseteq U_2$;

(4) $x, y \in A - \mathbb{N}$, $x < y$, and $2 \le i \le n \Rightarrow \exists z \in U_i(x < z < y)$.

Let $\mathfrak{A}_n = (A, <, U_2, \ldots, U_n, 0, 1, 2, \ldots)$, and let $T_n = Th_\forall(\mathfrak{A}_n)$. (An explicit recursive axiomatization for T_n can easily be supplied.) Obviously T_n has the joint-embedding property. Let $\phi(x)$ be the formula

$$\forall yz(x \le y < z \to \exists w(y < w < z)) .$$

For each $i \le n$ let σ_i be the sentence defined as follows:

$$\sigma_0 =_{df} \forall x \neg \phi(x) ,$$

$$\sigma_1 =_{df} \exists x \phi(x) \wedge \forall x(\exists y < x)(\phi(x) \to \phi(y)) ,$$

and for $2 \le i \le n$,

$$\sigma_i =_{df} \exists x[\phi(x) \wedge U_i(x) \wedge (\forall y < x) \neg \phi(y)] .$$

It is easy to see that for each $i \le n$ there is an existentially complete model of T_n which is also a model of σ_i . Furthermore, any two existentially complete models of σ_i are elementarily equivalent. Finally, any model of σ_i is not a model of σ_j whenever $i < j \le n$. Therefore $n_{\mathcal{E}}(T) = n+1$. ⊠

The above examples can be converted to ones in a finite language by, for example, "attaching" to each $x \in A$ an algebraically closed field K_x , where K_x has characteristic 0 unless $x \in \mathbb{N}$ in which case it has characteristic p_x (the x-th prime). It would be interesting to know whether or not finitely axiomatizable examples exist.

References

[1] Fisher, E., Simmons, H., and Wheeler, W., Elementary equivalence classes of generic structures and existentially complete structures, this volume.

[2] Vaught, R.L., Denumerable models of complete theories, Infinitistic Methods, (Proc. Sympos. on Found. of Math., Warsaw, 1959), New York, Pergamon Press, 1961, pp. 303-321.

FINITE FORCING AND GENERIC FILTERS IN ARITHMETIC

J.Hirschfeld

Abstract

We introduce the notions of forcing with respect to recursive function and recursive sets and forcing with respect to partial recursive function and r.e. sets. It is shown that the generic models are the same in both cases and coincide with the finitely generic model for a natural modification of arithmetic. This gives rise to a very interesting class of models which is investigated using the different ways in which it was obtained.

* * *

In [1] J. Barwise and A. Robinson proved that the only finitely generic model for full arithmetic is the standard model. This immediately raised the question of what kind of generic models are obtained if one excludes the standard model by adding a new constant and axioms that claim that this constant differs from all standard elements. This question looked more intriguing after S. Shelah proved that by adding an uncountable number of constants one obtains a theory without generic models [6].

This question seems to have no connection to the following construction which is of interest in its own sake: Let I be a set and $\mathfrak{A}$ a family of subsets of I which is closed under intersections. Let $\mathcal{F}$ be a countable family of functions from I into a model M . In this situation there is a natural definition of forcing: For an atomic formula $A \Vdash \phi(f_1,..f_n)$ if $M \models \phi(f_1(i),..f_n(i))$ for every i in A. $A \Vdash \exists x\phi(x)$ if there is a function h in $\mathcal{F}$ such that $A \Vdash \phi(h)$, and $A \Vdash \sim \phi$ if no subset of A which is in $\mathfrak{A}$ forces ϕ (disjunctions and conjunctions are handled like satisfaction). With this definition one obtains complete sequences and generic models with equivalence classes of functions as elements and in such a model $\phi([f_1],..[f_n])$ is satisfied iff there is a set in the sequence that forces $\phi(f_1,..f_n)$.

Of course, in this generality one could not expect interesting results, which depend on the appropriate choice of M, I, $\mathfrak{A}$ and $\mathcal{F}$.

However, it seemed, that beginning with the set of recursive functions and the family of recursive sets, an interesting class of homomorphic images of the semiring of recursive functions will emerge. A similar consideration suggested treatment of the set of partial recursive function and the family of r.e. sets.

As it turned out, recursively generic models are the same as r.e. generic models (section 3) and even the same as finitely generic models (section 4). This solves to some extent the original question about finite forcing, and points out a class of models of a fragment of arithmetic whose main properties and listed in section 5. Most of these properties cannot be deduced (as far as we can see) using only finite forcing and follow from the more detailed investigation of recursively generic models in section 2.

Throughout the paper we use some of results of [2] and [3] which are listed in section 1 both for completeness and to set the arithmetical backround.

We would like to thank L. Manivetz, who helped us to work out the proof of 2.13 and D.J.Brown, with whom we first discussed the idea of "generic filters".

1. Preliminaries.

The standard model of Arithmetic N is treated in the language that has names for all the numbers, for the functions + and · and for the order relation < .

1.1 Let R be the set of recursive functions and let F be a free filter of recursive sets. We defined the reduced recursive power R/F modifying slightly the usual approach. We define an equivalence relation: $f \equiv g$ if there is a set A in F such that $f(x) = g(x)$ a.e. (almost everywhere) on A. I.e.: $f(x) = g(x)$ for all but a finite number of elements of A. This yields the universe whose elements are the equivalence classes. For every function f let $[f]$ be its equivalence class. We define $[f]+[g] = [f+g]$, $[f]\cdot[g] = [fg]$ and $[f] < [g]$ if for some set $A \in F$, $f(x) < g(x)$ a.e. on A. Then R/F is a model in our language and it includes N as an initial segment if we identify n with the function $f(x) \equiv n$.

The modification that we made is not essential, but is rather natural if we wish to exclude principal filters. If F′ is obtained from F by adding all the sets that differ from some set in F only by a finite number of elements than R/F = R/F′ and for R/F′ the two definitions coincide. We say that F is maximal and R/F is a recursive ultrapower if for every recursive set A there is a set B in F such that B-A is finite or $B \cap A$ is finite. In this case, if F′ is as above then it is indeed a maximal filter of recursive sets.

Every recursive ultrapower is a model of T_{π_2} - the π_2 statements that hold in N [2]. We recall that a Σ_1 formula has the form $\exists \bar{x}\phi$ where ϕ has only bounded quantifiers (quantifiers of the form $\exists x<t$ and $\forall x<t$). A π_2 formula has the form $\forall \bar{y}\psi$ where ψ is Σ_1 .

1.2 Matijasevic's result [5] is essential to our discussion. It says that every r.e. predicate can be described by an existential formula. Adapted to our terminology, this means that every Σ_1 formula which we call also an r.e. formula) is equivalent in T_{π_2} to an existential formula. From this it follows that r.e. formulas are preserved under extensions among the models of T_{π_2}

and so are partial recursive functions if we think of their graph as given by a Σ_1 formula.

We call a formula ϕ <u>recursive</u> if ϕ and $\sim\phi$ are both equivalent in T_{π_2} to existential formulas. (Every such formula describes in N a recursive set and by Matijasevic's theorem every recursive set may be described by such a formula). Among models of T_{π_2} recursive predicates are absolute and so are recursive functions as their graph may be described by a recursive formula (we mention also that recursive or partial recursive function described by such graphs are indeed functions in all models of T_{π_2} and that their values are independent of the particular Σ_1 formula that describes them). For models of T_{π_2} we allow ourselves notations like "$a\varepsilon A$" or "$f(a)=b$" where A is a recursive set and f is a recursive function, meaning the formulas $\phi(a)$ and $\Psi(a,b)$ where ϕ and Ψ are some Σ_1 formulas that describe A and f respectively.

1.3 We summarize some of the results of [2]

a) If $M \models T_{\pi_2}$ and $M' \subset M$ then $M' \models T_{\pi_2}$ iff M' is closed in M under recursive functions.

b) The minimal submodel of M that contains an element a and is itself a model of T_{π_2} is the closure of a under recursive functions. We denote this model by $N[a]$.

c) $N[a]$ is isomorphic to R/F where F is the (maximal) filter of sets A such that $M \models$ "$a\varepsilon A$" . In other words, F is the filter of the recursive sets that may be described by a recursive formula $\phi(x)$ for which $M \models \phi(a)$. We call F the <u>recursive type</u> of a .

d) If $\phi(x_1,..x_n)$ is a Σ_1 formula then $R/F \models \phi([f_1],..[f_n])$ iff $\{x|N \models \phi(f_1(x),..f_n(x))\}$ almost includes a set in F . In particular, if I is the identify function then $R/F \models$ "$f([I]) = [f]$".

1.4 Let Σ be the class of submodels of models of Arithmetic (i.e. of T(N)). Let $\mathcal{E}$ be the subclass of models which are existentially complete in Σ . Let C be the set of partial recursive functions and F a filter of r.e. sets. We define the <u>reduced r.e.</u>

power C/F similarly to 1.1, except that we deal only with functions that are defined a.e. on some set of F . Then the r.e. ultrapowers play in $\mathcal{E}$ a role which is similar to that played by recursive ultrapowers in the class of models of T_{π_2} [3, chapters 8 and 9] .

a) Every model in $\mathcal{E}$ is a model of T_{π_2}.

b) R.e. predicates and partial recursive function are absolute among models in $\mathcal{E}$.

c) If $E \in \mathcal{E}$ and $M \subset E$ then M is existentially complete iff M is closed in E under partial recursive functions.

d) The minimal submodel of E that contains a given element a and is itself existentially complete is the closure of a under partial recursive functions (the existential closure of a) .

e) The existential closure of a is isomorphic to C/F where F is the existential type of a in E (i.e. the collection of r.e. sets which may be described by an existential formula which is satisfied by a in E).

f) Every r.e. ultrapower is existentially complete.

g) For every existential formula $\phi(x_1,..x_n)$, $C/F \models \phi([f_1],..[f_n])$ iff $\{x \mid N \models \phi(f_1(x),..f_n(x))\}$ is (almost) in F .

1.5 If F is an r.e. filter such that the collection F' of recursive sets in F is a base for F then R/F'=C/F. This is true since every partial recursive function is defined a.e. on some recursive set in F and can therefore be replaced by an equivalent total function.

1.6 There is a formula N(x) which is satisfied in every existentially complete model exactly by the standard elements (elements of N) [3,8.24]. Similarly- there is a (different) formula N(x) which describes N in every recursive ultrapower [2].

1.7 (Kleene's enumeration theorems, modified in view of 1.2). For every n there are existential formulas $V^n(i,\bar{x})$ and $F^n(i,\bar{x},y)$ such that

(i) For every n-place existential formula $\phi(\bar{x})$ there is some i such that

$T_{\pi_2} \models \forall\bar{x}(V(i,\bar{x}) \leftrightarrow \phi(\bar{x}))$

(ii) $T_{\pi_2} \models \forall\bar{x}\,\forall y\,\forall z(F(i,\bar{x},y) \wedge F(i,\bar{x},z) \rightarrow y=z)$

(iii) For every existential formula $\phi(\bar{x},y)$ which is the graph of a partial function there is some i such that $T_{\pi_2} \models \forall\bar{x}\,\forall y(F(i,\bar{x},y) \leftrightarrow \phi(\bar{x},y))$.

Thus V enumerates the r.e. sets and F the partial recursive functions.

1.8 1.6 and 1.7 give a powerful tool that enables us to state in the first order language properties which are of second order nature. We shall illustrate this by an example.

We define the formula

$$rec(i) \equiv \exists_j[N(j) \wedge \forall x(V(i,x) \leftrightarrow \sim V(j,x))]$$

Next, look at the statement

1.8.1 $\forall x[\phi(x) \rightarrow \exists i(N(i) \wedge rec(i) \wedge V(i,x) \wedge \forall n[N(n) \wedge V(i,n) \rightarrow \phi^N(n)])]$.

Here $\phi(x)$ is an arbitrary formula and ϕ^N is its relativization to the formula $N(x)$. Then in every existentially complete model (or recursive ultrapower) the statement above means:

1.8.2 $\phi(x)$ holds only if there is a recursive set A such that "$x \in A$" and $\phi(n)$ holds in N for every $n \in A$.

With this example in mind, we shall avoid later the formal statements of the form 1.8.1 and use informal descriptions of the form 1.8.2.

2. Generic Recursive Ultrapowers.

Let R be the set of recursive functions and S the set of infinite recursive sets. We define by induction when does a set A <u>force</u> a formula $\phi(f_1,\dots,f_n)$, where A is in S and $f_1,\dots,f_n$ are functions in R. The definition follows closely those in [1].

2.1 <u>Definition:</u>

i) If ϕ is atomic then $A \Vdash \phi(f_1,..f_n)$ iff $N \models \phi(f_1(x),..f_n(x))$ a.e.(almost everywhere) in A. In other words, if the set $\{x | N \models \phi(f_1(x),..f_n(x))\}$ contains all but a finite number of elements of A.

ii) $A \Vdash \phi \vee \Psi$ iff $A \Vdash \phi$ or $A \Vdash \Psi$

iii) $A \Vdash \phi \wedge \Psi$ iff $A \Vdash \phi$ and $A \Vdash \Psi$

iv) $A \Vdash \sim\phi$ iff there is no (infinite) set B, $B \varepsilon S$ and $B \subset A$, such that $B \Vdash \phi$.

v) $A \Vdash \exists x \phi(x)$ if for some $g \varepsilon R$ $A \Vdash \phi(g)$

It follows by a straightforward induction that:

2.2i) If $f_1(x) = g_1(x)$ a.e. in A for $i=1,\dots n$ then $A \Vdash \phi(f_1,..f_n)$ iff $A \Vdash \phi(g_1,..,g_n)$

ii) If $A-B \cup B-A$ is finite, then for every formula ϕ, $A \Vdash \phi$ iff $B \Vdash \phi$

A complete sequence and a generic model are now constructed as usual:

2.3i) A set cannot force a formula and its negation.

ii) If $A \Vdash \phi$ and $B \subset A$ then $B \Vdash \phi$

iii) For every set $A \varepsilon S$ and every formula ϕ there is a set B, $B \subset A$ and $B \varepsilon S$, such that $B \Vdash \phi$ or $B \Vdash \sim\phi$.

iv) Every set $A \varepsilon S$ is contained in a <u>complete sequence</u>; a decreasing sequence F of sets in S such that for every formula ϕ there is a set $B \varepsilon F$ such that $B \Vdash \phi$ or $B \Vdash \sim\phi$.

As usual a complete sequence yields a <u>generic model</u>; the complete sequence F is a filter base (we shall not distinguish between filters and filter bases). Thus we have the reduced power

R/F as defined in 1.1.

2.4 If F is complete then $R/F \models \phi([f_1],\ldots[f_n])$ iff $A \Vdash \phi(f_1,\ldots,f_n)$ for some set $A \varepsilon F$.

The prove is routine in view of the fact that for atomic formulas the notions of forcing and satisfaction coincide.

Weak Forcing is introduced as usual:

$$A \Vdash^* \phi \quad \text{if} \quad A \Vdash \sim\sim\phi$$

We have

2.5 i) If $A \Vdash \phi$ then $A \Vdash^* \phi$

ii) If $A \Vdash^* \sim\phi$ then $A \Vdash \sim\phi$

2.6 $A \Vdash^* \phi$ iff ϕ holds in every generic model R/F for which $A \varepsilon F$. From this we have the corollaries

2.7 If $A \Vdash^* \phi$ and Ψ is logically equivalent to ϕ then $A \Vdash^* \Psi$

2.8 ϕ holds in every generic model iff $N \Vdash^* \phi$.

The following lemma was proved by C. Coven for infinite forcing [1] but it is evident from our proof that it holds with all the natural definitions of forcing

2.9 Lemma:

Let R/F be a recursive reduced power with the following property: If $R/F \models \phi([f_1],\ldots,[f_n])$ then there is a set $A \varepsilon F$ such that $A \Vdash^* \phi(f_1,\ldots,f_n)$.
Then F is a complete sequence (and R/F is the corresponding generic model).

Proof:

We show by induction that if $R/F \models \phi([f_1],\ldots[f_n])$ then there is a set $A \varepsilon F$ such that $A \Vdash \phi(f_1,\ldots,f_n)$. This clearly suffices.
The assertion holds for atomic formulas as the definitions of forcing and satisfaction coincide.
The assertion follows immediately from the induction assumption

for $\phi \vee \Psi$, $\phi \wedge \Psi$ and $\exists x \phi$.

Finally - the assertion holds for the negation of a formula by the assumption of the lemma as forcing and weak forcing coincide for negations of formulas.

The following lemma is the key to a more concrete description of the class of generic models. We call a formula <u>conjunctive</u> if it does not contain disjunctions.

2.1o <u>Lemma:</u>

If $\phi(y_1,\ldots,y_n)$ is a conjunctive existential formula then $A \Vdash \phi(f_1,\ldots,f_n)$ iff $\phi(f_1(x),..f_n(x))$ holds a.e. in A.

<u>Proof:</u>

We first prove the lemma for quantifier free conjunctive formulas. This is done by induction on the complexity of the formula.

The assertion holds for atomic formulas by definition, and for conjunctions it follows immediately from the induction assumption.

Next, let ϕ be the formula $\sim\Psi(x_1,..,x_n)$. We assume first that $\sim\Psi(f_1(x),..,f_n(x))$ holds a.e. in A . Then clearly A has no infinite subset in which $\Psi(f_1(x),..f_n(x))$ holds a.e. By induction assumption- there is no infinite subset of A which forces Ψ so that $A \Vdash \phi(f_1,..f_n)$.

Assume on the other hand that $A-B$ is infinite, where $B = \{x \mid N \models \phi(f_1(x),..f_n(x))\}$. Then $A-B$ is recursive (as ϕ is quantifier free), and $\Psi(f_1(x),..f_n(x))$ holds for every x in $A-B$. By induction assumption $A-B \Vdash \Psi(f_1,..,f_n)$ so that A does not force the negation of this formula.

Finally - let ϕ be the formula $\exists \bar{z} \Psi(f_1,\ldots f_n, z_1,..z_k)$ where Ψ is a quantifier-free conjunctive formula. If $A \Vdash \phi$ then there are recursive functions $g_1,..g_k$ such that $A \Vdash \Psi(f_1,..f_n, g_1,..g_k)$. Ψ is quantifier free so that $\Psi(f_1(x),..f_n(x), g_1(x),..g_k(x))$ holds a.e. in A . It follows immediately that $\phi(f_1(x),..f_n(x))$ holds a.e. in A .

If, in the other hand, $\phi(f_1(x),\ldots,f_n(x))$ holds a.e. in A, then it holds for every x in a subset A' of A which is recursive and contains all but a finite number of elements of A. We define the function

$$h(x) = \begin{cases} \mu y[\Psi(f_1(x),\ldots f_n(x),\ (y)_1,\ldots(y)_x)] & \text{if } x\varepsilon A' \\ 1 & \text{otherwise.} \end{cases}$$

$(y)_i$ is the highest power to which the i'^{th} prime P_i divides y).

As Ψ is quantifier free and A' is recursive h is a recursive function. Let h_i be the function $h_i(x) = (h(x))_i$ then for every x in A', $\psi(f_y(x),..f_n(x),\ h_1(x),..h_k(x))$ holds. By the first part of the proof $A' \Vdash \Psi(f_1,..,f_n,h_1,..h_k)$. Thus $A' \Vdash \phi(f_1,..,f_n)$ and as mentioned in 2.2 also $A \Vdash \phi(f_1..,f_n)$.

2.11 Theorem:

Let F be a complete sequence and let B be an r.e. set. Then there is a set A in F such that either $A-B$ is finite or $A \cap B$ is finite.

Proof:

By 1.2 B may be described by an existential formula $\phi(x)$ and clearly the matrix may be assumed to be in a conjunctive form so that Lemma 2.1o applies to ϕ. F is complete and it must contain a set A such that $A \Vdash \phi(I)$ or $A \Vdash \sim\phi(I)$ where I is the identify function. In the first case $A-B$ is finite and in the second one $A \cap B$ is finite as it is an r.e. set and every infinite r.e. set includes an infinite recursive set which would force $\phi(I)$.

2.12 Corollaries:

Let F be a complete sequence

i) F is the base for a maximal filter. Moreover - F is the base for a maximal filter, even in the class of r.e. sets.

ii) $R/F \models T_{\pi_2}$

iii) R/F is existentially complete in the class Σ of submodels of models of arithmetic.

Proof:

i) is just a restatement of theorem 2.11. From this (iii) and (ii) follow by 1.5 and 1.4

More properties of the generic models will follow from the next lemma.

2.13 Lemma:

If g is a recursive permutation of N and $g(A) = B$ then $B \Vdash \phi(f_1,..f_n)$ iff $A \Vdash \phi(f_1 \circ g,...,f_n \circ g)$.

Proof:

Let ϕ be atomic and $B \Vdash \phi(f_1,..f_n)$. Then $\phi(f_1(x),..f_n(x))$ holds a.e. in B and since g is one-to-one $\phi(f_1(g(x)),.. ..f_n(g(x)))$ holds a.e. in A so that $A \Vdash \phi(f_1 o g,..f_n o g)$. If, on the other hand, $\phi(f_1(g(x)),...f_n(g(x)))$ holds a.e. in A then $\phi(f_1(x),..f_n(x))$ holds a.e. in B since g is onto B . For conjunctions and disjunctions the assertion follows directly from the induction assumption.
Assume next that $B \Vdash \exists x\phi(f_1,..f_n,x)$. Then there is a function h such that $B \Vdash \phi(f_1,..f_n,h)$. By the induction assumption $A \Vdash \phi(f_1 o g,..f_n o g, h o g)$ and $A \Vdash \exists x\phi(f_1 o g,..f_n ,x)$. And if $A \Vdash \exists x\phi(f_1 o g,..f_n o g,x)$ then there is a function h such that $A \Vdash \phi(f_1 o g,..f_n o g,h)$. As g is a permutation $h o g^{-1} = h'$ is a recursive function and $A \Vdash \phi(f_1 o g, f_n o g, h' o g)$. Again, by the induction assumption $B \Vdash \phi(f_1,..f_n,h')$ and $B \Vdash \exists x\phi(f_1,..f_n,x)$.

Finally, if not $A \Vdash \sim\phi(f_1 o g,..f_n o g)$ then there is an infinite recursive subset A' such that $A' \Vdash \phi(f_1 o g,..f_n o g)$. $g(A')$ is recursive and by assumption $g(A') \Vdash \phi(f_1,..f_n)$. But $g(A') \subset B$ so it is not true that $B \Vdash \sim\phi(f_1,..,f_n)$. A similar argument proves the other direction.

2.14 Theorem:

All the generic models are elementarily equivalent.

Proof:

In view of 2.8 we have to show that for every statement ϕ, $N \Vdash \sim\phi$ or $N \Vdash^{*} \phi$. Assume that not $N \Vdash \sim\phi$. Then there is an infinite recursive set A such that $A \Vdash \phi$. Let B be an infinite recursive subset of A such that its complement is infinite. Then $B \Vdash \phi$. We conclude from Lemma 2.13 that for every infinite recursive set B' with an infinite comlement $B' \Vdash \phi$, as any two such sets can be mapped on each other by a recursive permutation. Since any (infinite) recursive set includes such a set we have $N \Vdash^{*} \phi$.

Among the generic models many properties can be described in first order language. Thus we can prove structural theorems by constructing an example and using the fact that all the models are elementarily equivalent. We use this method to prove theorem 2.17.

2.15 Lemma:

There exists a complete sequence (filter) F such that every recursive function is either constant or monotone on some set of the sequence.

Proof:

We order all the formulas with parameters, $\langle\phi_i\rangle_{i<\omega}$ and all the recursive functions $\langle f_i\rangle_{i<\omega}$. F is constructed by induction: Assume that for every $i \leq n$ $A_n \Vdash \phi_i$ or $A_n \Vdash \sim\phi_i$ and that A_n satisfies the requirement for every f_i with $i \leq n$. First we extend the property to f_{n+1}: if A_n includes an infinite subset on which f_{n+1} is constant we make it A_n'. Otherwise the following definition describes a(total) monotone function:

$$h(0) = \mu x[x \varepsilon A_n] \;,\; h(n+1) = \mu x[x \varepsilon A_n \wedge h(n) < x \wedge f_{n+1}(h(n) < f(x)]$$

Let now A_n' be the range of h, then A_n' is a recursive subset of A_n on which f_{n+1} is monotone. To conclude the induction step let A_{n+1} be any subset of A'_n such that $A_{n+1} \Vdash \phi_{n+1}$ or $A_{n+1} \Vdash \sim\phi_{n+1}$. Clearly, this construction proves the lemma.

2.16 Let now F be the filter which was constructed in lemma 2.15 and let R/F by the corresponding generic model.

a) For every recursive function f, which is not constant on any set of F, there is a set $A \varepsilon F$ such that $f(B)$ is recursive for every $B \varepsilon F$ such that $B \subset A$. Moreover - there is a recursive permutation g such that $g(x) = f(x)$ for every $x \varepsilon A$.

b) For f and g, as in part (a), let G be the filter

$$G = \{B \varepsilon R \mid f^{-1}(B) \varepsilon F\} = \{B \varepsilon R \mid g^{-1}(B) \varepsilon F\}$$

It follows easily from 2.13 that G is a complete filter and R/G is a generic model. It is also easy to see that every recursive function h is either constant or monotone on some set of G (because $h \circ g$ has this property for F).

c) $R/F \Vdash$ "$g^{-1}(|f|) = |I|$" (see 1.2). Thus the minimal model of T_{π_2} which is generated by $[f]$ is R/F, as it is closed under recursive functions. But this minimal model is just R/G (1.2).

d) We conclude that R/F is generated by any one of its non standard elements and that the recursive type of every such element is a complete sequence, which again has the properties described in lemma 2.15.

2.17 <u>Theorem:</u>

a) Every generic model is minimal (does not include a non trivial submodel of T_{π_2})

b) Every nonstandard element realizes a recursive type which is a complete sequence.

(Thus the property of being a generic model does not depend on the choice of the filter by which it is constructed.)

<u>Proof:</u>

Using the enumerations $F(i,x,y)$ and $V(i,x)$ of the partial recursive functions and of r.e. sets and the predicate $N(x)$ which defines N the following becomes a first order predicate which holds in R/F by 2.16 (b), and has the same meaning in every generic model (compare with 1.8).

"For every non-standard element x and for every recursive function there is a recursive set which contains x and on which the function is either constant or monotone".

By 2.14 this is true in every generic model so that all of 2.16 holds there, and, in particular 2.16 (d).

3. Generic r.e. ultrapowers.

Let C be the set of partial recursive functions and S the set of infinite r.e. sets. The definition of $A \Vdash \phi(f_1,..f_n)$ follows closely that of forcing for recursive functions, except that at any stage one adds the requirement that all the functions which are mentioned are defined a.e. on A .

For a complete sequence F we require that for every formula $\phi(\bar{x})$ there exists a set A in F such that $A \Vdash \phi(f_1,..f_n)$ or $A \Vdash \sim\phi(f_1,..f_n)$ or $f_i(x)$ is almost nowhere defined on A for some $i \leq n$.

2.2-2.12 follow easily and 2.13 can be strengthened with a similar proof to

3.1 Lemma:

If g(x) is a partial recursive function which is defined for every element of A and is one-to-one there, and if $g(A) = B$ then $B \Vdash \phi(f_1,..f_n)$ iff $A \Vdash \phi(f_1 o g,..f_n o g)$.

Thus we have again (like 2.14)

3.2 Theorem:

All the r.e. generic models are elementarily equivalent.

Again we use a particular construction to obtain a property which is expressible in the first order language (theorem 3.5)

3.3 Lemma:

There is a complete sequence F all of whose members are recursive sets such that every recursive function is either constant or monotone on some set in F .

Proof:

The proof is exactly like 2.15 with one modification: After choosing a set A_{n+1} which forces either ϕ_{n+1} or $\sim\phi_{n+1}$ one replaces A_{n+1} (if necessary) by an infinite recursive subset.

3.4 For the sequence F of lemma 3.3 we have

a) R/F = C/F (1.5)

b) For every element $[f]$ in C/F the existential type of $[f]$ (viewed as an r.e. filter) has a base by recursive sets:

Assume that $\phi(x)$ is existential and $C/F \models \phi([f])$ f can be assumed to be a recursive permutation and $\{x|N \models \phi(f(x))\} = A$ is in F(by 2.1o for r.e. forcing). Thus $f(A) \subset \{y|N \models \phi(y)\}$, f(A) is recursive and since "$[I] \varepsilon A$" we have also "$f([I] \varepsilon f(A)$" which means that $[f]$ is in f(A).

3.5 Theorem:

If G is a complete r.e. filter the recursive sets in G are a base for the filter.

Proof:

By 3.4 there is a generic ultrapower that satisfies: "For every element the existential type has a base by recursive set". The statement is expressible in the first order language (compare again with 1.8) and means the same thing in C/G . As all the generic models are elementary equivalent the assertion for G follows.

To show that r.e. generic models and recursive generic models are the same, we need now only the following lemma.

3.6 Lemma:

Let A be a recursive set and $f_1,\ldots f_n$ recursive functions. Denote by $\Vdash_1$ forcing with respect to recursive sets and by $\Vdash_2$ forcing with respect to r.e. sets. Then $A \Vdash_1 \phi(f_1,..f_n)$ iff $A \Vdash_2 \phi(f_1,..f_n)$.

Proof:

The assertion holds for atomic formulas by definition and follows from the induction hypothesis for conjunctions and disjunctions.

Assume that $\phi = \exists x\psi(x)$ (we omit the parameters) and $A \Vdash_2 \phi$. Then for some partial recursive function g $A \Vdash_2 \psi(g)$. G can be replaced by a total function g' which equals to g on A and $A \Vdash_2 \psi(g')$. By the induction assumption $A \Vdash_1 \psi(g')$ so that $A \Vdash_1 \phi$. The opposite direction is easy. Finally, assume that not $A \Vdash_2 \sim\phi$. Then for some r.e. subset $B \subset A$, $B \Vdash_2 \phi$. Let C be an infinite recursive subset of B . $C \Vdash_2 \phi$ and by the induction assumption $C \Vdash_1 \phi$. As C is a subset of A it is not the case that $A \Vdash_1 \sim\phi$. Again, the opposite direction is easy.

3.7 Theorem:

Every recursive generic model is an r.e. generic model and every r.e. model is a recursive generic model.

Proof:

Let F be a complete r.e. filter. By 3.5, F has a base F' which is composed of recursive sets and for every formula ϕ with recursive functions as parameters there is a set $A \varepsilon F'$ such that $A \Vdash_2 \phi$ or $A \Vdash_2 \sim\phi$. By lemma 3.6 $A \Vdash_1 \phi$ or $A \Vdash_1 \sim\phi$.

Thus F' is a complete recursive sequence and $C/F = R/F'$.

Let now F be a complete recursive filter. By 2.12 F is also a maximal r.e. filter (base) . Let $\phi(f_1,..f_n)$ be a formula such that $f_1,..f_n$ are partial recursive functions, and assume that the domains of these functions intersect (by an infinite set) every element of F . By the maximality of F all these domains are in the r.e.filter generated by F and their intersection includes a recursive set A in F . On A these functions equal to total functions $f'_1,..f'_n$ and for some $B \subset A$, $B \varepsilon F$ and $B \Vdash_1 \phi(f'_1,..f'_n)$. By lemma 3.6 $B \Vdash_2 \phi(f'_1,..f'_n)$ and also $B \Vdash_2 \phi(f_1,..,f_n)$. Thus F is a complete r.e. sequence and again $C/F = R/F$.

4. Finitely generic models.

Let c be a new constant and let T be the theory $T(N) \cup \{c \neq n \mid n \in N\}$.

T is countable and has finitely generic models. (All the properties of finitely generic models that we use appear in sections 3 of [1]). Let M be such a model. Since $M \models T_{\forall}$ M is in the class Σ of submodels of models of $T(N)$ (if we ignore the new constant symbol). Moreover - M is existentially complete in Σ as it is existentially complete in the class of models of $T_{\forall}$ and every model of $T(N)$ which extends M becomes a model of T which extends M once c is determined in M. Clearly the notion of existential completeness is not effected by adding a name to the language.

Let $N[c]$ be the closure of c in M under partial recursive functions. Then $N[c]$ is existentially complete (1.4) in M, $N[c]$ itself is finitely generic and $N[c] \prec M$.

Let $N(x)$ be the formula that defines N in the existentially complete models and let $F(i,x,y)$ be the graph of the universal partial recursive function. Then $N[c] \models \forall y \exists i (N(i) \wedge F(i,c,y))$. The same statement holds in M. As $F(i,x,y)$ is existential and (at most) one valued we have $N[c] = M$. Finally - $N[c]$ is isomorphic to C/F where F is the r.e. ultrafilter (the existential type) which is realized by c in M(1.4).

We conclude from the discussion above:

4.1 **Theorem:**

Every finitely generic model for T is an r.e. ultrapower.

Theorem 4.1 implies that r.e. generic models and finitely generic models are the same (theorem 4.5). The proof however is quite tedious because of the differences in the definitions. To simplify the notations a little we shall not distinguish between a condition P and the conjunction of the formulas in P.

a) With every condition $P(\bar{a},c)$ we associate the r.e. set A_p, which is described by the formula $\exists \bar{y} P(\bar{y},x)$.

b) For every formula $\phi(\bar{x},y)$ we denote by $\phi(f_1(c),..f_n(c),c)$ the formula $\exists \bar{z}[\phi(\bar{z},c) \wedge f_1(c)=z_1 \wedge \ldots \wedge f_n(c)=z_n]$ where $f_i(c)=z_i$ is an arbitrary existential formula that describes f_i.

The following are easily verified:

c) $P(\bar{a},c)$ is a condition iff A_P is infinite, iff $\exists \bar{x} P(\bar{x},[I])$ holds in some nontrivial r.e. ultrapower.

d) If $Q \Vdash P$ then Q extends P and if Q extends P then $A_Q \subset A_P$.

4.2 Lemma:

Let B be an infinite r.e. subset of A_P . Then there exists a condition Q such that $Q \supset P$ and $A_Q \subset B$.

Proof:

Let P be the condition $P(\bar{a},c)$ where $\bar{a}$ is a string of new constants. B may be described by a predicate

$$\beta(x) = \exists \bar{z} \beta_1(\bar{z},x) \bigvee .. \bigvee \exists \bar{z} \beta_n(\bar{z},x)$$

where for $i=1,...,n$ β_i is a conjunction of atomic formulas and negations of atomic formulas. Each of the disjuncts describes a subset of B and at least one of them (say the first) describes an infinite r.e. subset B_1 . As $B_1 \subset A_P$ every r.e. ultrapower which has B_1 in its generating filter satisfies $\exists \bar{x} \beta_1(\bar{x},[I])$ and $\exists \bar{x} P(\bar{x},[I])$. Hence the following is a condition $Q = \beta_1(\bar{b},c) \bigwedge P(\bar{a},c)$. It is easy to see that $Q \supset P$ and $A_Q \subset B$.

4.3 Theorem:

For every condition P , for every formula $\phi(\bar{x})$ and for all partial recursive functions $f_1,..f_n$ the following are equivalent:

i) There is a condition Q which extends P such that

$$Q \Vdash \phi(f_1(c),...f_n(c)$$

ii) There is an infinite r.e. subset B of A_P such that

$$B \Vdash \phi(f_1,..f_n)$$

Proof:

Assume first that ϕ is atomic and that Q is an extensions of

P such that $Q \Vdash \phi(f_1(c),..f_n(c))$. Let M be a finitely generic model which is constructed from a complete sequence that contains Q . Then $M \models \phi(f_1(c),..f_n(c)$ and $M \models Q(c)$. But by 4.1 M is the generic ultrapower C/F where F is the existential type of c . Thus $M \models \phi\,[f_1],..[f_n])$ and the set $A = \{x | N \models \phi(f_1(x),..$ $..f_n(x))\}$ is in F . A_Q is also in F and so is the set $B = A \cap A_Q$. Now $B \subset A_p$, B is infinite like all the sets in F and $B \Vdash \phi(f_1,..,f_n)$ by the definition of forcing for atomic formulas, since $B \subset A$.

Assume, on the other hand, that $A_Q \Vdash \phi(f_1,..f_n)$ for some Q which extends P (note that whenever we assume that condition (ii) of the theorem holds, it may be stated in this form using lemma 4.2. This will be done throughout the proof). Let M be a finitely generic model obtained from a complete sequence that contains Q . Again by 4.1 M is an r.e. ultrapower which satisfies $Q(\bar{a},c)$. As $\phi(f_1(x),..f_n(x))$ holds a.e. in A_Q we have also $M \models (f_1(c),..f_n(c))$. But M is generic so that there exists a condition $Q'(\bar{a},c,\bar{b})$ in the diagram of M such that $Q' \Vdash Q \wedge \phi(f_1(c),..f_n(c))$. In particular $Q \subset Q'$ and $Q' \Vdash \phi(f_1(c),..$ $..f_n(c))$.

Next, we assume that $Q \supset P$ and $Q \Vdash \sim\phi$ (from now on we omit the parameters). Q has no extension that forces ϕ and by assumption A_Q does not include an infinite r.e. set that forces ϕ . Hence $A_Q \Vdash \sim\phi$.

If, on the other hand, $A_Q \Vdash \sim\phi$ for some $Q \supset P$ (using again 4.2) then $Q \Vdash \sim\phi$ for similar reasons.

Next we assume that for some condition Q , $Q \supset P$ and $Q \Vdash \phi \wedge \Psi$. Hence $Q \Vdash \phi$ and $Q \Vdash \Psi$. By assumption there is a condition Q' such that $Q' \supset Q$ and $A_{Q'} \Vdash \phi$. But we have also $Q' \Vdash \Psi$ so that there is also a condition Q'' such that $Q'' \supset Q'$ and $A_{Q''} = Q$ and $A_{Q''} \Vdash \Psi$. Clearly $A_{Q''} \Vdash \phi \wedge \Psi$. The other direction is proved similarly.

For $\phi \vee \Psi$ the proof is similar.

Finally assume that $Q \supset P$ and $Q \Vdash \exists x\phi(x)$. Then $Q \Vdash \phi(b)$ for some b . Let M be a generic model with Q in its diagram.

Then $M \models \phi(b)$. But M is an r.e. ultrapower so that there is a partial recursive function h such that $M \models \phi(h(c)) \wedge Q$. Thus there is a condition Q' in the diagram which forces $\phi(h(c))$ and extends Q. By induction assumption there is also a set $B \subset A_Q$ such that $B \Vdash \phi(h)$, and therefore $B \Vdash \exists x\phi(x)$.

Assume, on the other hand, that $Q \supset P$ and $A_Q \Vdash \exists x\phi(x)$. Then $A_Q \Vdash \phi(h)$ for some partial function h. By assumption $Q' \Vdash \phi(h(c))$ for some Q' which extends Q. Let M be a generic model with Q' in its diagram. Then $M \models \exists x\phi(x) \bigwedge Q$ and there exists a condition Q'' in the diagram which forces this statement. Therefore $Q'' \supset Q$ and $Q'' \Vdash \exists x\phi(x)$.

4.4 Corollary:

$P \Vdash^* \phi(f_1(c),..f_n(c))$ iff $A_p \Vdash^* \phi(f_1,..,f_n)$

Proof:

The left assertion claims that no extensions of P forces $\sim\phi$ which is equivalent by 4.3 to the claim that no infinite r.e. subset of A_p forces $\sim\phi$. But this is exactly the definition of $A_p \Vdash^* \phi$.

4.5 Theorem:

C/F is r.e. generic iff it is finitely generic (when c is interpreted as the class $[I]$).
(Note that since, in both cases, every generic model is an r.e. ultrapower, the theorem claims that all the generic models are the same).

Proof:

Assume first that C/F is finitely generic. We shall prove that if $C/F \models \phi([f_1],..[f_n])$ then there is a set $A \varepsilon F$ such that $A \Vdash^* \phi(f_1,..f_n)$. This suffices by 2.9 (for r.e. forcing). But if $C/F \models \phi([f_1],..[f_n])$ then $C/F \models \phi(f_1(c),..f_n(c))$. Therefore, there is a condition P in the diagram such that $P \Vdash \phi(f_1(c),..$ $..f_n(c))$. By 4.4 $A_p \Vdash^* \phi(f_1,..,f_n)$ and clearly $A_p \varepsilon F$.

Assume, on the other hand, that F is a complete r.e. filter. We shall prove that every formula which is satisfied in C/F is weakly (finitely) forced by a condition in the diagram. This suffices by the analogue of 2.9 for finite forcing (which is proved similarly). Assume that $C/F \models \phi([f_1],..[f_n])$. There is a set $B \varepsilon F$ such that $B \Vdash \phi(f_1,..f_n)$. B is described by a formula $\exists \bar{z}\beta_1(\bar{z},x) \bigvee .. \bigvee \exists \bar{z}\beta_k(\bar{z},x)$ where for $i=1,..n$ β_1 is a conjunction of atomic formulas and negations of atomic formulas. Each disjunct describes an r.e. set and their union is B. Therefore, one of these sets (say B_1-which is described by $\exists \bar{z}\beta_1(\bar{z},x)$ is in F. We conclude that $C/F \models \exists \bar{z}\beta_1(\bar{z},c)$ and there are elements $g_1 \dots g_m$ such that $C/F \models \beta_1([g_1],\dots,[g_m],c)$. We denote this statement by P so that P is a condition in the diagram. As $A_p = B_1$ we have by lemma 4.4 $P \Vdash^{*} \phi(f_1(c),..$ $..f_n(c))$.

In C/F we have also "$f_1(c) = [f_1]$" $\bigwedge .. \bigwedge$ "$f_n(c) = [f_n]$". This is an existential formula $\exists \bar{z}\alpha(\bar{z},[f_1],..[f_n],c)$. Therefore there are $h_1,\dots h_r$ such that

$$C/F \models \alpha([h_1],..[h_r],\ [f_1],..[f_n],\ c)$$

Replacing (if necessary) α by one of its disjuncts that holds in C/F we can assume that α is a condition Q in the diagram of C/F. Clearly

$$P \wedge Q \Vdash^{*} \phi(f_1(c),..f_n(c)) \wedge \exists \bar{z}\alpha(\bar{z},[f_1],..[f_n],c).$$

To conclude the proof we show that $P \wedge Q \Vdash^{*} \phi([f_1],..[f_n])$ i.e. - that ϕ holds in every generic model that has $P \wedge Q$ in its diagram. But this is clearly the case since every generic model is existential complete and $\exists \bar{z}\alpha(\bar{z},a_1,..a_n)$ actually means there that $f_1(c) = a_1 \bigwedge \dots \bigwedge f_n(c) = a_n$.

5. Properties of the generic models

Using three different kinds of forcing we arrived at the same class of generic models. We list some properties of the class and the models in it. Most of the proofs have already been established.

5.1 Every generic model is a recursive ultrapower, an r.e. ultrapower, a model of T_{π_2} and existentially complete in the class of submodels of models of Arithmetic.

5.2 In a generic model the existential type of any element has a base of recursive sets. Every non-standard element generates the whole model (as the closure under recursive functions). The genericity of the model does not depend on the choice of the generator.

5.3 The generic models are minimal models of T_{π_2} - they do not properly include any non-standard model of T_{π_2}.

5.4 a) All the generic models are elementarily equivalent.
b) Every existentially complete model and every recursive ultrapower which is elementarily equivalent to a generic model is also generic.

5.5 There are $2^{\aleph_0}$ generic models which are pairwise not isomorphic.

Proof:

5.1 is 2.12, and 5.2 follows from 2.17(b) and 2.12.
5.3 is 2.17(a). 5.4(a) is 2.14. To prove 5.4(b), we shall need the following:

5.6 For every formula $\phi(x)$ the following predicate is arithmetical:

$\text{Force}_\phi\ (i,a_1,..,a_n) \equiv$ "i is the Goedel number of a recursive set which forces $\phi(f_1,..,f_n)$ where for i=1...n a_1 is the Goedel number of f_i".

(For atomic formulas this has the form:
$\text{Rec}\ (i) \wedge \exists y \forall x \forall z_1,\ldots,z_n [(x>y \wedge V(i,x) \wedge F(a_1,x,z_1) \wedge \ldots F(a_r,x,z)) \rightarrow \phi(z_1,\ldots z_n)]$.

For more complex formulas the proof is by induction. The notations are explained in 1.8).

Let now $N(x)$ be a formula that describes N in the given model. As M is elementarily equivalent to a generic model it satisfies for every formula $\phi(\bar{x})$

$$\forall x_1 \ldots x_n [\sim N(x_1) \wedge \phi(x_1,\ldots x_n) \rightarrow \exists a_2 \ldots a_n (\bigwedge_{i=2}^{n} \mathrm{rec}(a_i)) \wedge (\bigwedge_{i=2}^{n} N(a_i))$$

$$\wedge (\bigwedge_{i=2}^{n} F(a_1,x_1,x_i)) \wedge \exists_i (N(i) \wedge V(i,x_1) \wedge \mathrm{Force}_{\phi}(i,j,a_2 \ldots a_n))]$$

(here j is a fixed index of the identify function).

The first conjuncts imply among other things that M is a minimal recursive ultrapower ($M \neq N$ since $M \models \sim \forall x N(x)$). Together with the last conjunct we can deduce easily that satisfaction in M implies weak forcing by some set in the recursive type of any (arbitrary) non-standard element. Thus M is generic.

To prove 5.5 one constructs simultaneously $2^{\aleph_0}$ complete sequences of recursive set by splitting every other step the recursive sets into two disjoint recursive sets. This yields $2^{\aleph_0}$ recursive types which are realized in generic models. Since all of them are countable, there must be $2^{\aleph_0}$ non isomorphic models to realize all the types.

5.7 From the point of view of finite forcing the following features are of interest.

a) The addition of a new constant leaves us with generic models that are elementarily equivalent in the original language. Genericity of a model does not depend on the interpretation of the new constant.

b) All the generic models are countable. Only the standard model of $T(N)$ can be extended to a generic model (since the generic models are minimal and they are not models of $T(N)$. As $T = T(N) \cup \{c \neq n \mid n \in N\}$, no model of T can be embedded in a generic model for T.

c) Every existentially complete model which is elementarily equivalent to a generic model is generic (but not all the existentially complete models are generic).

5.8 The class of generic models seems to be interesting also from the point of view of the theory of models of arithmetic. However, the only concrete result that we can point out is that there are elementarily equivalent recursive ultrapowers (homomorphic images of the semiring of the recursive functions) which are not isomorphic. In the previous discussions of this subject [2] and [4] there were given sufficient conditions under which elementary equivalence implies isomorphism.

The theory T^f : The new constant c does not play an essential role in the description of T^f. More precisely - let T' be the set of statements which are deducible from T^f and which do not contain any occurence of c. Then:

5.9 $$T^f \equiv T' \cup \{\sim N(c)\}$$

Proof:

Assume that $T^f \vdash \phi(c)$. Then $\phi(c)$ holds in every generic model M. Since every non-standard element of M can serve as generator for M and still keep M generic we have $M \models \forall x(\sim N(x) \to \phi(x))$. Therefore $T' \vdash \forall x(\sim N(x) \to \phi(x))$ and $T' \cup \{\sim N(c)\} \vdash \phi(c)$.

On the other hand T' itself is very complicated. We have $T(N) \vdash \phi$ iff $T' \vdash \phi^N$ (the relativization of ϕ to $N(x)$), so that T' is not an arithmetical set. It seems that the interesting question that remains open is not how to axiomatize T^f but what are necessary and sufficient model theoretic conditions for a model to be generic. Some necessary conditions which look relevant appear in 5.1-5.3.

6. Additional remarks

In the paper we used two ideas that are much more general than in our presentation: Adding constants and requirements to a theory, and constructing generic filters. Clearly, in the more general case there is no relation between the two things.

The addition of constants: This may have many variations (like adding a diagram of a model to T). To stay close to our work we shall consider only the case of complete theory $T(M)$ of a model (e.g. a generic model of a theory which has no proper extension that is generic). Unlike arithmetic, there is no reason to think that the addition of one new constant will yield a complete theory in $L(M)$. However:

6.1 If $T = T(M) \cup \{c_1 \neq a \mid i < \omega \ , \ a \in M\}$ then all the generic models of T are elementarily equivalent in $L(M)$. This follows immediately from the fact that $T(M)$ has the joint embedding property and from the following lemma:

6.2 If $P(\bar{a}, c_{i_1}, .. c_{i_n}) \Vdash \phi(\bar{a}, c_{i_1}, .. c_{i_n})$ then

$P(\bar{a}', c_{j_1}, .. c_{j_1}) \Vdash \phi(\bar{a}', c_{j_1}, .. c_{j_n})$. (the proof is by induction).

From this, one concludes that if ϕ is a statement in $L(M)$ then $\phi \Vdash^* \phi$ or $\phi \Vdash^* \sim\phi$.

Additional requirements on the new constants can be thought of which preserve a modified version of 6.2. E.g. - if we have order in the language we can order the new constants (if it is consistent) in the order $\omega^* + \omega$.

Generic filters: Let $\widetilde{F}$ be a countable set of functions from I to a model M. Let $\mathfrak{A}$ be a collection of sets of I. For atomic formulas define $A \Vdash \phi(f_1 .. f_n)$ if $M \models \phi(f_1(i), .. f_n(i))$ for every i in A. Conjunction, disjunction and the existential quantifier are treated as expected and $A \Vdash \sim\phi$ if no subset of A in $\mathfrak{A}$ forces ϕ. This construction yields models for which satisfaction and being forced by a set in the sequence coincide.

The only additional case that we checked brought about negative results.

6.3 If $\mathfrak{A}$ is the set of arithmetical sets and $\tilde{F}$ the class of arithmetical functions then every arithmetical ultrapower is generic and conversely. For this we use the following lemma

6.4 If ϕ does not contain conjunctions then $A \Vdash \phi(f_1..,f_n)$ iff $\phi(f_1(i),..f_n(i))$ holds for every i in A.

It follows that every complete filter is maximal and that for every arithmetical ultrapower satisfaction implies weak forcing.

References

[1] J. Barwise and A. Robinson: Completing theories by forcing, Ann. of Math. Logic 2 (197o) 119-142.

[2] J. Hirschfeld: Models of Arithmetic and recursive functions, Israel J. of Math. (to appear).

[3] J. Hirschfeld and W.H. Wheeler: Forcing, Arithmetic and division rings. Lecture Notes in Mathematics, Vol. 454. Springer Verlag (1975).

[4] M. Lerman: Recursive functions modulo Co-r-maximal sets. Trans. of the A.M.S. 148 (197o) 429-444.

[5] Yu. V. Matijasevic: Diophantine representation of recursive enumerable predicates. Proceedings of the second Scandinavian Logic symposium. North Holland (1971).

[6] S. Shelah: A note on model complete models and generic models. Proceeding of the A.M.S. 34 (1972) 5o9-514.

DENSE EMBEDDINGS I: A THEOREM OF ROBINSON IN A GENERAL SETTING*

Angus Macintyre

Yale University

TO THE MEMORY OF ABRAHAM ROBINSON

0. INTRODUCTION

In 1959 Robinson published "Solution of a Problem of Tarski". In this paper [12] he proved the completeness of the theory of a real closed field with a distinguished dense proper subfield. The problem of Tarski [16] had been to prove the decidability of the theory of the field of real numbers with a predicate distinguishing the field of real algebraic numbers. This problem is easily solved as a by-product of Robinson's method.

Robinson's technique in this paper is rather special, and no generalization is readily apparent. However, we know from conversations with Robinson that he was interested in finding generalizations. In 1966 we found a rather more direct proof [7] using saturated models. About the same time, P. J. Cohen had a proof using quantifier elimination. In 1968 [8] we extended Robinson's result to p-adic fields, using our 1966 method. At the same time, we were aware of various examples blocking a comprehensive generalization.

Lately, as a result of helpful and enjoyable conversations with Peter Winkler at Yale, we returned to this theme. In the present paper we put Robinson's result into a general setting, using mainly concepts that have become standard in connection with $\aleph_1$-categoricity.

---

*Partially supported by NSF - GP - 34088X

1. FORMULATION OF CONCEPTS

1.1. L will be a first order logic and T an L-theory. L_1 is a first order logic obtained from L by adding a single unary predicate A. We shall be interested in L_1-structures $\mathfrak{M}_1$ whose reduct to L is an L-structure $\mathfrak{M}$ with domain M, such that $\mathfrak{M} \models T$, and such that A denotes a subset of M which is the domain of a substructure $\mathfrak{A}$ of $\mathfrak{M}$, and $\mathfrak{A} \models T$. We shall denote this L_1-structure by $(\mathfrak{M}, \mathfrak{A})$ and refer to it as a pair of models of T.

For example, in [12] Robinson considered pairs of real closed fields.

1.2. What is a reasonable explication of the notion:
$\mathfrak{A}$ is dense in $\mathfrak{M}$?

We know exactly what we mean by this when $\mathfrak{A}$ and $\mathfrak{M}$ are real closed fields, namely that if x and y are elements of $\mathfrak{M}$, and $x < y$, then there is an element z of $\mathfrak{A}$ with $x < z < y$.

We propose the following definition. First extend L to $L^{\exists^\infty}$, with the generalized quantifier $\exists^\infty$ such that $\exists^\infty x\, \Phi(x)$ "means" that there are infinitely many x satisfying Φ. See [17]. Of course, any L-structure naturally becomes an $L^{\exists^\infty}$-structure, and we have a "natural equality"

$$(L_1)^{\exists^\infty} = (L^{\exists^\infty})_1 .$$

Let $\mathfrak{A}, \mathfrak{M}$ be L-structures with $\mathfrak{A} \subseteq \mathfrak{M}$. Let $\mathfrak{M}_1$ be $(\mathfrak{M}, \mathfrak{A})$. We say <u>$\mathfrak{A}$ is dense in $\mathfrak{M}$</u>

$$(\text{written } \mathfrak{A} \subseteq_d \mathfrak{M})$$

if $\mathfrak{M}_1$ satisfies the following $L_1^{\exists^\infty}$ - axioms:

$$\forall x_1 \cdots \forall x_n [\exists^\infty x\, \Phi(x, x_1, \cdots x_n) \longrightarrow \exists_y (A(y) \wedge \Phi(y, x_1, \cdots x_n))]$$

for each L-formula Φ.

Notice at once that there is no reason to think that for a given T the class of models $(\mathfrak{M},\mathfrak{A})$ of T with $\mathfrak{A} \subseteq_d \mathfrak{M}$ is an EC_Δ. See Appendix 1 for a counterexample.

It is an instructive exercise to show that when $\mathfrak{M}$ and $\mathfrak{A}$ are real closed fields then our notion of dense embedding agrees with the classical order-theoretic definition used in [12].

1.3. Now we come to some considerations about algebraic dependence. The concepts involved are basic to the Baldwin-Lachlan approach to $\aleph_1$-categoricity. We will use them here in a wider setting.

We refer to [2] for the notion $cl_{\mathfrak{M}}(X)$, the algebraic closure of X in $\mathfrak{M}$, and for the notion of independent set. More generally, if $\mathfrak{A} \subseteq \mathfrak{M}$, we say X is $\mathfrak{A}$-independent if no x in X is in $cl_{\mathfrak{M}}(\mathfrak{A} \cup (X-\{x\}))$. We have a corresponding notion of $\mathfrak{A}$-spanning set in $\mathfrak{M}$, and $\mathfrak{A}$-basis for $\mathfrak{M}$.

The general theory of all this is obscure. When $\mathfrak{M}$ is a model of a minimal theory [2], we have a precise analogue to the classical Steinitz theory [15] of transcendence base.

To get such an analogue, the decisive property is:

Steinitz Exchange Property for $\mathfrak{M}$:

If $y \in cl_{\mathfrak{M}}(X \cup \{x\})$, and $y \notin cl_{\mathfrak{M}}(X)$, then $x \in cl_{\mathfrak{M}}(X \cup \{y\})$.

Lemma 1 Suppose $\mathfrak{M}$ has the Steinitz Exchange Property and $\mathfrak{A} \subseteq \mathfrak{M}$. If X is a maximal $\mathfrak{A}$-independent subset of $\mathfrak{M}$ then X is an $\mathfrak{A}$-basis for $\mathfrak{M}$.

Proof Folklore.

Corollary If $\mathfrak{M}$ has Steinitz Exchange Property, and $\mathfrak{A} \subseteq \mathfrak{M}$, then $\mathfrak{M}$ has an $\mathfrak{A}$-basis.

Proof Zorn's Lemma.

Definition a) $\mathfrak{M}$ has transcendence degree κ if $\mathfrak{M}$ has a basis of cardinal κ. b) $\mathfrak{M}$ has transcendence degree κ over $\mathfrak{A}$ if there is an $\mathfrak{A}$-basis of $\mathfrak{M}$ of cardinal κ.

Notice that there is no reason to suppose that any two $\mathcal{A}$-bases of $\mathcal{M}$ have the same cardinal. However,

Lemma 2 Let $\mathcal{M}$ be an L-structure, where card(L) = λ. Let $\mathcal{A} \subseteq \mathcal{M}$, and suppose X,Y are $\mathcal{A}$-bases for $\mathcal{M}$. Suppose card(X) $\geq$ max(λ, card $\mathcal{A}$). Then card(X) = card (Y).

Proof Trivial counting argument.

We shall need also the following easy lemma.

Lemma 3 Suppose $\mathcal{M}$ has Steinitz Exchange Property. Suppose $\mathcal{A} \prec \mathcal{M}$, and X is a basis for $\mathcal{A}$, and Y is an $\mathcal{A}$-basis for $\mathcal{M}$. Then $X \cap Y = \emptyset$, and $X \cup Y$ is a basis for $\mathcal{M}$.

Proof Suppose $t \in X \cap Y$. Then $t \in \mathcal{A}$. Thus $Y - \{t\}$ is an $\mathcal{A}$-base for $\mathcal{M}$, contradiction.

Suppose $m \in \mathcal{M}$. Then $m \in cl_{\mathcal{M}}(\mathcal{A}' \cup Y')$ for some finite $\mathcal{A}', Y'$ with $\mathcal{A}' \subseteq \mathcal{A}$, $Y' \subseteq Y$. But $\mathcal{A}' \subseteq cl_{\mathcal{A}}(X)$, so $\mathcal{A}' \subseteq cl_{\mathcal{M}}(X)$, since $\mathcal{A} \prec \mathcal{M}$. Therefore, $m \in cl_{\mathcal{M}}(X \cup Y')$, so $m \in cl_{\mathcal{M}}(X \cup Y)$. Therefore, $X \cup Y$ spans $\mathcal{M}$. If $X \cup Y$ is not independent then either

a) $\exists x_0 \in X$, $x_0 \in cl_{\mathcal{M}}((X - \{x_0\}) \cup Y)$ or

b) $\exists y_0 \in Y$, $y_0 \in cl_{\mathcal{M}}(X \cup (Y - \{y_0\}))$.

In Case (b), Y is not $\mathcal{A}$-independent, contradiction.

In Case (a), select minimal finite $Y_0 \subseteq Y$ such that $x_0 \in cl_{\mathcal{M}}((X - \{x_0\}) \cup Y_0)$. If $Y_0 = \emptyset$, $x_0 \in cl_{\mathcal{M}}(X - \{x_0\})$, so since $\mathcal{A} \prec \mathcal{M}$, $x_0 \in cl_{\mathcal{A}}(X - \{x_0\})$, contradicting independence of X. If $Y_0 \neq \emptyset$, then the Exchange Property for $\mathcal{M}$ gives $y_1 \in Y_0$ such that $y_1 \in cl_{\mathcal{M}}(X \cup (Y_0 - \{y_1\}))$, contradicting $\mathcal{A}$-independence of Y.

Therefore $X \cup Y$ is independent.

Notation a) t.d. $\mathcal{M} = \varkappa$ if $\mathcal{M}$ has transcendence degree $\varkappa$.

b) t.d. $\mathcal{M}|\mathcal{A} = \varkappa$ if $\mathcal{M}$ has transcendence degree $\varkappa$ over $\mathcal{A}$.

Now we come to:

Assumption 1

a) All models of T have the Steinitz Exchange Property.

b) If $\mathfrak{M} \models T$, then $cl_{\mathfrak{M}}(X) \prec \mathfrak{M}$ if X is a subset of $\mathfrak{M}$.

c) If $\mathfrak{A}_i \subseteq \mathfrak{M}_i$ (i = 1,2), and $\mathfrak{M}_i \models T$ (i = 1,2), and $f: \mathfrak{A}_1 \cong \mathfrak{A}_2$ is an isomorphism then f extends to an isomorphism

$$\bar{f}: cl_{\mathfrak{M}_1}(\mathfrak{A}_1) \cong cl_{\mathfrak{M}_2}(\mathfrak{A}_2).$$

We propose to call a theory T satisfying Assumption 1 a Steinitz theory.

Lemma 4 If T satisfies 1(b) and 1(c) then T admits elimination of quantifiers.

Proof We shall apply the criterion of Shoenfield [14]. It is enough to prove:

Suppose $\mathfrak{M}_1, \mathfrak{M}_2 \models T$ and $\mathfrak{M}_2$ is $card(L)^+$ - saturated. Suppose $\mathfrak{A}_i \subseteq \mathfrak{M}_i$, i = 1,2. Suppose $card(\mathfrak{M}_1) \leq card(L)$. Suppose $f: \mathfrak{A}_1 \cong \mathfrak{A}_2$ is an isomorphism. Then f extends to a monomorphism of $\mathfrak{M}_1$ into $\mathfrak{M}_2$.

But by 1(c), f extends to $\bar{f}: cl_{\mathfrak{M}_1}(\mathfrak{A}_1) \cong cl_{\mathfrak{M}_2}(\mathfrak{A}_2)$. By 1(b), $cl_{\mathfrak{M}_i}(\mathfrak{A}_i) \prec \mathfrak{M}_i$. But then it is trivial, using $card(L)^+$ - saturation of $\mathfrak{M}_2$ to extend $\bar{f}$ to a monomorphism $\mathfrak{M}_1 \longrightarrow \mathfrak{M}_2$.

Corollary If T satisfies 1(b) and 1(c) then T is model complete.

1.4. The next concept is that of Vaughtian pair, which is now known [2] to be of fundamental importance in $\aleph_1$-categoricity.

Recall that T has a Vaughtian pair if there are $\mathfrak{A}, \mathfrak{M} \models T$ with $\mathfrak{A} \prec \mathfrak{M}$, $\mathfrak{A} \neq \mathfrak{M}$, and some $\Phi(v) \in L(\mathfrak{A})$ such that $\Phi^{\mathfrak{A}}$ is infinite and $\Phi^{\mathfrak{A}} = \Phi^{\mathfrak{M}}$. An $\aleph_1$-categorical theory in a countable language cannot have a Vaughtian pair [2].

<u>Assumption 2</u> T has no Vaughtian pair.

We claim that Assumption 2 is satisfied for real closed fields. Now it is well-known [16] that if a set Δ is definable over the real closed field $\mathfrak{A}$ then Δ is a finite union of non-overlapping intervals (open, closed, half-open, etc.). If now this definition remains the same over $\mathfrak{M}$, where $\mathfrak{A} \subsetneqq \mathfrak{M}$, then each interval involved gets no new member added. But then by [4], each interval is a singleton, so Δ is finite.

<u>Lemma 5</u> Suppose T has no Vaughtian pairs. Then the class of pairs of models $(\mathfrak{M}, \mathfrak{A})$ of T, with $\mathfrak{A} \subseteq_d \mathfrak{M}$, is an EC_Δ, provided it is non-empty.

<u>Proof</u> (This was pointed out to me by Winkler). If T has no Vaughtian pairs, T is algebraically bounded [17, 18] by the arguments of [2]. So [17, 18] T admits elimination of the $\exists^\infty$-quantifier, in the sense that for each L-formula φ there is an $n \in \omega$ such that

$$T \models \exists^\infty v\, \varphi(v) \longleftrightarrow \exists^{\geq n} v\, \varphi(v).$$

Then of course our axioms for density are first order.

This leads to our next assumption.

<u>Assumption 3</u> There is a pair $(\mathfrak{M}, \mathfrak{A})$ of models of T with $\mathfrak{A} \subsetneqq \mathfrak{M}$ and $\mathfrak{A} \subseteq_d \mathfrak{M}$.

<u>Remark</u> This assumption does not always hold. Winkler observed that an example in [11] does not satisfy Assumption 3.

1.5. To state the next assumption we need a definition.

<u>Definition</u> a) T cannot express bounded transcendence degree if there is no model $\mathfrak{M}$ of T such that there exists an L-formula $\Phi(v_0, v_1, -- v_{n-1})$ and an integer k such that

i) $\mathfrak{M} \models \forall v_1 -- \forall v_{n-1} \exists^{\leq k} v_0\ \Phi(v_0, v_1, -- v_{n-1})$ and

ii) $\mathfrak{M} \models \forall v_0 -- \forall v_{n-1}[\bigvee_{\sigma \in S_n} \Phi^\sigma(v_0, v_1, -- v_{n-1})]$

where S_n is the group of permutations of $\{0, -- n-1\}$ and

$$\Phi^{\sigma}(v_0, v_1, -- v_{n-1}) = \Phi(v_{\sigma(0)}, v_{\sigma(1)}, - v_{\sigma(n-1)}).$$

b) T cannot express bounded relative transcendence degree if there is no pair $(\mathcal{M}, \mathcal{A})$ of models of T with $\mathcal{A} \subsetneq \mathcal{M}$ such that there exists an L-formula $\Phi(v_0, v_1, -- v_{n-1}, v_n, -- v_{n+m})$ and an integer k such that

i) $\mathcal{M} \models \forall v_1 -- \forall v_{n+m} \exists^{\leq k} v_0 \Phi(v_0, -- v_{n+m})$ and

ii) $(\mathcal{M}, \mathcal{A}) \models \forall v_0 -- v_{n-1} \exists v_n -- \exists v_{n+m}$

$$[A(v_n) \wedge -- \wedge A(v_{n+m}) \wedge \bigvee_{\sigma \in S_n} \Phi(v_{\sigma(0)} -- v_{\sigma(n-1)}, v_n -- v_{n+m})]$$

Assumption 4 a) T cannot express bounded transcendence degree.

b) T cannot express bounded relative transcendence degree.

Remark For an example where Assumption 4 does not hold, take T as the theory of infinite sets. Consider a pair $(\mathcal{M}, \mathcal{A})$ where $\mathcal{M} - \mathcal{A}$ has cardinality 1. Take $\Phi(v_0, v_1, v_2)$ as $v_0 = v_1 \vee v_0 = v_2$. Let $k = 2$, $n = 2$, $m = 0$. 4(b) is violated. (Note that 4(a) fails only if T has a finite model.)

Lemma 6 Suppose T satisfies Assumption 4.

a) Suppose $\mathcal{M} \models T$ and $\mathcal{M}$ is α-saturated, where $\alpha > \text{card}(L)$. Then for any $\beta < \alpha$ $\mathcal{M}$ has an independent subset of cardinal β.

b) Suppose $(\mathcal{M}, \mathcal{A})$ is a pair of models of T with $\mathcal{A} \subsetneq \mathcal{M}$, and $(\mathcal{M}, \mathcal{A})$ is α-saturated where $\alpha > \text{card}(L)$. Then for any $\beta < \alpha$ $\mathcal{M}$ has an $\mathcal{A}$-independent subset of cardinal β.

Proof Trivial compactness argument.

Note It is well-known [6] that if T is the theory of real closed fields then T satisfies Assumption 4.

1.6. Now we come to something rather less standard. We shall call this the Interior Condition for Types. It has an obvious topological significance, as will be seen when we verify it for real closed fields.

<u>Definition</u> Let $\tau(v_0) \in S_1(\mathcal{M})$, and let $\mathcal{M} < \mathcal{N}$. We say τ has interior in $\mathcal{N}$ if there exists $\varphi(v_0) \in L(\mathcal{N})$ such that

i) $\mathcal{N} \models \exists^\infty v_0\, \varphi(v_0)$ and

ii) for all $\Psi(v_0)$ in τ, $\mathcal{N} \models (\forall v_0)[\varphi(v_0) \longrightarrow \Psi(v_0)]$.

<u>Definition</u> T satisfies the Interior Condition for Types if for every model $\mathcal{M}$ of T and every non-principal $\tau \in S_1(\mathcal{M})$, τ has interior in some $\mathcal{N}$ with $\mathcal{M} < \mathcal{N}$.

<u>Assumption 5</u> T has the Interior Condition for Types.

Let us verify this when T is the theory of real closed fields. Let τ be a non-principal 1-type over a real closed field $\mathcal{M}$. Then τ corresponds to a Dedekind cut over $\mathcal{M}$, not realized in $\mathcal{M}$. Realize this cut in $\mathcal{N} > \mathcal{M}$, where $\mathcal{N}$ is (card $\mathcal{M}$)-saturated. Choose α, β in $\mathcal{N}$ in this cut, $\alpha < \beta$. This is possible by saturation. Now take $\varphi(v_0)$ as $\alpha < v_0 \wedge v_0 < \beta$.

1.7. <u>Definition</u> Suppose X is a subset of $\mathcal{M}$. X is dense in $\mathcal{M}$ if $X \cap \varphi^{\mathcal{M}} \neq \emptyset$ for each infinite $\varphi^{\mathcal{M}}$, where φ is an $L(\mathcal{M})$-formula.

Our final assumption concerns transitivity of density.

<u>Assumption 6</u> If Y is dense in $\mathcal{A}$, where $\mathcal{A} \models T$, and $\mathcal{A} < \mathcal{M}$, and $\mathcal{A} \subseteq_d \mathcal{M}$, then Y is dense in $\mathcal{M}$.

It is easily verified that this holds when T is the theory of real closed fields.

2. THE MAIN THEOREM

<u>Theorem 1</u> Suppose T is complete, and satisfies Assumptions 1 - 6. Let T^d be the theory of pairs $(\mathcal{M}, \mathcal{A})$ of models of T, with $\mathcal{A} \subsetneq_d \mathcal{M}$. Then T^d is complete.

<u>Proof</u> By Assumption 3, T^d is consistent. So we have to prove that if $(\mathcal{M}_1, \mathcal{A}_1)$ and $(\mathcal{M}_2, \mathcal{A}_2)$ are models of T^d then $(\mathcal{M}_1, \mathcal{A}_1) \equiv (\mathcal{M}_2, \mathcal{A}_2)$.

By using Assumption 4 and Lemma 6, and taking suitable good ultrapowers [3], we can assume without loss of generality that for some $\mathcal{K}$,

a) $\mathrm{card}(\mathcal{M}_i) = \mathrm{card}(\mathcal{A}_i)$
$= \mathcal{K}$
$> \max(\aleph_0, \mathrm{card}(L)), \quad i = 1, 2$;

b) $\mathcal{M}_i$ and $\mathcal{A}_i$ are special [3] of cardinal $\mathcal{K}$, $i = 1, 2$;

c) t.d. $\mathcal{M}_i | \mathcal{A}_i = \mathcal{K}$, $i = 1, 2$;

d) t.d. $\mathcal{A}_i = \mathcal{K}$, $i = 1, 2$;

e) cofinality $\mathcal{K} > \mathrm{card}\,(L)$.

We will in due course deduce that $(\mathcal{M}_1, \mathcal{A}_1) \equiv (\mathcal{M}_2, \mathcal{A}_2)$, whence the theorem.

<u>Lemma 7</u> Suppose T satisfies Assumptions 1 - 6. Suppose $\mathcal{M}, \mathcal{A} \models T$, $\mathcal{A} \subseteq \mathcal{M}$. Suppose $\mathrm{card}\,\mathcal{M} = $ t.d. $\mathcal{M}|\mathcal{A} = \mathrm{card}\,\mathcal{A} \geq \mathrm{card}\ L$. Then there is an $\mathcal{A}$-basis X of $\mathcal{M}$ such that for every infinite set D definable in $\mathcal{M}$ (using constants from $\mathcal{M}$), $X \cap D \neq \emptyset$.

<u>Proof</u> Enumerate as D_α, $\alpha < \mathrm{card}\,\mathcal{M}$, the definitions of the infinite D definable in $\mathcal{M}$ using constants from $\mathcal{M}$. We construct an increasing chain X_α, $\alpha < \mathrm{card}\,\mathcal{M}$, of $\mathcal{A}$-independent subsets of $\mathcal{M}$, such that $X_\alpha \cap D_\alpha \neq \emptyset$, all α, and $\mathrm{card}\,X_\alpha < \mathrm{card}\,\mathcal{M}$, all α. Suppose we have the construction for all $\alpha < \mu < \mathrm{card}\,\mathcal{M}$. Select a finite subset E_μ of $\mathcal{M}$ such that D_μ is defined using constants from E_μ. Consider $\mathcal{A}_\mu = \mathrm{cl}_{\mathcal{M}}(\mathcal{A} \cup \bigcup_{\alpha<\mu} X_\alpha \cup E_\mu)$. By Assumption 1(b), $\mathcal{A}_\mu \prec \mathcal{M}$. By Assumption 1(a), t.d. $\mathcal{A}_\mu | \mathcal{A} < \mathrm{card}\,\mathcal{M}$. So if $\mathcal{A}_\mu = \mathcal{M}$, we have both t.d. $\mathcal{M}|\mathcal{A} = \mathrm{card}\,\mathcal{M}$, and t.d. $\mathcal{M}|\mathcal{A} < \mathrm{card}\,\mathcal{M}$, contradicting Lemma 2. $\therefore \mathcal{A}_\mu \subsetneq \mathcal{M}$.

If $D_\mu^{\mathcal{A}_\mu} = D_\mu^{\mathcal{M}}$, we have a Vaughtian pair for T, contradicting Assumption 2. Therefore $D_\mu^{\mathcal{A}_\mu} \subsetneq D_\mu^{\mathcal{M}}$. Select $x \in D_\mu^{\mathcal{M}}$, $x \notin D_\mu^{\mathcal{A}_\mu}$. So $x \notin \mathcal{A}_\mu$. Let $X_\mu = (\bigcup_{\alpha<\mu} X_\alpha) \cup \{x\}$. By Exchange Property, X_μ is $\mathcal{A}$-independent. This completes the construction.

Now let X be any $\mathcal{A}$-basis extending $\bigcup X_\alpha$. X exists by Exchange Property.

Lemma 7a Suppose T satisfies Assumptions 1 - 6. Suppose $\mathcal{M} \models T$, t.d. $\mathcal{M} \geq$ card L. Then there is a basis X of $\mathcal{M}$ such that for every infinite D definable over $\mathcal{M}$ (using constants from $\mathcal{M}$), $X \cap D \neq \emptyset$.

Proof Same as preceding (it is really the case $\mathcal{A} = \emptyset$).

Note The above lemmas select dense transcendence bases in the style of [4].

Lemma 8 Suppose T satisfies Assumptions 1 - 6. Suppose $(\mathcal{M}, \mathcal{A}) \models T^d$. Suppose card $\mathcal{M}$ = t.d. $\mathcal{M} | \mathcal{A}$ = t.d. $\mathcal{A} \geq$ card L. Then there is an $\mathcal{A}$-basis X of $\mathcal{M}$, and a basis Y of $\mathcal{A}$ such that for every infinite D definable over $\mathcal{M}$ (using constants from $\mathcal{M}$), $X \cap D \neq \emptyset$, and $Y \cap D \neq \emptyset$.

Proof Lemma 6 gives us X. Lemma 7(a) gives us Y dense in $\mathcal{A}$. But $\mathcal{A} \prec \mathcal{M}$, by Lemma 4. By Assumption 6, Y is dense in $\mathcal{M}$, as required.

Now we return to the proof of Theorem 1. Select $\mathcal{A}_i$-bases X_i of $\mathcal{M}_i$ (i = 1, 2), and bases Y_i of $\mathcal{A}_i$ (i = 1, 2), such that both X_i and Y_i are dense in $\mathcal{M}_i$. (We use Lemma 8).

Our strategy now is to get $(\mathcal{M}_1, \mathcal{A}_1) \equiv (\mathcal{M}_2, \mathcal{A}_2)$ by matching up X_1 to X_2, and Y_1 to Y_2 in a suitable way.

Lemma 9 Suppose $\mathcal{N}_i \subseteq \mathcal{M}_i$, i = 1, 2. Suppose $\mathcal{N}_i = cl_{\mathcal{M}_i}(X_i' \cup Y_i')$, i = 1, 2, where $X_i' \subseteq X_i$ and $Y_i' \subseteq Y_i$. Suppose f is an isomorphism $\mathcal{N}_1 \cong \mathcal{N}_2$ such that $f(X_1') = X_2'$, and $f(Y_1') = Y_2'$. Then f maps $\mathcal{N}_1 \cap \mathcal{A}_1$ isomorphically onto $\mathcal{N}_2 \cap \mathcal{A}_2$.

Proof Suppose not. Then, by symmetry, we can assume without loss of generality that for some α in $\mathcal{N}_1 \cap \mathcal{A}_1$, $f(\alpha) \notin \mathcal{A}_2$.

Now, for some finite $Y_1^* \subseteq Y_1'$, $\alpha \in cl_{\mathcal{A}_1}(Y_1^*)$. By model-completeness of T, and Assumption 1, $cl_{\mathcal{A}_1}(Y_1^*) \cap \mathcal{N}_1 = cl_{\mathcal{N}_1}(Y_1^*)$.

Therefore $\alpha \in cl_{\mathfrak{N}_1}(Y_1^*)$. Therefore $f(\alpha) \in cl_{\mathfrak{N}_2}(f(Y_1^*))$. But $f(Y_1^*) \subseteq Y_2'$, so $f(\alpha) \in cl_{\mathfrak{N}_2}(Y_2') \subseteq cl_{\mathfrak{M}_2}(Y_2) \subseteq \mathfrak{A}_2$, a contradiction.

This proves the lemma.

We are now ready to complete the proof of Theorem 1. We show that a certain back-and-forth argument can be done. The problem is to construct a system of local isomorphisms in the sense of [5].

Let $Z_i = X_i \cup Y_i$. By Lemma 3, $X_i \cap Y_i = \emptyset$, and Z_i is a basis for $\mathfrak{M}_i$. Also, X_i, Y_i and Z_i have cardinal $\aleph$. It is clearly enough, because of Lemma 9, and because of routine back and forth manipulations, to prove Lemmas 10 and 10(a) below. 10(a) will provide some local isomorphisms, and 10 will allow us to extend local automorphisms.

<u>Lemma 10</u> Let Z_i' be a finite subset of Z_i, $i = 1, 2$. Suppose f is an isomorphism

$$f: cl_{\mathfrak{M}_1}(Z_1') \cong cl_{\mathfrak{M}_2}(Z_2'),$$

such that $f(Z_1') = Z_2'$, and $f(Z_1' \cap Y_1) = Z_2' \cap Y_2$. Let $t_1 \in Z_1$. Then there exists $t_2 \in Z_2$, and an extension g of f such that g is an isomorphism

$cl_{\mathfrak{M}_1}(Z_1' \cup \{t_1\}) \cong cl_{\mathfrak{M}_2}(Z_2' \cup \{t_2\})$, $g(t_1) = t_2$, and

$$t_1 \in Y_1 \Longleftrightarrow t_2 \in Y_2 .$$

<u>Proof</u> Because of Assumption 1(c), it suffices to find $t_2 \in Z_2$, and $h \supseteq f$, such that h is an isomorphism of $\mathfrak{N}_1$ onto $\mathfrak{N}_2$, where $\mathfrak{N}_i$ is the substructure of $\mathfrak{M}_i$ generated by $cl_{\mathfrak{M}_i}(Z_i')$ and t_i, and such that $h(t_1) = t_2$, and $t_1 \in Y_1 \Longleftrightarrow t_2 \in Y_2$.

Let τ be the type of t_1 over $cl_{\mathfrak{M}_1}(Z_1')$. Let τ^f be the image of τ under f [10, 13], so τ^f is a set of formulas in one variable over $cl_{\mathfrak{M}_2}(Z_2')$. We have to find t_2 in Z_2 satisfying τ^f, and such that $t_1 \in Y_1 \Longleftrightarrow t_2 \in Y_2$. This will conclude the proof.

Case 1 $\mathcal{T}$ principal. Now, $\mathrm{cl}_{\mathfrak{M}_1}(Z_1')$ is infinite, since it is a model of T. So $t_1 \in \mathrm{cl}_{\mathfrak{M}_1}(Z_1')$, and the result is trivial.

Case 2 $\mathcal{T}$ non-principal. Since $\mathrm{cl}_{\mathfrak{M}_1}(Z_1') \prec \mathfrak{M}_1$, $\mathcal{T}$ is finitely satisfiable in $\mathrm{cl}_{\mathfrak{M}_1}(Z_1')$, so $\mathcal{T}^f$ is finitely satisfiable in $\mathrm{cl}_{\mathfrak{M}_2}(Z_2')$. Clearly $\mathcal{T}^f$ is then a non-principal type over $\mathrm{cl}_{\mathfrak{M}_2}(Z_2')$.

By Assumption 5, $\mathcal{T}^f$ has interior in some $\mathfrak{M}_2{}^*$ where $\mathfrak{M}_2{}^* \models T$ and $\mathrm{cl}_{\mathfrak{M}_2}(Z_2') \subseteq \mathfrak{M}_2{}^*$. Since K has cofinality $>$ card(L) $=$ card(L($\mathrm{cl}_{\mathfrak{M}_2}(Z_2')$)), $\mathfrak{M}_2$ is card(L)-saturated, and so by model-completeness of T we may take $\mathfrak{M}_2{}^* = \mathfrak{M}_2$.

So, there exists $\varphi(v_0)$ in $L(\mathfrak{M}_2)$ such that

$$\mathfrak{M}_2 \models \exists^{\infty} v_0\, \varphi(v_0) \quad \text{and}$$

$$\mathfrak{M}_2 \models (\forall v_0)[\varphi(v_0) \longrightarrow \Psi(v_0)]$$

for all Ψ in $\mathcal{T}^f$.

Subcase 1 $t_1 \in X_1$

Now $\varphi^{\mathfrak{M}_2}$ is infinite, so $X_2 \cap \varphi^{\mathfrak{M}_2} \neq \emptyset$. Select t_2 in $X_2 \cap \varphi^{\mathfrak{M}_2}$. Then t_2 satisfies $\mathcal{T}^f$ and $t_2 \in X_2$.

Subcase 2 $t_1 \in Y_1$

Y_2 is dense in $\mathfrak{M}_2$, by assumption, so, as above, $Y_2 \cap \varphi^{\mathfrak{M}_2} \neq \emptyset$. Select t_2 in $Y_2 \cap \varphi^{\mathfrak{M}_2}$. Then t_2 satisfies $\mathcal{T}^f$, and $t_2 \in Y_2$.

This concludes the proof.

Lemma 10(a) Let $t_1 \in Y_1$. Then there exists $t_2 \in Y_2$, and an isomorphism $f\colon \mathrm{cl}_{\mathfrak{M}_1}(\{t_1\}) \cong \mathrm{cl}_{\mathfrak{M}_2}(\{t_2\})$ with $f(t_1) = t_2$.

Proof Let $\mathcal{T}$ be the pure type of t_1 in $\mathfrak{M}_1$. As in Lemma 10, all we have to do is to find t_2 in Y_2 such that $\mathcal{T}$ is the pure type of t_2 in $\mathfrak{M}_2$. First we note that the completeness of T gives that $\mathcal{T}$ is finitely satisfiable in $\mathfrak{M}_2$.

Case 1 $\mathcal{T}$ principal. Then let φ generate $\mathcal{T}$. Since $t_1 \in \varphi^{\mathcal{M}_1}$, $\varphi^{\mathcal{M}_1}$ is infinite. So, by completeness, $\varphi^{\mathcal{M}_2}$ is infinite. Thus $\varphi^{\mathcal{M}_2} \cap Y_2 \neq \emptyset$. This gives t_2 as required.

Case 2 $\mathcal{T}$ non-principal. We can argue as in Case 2 of Lemma 10.

This concludes the proof of Lemma 10a and Theorem 1.

Remark In view of the importance of the Interior Condition for Types in the above, we want to remark that the Interior Condition implies that α_T, the Morley rank of T [10, 13], is ≤ 2.

3. APPLICATIONS

3.1. Real closed fields. When T is the theory of real closed ordered fields, we have verified all the assumptions, and this gives Robinson's theorem that the theory of pairs $(\mathcal{M}, \mathcal{A})$ of real closed fields with $\mathcal{A} \subsetneq \mathcal{M}$ and $\mathcal{A} \subseteq_d \mathcal{M}$ is complete.

3.2. P-adically closed fields. We prefer the notation p-adically closed fields for what Ax-Kochen [1] call formally p-adic fields.

Suppose $\mathcal{A}$ and $\mathcal{M}$ are p-adically closed fields and $\mathcal{A} \subseteq \mathcal{M}$. $\mathcal{A}$ and $\mathcal{M}$ are naturally topological fields, and so we have a notion $\mathcal{A}$ is dense in $\mathcal{M}$.

We shall verify that this coincides with the general notion of our paper, and prove that the theory of pairs $(\mathcal{M}, \mathcal{A})$ with $\mathcal{A} \subsetneq \mathcal{M}$ and $\mathcal{A} \subseteq_d \mathcal{M}$ is complete.

The central theme of our analysis is the structure of definable subsets of p-adically closed fields. We do this in detail in another publication [9], and we shall quote the main results here.

The theory of p-adically closed fields is model complete [1], so the notion of extension will not be changed if we consider p-adically closed fields as fields K with valuation subring V and subsets P_n where P_n is the set of nth powers in K. So, let $\mathcal{L}$ be the language of field theory together with predicates

V and P_n $(n \geq 2)$. T is the theory of p-adically closed fields construed as $\mathfrak{L}$-structures. (It is routine to interpret the valuation and value group in terms of the valuation ring V).

In [9] we proved:

i) T admits elimination of quantifiers;

ii) if $\mathfrak{M} \models T$ and $\alpha \in cl_{\mathfrak{M}}(X)$ then α is algebraic over $\mathbb{Q}(X)$ in the classical field-theoretic sense;

iii) if $\mathfrak{M} \models T$ and $\varphi^{\mathfrak{M}}$ is infinite, then $\varphi^{\mathfrak{M}}$ has non-empty interior in the valuation topology on $\mathfrak{M}$.

From (ii) we can immediately deduce that if $\mathfrak{M} \models T$ the $cl_{\mathfrak{M}}(X)$ is exactly the relative algebraic closure of X in $\mathfrak{M}$, in the field-theoretic sense. From this Assumption 1(a) follows. Assumption 1(b) then comes from the result [1] that a relatively algebraically closed subfield of a p-adically closed field is an elementary submodel.

Assumption 1(c) is proved in [9].

Assumptions 3 and 4 are clear.

For Assumption 2, suppose we have $\mathfrak{A} \subsetneqq \mathfrak{M} \models T$ and $\varphi^{\mathfrak{A}} = \varphi^{\mathfrak{M}}$, $\varphi^{\mathfrak{A}}$ infinite. By (iii), $\varphi^{\mathfrak{A}}$ has interior. So $\exists\, \alpha \in \mathfrak{A}$, and γ in the value group of $\mathfrak{A}$ such that

$$\mathfrak{A} \models \forall x[v(x-\alpha) > \gamma \longrightarrow \varphi(x)].$$

Since $\varphi^{\mathfrak{A}} = \varphi^{\mathfrak{M}}$, it follows that $\{x \in \mathfrak{M} : v(x-\alpha) > \gamma\}$

$$= \{x \in \mathfrak{A} : v(x-\alpha) > \gamma\}.$$

Since $\mathfrak{M} \neq \mathfrak{A}$, there exists $y \in \mathfrak{M} - \mathfrak{A}$, $v(y) > 0$. Select $x_0 \in \mathfrak{A}$ with $v(x_0-\alpha) = \delta > \gamma$. Select $x_1 \in \mathfrak{A}$ with $v(x_1) = \delta$. Then $v(\beta x_1) > \delta$, so $v(\beta x_1 + x_0 - \alpha) = \delta > \gamma$, and $\beta x_1 + x_0 \notin \mathfrak{A}$. This contradiction proves Assumption 2.

The next item is the Interior Condition for Types. Let $\mathfrak{M} \models T$, and let p be a non-principal member of $S_1(\mathfrak{M})$. Realize p by α in some elementary extension $\mathfrak{N}$ of $\mathfrak{M}$. Since p is non-principal, $\varphi^{\mathfrak{M}}$ is infinite for each φ in p, and so $\varphi^{\mathfrak{M}}$ has interior for each φ in p. It follows that $\varphi^{\mathfrak{N}}$ has interior for each φ in p. Now we show that a suitable choice of $\mathfrak{N}$ will guarantee that p has interior in $\mathfrak{N}$. Select $\mathfrak{N}$ so that $\mathfrak{N}$ has some element $\beta \neq 0$ with $v(\beta) > v(x)$ for all x in $\mathfrak{M}$ with $x \neq 0$. Then a trivial argument shows that

$$\mathfrak{N} \models \exists^{\infty} y\, v(y-\alpha) > v(\beta) \quad \text{and}$$
$$\mathfrak{N} \models (\forall y)[v(y-\alpha) > v(\beta) \longrightarrow \varphi(y)]$$

for all φ in p.

Finally, we have to show that density is transitive. This follows if we can show that our model-theoretic density agrees with the usual topological density. But this is immediate from our result that if $\mathfrak{M} \models T$ and $\varphi^{\mathfrak{M}}$ is infinite then $\varphi^{\mathfrak{M}}$ has interior.

This completes the proof.

4. DECIDABILITY RESULTS

Let us continue to assume that T is complete with an infinite model, and satisfies Assumptions 1 - 6. Then T is algebraically bounded, and has elimination of quantifiers.

Theorem 2 T^d has the same Turing degree as T.

Proof (We are very indebted to Peter Winkler for the key idea of using Vaughtian pairs below).

Clearly T is recursive in T^d.

As we remarked before, T allows elimination of the $\exists^{\infty}$-quantifier. That is, for each $\varphi(\vec{v},w)$ in L, there is $\Psi(\vec{v})$ in L such that $T \models [\exists^{\infty} w\, \varphi(\vec{v},w)] \longleftrightarrow [\Psi(\vec{v})]$. Moreover, for each such φ there is an integer n such that

$$T \models \exists^{\geq n} w\, \varphi(\vec{v},w) \longrightarrow \exists^{\infty} w\, \varphi(\vec{v},w).$$

Our problem is: Given φ, to find n.

Since T^d is complete, we can prove T^d is recursive in T provided we show that there is a process recursively enumerable in T which to an L-formula $\varphi(v,v_1,--v_m)$ finds an n such that $T \models \exists^{\infty} v\, \varphi(v,v_1,--v_m) \longleftrightarrow \exists^{\geq n} v\, \varphi(v,v_1,--v_m)$.

Here is the process. Consider a set T_D of L_1-sentences recursively enumerable in T and whose models are the pairs $(\mathfrak{M},\mathfrak{A})$ where $\mathfrak{M} \models T$ and $\mathfrak{A} \prec \mathfrak{M}$, $\mathfrak{A} \neq \mathfrak{M}$. (Clearly such a T_D exists).

Now for any L-formula $\varphi(v,v_1,--v_m)$,

$$T_D \cup \{\exists^{\geq n} v\, \varphi(v,v_1,--v_m) : n \in \omega\} \cup \{A(v_j) : 1 \leq j \leq m\} \models (\exists v)(\neg A(v) \wedge \varphi(v,v_1,--v_m)).$$

This is because T has no Vaughtian pairs.

But then by compactness there exists n such that

$$T_D \cup \{\exists^{\geq n} v\, \varphi(v,v_1,--v_m)\} \cup \{A(v_j) : 1 \leq j \leq m\} \models (\exists v)[\neg A(v) \wedge \varphi(v,v_1,--v_m)].$$

Clearly there is a process recursively enumerable in T for finding such an n. Then, since T_D implies $\mathfrak{A} \prec \mathfrak{M}$, we have

$$T \models \exists^{\geq n} v\, \varphi(v,v_1,--v_m) \longleftrightarrow \exists^{\infty} v\, \varphi(v,v_1--v_m).$$

This proves the theorem.

<u>Corollary</u> The theory of pairs $(\mathfrak{M},\mathfrak{A})$ where $\mathfrak{A} \subsetneq_d \mathfrak{M}$ and $\mathfrak{A},\mathfrak{M} \models T$ is decidable, when T is either the theory of real closed fields or p-adically closed fields.

5. THE $\aleph_1$-CATEGORICAL CASE

In [12], Robinson also proved that the theory of pairs $(\mathfrak{M},\mathfrak{A})$ of algebraically closed fields, of prescribed characteristic, with $\mathfrak{A} \subsetneq \mathfrak{M}$, is complete and decidable. His method for this resembles his proof for real closed fields. Keisler [6] gave a much simpler proof. In this section we shall give a wide generalization of Robinson's result, extending Keisler's proof by using the Baldwin-Lachlan theorem.

We remark first that if T is the theory of algebraically closed fields of some fixed characteristic then T does not satisfy Assumptions 1-6. The culprit is the Interior Condition for Types, as is easily seen.

Suppose T is a complete theory. Let $T^{(2)}$ be the theory of pairs $(\mathfrak{M},\mathfrak{A})$ where $\mathfrak{M},\mathfrak{A} \models T$ and $\mathfrak{A} \subsetneq \mathfrak{M}$. We shall prove that $T^{(2)}$ is complete under certain circumstances. (Note that $T^{(2)}$ is consistent if T has an infinite model). Suppose T has an infinite model. Then obviously $T^{(2)}$ is not complete unless T is model complete.

There is another less obvious necessary condition on T in order for $T^{(2)}$ to be complete. Let $(\mathfrak{M},\mathfrak{A}) \models T^{(2)}$. Let $\varphi(v_0) \in L(\mathfrak{A})$, and suppose $\varphi^{\mathfrak{A}}$ is infinite. We now assume T is model complete. So $\varphi^{\mathfrak{M}} \cap \mathfrak{A} = \varphi^{\mathfrak{A}}$, and $\varphi^{\mathfrak{M}}$ is infinite, Suppose there is some $\Psi(v_0,v_1 - - v_{n-1},v_n, - - v_{n+m})$ in $\mathcal{L}$ and some $k \in \omega$ such that

$$\mathfrak{M} \models \forall v_1 - - \forall v_{n+m} \exists \leq k \; v_0 \Psi(v_0, - - v_{n+m})$$

and

$$(\mathfrak{M},\mathfrak{A}) \models \forall v_0 - - \forall v_{n-1}[\Phi(v_0) \wedge - - \wedge \Phi(v_{n-1})$$
$$\longrightarrow \exists v_n - - \exists v_{n+m}(A(v_n) \wedge - - \wedge A(v_{n+m})$$
$$\wedge \bigvee_{\sigma \in S_n} \Psi(v_{\sigma(0)} - - v_{\sigma(n-1)}, v_n - - v_{n+m})].$$

Then $T^{(2)}$ is not complete. For clearly there is some $\mathfrak{M}' \succ \mathfrak{M}$ with $\mathfrak{M} \models T$ such that for all $\Phi' \in L(\mathfrak{A})$ such that $\Phi'^{\mathfrak{A}}$ is infinite, and all $\Psi(v_0,v_1, - - v_{n-1},v_n, - - v_{n+m})$ in L for which there exists $k \in \omega$ such that

$\mathfrak{M}' \models \forall v_1 - - \forall v_{n+m} \exists^{\leq k} v_0 \Psi(v_0, - - v_{n+m})$ we have

$(\mathfrak{M}',\mathcal{A}) \models \exists v_0 \cdots \exists v_{n-1}[\Phi'(v_0) \wedge \cdots \wedge \Phi'(v_{n-1})$
$\wedge \forall v_n \cdots \forall v_{n+m}(A(v_n) \wedge \cdots \wedge A(v_{n+m}) \rightarrow \neg \bigvee_{\sigma \in S_n} \Psi(v_{\sigma(0)} \cdots v_{\sigma(n-1)}, v_n \cdots v_{n+m})]$

The existence of $\mathfrak{M}'$ is an easy compactness argument. Then $(\mathfrak{M}',\mathcal{A}) \not\equiv (\mathfrak{M},\mathcal{A})$.

This leads us to :

<u>Assumption 4$^{\#}$</u> Suppose $(\mathfrak{M},\mathcal{A}) \models T^{(2)}$. Suppose $\Phi(v_0) \in L(\mathcal{A})$ and $\Phi^{\mathcal{A}}$ is infinite. Suppose $\Psi(v_0, v_1, \cdots v_{n-1}, v_n, \cdots v_{n+m})$ is an L-formula for which there exists $k \in \omega$ such that $\mathfrak{M} \models \forall v_1 \cdots v_{n+m} \exists^{\leq k} v_0 \ \Psi(v_0, v_1, \cdots v_n, \cdots v_{n+m})$. Then

$(\mathfrak{M},\mathcal{A}) \models \exists v_0 \cdots \exists v_{n-1}[\Phi(v_0) \wedge \cdots \wedge \Phi(v_{n-1})$
$\wedge \forall v_n \cdots \forall v_{n+m}(A(v_n) \wedge \cdots \wedge A(v_{n+m})$
$\longrightarrow \neg \bigvee_{\sigma \in S_n} \Psi(v_{\sigma(0)} \cdots v_{\sigma(n-1)}, v_n \cdots v_{n+m})]$

<u>Theorem 3</u> Suppose L is countable, and T is complete, model complete and satisfies Assumption 4$^{\#}$. Then $T^{(2)}$ is complete, and has the same Turing degree as T.

<u>Proof</u> The second part is trivial, given the first.

By [2], some principal extension T^* of T has a strongly minimal formula $\Phi(v_0)$. T^* has as axioms $T \cup \Theta(\vec{c})$, where Θ is an L-formula and the $\vec{c}$ are some new constants added to L. Since T is complete and model-complete, and Assumption 4$^{\#}$ transfers to T^*, we may assume without loss of generality that $T^* = T$, i.e. that $\Phi(v_0)$ is strongly minimal for T.

Now, by [2] any model of T has a well-defined Φ-dimension. It is clear that any α-saturated model of T has Φ - dimension $\geq \alpha$ if $\alpha \geq \omega$. From Assumption 4$^{\#}$ we see that if $(\mathfrak{M},\mathcal{A}) \models T^{(2)}$ and is α-saturated, and if X is a Φ -basis for $\mathcal{A}$, then there exists Y of cardinal $\geq \alpha$ such that $X \cap Y = \emptyset$ and $X \cup Y$ is a Φ -basis for $\mathfrak{M}$.

So if $(\mathfrak{M},\mathfrak{A}) \models T^{(2)}$ and is special of cardinal α, then there exist X,Y each of cardinal α, such that $X \cap Y = \emptyset$, X is a Φ - basis for $\mathfrak{A}$, and $X \cup Y$ is a Φ -basis for $\mathfrak{M}$.

Now suppose $(\mathfrak{M}_i,\mathfrak{A}_i)$, $i = 1, 2$, are special models of $T^{(2)}$ of cardinal α, and that X_i,Y_i are chosen as above. By [2], any bijection of $(X_1 \cup Y_1)$ onto $(X_2 \cup Y_2)$ extends to an isomorphism $f: \mathfrak{M}_1 \cong \mathfrak{M}_2$. If we make sure that the bijection maps X_1 onto X_2, then a routine Vaughtian pair argument as in [2] shows that $f(\mathfrak{A}_1) = \mathfrak{A}_2$.

This proves the theorem.

Notes 1 It is not the case in general for $\aleph_1$-categorical T that $T^d = T^{(2)}$. Indeed (cf. Assumption 3) T^d may be inconsistent.

2. Assumption $4^{\#}$ is needed. To see this take T as the theory of infinite sets. Then $\mathfrak{M} - \mathfrak{A}$ may be finite or infinite.

Appendix 1 Let T be the theory of discrete linear order with first element 0, no last element, and such that every non-zero element is a successor.

Suppose $\mathfrak{A} \subseteq \mathfrak{M}$, where $\mathfrak{A}, \mathfrak{M} \models T$. It is easy to see that $\mathfrak{A} \subseteq_d \mathfrak{M}$ if and only if whenever $[x, y]$ is an interval in $\mathfrak{M}$ and $[x, y] \cap \mathfrak{A} = \emptyset$ then $[x, y]$ has only finitely many members. But then a routine compactness argument shows that in this case density is not elementary.

Appendix 2 Dense linear order. Let T be the theory of dense linear order without end points. Neither Theorem 1 nor 3 applies. However, the following is easy.

Theorem 4 The theory of pairs $(\mathfrak{M},\mathfrak{A})$ of dense linear orders, where $\mathfrak{A}$ and $\mathfrak{M} - \mathfrak{A}$ are dense in $\mathfrak{M}$, is complete and decidable.

Problem Put this in a general setting.

REFERENCES

[1] J. Ax and S. Kochen, Diophantine problems over local fields: III. Decidable Fields, Annals of Math., 83 (1966), 437-456.

[2] J. Baldwin and A. Lachlan, On strongly minimal sets, J. S. L. 36 (1971), 79-96.

[3] C. C. Chang and H. J. Keisler, Model Theory, North Holland, 1973.

[4] P. Erdös, L. Gillman and M. Henriksen, An isomorphism theorem for real closed fields, Annals of Math., 61 (1955), 542-554.

[5] C. Karp, Languages with Expressions of Infinite Length, North Holland, 1964.

[6] H. J. Keisler, Complete theories of algebraically closed fields with distinguished subfields, Michigan Mathematical Journal, 11 (1964), 71-81.

[7] A. Macintyre, Classifying Pairs of Real Closed Fields, Ph. D. Thesis, Stanford, 1968.

[8] __________, Complete theories of topological fields with distinguished dense proper subfields, J. S. L. 34 (1969), 538.

[9] __________, Definable subsets of valued fields, in preparation.

[10] M. D. Morley, Categoricity in power, Transactions A. M. S. 114 (1965), 514-538.

[11] M. Mortimer, Ph. D. Thesis, Bedford College, London, 1973.

[12] A. Robinson, Solution of a problem of Tarski, Fundamenta Math. 47 (1959), 179-204.

[13] G. Sacks, Saturated Model Theory, Benjamin, 1972.

[14] J. R. Shoenfield, A theorem on quantifier elimination, Symposia Mathematica 5, 1971, 173-176.

[15] E. Steinitz, Algebraische Theorie der Korper, Berlin, 1930.

[16] A. Tarski and J. C. C. McKinsey, A Decision Method for Elementary Algebra and Geometry, Rand Corporation, Santa Monica, 1948.

[17] P. Winkler, This volume.

[18] __________, Ph. D. Thesis, Yale, 1975.

NEW FACTS ABOUT HILBERT'S SEVENTEENTH PROBLEM

Kenneth McKenna, Yale College

Hilbert originally stated his seventeenth problem as the question: Is a rational function in n variables with rational coefficients which is everywhere non-negative on the rationals necessarily a sum of squares of rational functions with rational coefficients? Artin proved the following stronger result: Let K be a uniquely orderable field which is Archimedean. Then if f is a rational function with coefficients from K and f is non-negative on K then f is a sum of squares of rational functions with coefficients in K. The proof given by Artin naturally depends heavily on the fact that K is Archimedean. In conversations with Angus MacIntyre the following question was raised: What exactly characterizes the ordered fields for which Hilbert's conjecture holds?[1] This paper deals with this and related questions. In particular, the principal result in this direction will be a proof that Hilbert's conjecture holds on an ordered field, K, if and only if K is dense in its real closure and K is uniquely orderable. The fact that all ordered fields are not dense in their real closures can easily be seen by considering the field Q(t), where t is a transcendental which is placed greater than all the rationals. There is nothing from Q(t) in the interval $(\sqrt{t}, 2\sqrt{t})$. In fact, if A is the field of real algebraic numbers, no non-rational element of A is a limit point of Q(t).[2]

We begin with a definition:

<u>Definition:</u> Let K be an ordered field and f a rational function in n variables defined on K. Then we will say that f is <u>definite</u> on K (or simply <u>definite</u>) if and only if f is non-negative everywhere it is defined on K.

If Hilbert's conjecture holds on an ordered field, K, we will say "K has Hilbert's Property, (HP)". Critical to our purposes is a property we will call "*".

---

1 For a very nice example of a non-Archimedean, uniquely orderable ordered field on which Hilbert's conjecture is not true see Dubois.

2 This was essentially noticed by Keisler in a conversation with the author.

Definition: Let K be an ordered field with real closure $\overline{K}$. Then we will say "K has *" if and only if every function which is definite on K with coefficients in K is definite on $\overline{K}$.

The importance of * in our considerations will be clear from

Theorem 1. Let K be an ordered field. Then K has HP if and only if K is rigid (i.e. uniquely orderable up to isomorphism) and has *.

Proof. Suppose f is a function definite on K with coefficients in K. If f is not a sum of squares of rational functions then by a well known theorem of Artin (cf. Jacobson, vol. 3) there is an ordering of the rational function field, $K(\overline{x})$ that puts f negative. Since K is rigid this ordering extends the ordering on K. By taking the relative algebraic closure of K in L, the real closure of $K(\overline{x})$, we note the following diagram commutes.

$$\begin{array}{ccc} \overline{K} & \longrightarrow & L \\ \uparrow & & \uparrow \\ K & \longrightarrow & K(\overline{x}) \end{array}$$

$K(\overline{x})$ models the sentence $E\overline{x}\, f(\overline{x}) < 0$. Since this is an existential sentence, L also models it. Since L and $\overline{K}$ are real closed fields and the theory of real closed fields is model complete we see that $L > K$. Since f has coefficients from K we know that $\overline{K}$ then models $E\overline{x}\, f(\overline{x}) < 0$. But this contradicts the fact that K has *.

Conversely, if K has HP every positive element is expressible as the sum of squares. Hence K is rigid. Further, every definite function with coefficients from K is the sum of squares of functions and so is definite on $\overline{K}$.

Definition: Let K and L be ordered fields such that K is a subfield of L. Then we say "K is dense in L" if and only if for every two distinct elements of L there is an element of K lying between them.

We now need three lemmas whose use will be apparent later.

Lemma 1. If K is an ordered field which is not dense in $\overline{K}$ there is an element, p, of $\overline{K}$ such that K is not dense in $K(p)$. Furthermore there is an element of K^+, h, so that the interval $(p - h, p + h)$ is disjoint from K, where $K^+ = \{k \in K \mid k > 0\}$.

Proof. Before beginning the proof itself we notice that K is cofinal in $\overline{K}$. This is clear from the observation that if r is an element of $\overline{K}$ with minimum polynomial $g(x)$, then if $g(x) = x^n + a_{n-1}x^{n-1} + \ldots a_0$ where a_i is in K then we have the estimate $|r| < |1| + |a_{n-1}| + \ldots + |a_0|$ if $|r| > 1$.

From this it follows that $\overline{K}$ contains no infinitesimals with respect to K.

Since K is not dense in $\overline{K}$ there are elements p and q from $\overline{K}$ such that (p, q) is disjoint from K. Since $\overline{K}$ contains no infinitesimals we can find h in K^+ so that $h < q - p$. Thus the interval $(p, p + h)$ is disjoint from K. But also $(p - h, p)$ must be disjoint from K, since if k were in $(p - h, p)$ $k + h$ would be in $(p, p + h)$.

Definition: If K is an ordered field and k is an element of $\overline{K}$ then we say that k is a limit point of K if and only if for every h in K^+ there is a g in K so that g is in the interval $(k - h, k + h)$.

We immediately note that this definition implies that if h is in $\overline{K}^+$ and k is a limit point of K there is an element, g, in K that lies in the interval $(k - h, k + h)$.

Definition: Let K be an ordered field. Then we define $K!$ to be the set of all K limit points.[1]

We notice that $K!$ is closed under addition, subtraction, multiplication and inverses and hence that $K!$ is a field. Clearly $K!$ contains K.

Lemma 2. Let f be a polynomial in one variable with coefficients in an ordered field, K. If p is a root of f and p is in $K!$ then the polynomial $f(x)/x - p$ has coefficients in $K!$.

1 For comparison, see Scott, note especially that all of our limit points are algebraic elements over K.

Proof. K! is a field and p and K are in it.

Lemma 3. If $f(x)$ is a polynomial with coefficients in K! and root p in $\overline{K}$ then there exists a polynomial $m(x)$ with coefficients in K such that if h is in K^+ then:

i) $\deg(f) = \deg(m)$ and

ii) m has a root in the interval $(p - h, p + h)$.

Proof. This lemma says nothing more than the roots of a polynomial are continuous functions of the coefficients. We give a proof much like one presented in Scott for a similar fact.

Let $f(x) = x^n + a_{n-1}x^{n-1} + \ldots + a_0$. We can assume that f is irreducible over K! and hence has only simple roots. Hence f is monotonic in a neighborhood of all its roots, in K, say where $|p_i - x| < d \in K^+$ where $p_1, p_2, \ldots p_t$ are all the roots of f in $\overline{K}$. Hence $f(p_i - d)$ and $f(p_i + d)$ have different signs for $i = 1, \ldots t$. Next choose e from K^+ so that $e < \min(|f(p_i + d)|)$ $i = 1, \ldots t$. Next choose k in K^+ so that $|p_i + d|^j < k$ for $j = 0, \ldots n$. Since every a_j is in K!, for every j we can pick b_j from K so that $|a_j - b_j| < e/nk$. Let $m(x) = x^n + b_{n-1}x^{n-1} + \ldots + b_0$. We now note that

$$|f(p \pm d) - m(p \pm d)| \leq \sum |a_j - b_j| \, |p \pm d|^j$$

$$\leq \sum e/nk \cdot k = e .$$

Hence we see that $m(p - d)$ and $m(p + d)$ are of opposite signs. It follows that m has a root on $(p - d, p + d)$. Since we have shown this for all sufficiently small d, we are done.

We now have enough to prove the main theorem.

Theorem 2. Let K be an ordered field. Then K has * if and only if K is dense in $\overline{K}$.

Proof. Let K have $*$ and suppose K is not dense in $\overline{K}$. Then by Lemma 1 we know there is an element of $\overline{K}$, p, so that there is h in K^+ such that the interval $(p - h, p + h)$ is disjoint from K. Let p be of smallest degree so that this occurs.

Let p have minimum polynomial f over K. We claim every real root of f is isolated from K. Suppose this claim is not true. Then there is p' in $K!$ such that $f(p') = 0$. By Lemma 2 $g(x) = f(x)/x - p'$ has coefficients in $K!$. Clearly $p \neq p'$ so $g(p) = 0$. By Lemma 3 we can choose a polynomial with coefficients in K, $m(x)$, so that m and g have equal degree and m has a root, r, that lies in the interval $(p - h/2, p + h/2)$. The degree of r is at most the degree of m, which is one less than the degree of f, which is equal to the degree of p. But r is isolated from K, which violates the choice of p, which was taken of minimum degree to be isolated. This proves that all the roots of f in $\overline{K}$ are isolated from K. We can assume we have chosen h from K^+ so small that the intervals $(p_1 - h, p_1 + h), \ldots, (p_t - h, p_t + h)$ are all disjoint from K, where $p_1, \ldots, p_t$ are all the roots of f in $\overline{K}$ in the order in which they occur in $\overline{K}$.

Consider the element $p_1 + h/2$. Let it have minimum polynomial $s(x)$ over K. Observe that $s(x + h/2)$ has a root at p_1. Since $p_1 + h/2$ is an element of $K(p_1)$ it follows that $s(x + h/2) = f(x)$. Let k be an element of K. We claim that the sign of $s(k)$ is the same as the sign of $f(k)$.

We assume both $s(x)$ and $f(x)$ have leading coefficient of 1. It follows that the sign of $f(k)$ is the same as the sign of $s(k)$ for all k less than p_1. Since both $s(x)$ and $f(x)$ are irreducible, they have only simple roots. Hence they must change sign at their roots and only at their roots. The roots of $f(x)$ are $p_1, \ldots, p_t$ in $\overline{K}$. From our above observation that $s(x + h/2) = f(x)$ we see the roots of $s(x)$ are $p_1 + h/2, \ldots, p_t + h/2$. Thus $f(x)$ changes sign at each root and stays of constant sign on (p_i, p_{i+1}). Likewise, $s(x)$ changes sign at each one of its roots and stays of constant sign on $(p_1 + h/2, p_{i+1} + h/2)$. But p_i and $p_i + h/2$ are both in the interval $(p_i - h, p_i + h)$, which is disjoint from K. It follows from this and the fact that $f(x)$ and $s(x)$ have the same sign on $(-\infty, p_1)$ that the sign of $f(k)$ is the same as the sign of $s(k)$ for all k in K.

We now consider the polynomial $F(x) = f(x)\, s(x)$. $F(k)$ is positive for all k in K. However, it is clear from the choice of $f(x)$ and $s(x)$ that if h is taken small enough then $F(x)$ will have only simple roots. It follows that $F(x)$ changes sign at p_1 and hence is somewhere negative on $\overline{K}$. But this contradicts the assumption that K had $*$. Hence if K has $*$, K is dense in $\overline{K}$.

Conversely, suppose K is dense in $\overline{K}$. Then if $f(x)$ is a rational function with coefficients in K and h is in K^+ we can write the formal Taylor series for f as $f(\overline{x} + \overline{h}) = f(\overline{x}) + e(\overline{x}, \overline{h})$. Where if $\overline{k}'$ is in $\overline{K}$ and f is defined at $\overline{k}$, $e(\overline{k}', h)$ gets arbitrarily small as h does. Hence if $f(\overline{k}')$ is negative we can find $\overline{k}$ in K close enough to $\overline{k}'$ so that $f(\overline{k})$ is also negative. Hence K has $*$.

As we have characterized it so far, $*$ is a property of K and $\overline{K}$ together. We now characterize $*$ algebraically in terms of K alone.

<u>Definition</u>: We say an ordered field, K, has the <u>Weak Hilbert Property</u> (WHP) if and only if every definite function on K is expressible as a sum of the form $a_1 g_1^2(\overline{x}) + \ldots + a_m g_m^2(\overline{x})$ where each a_j is in K^+ and each g_j is a rational function with coefficients in K.

<u>Theorem 3</u>. Let K be an ordered field. Then K has $*$ if and only if K has the WHP.

<u>Proof</u>. If K has the WHP and f is definite on K then $f(\overline{x}) = a_1 g_1^2(\overline{x}) + \ldots + a_m g_m^2(\overline{x})$ where a_i is in K^+. Hence f is definite on $\overline{K}$.

Conversely, suppose K has $*$ and f is definite on K with coefficients in K. If f is not a sum of the above form, then by a slightly modified version of Artin's Theorem there is an ordering of the rational function field which preserves the ordering on K and puts f negative. As before, the following diagram commutes. (L is again the real closure of $K(\overline{x})$.)

$$\begin{array}{ccc} \overline{K} & \longrightarrow & L \\ \uparrow & & \uparrow \\ K & \longrightarrow & K(\overline{x}) \end{array}$$

So again we get a contradiction in the fact that $\overline{K}$ must model the sentence $E\overline{x}\ f(\overline{x}) < 0$ by model completeness of real closed fields.

Now it follows from inspection of the proof of Theorem 2 that an ordered field, K, has HP if and only if every polynomial in one variable with coefficients in K which is definite on K is definite on $\overline{K}$ and K is rigid. If follows from an argument similar to the ones used in proving Theorems 1 and 3 that K has the property that every polynomial with coefficients in K that is definite on K is definite on $\overline{K}$ if and only if every such polynomial is the sum of squares of rational functions in one variable with coefficients in K, provided K is rigid. Now, Artin proved that a polynomial which is the sum of squares of elements in K(x) is already the sum of squares of polynomials in one variable with coefficients in K. (cf. Artin). It is therefore possible to state:

Theorem 4. Let K be an ordered field. Then Hilbert's conjecture holds on K if and only if every polynomial in one variable which is definite on K with coefficients from K is the sum of squares of polynomials in one variable and coefficients in K.

It is also clear from the inspection of the proof of Theorem 2 that the following is true.

Corollary. If K is an ordered field which is not dense in $\overline{K}$ then there is a polynomial in one variable with coefficients from K that has only isolated roots in $\overline{K}$.

The question naturally arises: Is the property of being dense in its real closure a first order property of an ordered field? We answer this question in the affirmative with:

Theorem 5. An ordered field, K, is dense in its real closure if and only if K models the following first order set of sentences:

For each natural number, n, we write:

$$(x_0)\ldots(x_n)(x)(y)(z)\,\exists w[(x < y \;\wedge\; x_n x^n + \ldots + x_0 > 0$$

$$\wedge\; x_n y^n + \ldots x_0 < 0) \rightarrow (x < w < y$$

$$\wedge\; (x_n w^n + \ldots + x_0)^2 < z^2)] \;.$$

We call the set of all these axioms "S" and fix this name throughout the rest of this paper. These axioms are intended to express nothing more than the fact that a K-polynomial that changes sign on an interval in K must come arbitrarily close to 0 on that interval in K.

Proof. Suppose K is dense in $\overline{K}$. Then if f(x) is positive and f(y) is negative f has a root in (x, y) (in $\overline{K}$). Since K is dense in $\overline{K}$ this root is a limit point of K. Since f is continuous, it must get arbitrarily small as the root is approached from the right.

Conversely, suppose K models S, but is not dense in $\overline{K}$. Then choose, as before, p in $\overline{K}$ of minimum degree so that the interval (p - h, p + h) is disjoint from K for some h in K^+. Let p have minimum polynomial f(x). Then f' has no isolated roots. We know that the mean value theorem for polynomials holds for real closed fields, hence we can choose k and m from K so that f' has no roots in (k, m) and p lies in (k, m) and know that f must be monotonic in (k, m). We can assume f(k) is positive. Since K models S, f must get arbitrarily small on (k, m). But since f is monotonic the only place this can happen is arbitrarily close to p.

Corollary. The family of fields with HP is inductive. (Closed under union of chains.)

Proof. Let $K_1 \subset K_2 \subset K, \ldots$ be a chain of ordered fields with HP. Then, since S is $\forall\exists$, $K = \bigcup K_i \models S$ and thus is dense in $\overline{K}$. If $k \in K^+$ then $k \in K_i$ for some i so k is a sum of squares in K_i, hence in K.

Corollary. The family of ordered fields with the WHP is inductive and closed under ultrapowers.

Corollary. If K is an ordered field the following are equivalent:

(i) K is dense in $\overline{K}$.

(ii) K has *.

(iii) K has the WHP.

(iv) $K \models S$.

We now give a definition apparently due to Scott.

Definition: Let K be an ordered field. K is complete if and only if K has no proper ordered field extensions in which it is dense.

Scott proves that for every ordered field, K, there is a unique complete ordered field, K#, in which K is dense. This K# is the completion of K. Scott then proves that K# is real closed if and only if K is dense in $\overline{K}$. Using these facts and noting that if K is complete $K = K^{\#}$ we have the following.

Theorem 6. Let K be a complete ordered field. Then K is real closed if and only if K has HP.

It is interesting to note (and easy to see) that if we assume GCH there are no saturated, real closed, complete ordered fields.

In order to avoid some possible confusion, the reader should take note of the following facts. We have defined a rigid field to be a field which, up to isomorphism, admits only one ordering. This is equivalent to saying that for every non-zero element in the field exactly it or its negative is totally positive in the Hilbert sense, and hence is a sum of squares. It does not follow that a rigid field does not admit non-trivial automorphisms. As we will show, every elementary extension of the rationals is rigid. However, the work of Ehrenfeuct and Mostowski (cf. Sacks, ch. 34) proves that every infinite structure has an elementary extension that admits non-trivial automorphisms. This is an important difference between the field theoretic and model theoretic use of the word "rigid".

Examples:

1. Consider the field of formal power series in one indeterminate and integral exponents, $Q((t))$, where t^{-1} is placed larger than all the rationals. Then there is nothing between $\sqrt{t}$ and $2\sqrt{t}$, hence $Q((t))$ does not have the WHP. (This example is due to Scott.)

2. $Q(t)$ can be ordered to satisfy S by making t an Archimedean transcendental.

3. Let K be an ordered field with real closure $\overline{K}$. Recall that the axioms, S, are $\forall\exists$ sentences of the form $(\overline{x})\, Ew(p(\overline{x}, w))$. We choose a Skolem function for each axiom and adjoin new axioms of the form $(\overline{x})\, (Ew(p(\overline{x}, w)) \rightarrow p(\overline{x}, f(\overline{x})))$ to form the new set of axioms, S'. Using the usual Skolem techniques we can expand $\overline{K}$ to a model of $S', \overline{K}'$. We can then close K under the field operations and these new Skolem functions to form a hull that satisfies S, but is not necessarily equal to $\overline{K}$ and which might contain other models of S that contain K.

4. Let K_i $i \in \mathbb{N}$ be a family of ordered, rigid fields. Then every element of K_i is a sum of squares if and only if it is non-negative. Suppose further that there is a uniform bound on the number of squares needed to express any positive element of any K_i. If each K_i models S, then Hilbert's conjecture holds on $\Pi_{\mathbb{N}} K_i/D$. Furthermore, if K_i' is elementary equivalent to K_i then Hilbert's conjecture holds on K_i', since the order relation is definable in terms of a finite number of squares. Thus, all ordered fields elementary equivalent to Q have HP, since on Q this bound is 4.

5. On the other hand, if K_i $i \in \mathbb{N}$ is as in the preceding example but there is no uniform bound on the number of squares needed to express a given positive element then there exists an ultrafilter over $\mathbb{N}$ so that, Hilbert's conjecture will not hold on the ultraproduct, since it will not be rigid. The product will still, of course, have WHP.

6. In particular, if Hilbert's conjecture holds for K but there is no bound on the number of squares needed to express positive elements, K has a non-rigid elementary extension.

7. Pfister has proved that if K is a field of transcendency degree n over a real closed field, then every element of K that is a sum of squares is a sum of 2^n squares. Thus, if K is such a field with HP and $L \equiv K$ then L has HP.

The author is greatly indebted to Angus MacIntyre both for the many helpful suggestions he made on this paper and for the seemingly endless quantities of patience, intelligence, and good humor he has supplied over the last year.

Kenneth McKenna
Yale College, 1975

Bibliography

Artin, E., Über die Zerlegung definiter Funktionen in Quadrate, Abh. Math. Sem. Hamburg 5 (1927), pp. 100-115.

Dubois, D. W., Note on Artin's Solution to Hilbert's 17th Problem, Bull. Am. Math. Soc. 73 (1967), 540-541.

Chang and Keisler, Model Theory, North Holland, 1973.

Jacobson, Lectures in Abstract Algebra, Van Nostrand, 1964.

Pfister, Zur Darstellung definiten Funktionen als Summe von Quadraten, Inv. Math. 4 (1967), 229-237.

Robinson, A. Model Theory, North Holland, 1953.

Sacks, Saturated Model Theory, Benjamin, 1972.

Scott, "On Completing Ordered Fields" International Symposium on the Applications of Model Theory to Algebra, Analysis and Probability, Ed. by W. A. J. Luxemburg New York, Rinehart and Winston, 1969.

NONSTANDARD ASPECTS OF HILBERT'S IRREDUCIBILITY THEOREM.

Peter Roquette

University of Heidelberg

1. INTRODUCTION.

In personal conversations and discussions, Abraham Robinson would often return to Hilbert's irreducibility theorem, emphasizing its central role in what is called diophantine geometry, but also with a view on general algebra und model theory. This note is directly influenced by these conversations and his stimulating remarks. It is to be regarded as a comment and a supplement to the article [7] by Gilmore and Robinson concerning Hilbert's irreducibility theorem, published twenty years ago.

That article, together with the other article by Robinson of the same year concerning Hilbert's 17th problem, has been said to "mark a watershed in the development of model theory" (S.Kochen). Since then it has become increasingly clear that model theory provides a new way of mathematical reasoning, capable of applications to mathematical problems of widespread interest and universal significance. The appeal of model theoretic arguments often lies in their extreme elegance and simplicity. For many of us, simplification means a better insight into the nature of the problem and thus an increase of mathematical knowledge.

Let K be a field. As usual, K is called <u>Hilbertian</u> if Hilbert's irreducibility theorem holds over K. The main result of Gilmore-Robinson [7] gives a necessary and sufficient condition for the field

K to be Hilbertian. This condition is "metamathematical" in the sense that it refers to an enlargement of K. Let *K denote such an enlargement, for a higher order language. *K will be fixed throughout the following discussion. It is well known and easy to see that K is algebraically closed in *K. Hence every nonstandard element $t \in {}^*K$ is transcendental over K; the field K(t) is isomorphic to the field of rational functions in one variable over K. Now, the main theorem of [7] can be stated as follows:

THEOREM of GILMORE and ROBINSON. <u>The field</u> K <u>is Hilbertian if and only if there exists a nonstandard element</u> $t \in {}^*K$ <u>such that</u> K(t) <u>is algebraically closed in</u> *K.

In our opinion, the significance of the Gilmore-Robinson condition lies in the fact that it is structural, concerning the field structure of *K in relation to its rational subfields K(t). Therefore, this condition is susceptible to investigations with the general structural methods of algebra and field theory: either in cases where one wants to prove Hilbert's irreducibility theorem for special classes of fields K, or if one wants to apply or to amend it in special situations.

In this note, we intend to exemplify these ideas while discussing the following classes of fields:

Section 2: Number fields,

Section 3: Function fields,

Section 4: Finitely generated extensions of Hilbertian fields.

In each of these cases, our discussion will yield a new "metamathematical" proof of Hilbert's irreducibility theorem. Moreover, we shall exhibit explicit constructions of elements $t \in {}^*K$ which satisfy the Gilmore-Robinson condition. <u>It is the nature of these constructions which is the main object of this note,</u> and which seems significant in various respects. Let us explain this in more detail.

DEFINITION. An element $t \in {}^*K$ is called Hilbertian for K if it satisfies the Gilmore-Robinson conditions, i.e. if t is nonstandard and $K(t)$ is algebraically closed in ${}^*K$.

Thus the theorem of Gilmore-Robinson can be expressed by saying that the existence of Hilbertian elements implies the field to be Hilbertian, and conversely.

As said above, our aim is to give explicit constructions of Hilbertian elements in the cases mentioned above. Our motivation and guide line will be the methods of algebraic geometry, which we want to apply to our given situation. The idea is to use the enlargement as some kind of "universal field" which contains the coefficients of the algebraic objects to be considered. Usually, the universal field of algebraic geometry is taken to be algebraically closed, of large degree of transcendency. However, in the study of rationality questions over particular fields, it is often advisable to use a smaller universal field, which somehow is adapted to the structure of the ground field. This has been done, for instance, in the p-adic case by S.Lang [14] who used the p-adic completion as universal field of p-adic geometry. From this point of view it seems quite natural to try to use the enlargement ${}^*K$ as a "universal field" for the geometry over K. We will not attempt here a systematic development of such "nonstandard geometry". We have mentioned this only to give the reader an idea of the background and of our motivation for this note.

Specifically, our construction of Hilbertian elements will be similar to those geometric constructions which appear in the context of the theorem of Bertini. In fact, the starting point of this work was the observation that in the case of function fields, Hilbertian elements can be constructed as generic hyperplane sections, i.e. in the form

$$t = t_o + t_1u_1 + \dots + t_nu_n$$

where the coefficients $t_o, t_1, \dots, t_n$ are algebraically independent over K. Later, after completing the first draft of the manuscript, it turned out that the following shorter expression is already sufficient:

$$t = t_o + t_1 u$$

with t_o, t_1 algebraically independent over K. We shall see that with various conditions on u in the respective cases, such expressions yield Hilbertian elements.

In every case, the proof of the Hilbertian property of t consists in the reduction to a well known lemma from field theory and algebraic geometry. This lemma is known as "Matsusaka's lemma", and in algebraic geometry it plays a central role in the proof of Bertini's theorem. For the convenience of the reader we have included a proof of Matsusaka's lemma in section 5 of this note.

It seems remarkable that the above construction works, not only in the case of function fields where it seems natural, but also for number fields. Nonstandard methods seem to be suited to simulate "geometric" situations in number fields, an experience which we made earlier in our paper [21] on the Siegel-Mahler theorem (in collaboration with A.Robinson). By the way, we shall use the methods and results of [21] in the discussion in section 2, which deals with number fields. The other sections are independent of [21] and self-contained.

In some sense, the study of Hilbertian elements $t \in {}^*K$ is equivalent to the study of Hilbert subsets $H \subset K$. Recall that a basic Hilbert set H_f is given by an irreducible polynomial $f = f(T,X) \in K[T,X]$ in two variables X,T over K; the set H_f consists of those $t \in K$ for which $f(t,X)$ is irreducible in $K[X]$. An arbitrary Hilbert subset H of K is then the intersection of finitely many basic Hilbert sets, i.e.

$$H = H_{f_1} \cap H_{f_2} \cap \dots \cap H_{f_n} .$$

Hilbert's irreducibility theorem can be stated as saying that every Hilbert set $H \subset K$ is nonempty.

Now, the connection between Hilbertian elements $t \in {}^*K$ and Hilbert subsets $H \subset K$ is given by the following theorem. In this theorem, $\mathfrak{E}$ denotes any property of field elements, defined in the language of K.

GENERALIZED THEOREM of GILMORE and ROBINSON. *If there exists a Hilbertian element* $t \in {}^*K$ *with property* $\mathfrak{E}$ *then every Hilbert subset* $H \subset K$ *contains an element with property* $\mathfrak{E}$ *and conversely.*

The original theorem of Gilmore-Robinson can be viewed as a special case, by taking for $\mathfrak{E}$ the trivial property which holds for every field element.

The above theorem may be regarded as a "translation principle" which allows the transition from the language of Hilbertian elements to the language of Hilbert sets. Using this principle, it is possible to translate our results about explicit constructions of Hilbertian elements, into statements about Hilbert sets. We shall leave the details as an exercise to the reader.

By the way, Gilmore and Robinson [7] have already given a beautiful example of this translation principle: For each nontrivial valuation of the Hilbertian field K, they proved that every Hilbert set is dense in K, with respect to the topology defined by that valuation. In other words: given $a,b \in K$ (with $b \neq 0$) then every Hilbert set contains an element $t \in K$ such that $|t-a| \leqq |b|$. (We write the valuation multiplicatively.) Due to the above theorem, this is equivalent to saying that there exists a Hilbertian element $t \in {}^*K$ such that $|t-a| \leqq |b|$. But this is trivially verified: starting from an arbitrary Hilbertian element $u \in {}^*K$ we put $t = bu+a$ or $t = bu^{-1}+a$ according to whether $|u| \leqq 1$ or $|u| > 1$.

For convenience, let us close this section by giving a proof of the generalized Gilmore-Robinson theorem. Let us remark, however, that this proof is not new and essentially contained in [7] already.

Proof of the generalized Gilmore-Robinson theorem.

(i) The first step in the proof consists of rewriting the defining property of Hilbertian elements in terms of irreducible polynomials. Let $t \in {}^*K$ be transcendental over K. The field K(t) is algebraically closed in *K if and only if every irreducible polynomial over K(t) remains irreducible over *K. That is, the inclusion map $K(t)[X] \subset {}^*K[X]$ should preserve irreducibility. In this statement, the ring K(t)[X] can be replaced by K[t,X]. This can be done because, by Gauss'lemma, every irreducible polynomial of K(t)[X] splits into a product of the form

$$g(t).f(t,X)$$

where f(t,X) is irreducible in K[t,X] and where g(t) is a factor from K(t); conversely, every such product is irreducible in K(t)[X]. We conclude: t is Hilbertian if and only if the inclusion map $K[t,X] \subset {}^*K[X]$ preserves irreducibility.

Since t is transcendental over K, the ring K[t,X] is isomorphic to the polynomial ring K[T,X] in two independent variables T and X. This isomorphism is given by the specialization $T \to t$ and it preserves irreducibility. Hence: $t \in {}^*K$ is Hilbertian if and only if it satisfies the following irreducibility condition:

(I) The specialization map $K[T,X] \to {}^*K[X]$ given by $T \to t$ preserves irreducibility. That is, if f(T,X) is irreducible in K[T,X] then f(t,X) is irreducible in *K[X].

In the foregoing discussion we had assumed from the start that t is transcendental over K. However, this is not necessary since the condition (I) implies automatically that this is the case. For, if t would satisfy an irreducible equation g(t) = 0 over K then we consider the polynomial $f(T,X) = g(T) + X^2$ which is irreducible in K[T,X]; from (I) it would follow that $f(t,X) = X^2$ is irreducible in *K[X] which is absurd. Thus we see that condition (I) is necessary and sufficient for an element $t \in {}^*K$ to be Hilbertian.

This condition can be expressed in terms of Hilbert subset, as follows. As above, H_f denotes the basic Hilbert subset of K defined by f; it consists of all $t \in K$ for which $f(t,X)$ is irreducible in $K[X]$. Let $^*H_f$ denote its enlargement in $^*K$; it consists of those $t \in {}^*K$ for which $f(t,X)$ is irreducible in $^*K[X]$. With this notation, condition (I) can be put into the form

$$t \in \bigcap_f {}^*H_f$$

where f ranges over all irreducible polynomials of $K[T,X]$.

(ii) This being said, we now conclude the proof as follows: Let E denote the subset of K which is defined by the property $\mathfrak{E}$. That is, E consists of those elements of K which have property $\mathfrak{E}$. The enlargement $^*E$ consists of those elements in $^*K$ which have property $\mathfrak{E}$. In view of (i), the intersection

$$D = \bigcap_f {}^*H_f \cap {}^*E$$

consists of all Hilbertian elements $t \in {}^*K$ which have property $\mathfrak{E}$. In other words: the existence of a Hilbertian element with property $\mathfrak{E}$ is equivalent to D being nonempty. On the other hand, it follows from general enlargement principles [20] that the intersection D is nonempty in $^*K$ if and only if every finite sub-intersection is nonempty in K, which is to say that

$$H_{f_1} \cap \ldots \cap H_{f_n} \cap E \neq \emptyset ,$$

for every finite system $f_1,\ldots,f_n$ of irreducible polynomials in $K[T,X]$. If we put

$$H = H_{f_1} \cap \ldots \cap H_{f_n}$$

then the above condition says that

$$H \cap E \neq \emptyset$$

for every Hilbertian subset $H \subset K$. In other words: every Hilbertian set should contain an element with property $\mathfrak{E}$.

QED.

2. NUMBER FIELDS.

In this section, K denotes an algebraic number field of finite degree.

Our first result will be negative: it says that certain nonstandard elements $u \in {}^*K$ are not Hilbertian. On the other hand, we shall see that the algebraic closure of K(u) in *K is not too large and, moreover, it can be explicitely described. This will lead us then to the construction of Hilbertian elements.

The elements u in question are those whose pricipal divisor (u) is composed of standard primes only. The existence of such u is easily established: for instance, let $a \neq 0$ be an element in K, and put

$$u = a^{\omega} \qquad \text{with } \omega \in {}^*\underline{\underline{N}} .$$

The principal divisor (u) contains exactly those primes which appear in the principal divisor (a); they are all standard since $a \in K$. If a is not a root of unity and if ω is infinite then u is nonstandard.

THEOREM 2.1. Let $u \in {}^*K$ be a nonstandard element whose principal divisor (u) is composed of standard primes only.

Then u is not Hilbertian, i.e. K(u) is not algebraically closed in *K. In fact, for every natural number $n \in \underline{\underline{N}}$ there is one and only one extension F_n of K(u) within *K, such that $[F_n:K(u)] = n$. The field F_n is rational over K. It can be generated by an element z_n such that

$$F_n = K(z_n) \quad \text{and} \quad z_n^n = c_n u$$

with a suitable constant $c_n \in K$.

In the proof, we shall use the notations and results of [21].

Proof.

(i) Existence: Let $n \in \underline{\underline{N}}$. Since n will remain fixed we omit the index n in the following proof. We claim that there exists an extension F of K(u) within *K such that

$$[F:K(u)] = n.$$

We try to construct F in the form F = K(z) where the nonstandard element z ∈ *K is chosen such that

$$(*) \qquad z^n = c\dot{u}$$

with some c ∈ K. In fact, this relation shows that K(z) is of degree n over K(u). Thus we are faced with proving the existence of z ∈ *K and c ∈ K such that (*) holds. This relation says that u is an n-th power in *K, up to a constant factor.

First, we shall prove the analogous statement for the principal divisor (u). By definition, (u) is an element of the group $^*\mathfrak{D}$ of internal divisors of *K. The group operation in $^*\mathfrak{D}$ is written additively. We claim that there exists an internal divisor $\mathfrak{a} \in {}^*\mathfrak{D}$ such that

$$(**) \qquad (u) \doteq n.\mathfrak{a} .$$

That is, (u) ∈ $^*\mathfrak{D}$ is divisible by n up to a finite summand. (As in [21] the symbol ≐ indicates the same order of magnitude, i.e. equality up to a finite quantity.)

Let S denote the set of those internal primes $\mathfrak{p}$ which appear in the principal divisor (u). By definition, S is an <u>internal</u> set. On the other hand, the hypothesis of theorem 2.1 implies that S consists of <u>standard</u> primes only. Now, it follows from general enlargement principles that every internal set which consists only of standard quantities is necessarily a <u>finite</u> set. We conclude that S is finite. That is, there are only finitely many primes $\mathfrak{p}$ such that $v_\mathfrak{p}(u) \neq 0$. (As in [21] the symbol $v_\mathfrak{p}(u)$ denotes the $\mathfrak{p}$-adic ordinal of u.)

We try to construct the divisor $\mathfrak{a} \in {}^*\mathfrak{D}$ such that it is composed of primes from S only. That is, $\mathfrak{a}$ should be a finite sum of the form

$$\mathfrak{a} = \sum_{\mathfrak{p} \in S} \alpha_\mathfrak{p} \cdot \mathfrak{p}$$

where the coefficients $\alpha_{\mathfrak{p}}$ have to be determined such that (**) holds. Let us put

$$\mathfrak{b} = (u) - n.\mathfrak{a} = \sum_{\mathfrak{p}\in S} \beta_{\mathfrak{p}}.\mathfrak{p}$$

where

$$\beta_{\mathfrak{p}} = v_{\mathfrak{p}}(u) - n.\alpha_{\mathfrak{p}}.$$

The condition (**) requires that $\mathfrak{b}$ is a finite divisor; since all $\mathfrak{p} \in S$ are standard this is equivalent to saying that every coefficient $\beta_{\mathfrak{p}}$ is a finite number.

Let us recall that the notion of "divisor" and "prime divisor" also includes the archimedean primes of *K. If $\mathfrak{p}$ is archimedean then the $\mathfrak{p}$-adic coefficient $\alpha_{\mathfrak{p}}$ of an internal divisor is a real number (standard or nonstandard). If $\mathfrak{p}$ is nonarchimedean then this coefficient is required to be an integer (standard or nonstandard). For the general definitions we refer to [21].

Now, if $\mathfrak{p}$ is nonarchimedean, then we use the Euclidean algorithm for $^*\underline{\underline{Z}}$ to define integers $\alpha_{\mathfrak{p}}, \beta_{\mathfrak{p}} \in {}^*\underline{\underline{Z}}$ such that

$$v_{\mathfrak{p}}(u) = n.\alpha_{\mathfrak{p}} + \beta_{\mathfrak{p}} \quad \text{and} \quad 0 \leqq \beta_{\mathfrak{p}} < n.$$

If $\mathfrak{p}$ is archimedean, we define $\alpha_{\mathfrak{p}}, \beta_{\mathfrak{p}} \in {}^*\underline{\underline{R}}$ by the formula

$$v_{\mathfrak{p}}(u) = n.\alpha_{\mathfrak{p}} + \beta_{\mathfrak{p}} \quad \text{and} \quad \beta_{\mathfrak{p}} = 0.$$

We obtain two internal divisors

$$\mathfrak{a} = \sum_{\mathfrak{p}\in S} \alpha_{\mathfrak{p}}.\mathfrak{p} \quad , \quad \mathfrak{b} = \sum_{\mathfrak{p}\in S} \beta_{\mathfrak{p}}.\mathfrak{p}$$

such that

$$(u) = n.\mathfrak{a} + \mathfrak{b} .$$

By construction, the coefficients $\beta_{\mathfrak{p}}$ are contained in the interval

$$0 \leqq \beta_{\mathfrak{p}} < n$$

and hence they are finite. As said above already, this implies that $\mathfrak{b}$ is a finite divisor. That is, we have $\mathfrak{b} \doteq 0$ and therefore

$$(u) \doteq n.\mathfrak{a}$$

as required.

Now let us consider the size $\sigma(\mathfrak{a})$. According to its definition [21] we have

$$\sigma(\mathfrak{a}) = \sum_{\mathfrak{p}\in S} \alpha_{\mathfrak{p}}.\log(N\mathfrak{p})$$

where $N\mathfrak{p}$ denotes the norm of $\mathfrak{p}$. (If $\mathfrak{p}$ is archimedean then we put $\log(N\mathfrak{p}) = 1$.) The size of a principal divisor vanishes; this is the additive way of stating the well known product formula for valuations. Hence, from (**) we infer that $\sigma(n.\mathfrak{a}) = n.\sigma(\mathfrak{a}) \doteq 0$ and therefore

$$\sigma(\mathfrak{a}) \doteq 0 .$$

That is, the size $\sigma(\mathfrak{a})$ is finite.

Now we use theorem 3.4 of [21] (the principal divisor theorem) which says that every divisor of finite size is principal, up to a finite summand. We conclude that there exists an element $z \in {}^*K$ such that

$$(z) \doteq \mathfrak{a} .$$

It follows $(z^n) \doteq n.\mathfrak{a}$ and therefore from (**):

$$(z^n) \doteq (u) .$$

In other words: the principal divisors of z^n and of u coincide, up to a finite term. Again referring to theorem 3.4 of [21] we conclude that z^n and u differ by a factor $c \in K$ only:

$$z^n = cu$$

This proves (*).

(ii) <u>Uniqueness:</u> Let F be any finite algebraic extension of $K(u)$ within $^*K$. We claim that F is of the form as specified in theorem 2.1.

We regard F as an algebraic function field of one variable over K. Then we have

$$K \subset F \subset {}^*K$$

which is just the situation studied in [21]. We know that F contains the nonstandard element u whose denominator (and numerator) is composed of standard primes only. Using theorem 1.1 of [21] (the Siegel-Mahler theorem) we conclude that F is of genus zero.

The subfields of *K of genus zero have been classified in [21]. This classification can be described as follows: An element of F is called "exceptional" if its denominator is divisible by standard primes only. For instance, both u and u^{-1} are exceptional in F. Let R denote the ring of all exceptional elements in F. The said classification describes the structure of R as follows: there is a generator z of $F|K$ such that

$$F = K(z)$$

and that one of the following four cases applies:

<u>Case 0.</u> $R = K$

<u>Case 1.</u> $R = K[z]$

<u>Case 2.</u> $R = K[z,z^{-1}]$

<u>Case 2a.</u> $R = K[\varphi^{-1}, z\varphi^{-1}]$ where $\varphi = z^2-a$.

Here, a denotes an element in K which is not a square in K.

In our present situation, we claim that case 2 applies. This is because the ring R contains both u and u^{-1}, i.e. the given element u is a <u>unit</u> of R. Moreover, $u \notin K$, i.e. u is nonconstant. On the other hand, it is easily verified that in cases 0,1 and 2a every unit of R is constant. This is clear in case 0. In cases 1 and 2a, we remark that every element in R has only one pole, namely the pole of z (in case 1) resp. the zero of $\varphi = z^2-a$ (in case 2a). Therefore a unit of R has no pole at all and hence is constant in these cases.

We have proved that in our case, $R=K[z,z^{-1}]$. The units of this ring are the constant multiples of the powers of z. Hence $u = az^n$ with suitable $a \in K$ and $n \in \underline{\underline{Z}}$. After replacing z by z^{-1} if necessary, we may assume that $n > 0$. If we put $c = a^{-1}$, then we have

$$F = K(z) \quad , z^n = cu \quad .$$

This shows that $[F:K(z)] = n$, and that F is of the form as specified in theorem 2.1.

We have to show that F is uniquely determined by its degree n. For, let F' be any other extension of K(u) within *K, of the same degree n over K(u). Then, by what has been proved above, we have

$$F' = K(z') \quad , \quad z'^n = c'u$$

with z' ∈ *K and c' ∈ K. We conclude

$$(z'z^{-1})^n = c'c^{-1} \in K \quad .$$

Since K is algebraically closed in *K it follows that $z'z^{-1} \in K$. Hence there exists b ∈ K such that

$$z' = bz$$

$$F' = K(z') = K(z) = F$$

QED.

Now, the idea of the following construction is the following: starting from an element u as in theorem 2.1 we try to obtain a Hilbertian element t as a linear polynomial of the form

$$t = t_o + t_1 u$$

where the coefficients t_o, t_1 are in some sense "generic". This notion of "generic" has to be made precise; to this end we perform the following construction.

As a mathematical structure of its own, *K also has an enlargement. Let **K denote an enlargement of *K. Then we have

$$K \subset {}^*K \subset {}^{**}K \ .$$

Referring to the definition of enlargement [20] it is immediate that **K can also be regarded as an enlargement of K, naturally. For brevity, we speak of the enlargements **K| *K and **K| K respectively. Accordingly, we have two notions of "standard" entities defined in **K, those which are standard over *K and those standard over K. For instance, every individual in *K is standard over *K, but it is K-standard only if it is contained in K. In general, every K-standard entity is also *K-standard but not conversely. The K-standard prime divisors of **K correspond 1-1 to the prime divisors of K, whereas the *K-standard primes correspond 1-1 to the internal primes of $^*K|K$.

Now let $u \in {}^{**}K$ be *K-nonstandard, i.e. $u \notin {}^*K$. We consider linear polynomials in u of the form

$$t = t_o + t_1 u$$

with coefficients $t_o, t_1 \in {}^*K$. Such a polynomial is called _generic over_ K if the coefficients $t_o, t_1 \in {}^*K$ are algebraically independent over K. It is immediate from the definition of enlargement that *K is of infinite degree of transcendency over K (see lemma 3.3 below); hence there exist generic linear polynomials in the above sense.

In particular, we are concerned with such elements $u \in {}^{**}K \setminus {}^*K$ whose principal divisor (u) is composed of K-standard primes only. The existence of such u is easily established: we may take

$$u = a^{\omega}$$

where $a \in K$ is not a root of unity, and where $\omega \in {}^{**}\underline{\underline{N}} \setminus {}^*\underline{\underline{N}}$. Such elements u lead to Hilbertian elements, according to the following theorem.

THEOREM 2.2. _Let_ $u \in {}^{**}K \setminus {}^*K$ _be such that its principal divisor_ (u) _is composed of_ K-_standard primes only. Consider a linear polynomial_

$$t = t_o + t_1 u \quad , \quad t_o, t_1 \in {}^*K$$

which is generic in the sense that t_o, t_1 _are algebraically independent over_ K.

Then K(t) _is algebraically closed in_ **K. _That is_, t _is Hilbertian in the enlargement_ $^{**}K|K$.

REMARK. Using the theorem of Gilmore-Robinson it follows from this that K is a Hilbertian field. Moreover, we have explicitely constructed a Hilbertian element t in the enlargement $^{**}K|K$.

Proof of theorem 2.2.

First we apply theorem 2.1 to the enlargement $^{**}K|K$ and the element $u \in {}^{**}K$. We conclude that for each $n \in \underline{\underline{N}}$ there is one and only one extension F_n of K(u) in **K such that $[F_n : K(u)] = n$.

Therefore the union of these fields F_n is the algebraic closure of $K(u)$ in $^{**}K$. This union will be denoted by F; thus we have

$$F = \bigcup_n F_n$$

and F is algebraically closed in $^{**}K$.

Now we apply theorem 2.1 to the enlargement $^{**}K | ^{*}K$. Although this theorem has been proved only for an algebraic number field as ground field, it is clear from general model theory that the theorem remains true if K is replaced by some other model of K, e.g. by $^{*}K$. It follows that $^{*}K(u)$ too has exactly one extension of degree n inside $^{**}K$, for each $n \in \underline{\underline{N}}$. This extension can be identified to be the field compositum $^{*}KF_n$. To see this, we generate F_n according to theorem 2.1 in the form

$$F_n = K(z_n) \quad , \quad z_n^n = c_n u \quad .$$

We conclude that

$$^{*}KF_n = {}^{*}K(z_n) \ , \quad z_n^n = c_n u \quad .$$

Since u is transcendental over $^{*}K$, the above equation for z_n is irreducible over $^{*}K(u)$ and thus

$$[\,^{*}KF_n : {}^{*}K(u)] = n$$

as contended.

It follows that the union of the fields $^{*}KF_n$ is the algebraic closure of $^{*}K(u)$ in $^{**}K$. Now we have

$$^{*}KF = \bigcup_n {}^{*}KF_n \quad .$$

<u>We conclude that the field compositum</u> $^{*}KF$ <u>is algebraically closed in</u> $^{**}K$.

By construction, F contains the element u. On the other hand, $^{*}K$ contains t_o, t_1. Therefore the compositum $^{*}KF$ contains the element $t = t_o + t_1 u$. Thus we have the field tower

$$K(t) \subset {}^{*}KF \subset {}^{**}K.$$

We know from above that $^{*}KF$ is algebraically closed in $^{**}K$. <u>Therefore,</u>

<u>in order to prove theorem 2.2 it suffices to show that</u> K(t) is <u>algebraically closed in</u> *KF.

This reduction from **K to *KF is the main step of our proof. The rest of proof will be purely algebraic, using well known arguments from algebraic geometry resp. general field theory.

By construction, the element u is transcendental over *K, and t_o, t_1 are elements of *K which are algebraically independent over K. Hence, the three elements t_o, t_1, u are algebraically independent over K. It follows that $t = t_o + t_1 u$ is transcendental over K(u) and hence also over the field F (which is algebraic over K(u)). Consider the field tower

$$K(t) \subset F(t) \subset {}^*KF \ .$$

We know that K is algebraically closed in F (because K is algebraically closed in its enlargement **K). This property of relative algebraic closure is preserved if we adjoin a transcendental element to the respective fields. We conclude that K(t) is algebraically closed in F(t). <u>Hence, it remains to verify that</u> F(t) <u>is algebraically closed in</u> *KF.

The situation of the fields involved is shown in the following diagram:

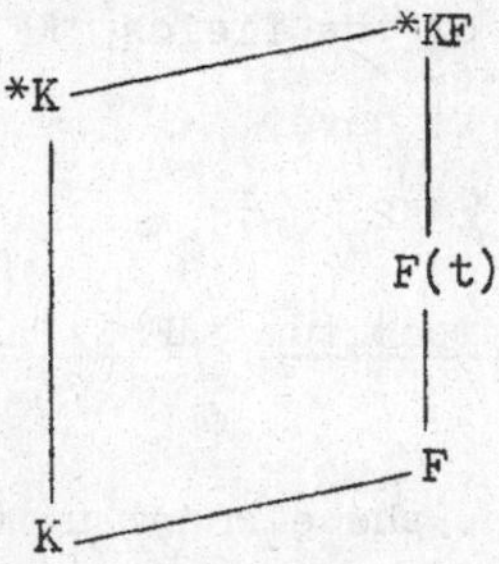

Concerning these fields, we have the following information available.

(1) K <u>is algebraically closed in</u> *K (because *K is the enlargement of K). Since K is a number field and hence of characteristic 0, this can also be expressed by saying that *K is <u>regular</u> over K, in the

sense of general field theory.

(2) F <u>is linearly disjoint from</u> *K <u>over</u> K. (This is because u is transcendental over *K and F algebraic over K(u); hence F is algebraically free from *K over K. Since *K|K is regular we conclude the linear disjointness.)

(3) $t = t_o + t_1 u$, where t_o, t_1 are in *K and algebraically independent over K, and where u is in F and transcendental over K.

From these informations (1)-(3) it follows that F(t) is algebraically closed in *KF. This is the content of a field theoretical lemma, known as "Matsusaka's lemma". We shall state this lemma below; a proof can be found in section 5.

QED.

The said "lemma of Matsusaka" refers to an arbitrary ground field K. We shall state it in full generality, although at present we have applied it only in the case of characteristic zero, and hence we could omit the discussion of inseparabilities. But in the next sections we want to apply this lemma also in other cases, including characteristic $p > 0$.

We consider the following situation: K is an arbitrary field and E, F are two extensions of K, both contained in a common overfield. The following conditions are imposed:

(1) E is a regular extension of K. That is, K is algebraically closed in E, and E is separable over K. (This separability condition is always satisfied if K has characteristic zero.)

(2) F is linearly disjoint from E over K.

Now let u be an element of F. We consider a linear polynomial

$$t = t_o + t_1 u$$

where the coefficients t_o, t_1 are in E. It is assumed that this polynomial is "generic" in the following sense:

(3) t_o, t_1 are algebraically independent over K.

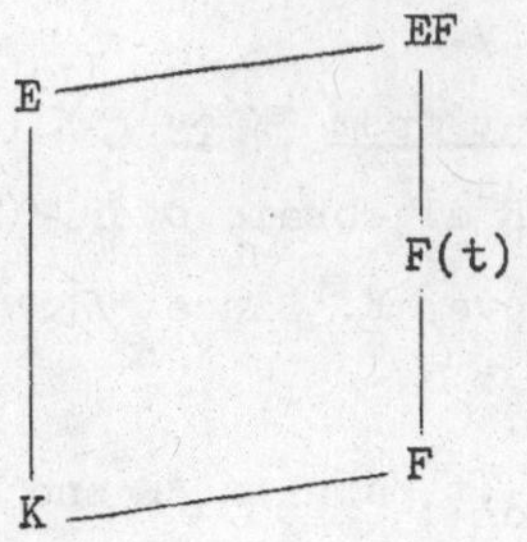

In this situation we have the following

LEMMA of MATSUSAKA.

(A) _If_ u _is transcendental over_ K _then_ F(t) _is separable-algebraically closed in_ EF.

(B) _Let_ char(K) = p > 0. _If_ u _is_ p-_free over_ K _then_ F(t) _is closed with respect to purely inseparable extensions within EF._

(C) _Hence, if_ u _is both transcendental and_ p-_free over_ K _then_ F(t) _is algebraically closed in_ EF.

As usual, u is called p-_free over_ K if $u \notin F^pK$. If char(K) = 0 then statement (B) is void and (A) is identical with (C).

A proof of this lemma can be found in section 5.

3. FUNCTION FIELDS.

Let us start with some preliminary remarks.

Let K be an infinite field, and let F be an extension field of K. The inclusion $K \subset F$ induces naturally an inclusion $*K \subset *F$ of their enlargements. Thus we have the following diagram of fields.

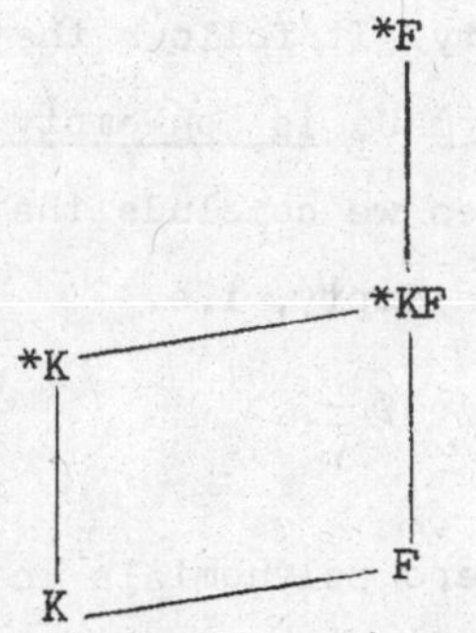

LEMMA 3.1 *K <u>and</u> F <u>are linearly disjoint over</u> K.

<u>Proof.</u>

Let $u_1,\dots,u_n$ be finitely many elements in F which are linearly independent over K. The statement that the u_i are K-linearly independent, remains true in the enlargement. Hence the $u_1,\dots,u_n$ are linearly independent over *K.

QED.

In lemma 3.1 we may take for F the algebraic closure of K. It follows that *K is K-linearly disjoint to the algebraic closure of K, which is to say:

COROLLARY 3.2. *K <u>is a regular extension of</u> K.

LEMMA 3.3. *K <u>is of infinite degree of transcendency over</u> K.

<u>Proof.</u>

Let $n \in \underline{\underline{N}}$ be a standard natural number. We have to show that *K contains n elements which are algebraically independent over K. Let $\underline{T} = (T_1,\dots,T_n)$ be a system of n independent variables, and let $f(\underline{T}) \in K[\underline{T}]$ be a nonzero polynomial. Since K is infinite there exists $\underline{t} = (t_1,\dots,t_n)$ in K^n such that $f(\underline{t}) \neq 0$. Let $U_f \subset K^n$ denote the Zariski-open set of those $\underline{t} \in K^n$ for which $f(\underline{t}) \neq 0$. Then U_f is non-empty; this holds for every $f \neq 0$. We have the formula

$$U_f \cap U_g = U_{fg}$$

which shows that the intersection of two such sets is again of the same form, and hence non-empty. It follows that <u>the intersection of finitely many sets of the form</u> U_f <u>is non-empty.</u> Consequently, from general enlargement principles we conclude that the intersection of all their enlargements is non-empty, i.e.

$$\bigcap_f {}^*U_f \neq \emptyset \quad ,$$

where f ranges over the nonzero polynomials in $K[\underline{T}]$.
Let $\underline{t} = (t_1,\dots,t_n) \in {}^*K^n$ be in this intersection. For every $f \neq 0$ in $K[\underline{T}]$ we know that $\underline{t} \in {}^*U_f$, which is to say that $f(\underline{t}) \neq 0$. This means that the $t_1,\dots,t_n \in {}^*K$ are algebraically independent over K.

QED.

This being said, we now consider finitely generated extension fields F of K. Following an accepted terminology, every such extension $F|K$ is called a <u>function field</u> over K. The <u>dimension</u> of that function is defined to be the degree of transcendency of F over K.

It has been proved by Franz [4] that every separable function field F of dimension ≥ 1 over K is Hilbertian, regardless of whether the ground field K is Hilbertian or not. The validity of this theorem is however not restricted to the separable case, as Inaba [9] has shown. As said in the introduction, we now proceed to give a nonstandard proof of this fact by explicitely constructing a Hilbertian element $t \in {}^*F$ for F. We shall not distinguish between the separable and the inseparable case.

Let $u \in F$ and consider a linear polynomial

$$t = t_o + t_1 u$$

where the coefficients t_o, t_1 are in *K, and are assumed to be algebraically independent over K. As above, t is called a "generic" linear polynomial in u. It follows from lemma 3.3 that there exist such generic polynomials. By definition, t is an element of *F. It follows from lemma 3.1 that t is not contained in F, i.e. t is nonstandard. Under certain conditions about u, we now claim that t is Hilbertian. These conditions are the following:

(A) u is transcendental over K, and

(B) u is p-free over K if $\mathrm{char}(K) = p > 0$.

The existence of such elements $u \in F$ is easily established, for every function field $F|K$ of dimension $\geqq 1$. This is clear for condition (A), since by definition the dimension equals the degree of transcendency. If $\mathrm{char}(K) = p > 0$ then we observe that every p-basis $u_1, \ldots, u_n$ of $F|K$ contains a basis of transcendeny; since $F|K$ has dimension $\geqq 1$ we conclude that there is at least one of the u_i which is transcendental over K. As a member of a p-basis, this u_i is at the same time p-free over K.

THEOREM 3.4. <u>Let</u> $F|K$ <u>be a function field of dimension</u> $\geqq 1$.

<u>Let</u> u <u>be an element of</u> F <u>which is transcendental and</u> p-<u>free over</u> K. <u>Consider a linear polynomial</u>

$$t = t_o + t_1 u$$

<u>whose coefficients</u> t_o, t_1 <u>are in</u> *K, <u>and generic in the sense that</u> t_o, t_1 <u>are algebraically independent over</u> K.

<u>Then</u> F(t) <u>is algebraically closed in</u> *F. <u>Hence the element</u> $t \in {}^*F$ <u>is Hilbertian for</u> F.

Again, using the theorem of Gilmore-Robinson it follows that every function field of dimension $\geqq 1$ is Hilbertian.

REMARK. In theorem 3.4 the ground field K is assumed to be <u>infinite</u>, according to the general hypothesis stated at the beginning of this section. The proof does not apply to function fields with finite ground fields (if K is finite then *K = K). Function fields with finite ground fields can be dealt with by the arithmetic methods of section 2.

<u>Proof of theorem 3.4.</u>

We consider the following diagram of fields:

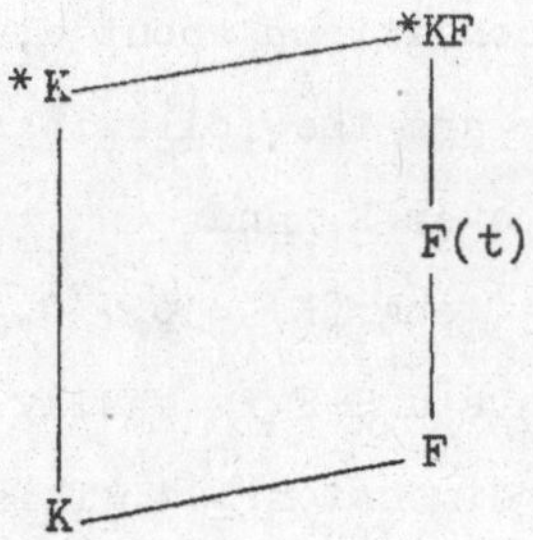

Here, *K is regular over K (corollary 3.2), and F is linearly disjoint to *K over K (lemma 3.1). Therefore, the assumptions of Matsusaka's lemma are satisfied (see the end of section 2 for a statement of this lemma). We conclude that F(t) is algebraically closed in *KF.

<u>It remains to show that</u> *KF <u>is algebraically closed in</u> *F. This is contained in the following.

THEOREM 3.5. <u>Let</u> F <u>be any finitely generated field extension of</u> K <u>(dimension zero included). Then</u> *F <u>is a regular extension of</u> *KF. <u>That is,</u> *KF <u>is algebraically closed in</u> *F, <u>and</u> *F <u>is separable over</u> *KF.

<u>Proof.</u>

We use induction with respect to the number r of generators of $F|K$, in order to reduce the proof to the case where $F|K$ is generated by a single element. This reduction step is carried out as follows: Let L be an intermediate field

$$K \subset L \subset F$$

and assume that $F|L$ is generated by one element, whereas $L|K$ is generated by r-1 elements. By induction, *L is then regular over *KL. Now we know that F is L-linearly disjoint to *L (lemma 3.1), hence *KF is *KL-linearly disjoint to *L (see diagram next page).

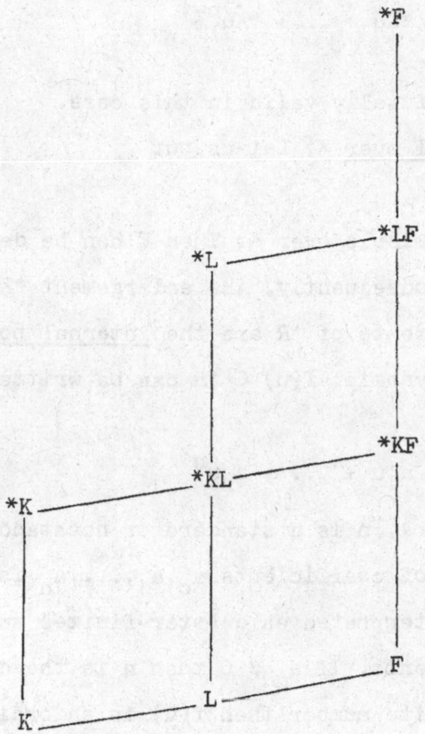

Therefore, from the regularity of *L| *KL we conclude the regularity of *LF| *KF.

Now, if we assume the theorem proved for one-generator extensions, then we apply this to F| L and conclude that *F| *LF is regular. Since a tower of regular extensions is again regular, it follows that *L| *KF is regular, as contended.

<u>Thus it suffices to prove theorem 3.5 under the additional hypothesis that</u> F = K(u) <u>is generated by a single element</u> u.

We distinguish two cases:

<u>Case 1:</u> u <u>is algebraic over</u> K. Let n = [F:K]. The elements $1, u, \ldots, u^{n-1}$ form a K-basis of F. The statement that these elements form a basis, remains true in the enlargement. Hence $1, u, \ldots, u^{n-1}$ form a *K-basis of *F. It follows [*F:*K] = n and

$$*F = *K + *Ku + \ldots + *Ku^{n-1}$$
$$= *KF \ .$$

We see that theorem 3.5 is trivially valid in this case.

Case 2: u is transcendental over K. Let us put

$$R = K[u] \ ,$$

the polynomial ring in one variable over K. Then F can be described as the quotient field of R. Consequently, the enlargement *F is the quotient field of *R. The elements of *R are the internal polynomials over *K. Such an internal polynomial $f(u) \in *R$ can be written in the form

$$f(u) = a_o + a_1 u + \ldots + a_n u^n$$

with coefficients $a_i \in *K$. Here, n is a standard or nonstandard natural number, the sequence of coefficients $a_o, a_1, \ldots, a_n$ is internal, and the above sum is to be interpreted as a "star-finite" sum as generally defined in enlargements. If $a_n \neq 0$ then n is the degree of f(u). If this degree is a finite number then f(u) is an ordinary polynomial, i.e. $f(u) \in *K[u]$. Thus, *R contains the ordinary polynomial ring *K[u] as a proper subring.

We also remark that *R is integrally closed in its quotient field *F. (This is so because R = K[u] is integrally closed in F = K(u).)

This being said, we proceed to show that *KF is algebraically closed in *F. Since F = K(u), we have

$$*KF = *K(u) \ ,$$

which is to say that *KF consists of the ordinary rational functions in u with coefficients in *K. Let $0 \neq x \in *F$ be algebraic over *K(u); we have to show that x is contained in *K(u). After multiplying x with a suitable element from *K[u] we may assume that x is integral over *K[u]; our aim is to show that $x \in *K[u]$.

We have seen above that $*K[u] \subset *R$ and that *R is integrally closed in *F. Hence x, being integral over *K[u], is contained in *R. This means that x can be represented in the form

$$x = f(u)$$

where f(u) is an internal polynomial as described above. We have to show that its degree is a <u>finite</u> number; this will imply that $x = f(u) \in {}^*K[u]$.

Consider the element x^{-1} which is also algebraic over $^*K(u)$. Again, after multiplying x^{-1} with a suitable factor $h(u) \in {}^*K[u]$ we obtain an element $y = h(u)x^{-1}$ which is integral over $^*K[u]$, hence

$$y = g(u)$$

where g(u) is an internal polynomial. It follows

$$xy = f(u)g(u) = h(u) ,$$

$$\deg f(u) \leqq \deg h(u) .$$

By construction h(u) is an ordinary polynomial, of finite degree. Therefore deg f(u) is finite too, hence $x = f(u) \in {}^*K[u]$ as contended.

We have shown that $^*K(u) = {}^*KF$ is algebraically closed in $^*F$. It remains to prove that $^*F$ is separable over $^*K(u)$. This assertion is nontrivial only if $\mathrm{char}(K) = p > 0$; it is equivalent to saying that p-independent elements of $^*K(u)$ remain p-independent in $^*F$. It suffices to consider p-independent elements from a fixed p-basis of $^*K(u)$. Now, a p-basis of the rational function field $^*K(u)$ is obtained by first choosing a p-basis of $^*K$ and then adding the single element u. Let

$$a_1,\dots,a_n, u$$

be a typical finite subset of that p-basis, the a_i being p-independent in $^*K$. We have to show that this set remains p-independent in $^*F$. In other words: we have to show the validity of the following statement:

" If $a_1,\dots,a_n$ are p-independent elements in $^*K$ then $a_1,\dots,a_n$, u are p-independent elements in $^*F$. "

Here n denotes a given standard natural number. The above statement refers to the enlargement $^*F$, regarded as a field extension of *K. This statement is true in $*F|*K$ if and only if it is true in $F|K$. But its validity in $F|K$ is obsious, since $F = K(u)$ is a purely transcendental extension. Namely, if $a_1,\dots,a_n$ are p-independent elements in K then $a_1,\dots,a_n,u$ are p-independent in $K(u) = F$.

QED.

4. FINITELY GENERATED EXTENSIONS OF HILBERTIAN FIELDS.

As Hilbert [8] has shown, every finite separable algebraic extension of a Hilbertian field is again Hilbertian. Actually Hilbert considered only the case of algebraic number fields, which are finite algebraic extensions of $\underline{\underline{Q}}$. However his proof is valid in general, at least in the separable case, as Franz [4] has observed. The inseparable case has been treated by Inaba [9].

We are now going to construct a Hilbertian element $t \in *F$, for every finitely generated extension F of the Hilbertian field K. We do not distinguish between the separable and the inseparable case. Also, we include the case where $F|K$ is of dimension > 0 although this has been treated in section 3 already. For, our aim is not only to prove the Hilbertian property of F but also to exhibit a certain universal construction for Hilbertian elements; in this respect we shall obtain new information also if $\dim F|K > 0$.

In theorem 3.4 we had seen that a "generic linear polynomial" could be used as Hilbertian element for F. Is this possible also in our present situation? Thus, if $u \in F$ is a suitable element, we try to find Hilbertian elements $t \in *F$ which are of the form

$$t = t_o + t_1 u$$

with $t_o, t_1 \in {}^*K$, algebraically independent over K. This time, however, we cannot expect that this construction is successful for arbitrary choice of K-algebraically independent elements $t_o, t_1 \in {}^*K$. For in our present situation we have to use the fact that K is Hilbertian; hence the choice of t_o, t_1 should be somehow adapted to the Hilbertian structure of K.

We need the following

LEMMA 4.1. <u>Assume</u> K <u>is Hilbertian. For every standard natural number</u> n <u>there exist</u> n <u>elements</u> $t_1, \ldots, t_n \in {}^*K$, <u>algebraically independent over</u> K, <u>such that</u> $K(t_1, \ldots, t_n)$ <u>is algebraically closed in</u> $^*K$.

Such a system $\underline{t} = (t_1, \ldots, t_n) \in {}^*K^n$ will be called a <u>Hilbertian system</u> of length n. (We avoid the name "Hilbertian set" since this is used in another sense; see section 1.)

REMARK 4.2. The statement of lemma 4.1 can also be expressed in terms of the field K, without referring to the enlargement. To see this let $\underline{T} = (T_1, \ldots, T_n)$ denote an n-tuple of independent variables. We consider irreducible polynomials $f(\underline{T}, X)$ in n+1 variables $\underline{T}$, X over K. For any such $f \in K[\underline{T}, X]$, let H_f denote the corresponding Hilbert set in K^n, consisting of all $\underline{t} \in K^n$ such that $f(\underline{t}, X) \in K[X]$ is irreducible. Let $^*H_f$ be its enlargement in $^*K^n$. Now, we have that $\underline{t} \in {}^*K^n$ <u>is Hilbertian if and only if</u> $\underline{t}$ <u>is contained in the intersection</u>

$$D = \bigcap_f {}^*H_f \quad .$$

The proof of this is exactly the same as the similar proof for n=1 in section 1 (see part (i) of the proof of the generalized Gilmore-Robinson theorem.) Hence, lemma 4.1 can also be expressed by saying that the above intersection D is non-empty. On the other hand, it follows from general enlargement principles that D is non-empty in $^*K^n$ if and only if every finite sub-intersection is non-empty in K^n, which is to say that

$$H_{f_1} \cap \ldots \cap H_{f_m} \neq \emptyset$$

for every finite system of irreducible polynomials $f_1,\ldots,f_m \in K[\underline{T},X]$. By definition, the finite intersections of the above form are the <u>Hilbert subsets of</u> K^n. Thus, the statement of lemma 4.1 can be expressed in terms of K as follows:

<u>If</u> K <u>is Hilbertian then for any natural number</u> $n \in \underline{N}$, <u>the Hilbert subsets of</u> K^n <u>are nonempty.</u>

This statement reflects the observation, already made by Hilbert [8], that the validity of the irreducibility theorem for one parametric variable T implies its validity for n parametric variables $\underline{T} = (T_1,\ldots,T_n)$, for any $n \in \underline{N}$. And we see that lemma 4.1 is the nonstandard version of this fact.

<u>Proof of lemma 4.1.</u>

We assume $n > 1$ and use induction with respect to n. (The case n=1 has been settled by the theorem of Gilmore and Robinson.) Thus we have n-1 elements $t_1,\ldots,t_{n-1} \in {}^*K$ which form a Hilbert system. For brevity, let us put

$$K' = K(t_1,\ldots,t_{n-1}) \quad .$$

Then K' is algebraically closed in *K.

As in section 2, we regard *K as a mathematical structure of its own and consider an enlargement **K| *K. Since *K is a model of K, the Hilbertian property of K is inherited by *K; hence there exists a Hilbertian element in **K for *K. Let $t_n \in {}^{**}K$ be such an element; then t_n is transcendental over *K and $^*K(t_n)$ is algebraically closed in **K. By construction, t_n is transcendental over K' and hence the system $\underline{t} = (t_1,\ldots,t_{n-1},t_n)$ is algebraically independent over K. We claim that the field

$$K(\underline{t}) = K'(t_n)$$

is algebraically closed in **K. To this end, consider the following field diagram.

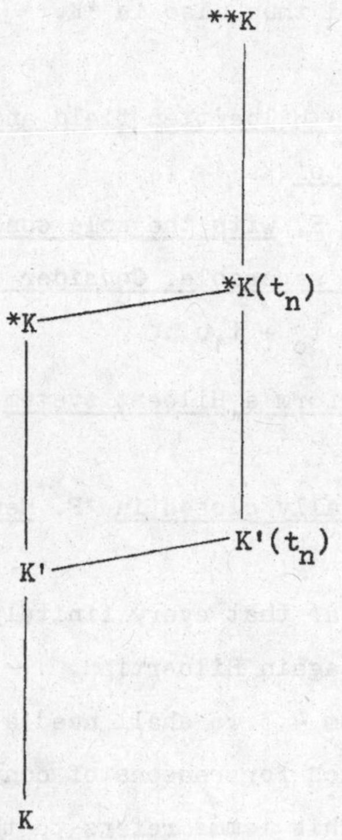

We know that K' is algebraically closed in *K. This property of relative algebraic closure is preserved if we adjoin a transcendental element to the respective fields; hence $K'(t_n)$ is algebraically closed in $*K(t_n)$. On the other hand, $*K(t_n)$ is algebraically closed in **K by construction of t_n. We conclude that, indeed, $K'(t_n)$ is algebraically closed in **K.

We have found a system $\underline{t} = (t_1,\ldots,t_n)$ of K-algebraically independent elements in **K, such that $K(\underline{t})$ is algebraically closed in **K. That is, we have found a Hilbert system of length n in **K.

We now regard **K as an enlargement of K, in the natural way. Thus we have shown that lemma 4.1 holds in some enlargement of K. On the other hand, we have seen in remark 4.2 that this lemma can be stated in terms of K only, without referring to any definite enlargement.

Therefore, since lemma 4.1 holds in some enlargement, it holds in every enlargement of K and thus also in *K.

QED.

THEOREM 4.3. *Let* K *be a Hilbertian field and* F *a finitely generated field extension of* K.

Let u *be an element of* F, *with the sole condition that* u *is* p-*free over* K *in case* F|K *is not separable. Consider a linear polynomial*

$$t = t_o + t_1 u$$

whose coefficients t_o, t_1 *form a Hilbert system (of length 2) in* *K *(see lemma 4.1).*

Then F(t) *is algebraically closed in* *F. *Hence* $t \in$ *F *is Hilbertian for* F.

This shows in particular that every finitely generated extension of a Hilbertian field is again Hilbertian.

In the proof of theorem 4.3 we shall need a well known fact from general field theory, which for reasons of convenient reference we shall state as a lemma. This lemma refers to the following field theoretic situation:

K is a field, E and F are extension fields of K, both contained in a common overfield. We assume:

E *and* F *are linearly disjoint over* K.

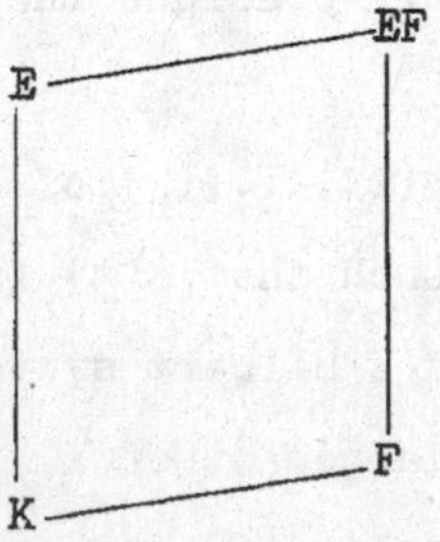

In this situation, we have

LEMMA 4.4. *If* K *is algebraically closed in* E *then* F *is* algebraically closed in EF, provided at least one of the extensions $E|K$ or $F|K$ is separable.

If this separability condition is not satisfied then we can only deduce that F is separable-algebraically closed in EF.

For the proof, we refer to Lang [15] page 58, theorem 4 (separable case) and page 61, corollary 6 (inseparable case).

Proof of theorem 4.3.

The contention is that F(t) is algebraically closed in its extension *F. We split this extension into three levels:

$$F(t) \subset F(t_o,t_1) \subset {}^*KF \subset {}^*F \ .$$

It suffices to show that in each of these three levels, the ground field is algebraically closed in its extension. As to the highest level:

$$^*KF \subset {}^*F$$

this has been proved in theorem 3.5 already.

As to the lowest level:

$$F(t) \subset F(t_o,t_1)$$

our contention follows immediately from the definition of $t = t_o + t_1u$ which shows that

$$F(t,t_1) = F(t_o,t_1) \ .$$

Since t_o, t_1 are algebraically independent over K, hence also over F, it follows that t_1 is transcendental over F(t). Hence $F(t_o,t_1)$ is a purely transcendental extension of F(t) which implies F(t) to be algebraically closed in $F(t_o,t_1)$.

It remains to discuss the intermediate level:

$$F(t_o,t_1) \subset {}^*KF.$$

Consider the following diagram of fields:

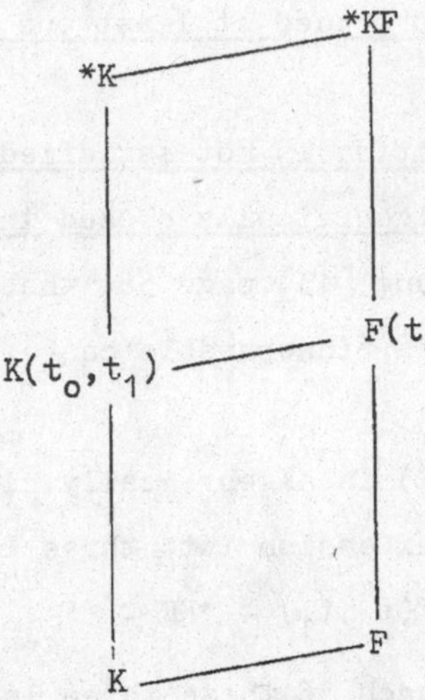

We know that *K and F are K-linearly disjoint (lemma 3.1). Hence *K and $F(t_o,t_1)$ are $K(t_o,t_1)$-linearly disjoint (see diagram). Also, we know that $K(t_o,t_1)$ is algebraically closed in *K (since by assumption t_o,t_1 is a Hilbert system for *K|K). Therefore, applying lemma 4.4 we conclude that $F(t_o,t_1)$ <u>is separable-algebraically closed in</u> *KF. If in addition F|K is separable then $F(t_o,t_1)$ is separable over $K(t_o,t_1)$ and therefore lemma 4.4 shows $F(t_o,t_1)$ algebraically closed in *KF.

If F|K is inseparable then the above proof shows only that F(t) is separable-algebraically closed in *KF. It remains to verify that F(t) is closed with respect to purely inseparable extensions within *KF. But this follows from part (B) of Matsusaka's lemma (see the end of section 2.) Observe that the hypothesis of that lemma, namely that u is p-free over K, is part of the hypothesis of theorem 4.3.

QED.

If F|K is separable then theorem 4.3 does not impose any condition on u; hence we may take u = 0 and conclude that $t = t_o$ is Hilbertian for F. In this case, the above proof can be modified such that t_1 does not appear at all; it is immediate from lemma 4.4 that $F(t_o) = F(t)$ is algebraically closed in *KF provided $K(t_o)$ is algebraically closed in *K. In other words, we obtain the following

COROLLARY 4.5. *In the same situation as in theorem 4.3, assume in addition that* F *is separable over* K. *Then every Hilbertian element* $t_o \in {}^*K$ *for* K *is at the same time a Hilbertian element for* F.

If one analyses the proof by Franz [4] for the separable-algebraic case, then it turns out that his proof yields precisely the standard equivalent of our corollary 4.5. Franz was not able to decide, by his methods, the question about inseparable extensions. These methods had to fail because, as we shall see now, corollary 4.5 is no longer true for inseparable extensions.

We assume that char(K) = p > 0. Recall that the p-*degree* of K is defined to be the number of elements in a p-basis of K. If K is Hilbertian then its p-degree is > 0, i.e. K is imperfect. For, if K would be perfect then *K too would be perfect, contradicting the fact that a Hilbertian element $t \in {}^*K$ is not a p-th power in *K by its very definition.

There is a difference of behavior of fields with finite and those with infinite p-degree. Those with finite p-degree show the worst deviation from corollary 4.5.

THEOREM 4.6. *Let* K *be a Hilbertian field of infinite* p-*degree. Then there exists a Hilbertian element* $t_o \in {}^*K$ *such that* *K *is separable over* $K(t_o)$. *Consequently,* *K *is then a regular extension of* $K(t_o)$.

Such $t_o \in {}^*K$ *remains Hilbertian for every finitely generated extension* F *of* K, *including inseparable ones.*

THEOREM 4.7. *Let* K *be Hilbertian of finite* p-*degree. Then* *K *is inseparable over* $K(t_o)$, *for every Hilbertian element* $t_o \in {}^*K$.

There exists a finite extension F *of* K *such that no Hilbertian element for* K *is also Hilbertian for* F.

Proof of theorem 4.6.

Let $a_1,\dots,a_n$ be a finite subset of K. First we try to find a

Hilbertian element in $*K$ which is p-independent from the a_i. Let $t_o \in *K$ be an arbitrary Hilbertian element. If t_o should be p-dependent from the a_i in $*K$ then we choose $b \in K$ which is p-independent from the a_i, and we replace t_o by t_o-b. (The existence of b follows from our assumption about K being of infinite p-degree.) It is clear that t_o-b is p-independent from the a_i in $*K$ (otherwise $b = t_o-(t_o-b)$ would be p-dependent from the a_i in $*K$ and hence also in K, contrary to the choice of b). Moreover t_o-b is Hilbertian since $K(t_o-b) = K(t_o)$. Writing again t_o instead of t_o-b, we have proved the existence of a Hilbertian element $t_o \in *K$ which is p-independent from $a_1,\dots,a_n$.

This holds for every finite subset $a_1,\dots,a_n$ of K. From this we can conclude the existence of a Hilbertian element $t_o \in *K$ which is p-independent from the whole field K. The argument is as follows:

Let $a_1,\dots,a_n$ as above, and $f_1,\dots,f_m$ be finitely many irreducible polynomials in $K[T,X]$. Since $t_o \in *K$ is Hilbertian we know that the $f_j(t_o,X)$ are irreducible in $*K[X]$. Thus the following statement is true in $*K$ and hence in K:

"There exists t_o such that the polynomials $f_1(t_o,X),\dots,f_m(t_o,X)$ are irreducible, and that t_o is p-independent from $a_1,\dots,a_n$."

Using fundamental enlargement principles we conclude that there exists a nonstandard $t_o \in *K$ which satisfies these conditions, simultaneously with respect to all irreducible polynomials and all finite subsets of K. In other words: this element $t_o \in *K$ is Hilbertian, and it is p-independent from K.

If t_o is chosen in this way, we claim that $*K$ is separable over $K(t_o)$. This is equivalent to saying that p-independent elements in $K(t_o)$ remain p-independent in $*K$. It suffices to consider p-independent elements of a fixed p-basis of $K(t_o)$. Now, a p-basis of $K(t_o)$ is obtained by first choosing a p-basis of K and then adding the single element t_o. Let $a_1,\dots,a_n,t_o$ be a typical finite subset of that p-basis, the a_i being p-independent in K. We have to show that $a_1,\dots,a_n,t_o$ are p-independent in *K. By construction, t_o is p-independent in *K from the a_i. On the other hand, since the a_i are p-independent in K, the same statement is true in the enlargement *K, i.e. the a_i are p-independent in *K.

<u>We have shown that</u> t_o <u>is a Hilbertian element such that</u> *K <u>is separable over</u> $K(t_o)$.

Now let F be an arbitrary finitely generated extension of K, not necessarily separable. We claim that t_o is Hilbertian also for F, which is to say that $F(t_o)$ is algebraically closed in *F. We know from theorem 3.5 that *KF is algebraically closed in *F; thus again it suffices to prove $F(t_o)$ to be algebraically closed in *KF. We apply lemma 4.4 to the following situation:

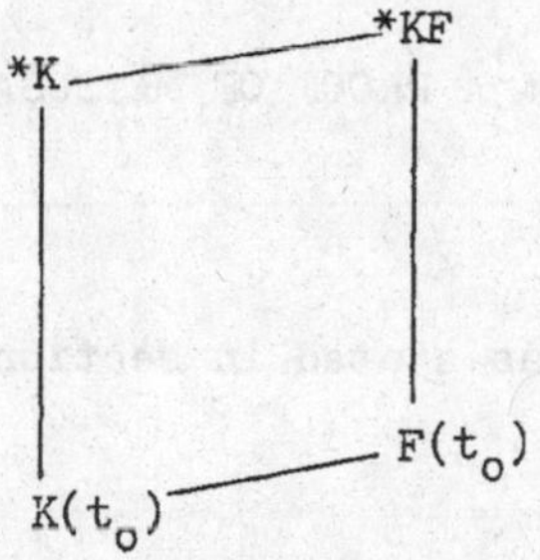

This time we know not only that $K(t_o)$ is algebraically closed in *K, but also that *K is separable over $K(t_o)$. In view of lemma 4.4, which only requires one of the two extensions involved to be separable, we conclude that $F(t_o)$ is algebraically closed in *KF.

QED.

REMARK 4.8. The same arguments lead to the conclusion that *F is regular over $F(t_o)$.

<u>Proof of theorem 4.7.</u>

Let n denote the p-degree of K. Let $a_1,\dots,a_n$ be a p-basis of K. The statement that the a_i form a p-basis of K, remains true in the enlargement; hence $a_1,\dots,a_n$ is also a p-basis of *K, and n is the p-degree of *K. In particular, every nonstandard element $t_o \in$ *K is p-dependent from the a_i in *K. On the other hand, since t_o is transcendental over K, the n+1 elements $a_1,\dots,a_n,t_o$ constitute a p-basis of $K(t_o)$. Thus a p-basis of $K(t_o)$ becomes p-dependent in *K, which implies inseparability of *K over $K(t_o)$. The foregoing argument holds for every nonstandard $t_o \in$ *K, in particular for every Hilbertian t_o.

Now let us put $F = K^{1/p}$. We have $[F:K] = p^n$, hence F is finite over K. The statement $F = K^{1/p}$ remains true in the enlargement; hence $*F = *K^{1/p}$. Thus every element $t_o \in$ *K is a p-th power in *F, and therefore t_o cannot be Hilbertian for F.

QED.

5. APPENDIX: A PROOF OF MATSUSAKA'S LEMMA.

The lemma of Matsusaka as stated in section 2 is concerned with the following situation:

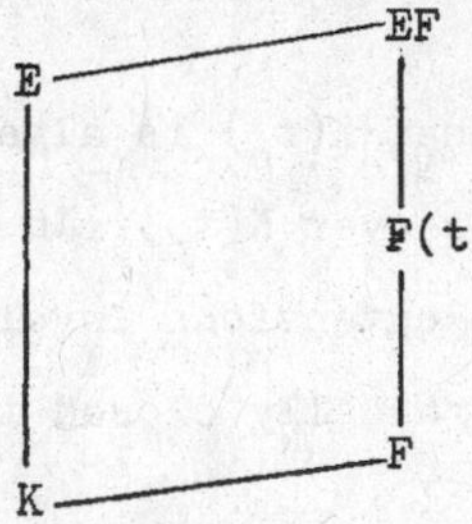

The fields E and F are linearly disjoint over K, and E|K is assumed to be regular. We have

$$t = t_o + t_1 u$$

where $t_o, t_1 \in$ E are assumed to be algebraically independent over K.

The element u is contained in F. Under certain assumptions about u it is claimed that F(t) is algebraically closed in EF. More precisely, the lemma consists of two parts (A) and (B), one dealing with separable extensions, the other with purely inseparable ones.

(A) If $u \in F$ is transcendental over K then F(t) is separable-algebraically closed in EF.

(B) If $u \in F$ is p-free over K then F(t) is closed with respect to purely inseparable extensions within EF.

It is a general phenomenon observed in field theory that the discussion of separable and of purely inseparable extensions requires different methods. So it will be in the following proof. In the inseparable part we shall use field derivations, whereas in the separable part we shall use field automorphisms. Let us start with the

Proof of (B).

Let $x \in EF$ such that $x^p \in F(t)$; we have to show that $x \in F(t)$. We write x^p as a quotient of two polynomials in F(t); after multiplying x with the denominator of this quotient we may assume without loss that $x^p \in F[t]$. We write

$$x^p = f(t).$$

Our aim is to show that f(t) is a p-th power in F[t].

Let f'(t) denote the polynomial derivative of f(t). First we claim that

$$f'(t) = 0.$$

To this end, we use the K-derivations of F. If an element of F is annihilated by all K-derivations of F then this element is contained in F^pK, and conversely. The hypothesis of part (B) implies that $u \notin F^pK$; hence there exists at least one K-derivation D of F which does not annihilate u. After multiplying D with a suitable factor from F we may assume that

$$Du = 1.$$

We extend D to an E-derivation of EF; this is possible and unique since E is K-linearly disjoint to F. The extended derivation will be denoted by the same symbol D. Since $f(t) = x^p$ is a p-th power in EF we then have

$$0 = Df(t) = f^D(t) + f'(t).Dt \ .$$

Here, $f^D(t) \in F[t]$ is the polynomial obtained from f(t) by applying D to its coefficients. As to the derivative Dt we compute

$$Dt = D(t_o+t_1u) = t_1 \quad .$$

(Notice that t_o,t_1 are in E and hence are annihilated by D.) It follows that

$$0 = f^D(t) + f'(t).t_1 \ .$$

Now, if $f'(t) \neq 0$ then this would be a <u>nontrivial</u> algebraic relation between t and t_1 over F. Hence t and t_1 would be algebraically dependent over F. But this is not the case: By definition of $t = t_o+t_1u$ we have

$$F(t,t_1) = F(t_o,t_1) \ .$$

Since t_o,t_1 are algebraically independent over K, hence over F, we conclude that t and t_1 are algebraically independent over F.

We have shown that

$$f'(t) = 0.$$

Now let D be an <u>arbitrary</u> derivation of F, not necessarily the one considered above, and not necessarily a K-derivation. Still, this derivation of F can be extended to a derivation of EF; this is so because EF is regular, hence separable over F. In general there will be many derivations of EF which extend the given derivation of F. We choose one such extension and denote it again with the symbol D. Then again, we have

$$0 = Df(t) = f^D(t) + f'(t).Dt \ .$$

This time we know already that $f'(t) = 0$; hence we conclude

$$f^D(t) = 0 \ .$$

That is, the coefficients of f(t) are annihilated by D. Since this is true for every derivation D of F, these coefficients are contained in

F^p. Hence

$$f(t) \in F^p[t] .$$

On the other hand, we know that $f'(t) = 0$ which means that $f(t)$ is a polynomial in t^p. Hence

$$f(t) \in F^p[t^p] = F[t]^p$$

as contended.

QED.

Before turning of the proof of (A) let us carry out the following reduction steps which will help to simplify the proof. We claim: <u>In order to prove (A) we may assume without loss that the following additional conditions are satisfied:</u>

(i) E <u>is finitely generated over</u> K,

(ii) F <u>is algebraically closed.</u>

As to (i), we consider subfields $E_o \subset E$ which contain $K(t_o,t_1)$ and are finitely generated over K. Every $x \in EF$ is contained in E_oF for some such E_o. Hence, if we know that $F(t)$ is separable-algebraically closed in E_oF for every such E_o, we conclude that $F(t)$ is separable-algebraically closed in EF.

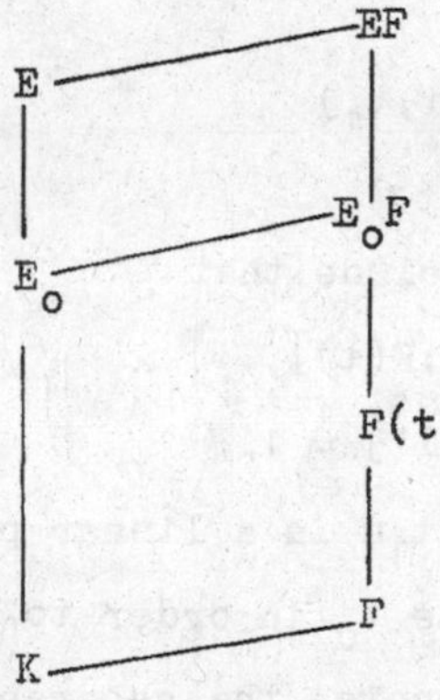

As to (ii), let $\tilde{F}$ denote the algebraic closure of F. Then $\tilde{F}$ is K-linearly disjoint to E. Hence $\tilde{F}(t)$ is $F(t)$-linearly disjoint to EF (see diagram next page).

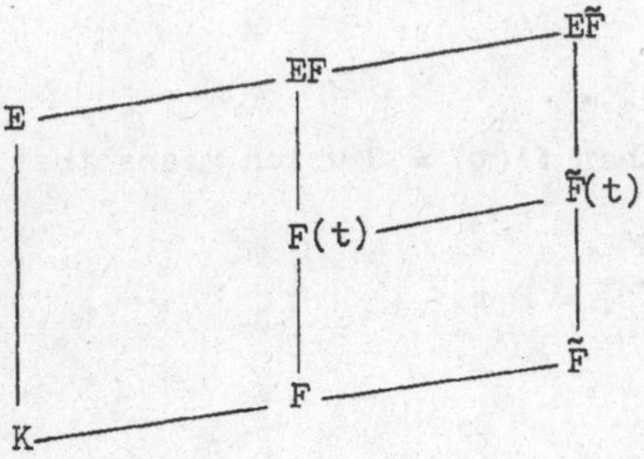

In particular, we conclude that

$$EF \cap \tilde{F}(t) = F(t).$$

Therefore, if we know that $\tilde{F}(t)$ is separable-algebraically closed in $E\tilde{F}$ then it follows $F(t)$ is separable-algebraically closed in EF.

<u>Proof of (A) under the additional hypotheses (i) and (ii).</u>

Let A denote the separable-algebraic closure of $F(t)$ in EF; we have to show that $A = F(t)$, i.e. $[A:F(t)] = 1$. The field degree $[A:F(t)]$ is not affected if we adjoin a transcendental element to both fields. Now, we have seen above already that t_1 is transcendental over $K(t)$ since

$$F(t,t_1) = F(t_o,t_1) .$$

For brevity, let us put

$$F' = F(t_o,t_1) = F(t,t_1)$$

$$A' = F'A = A(t_1) \quad .$$

By what we have said, we conclude that

$$[A':F'] = [A:F(t)] .$$

Our aim is to show that $[A':F'] = 1$.

Now recall that $t = t_o + t_1 u$ is a linear polynomial in u. If u varies inside F then we write t_u in order to indicate the dependence on u. Also, A_u denotes the separable-algebraic closure of $F(t_u)$ in EF, and $A'_u = F'A_u$.

Now let $v \in F$ such that $u \neq v$. From the formulas

$$t_u = t_o + t_1 u$$
$$t_v = t_o + t_1 v$$

we infer that

$$F(t_u, t_v) = F(t_o, t_1) = F'.$$

In particular, it follows that t_u and t_v are algebraically independent over F. Hence the fields A_u and A_v are algebraically free over F. On the other hand, both extension $A_u|F$ and $A_v|F$ are regular. (This is because they are subextensions of $EF|F$ which is regular.) We conclude that A_u and A_v are linearly disjoint over F. Consequently, A'_u <u>and</u> A'_v <u>are linearly disjoint over</u> F' (see diagram). This holds whenever $u \neq v$.

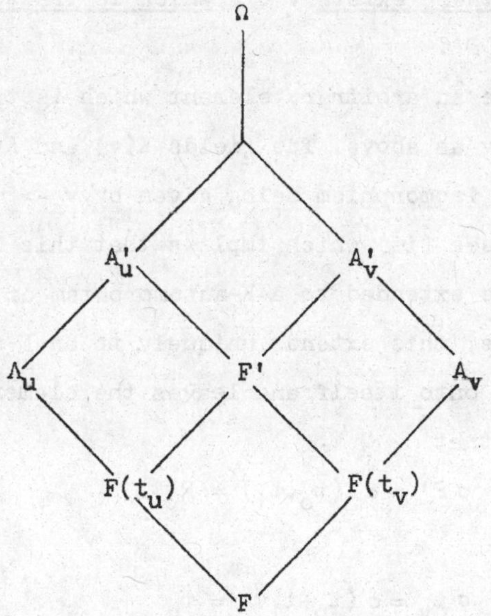

By construction, A'_u and A'_v are separable-algebraic field extensions of F'. That is, they are contained in the separable-algebraic closure Ω of F' in EF. Now we use the additional hypothesis (i) which implies that $EF|F$ and hence $EF|F'$ are finitely generated. It follows that $\Omega|F'$ is of finite degree. Now, every separable-algebraic extension of finite degree admits only finitely many intermediate fields. Hence there are only finitely many fields of the form A'_u with $u \in F$. Let these fields be $A'_{u_1},\dots,A'_{u_n}$ where $u_1,\dots,u_n \in F$. For every $v \in F$, the corresponding field A'_v coincides with one of these fields, say $A'_v = A'_{u_i}$. On the other hand, if $v \neq u_1,\dots,u_n$ then we know from above that A'_v is F'-linearly disjoint to A'_{u_i}. Hence A'_v is F'-linearly disjoint to itself, which implies that $A'_v = F'$.

We have proved that $A'_v = F'$ for all $v \in F$ except possibly a finite number. Since there are infinitely many transcendental elements in $F|K$, we conclude: there exists $v \in F$ which is transcendental over K and such that $A'_v = F'$.

Now let $u \in F$ be an arbitrary element which is transcendental over K. We choose v as above. The fields $K(v)$ and $K(u)$ are naturally K-isomorphic, such isomorphism being given by $v \to u$. Now we use the additional hypotheses (ii) which implies that this isomorphism $K(v) \to K(u)$ can be extended to a K-automorphism of F, say σ. By linear disjointness, this extends uniquely to an E-automorphism of EF. Note that σ maps F onto itself and leaves the elements $t_o, t_1 \in F$ fixed; it follows that

$$\sigma F' = \sigma F(t_o,t_1) = F' .$$

Moreover, we have

$$\sigma t_v = \sigma(t_o + t_1 v) = t_u$$

and therefore

$$\sigma A_v = A_u$$

by definition of A_v and A_u. We conclude:

$$\sigma A'_v = \sigma(F'A_v) = F'A_u = A'_u .$$

Since $A'_v = F'$, it follows $A'_u = F'$. As shown above, this implies

$A_u = F(t_u)$, i.e. $F(t_u)$ is separable-algebraically closed in EF.

QED.

REMARK. Our version of Matsusaka's lemma is slightly different from its usual version as concerns the hypothesis in part (B). The reader may compare our version with that given in the book of S.Lang [15], page 213-214.

BIBLIOGRAPHY.

[1] DÖRGE, K. Zum Hilbertschen Irreduzibilitätssatz. Math.Ann. 95 (1926) 84-97

[2] DÖRGE, K. Einfacher Beweis des Hilbertschen Irreduzibilitäts=satzes. Math.Ann. 96 (1926), 176-182.

[3] EICHLER, M. Zum Hilbertschen Irreduzibilitätssatz. Math.Ann. 116 (1939), 742-748

[4] FRANZ, W. Untersuchungen zum Hilbertschen Irreduzibilitätssatz. Math. Z. 33 (1931), 275-293

[5] FRIED, M. On Hilbert's irreducibility theorem. J. Number Theory 6 (1974), 211-231

[6] FRIED, M. - LEWIS, D.J. Solution spaces for Diophantine Problems, Chap.IV. (Mimeographed Notes).

[7] GILMORE, P.C. - ROBINSON, A. Metamathematical considerations on the relative irreducibility of polynomials. Canad. J.Math. 7 (1955), 483-489

[8] HILBERT, D. Über die Irreduzibilität ganzer rationaler Funktionen mit ganzzahligen Koeffizienten. Crelles J. 110 (1892), 104-129

[9] INABA, E. Über den Hilbertschen Irreduzibilitätssatz. Japan. J. Math. 19 (1944), 1-25.

[10] KNOBLOCH, H.W. Zum Hilbertschen Irreduzibilitätssatz. Abh. Math. Sem. Univ. Hamburg 19 (1955), 176-190

[11] KOCHEN, S. Abraham Robinson, Memorial Service, Sep.15(1974). (Mimeographed Copy).

[12] KUYK, W. Generic approach to the Galois embedding and extension problem. J. Algebra 9 (1968), 393-407

[13] KUYK, W. Extensions de corps hilbertiens. J. Algebra 14 (1970), 112-124

[14] LANG, S. Some applications of the local uniformization theorem. Amer.J. Math.76 (1954), 362-374

[15] LANG, S. Introduction to Algebraic Geometry (New York-London 1958)

[16] LANG, S. Le théorème d'irréductibilité de Hilbert. Sém. Bourbaki (1959/60), no. 201.

[17] LANG, S. Diophantine Geometry. (New York-London 1962).

[18] ROBINSON, A. On ordered fields and definite functions. Math.Ann.130 (1955), 257-271.

[19] ROBINSON, A. On Hilbert's Irreducibility Theorem. Unpublished Manuscript.

[20] ROBINSON, A. Nonstandard Analysis. (Amsterdam 1966)

[21] ROBINSON, A. - ROQUETTE, P. On the Finiteness Theorem of Siegel and Mahler concerning Diophantine Equations. J. Number Theory 7 (1975), 121-176.

[22] SCHINZEL, A. On Hilbert's Irreducibility theorem. Ann. Polon. Math. 16 (1965), 333-340

[23] SIEGEL, C.L. Über einige Anwendungen diophantischer Approximationen. Abh. Preuss. Akad.Wiss. Phys.Math. Kl. (1929), Nr.1

PROJECTIVE MODEL THEORY AND COFORCING

In Memory of Abraham Robinson (1918-1974)

George S. Sacerdote

Institute for Advanced Study

In ordinary model theory, such as in Bell and Slomson [2] or the first half of Robinson's book [12], the essential algebraic object of study is the notion of injection. In projective model theory, we reverse the arrows and study surjections between relational structures.

The fundamental theorems of projective model theory are Lyndon's Interpolation Lemma [9] and Lyndon's Homomorphism Theorem [10]; in the latter he proves that a sentence of the lower predicate calculus is equivalent to a positive sentence if and only if its truth is preserved under all surjections. These results and related ones from [17] are summarized in §1.

Projective model completeness was introduced by this author in [17]. Roughly speaking, a theory K is projectively model complete if and only if whenever $\alpha : M \twoheadrightarrow M'$ is a surjection and $M, M' \models K$, then any sentence in the vocabulary of M is true in M' if and only if it is true in M. (What is meant by the truth in M of a sentence in the vocabulary of M' will be explained in §1.) Projective model completeness can be used to prove completeness theorems, much as ordinary model completeness. In §2 we give another characterization of this notion; we also introduce the projective model completion of a consistent theory K and prove its uniqueness.

In §3 we introduce the notion of the permeability of a theory K' to a theory K. In general terms, K' is permeable to K if for each μ^+-saturated model N of K there exist μ^+-saturated models $N^* \models K$, and $N' \models K'$, and surjections $\beta : N^* \longrightarrow N'$ and $\alpha : N' \longrightarrow N$ such that any sentence is true in N if and only if it is true in N^*. We prove that K' is permeable to K if and only if for any pair $\{\varphi, \sim\theta\}$ where φ and θ are positive sentences, if $K \cup \{\varphi, \sim\theta\}$ is consistent, then $K' \cup \{\varphi, \sim\theta\}$ is consistent, provided that K has no trivial models. (A structure is trivial if it has one element λ and for each relation R in the language, $R(\lambda,\ldots,\lambda)$ holds.)

In §4 we introduce a "summing" construction for chains of elementary surjections, which plays a role similar to that of unions of chains in ordinary model theory. This construction, $\overleftarrow{\lim}^*(M_i)$, gives a special substructure of the ordinary projective limit which is a pre-image of each M_i with the following three properties: (i) If M_{i+1} is an elementary pre-image of M_i for each i, then $\overleftarrow{\lim}^* M_i$ is an elementary pre-image of each M_i; (ii) The sentences whose truth is preserved under the $\overleftarrow{\lim}^*$ construction are of form $\varphi \supset \psi$ where φ and ψ are both positive; and (iii) If K is a projectively model complete theory, then $\mathrm{Mod}(K)$ is closed under $\overleftarrow{\lim}^*$.

In §5 and §6 we introduce the notion of coforcing. This notion formalizes the idea that for free groups, all of their properties can be derived from the fact that their elements satisfy no relations other than those imposed by the axioms (i.e., associativity, $1 \cdot a = a$, etc.). In §7 we construct complete sequences of negative sentences which determine all properties of the corresponding cogeneric structures via the coforcing relation. In §8 we use the coforcing construction to show the completeness of various theories. Finally in §9 we specialize our theory to the study of free groups.

§1 Languages, Structures, and Homomorphisms

Let L be a first order language with individual variables $\{x_0, x_1, \ldots\}$, individual constants $\{a_0, a_1, \ldots, a_\alpha, \ldots\}$, finitary function symbols $\{f_0, f_1, \ldots\}$, finitary relation symbols $\{=, R_0, R_1, \ldots\}$, and the logical symbols &, $\vee$, $\sim$, $\forall$, and $\exists$. The formulas and sentences of L are obtained by the usual formation rules. An L-structure S consists of a set $|S|$ and a function F which assigns to constant symbols a in a subset of the constants, elements a^S of $|S|$, to n-ary function symbols f, operations f^S on $|S|$, and to n-ary relation symbols R, relations R^S on $|S|$. Often when one is given a structure S, the interpretations of the function and relation symbols are clear from the context and need not be further specified; consequently, we will usually denote structures by pairs (S,B) where $B = \{b_{i_0}, b_{i_1}, b_{i_2}, \ldots\}$ is an enumeration of the elements of $|S|$ which correspond to the set of constants $\{a_{i_0}, a_{i_1}, a_{i_2}, \ldots\}$ which are interpreted in S. Also, we will denote the set of constants interpreted in (S,B) by C_B, the element of C_B corresponding to $b \in B$ by c_b, and the element of B corresponding to $c \in C_B$ by b_c. By possibly extending the set B we will always assume that B generates S (i.e., S is the least substructure of S which contains B and is closed under the operations). A sentence φ of L is <u>defined</u> in (S,B) if all of the constants in φ are in C_B.

We shall often have need of a related language $\bar{L}$ which is obtained from L by replacing each relation symbol R by a new symbol $\bar{R}$. We will regard an L-structure M as an $\bar{L}$-structure by letting $\bar{R}^S$ be the complement of the relation R^S. There is an obvious equivalence between sentences of L and $\bar{L}$; we shall often abuse this equivalence by referring to sentences of both of these languages as if they were sentences of the same language. When we wish to be explicit about which language contains a particular sentence φ, we will refer to φ as an L-sentence or an $\bar{L}$-sentence.

Throughout this paper the sentential connective $\supset$, as in $\varphi \supset \theta$, is to be regarded as an abbreviation for $(\sim\varphi) \vee \theta$.

One can introduce the universal quantifier $\forall$ as $\sim\exists\sim$. If φ is a sentence of L (or $\bar{L}$), φ will be called a positive L-sentence (or $\bar{L}$-sentence) if φ is logically equivalent to a sentence of the same language, which can be written with the logical symbols &, $\bigvee$, $\exists$, and $\forall$, but not $\sim$ (other than those uses of $\sim$ implicit in $\forall$). A negative L- or $\bar{L}$-sentence is one whose negation is a positive L- or $\bar{L}$-sentence. In particular, a positive $\bar{L}$-sentence is (equivalent to) a negative L-sentence, and vice versa.

We will need two measures of the complexity of formulas of L and $\bar{L}$ which are called σ and ρ, respectively: the rank function σ is defined inductively on formulae involving the logical symbols &, $\bigvee$, $\sim$, and $\exists$. The symbol $\forall$ is treated as defined. If φ is an atomic sentence, $\sigma(\varphi) = 0$. If φ is $\sim\theta$, $\sigma(\varphi)$ is $\sigma(\theta) + 1$. If φ is $\exists x\theta$, $\sigma(\varphi) = \sigma(\theta) + 1$. If φ is $\theta \bigvee \psi$ or $\theta \,\&\, \psi$, $\sigma(\varphi) = \sigma(\theta) + \sigma(\psi) + 1$. The rank function ρ is defined on formulae involving the logical symbols &, $\bigvee$, $\sim$, $\exists$ and $\forall$. If φ is atomic, $\rho(\varphi) = 0$. If φ is $\theta \bigvee \psi$ or $\theta \,\&\, \psi$, $\rho(\varphi) = 2(\rho(\theta) + \rho(\psi))$. If φ is $\sim\theta$, $\rho(\varphi) = \rho(\theta) + 1$. If φ is $\exists x\theta$, $\rho(\varphi) = 2(\rho(\theta))$. If φ is $\forall x\theta$, $\rho(\varphi) = 6(\rho(\theta))$. Observe that if φ does not involve the symbol $\sim$, $\rho(\varphi) = 0$.

For the rest of §1, and all of §2, §3 and §4, we suppose that L and $\bar{L}$ have no function symbols. Alternatively, we may suppose that each function symbol f of L is replaced by a relation symbol R_f, and for each L-structure M, R_f^S is interpreted as the graph of f^S. Observe that $\bar{R}_f^S$ need not be a functional relation.

Given two structures (S,B) and (S',B') with $C_{B'} \subset C_{B'}$, and a function g with domain $|S|$ and range $|S'|$, we will call g a surjection if (i) for each constant a in $C_{B'}$, $g(a^S) = a^{S'}$; (ii) for each n-ary function symbol f and all elements $s_1,\ldots,s_n$ of $|S|$, $g(f^S(s_1,\ldots,s_n) = f^{S'}(g(s_1),\ldots,g(s_n))$, and (iii) for each relation symbol R and all elements $s_1,\ldots,s_n$ of $|S|$, if $R^S(s_1,\ldots,s_n)$ holds, then so does $R^{S'}(g(s_1),\ldots,g(s_n))$. A pre-image of (S',B') is a structure (S,B) for which there exists a surjection $g : (S,B) \longrightarrow (S',B')$.

The next three theorems are due to Lyndon [9,10].

<u>Theorem</u> 1.1: An L-sentence φ is positive if and only if for any L-structure (S',B') such that $(S',B') \models \varphi$, and for any L-structure (S,B) which is a surjective image of (S',B'), $(S,B) \models \varphi$.

<u>Theorem</u> 1.2: Let K be a consistent set of sentences of L. A sentence φ is true in all surjective images of models of K if and only if φ is a consequence of a positive L-sentence which is a consequence of K.

<u>Theorem</u> 1.3: Let K be a consistent set of sentences of L. Any L-structure (S,B) which satisfies all positive L-sentences which are consequences of K has an elementary extension which is a surjective image of a model of K.

The following four corollaries of 1.1-3 appear in [17]:

<u>Corollary</u> 1.4: An $\bar{L}$-sentence φ is positive if and only if whenever φ holds in an $\bar{L}$-structure (S,B) and (S',B') is a pre-image of (S,B), then $(S',B') \models \varphi$.

<u>Corollary</u> 1.5: Let K be a consistent set of sentences. A sentence φ is a consequence of a positive $\bar{L}$-sentence θ which in turn is a consequence of K if and only if for any model (S,B) of φ and any pre-image (S',B') of (S,B), if $(S,B) \models K$, then $(S',B') \models \varphi$.

<u>Corollary</u> 1.6: Let K be a consistent set of sentences. If (S,B) satisfies all positive $\bar{L}$-sentences which are consequences of K, there exists an elementary extension (S",B") of (S',B') which is a pre-image of a model of K.

Let T(S,B) be the set of sentences defined and true in (S,B). Let $D^{-}_{(S,B)}$ be the set of positive $\bar{L}'$-sentences in T(S,B).

<u>Corollary</u> 1.7: Let K be a consistent set of sentences and let (S,B) be an L-structure. If $K \cup D^{-}_{(S,B)}$ is consistent, then some elementary extension (S",B") of (S,B) is a surjective image of a model of $K \cup D^{-}_{(S,B)}$.

§2 Projective Model Completeness

Projective model completeness as such is discussed in detail in the author's paper [17]. In this section we will be concerned chiefly with a relativized version of projective model completeness because this new notion is intimately connected with coforcing. First we summarize the results of [17] relevant to this paper.

Let (S,B) and (S',B') be L-structures, where (S',B') is a pre-image of (S,B). (S',B') is an <u>elementary pre-image</u> of (S,B), if for any sentence φ in the vocabulary of (S,B), $(S,B) \models \varphi$ if and only if $(S',B') \models \varphi$. If g is the surjection from (S',B') onto (S,B), we call g an <u>elementary surjection</u>.

A consistent theory K is <u>projectively model complete</u> if each surjection between models of K is an elementary surjection.

<u>Lemma</u> 2.1: Let $g : (S',B') \longrightarrow (S,B)$ be an elementary surjection, and let (S',B'') be the substructure of (S',B') generated by the subset B'' of B' of elements corresponding to constants in C_B. Then the restriction of g to (S'',B'') is an isomorphism and $(S'',B'') \prec (S',B')$.

<u>Proof</u>: Since $g(B'') = B$ the restriction of g to (S'',B'') is a map onto (S,B). Since both B'' and B are in 1-1 correspondence with C_B, $g|(S'',B'')$ is also 1-1. Moreover for any function symbol since $g|(S'',B'')$ is a homomorphism, $g|(S'',B'')$ is an isomorphism onto (S,B). To see that $(S'',B'') \prec (S',B')$, observe that any sentence φ in the vocabulary of (S'',B'') is also in the vocabulary of (S,B). Thus $(S'',B'') \models \varphi$ if and only if $(S,B) \models \varphi$, if and only if $(S',B') \models \varphi$.

<u>Theorem</u> 2.2: Let K be a consistent theory. Then the following are equivalent.

(i) K is projectively model complete.

(ii) For any pair of models (S',B') and (S,B) of K, where (S,B) is a surjective image of (S',B') via a surjection α, and for any positive L-sentence φ defined in (S,B), $(S,B) \models \varphi$ only if $(S',B') \models \varphi$.

(iii) For any model (S,B) of K, $K \cup D^{-}_{(S,B)}$ is complete.

<u>Proof</u>: (i) $\Longleftrightarrow$ (ii) is Theorem 2.4 of [17].

(iii) $\Longrightarrow$ (i) is clear.

To see that (i) $\Longrightarrow$ (iii), suppose that K is projectively model complete. Let φ be a sentence such that both $K_0 = K \cup D^-_{(S,B)} \cup \{\varphi\}$ and $K_1 = K \cup D^-_{(S,B)} \cup \{\sim\varphi\}$ are consistent. Then we have models (S_0,B_0) and (S_1,B_1) of K_0 and K_1, respectively, elementary extensions (S'_0,B'_0) and (S'_1,B'_0) of (S,B) and surjections $\alpha : (S_0,B_0) \longrightarrow (S'_0,B'_0)$ and $\beta : (S_1 B_1) \longrightarrow (S'_1 B'_1)$ by 1.7.

$$\begin{array}{ccccc} (S_0,B_0) & \xrightarrow{\alpha} & (S'_0,B'_0) & & \\ & & & \hookleftarrow & (S,B) \\ (S_1,B_1) & \xrightarrow{\beta} & (S'_1,B'_1) & \hookleftarrow & \end{array}$$

Either $(S,B) \models \varphi$ or $(S,B) \models \sim\varphi$, but either conclusion leads to an immediate contradiction, since α and β must be elementary. Thus (iii) holds.

Let K be a non-empty consistent set of sentences. A consistent set of sentences K^* is <u>projectively model consistent relative to</u> (S,B) if (S,B) has an elementary extension (S',B') which is a surjective image of a model of K^*. Equivalently $K^* \cup D^-_{(S,B)}$ is consistent. If $K^* \supseteq K$ and K^* is projectively model consistent with every model of K, then K^* is <u>projectively model consistent with</u> K. K^* is <u>projectively model complete relative to</u> K if for every model (S,B) of K, every sentence φ defined in (S,B) either holds in all pre-images (S^*,B^*) of (S,B) among the models (S^*,B^*) of K^*, or else φ holds in no such pre-images. Alternatively, $K^* \cup D^-_{(S,B)}$ is complete. In particular, K is projectively model complete relative to itself precisely if K is projectively model complete.

The following theorem shows that there is at most one theory (up to logical equivalence) projectively model complete relative to a given theory K:

<u>Theorem</u> 2.3: Let K' and K^* be projectively model consistent with respect

to a consistent theory K. If K' and K^* are projectively model complete with respect to K, then K' and K^* are logically equivalent.

<u>Proof</u>: By symmetry, it suffices to show that $K^* \vdash K'$. For the sake of a contradiction, suppose that φ is a sentence in K' such that $K^* \cup \{\sim\varphi\}$ is consistent.

Let $(S,B) \models K^* \cup \{\sim\varphi\}$. Then $(S,B) \models K$ (since $K \subseteq K^*$), and since K^* is a projective model completion of K, $K^* \cup D^-_{(S,B)}$ is complete. Moreover, $K^* \cup D^-_{(S,B)} \vdash \sim\varphi$. Trivially $K' \cup D^-_{(S,B)} \vdash \varphi$ since $\varphi \in K'$. Thus, to obtain the desired contradiction it suffices to prove that for any model (S,B) of K and for any sentence φ defined in (S,B), $K^* \cup D^-_{(S,B)} \vdash \varphi$ if and only if $K' \cup D^-_{(S,B)} \vdash \varphi$.

If, on the contrary, we can find a model $(S,B) \models K$ and a sentence φ of L defined in (S,B) such that $K^* \cup D^-_{(S,B)} \vdash \sim\varphi$ and $K' \cup D^-_{(S,B)} \vdash \varphi$, choose (S,B) and φ of minimal ρ-rank. Clearly, $\rho(\varphi) > 0$. Furthermore, it is evident that φ cannot be of form $\theta \vee \psi$, $\theta \& \psi$, or $\sim\theta$. Therefore, φ is $\exists x\theta(x)$ or $\forall x\theta(x)$. Suppose the former case holds.

Choose elementary extensions (S'_0,B'_0) and (S^*_0,B^*_0) of (S,B), models $(S',B') \models K' \cup D^-_{(S,B)} \cup \{\varphi\}$ and $(S^*,B^*) \models K^* \cup D^-_{(S,B)} \cup \{\sim\varphi\}$ and surjections $\alpha : (S',B') \longrightarrow (S'_0,B'_0)$ and $\beta : (S^*,B^*) \longrightarrow (S^*_0,B^*_0)$ such that the following diagram holds.

$$(S',B') \xrightarrow{\alpha} (S'_0,B'_0) \succ (S,B) \prec (S^*_0,B^*_0) \xleftarrow{\beta} (S^*,B^*).$$

Now $(S^*,B^*) \models K$, $(S',B') \models K$. Since $(S',B') \models \varphi$, $(S',B') \models \theta(a)$ for some individual a. Since $K' \cup D^-_{(S,B)}$ is complete, $K' \cup D^-_{(S,B)} \vdash \theta(a)$. Since $\rho(\varphi)$ is minimal, and $\rho(\theta(a)) < \rho(\varphi)$, $K^* \cup D^-_{(S,B)} \vdash \theta(a)$. But then $K^* \cup D^-_{(S,B)} \vdash \varphi$, a contradiction. Thus, in this case we must conclude that $K^* \cup D^-_{(S,B)} \vdash \varphi$ if and only if $K' \cup D^-_{(S,B)} \vdash \varphi$.

A similar argument prevails if φ is $\forall x\theta(x)$. The proof of Theorem 2.3 is now complete.

In later sections we will return to the projective model completion K^* of a theory K. Specifically, in §4 it will be shown that K^*, if it exists, is equivalent to a set of sentences of form $\theta \supset \psi$ where θ and ψ are positive L-sentences. Alternatively, the class of models of K^* is closed under a special summing operation for chains of surjections.

§3 Permeability

Throughout this paper, μ is an infinite cardinal. Suppose that we are given a μ^+-saturated structure (S,B) of cardinal $\leq \mu^+$. If (S',B') is another such structure which is also a model of $D^-_{(S,B)}$, then there is a surjection α from M' onto M, as was noted by H. J. Keisler [20]. Let (S^*,B^*) be a μ^+-saturated model of $D^-_{(S',B')}$ of cardinal $\leq \mu^+$. Then there is a surjection from (S^*,B^*) onto (S',B').

$$(S^*,B^*) \xrightarrow{\beta} (S',B') \xrightarrow{\alpha} (S,B).$$

The μ^+-saturated structure (S',B') is said to be <u>permeable</u> to the μ^+-saturated structure (S,B) if (S^*,B^*) and the surjections α and β can be chosen so that $\alpha\beta$ is an elementary surjection. Otherwise (S',B') is said to be <u>impermeable</u> to (S,B).

An L-structure (S,B) is a <u>trivial structure</u> if it has only one element λ and for each relation symbol R, $R(\lambda,\lambda,\lambda,\ldots,\lambda)$ holds.

<u>Lemma</u> 3.1: (S,B) is a trivial structure if and only if $D^-_{(S,B)}$ is empty.

<u>Lemma</u> 3.2: If (S,B) is a trivial structure, then (S',B') is permeable to (S,B) if and only if $(S',B') \simeq (S,B)$.

For the rest of this section we consider only μ^+-saturated structures of cardinality $\leq \mu^+$.

<u>Theorem</u> 3.3: Suppose that (S,B) is not a trivial structure. An L-structure (S',B') is impermeable to (S,B) if and only if there exist positive L-sentences θ and φ which are defined in (S,B) such that $(S,B) \models \theta \mathbin{\&} {\sim}\varphi$ and $(S',B') \models \theta \supset \varphi$.

Proof: First suppose that (S',B') is impermeable to (S,B). If there are no surjections from (S',B') onto (S,B), then there is a sentence φ' in $D^-_{(S,B)}$ such that $(S',B') \models \sim\varphi'$. Let φ be $\sim\varphi'$. Then $(S,B) \models \forall x(x = x) \& \sim\varphi$, but $(S',B') \models \forall x(x = x) \supset \varphi$. Now suppose $(S',B') \models D^-_{(S,B)}$. There must be a surjection $\alpha : (S',B') \longrightarrow (S,B)$. Let T be the set of sentences true in (S,B). If (S'',B'') is a model of $D^-_{(S',B')}$, there is a surjection $\beta : (S'',B'') \longrightarrow (S',B')$. If $(S'',B'') \models T$ also, then $\alpha\beta$ is an elementary surjection; this is impossible in view of the impermeability of M' to M.

Choose a sentence $\psi(a_1,\ldots,a_n)$ in $D^-_{(S',B')}$ such that $T \vdash \sim\psi(a_1,\ldots,a_n)$; here, $a_1,\ldots,$ and a_n are the constants in ψ which lie in $C_{B'} - C_B$. By elementary logic, $T \vdash \sim\exists x_1,\ldots,x_n\psi(x_1,\ldots,x_n)$, where $x_1,\ldots,$ and x_n are new variables. Let θ be $\sim\exists x_1,\ldots,x_n\psi(x_1,\ldots,x_n)$. The positive L-sentence θ is defined and true in (S,B), and $(S',B') \models \sim\theta$. Now let φ be a positive L-sentence such that $\sim\varphi$ is in $D^-_{(S,B)}$ (recall that $D^-_{(S,B)}$ is non-empty). Then $(S,B) \models \theta \& \sim\varphi$ and $(S',B') \models \theta \supset \varphi$.

Conversely, if (S',B') is permeable to (S,B), then certainly for every positive L-sentence φ which is defined in (S,B), if $\sim\varphi$ is in $D^-_{(S,B)}$, then $(S',B') \models \sim\varphi$. Now let θ be a positive L-sentence which is defined and true in (S,B). Since $(S,B) \models \theta$, and $\alpha\beta$ is an elementary surjection, $(S^*,B^*) \models \theta$. Since θ is a positive L-sentence and β is a surjection, $(S',B') \models \theta$.

Let K' be a consistent theory. K' is <u>permeable to a μ^+-saturated structure</u> (S,B) of power $\leq \mu^+$, provided that there is a μ^+-saturated model (S',B') of K' of power $\leq \mu^+$ which is permeable to (S,B). Otherwise K' is <u>impermeable</u> to (S,B).

Theorem 3.4: A consistent theory K' is impermeable to a non-trivial structure (S,B) if and only if there is a pair (θ,φ) of positive L-sentences which are defined in (S,B) such that $(S,B) \models \theta \& \sim\varphi$, and $K' \vdash \theta \supset \varphi$.

Proof: Suppose that there is a pair (θ,φ) of positive L-sentences which are defined in (S,B) such that $(S,B) \models \theta \,\&\, \sim\varphi$, and $K' \vdash \theta \supset \varphi$. Then for any model (S',B') of K', $(S',B') \models \theta \supset \varphi$. Hence, by the previous theorem (S',B') is impermeable to (S,B). Hence K' is impermeable to (S,B).

Conversely, suppose that for every pair (θ,φ) of positive L-sentences which are defined in (S,B), $K' \cup \{\theta \,\&\, \sim\varphi\}$ is consistent. Then, by the compactness theorem, $K' \cup D^-_{(S,B)}$ is consistent. Furthermore, if $D^+_{(S,B)}$ is the set of positive L-sentences which are defined and true in (S,B), compactness also yields that $K' \cup D^-_{(S,B)} \cup D^+_{(S,B)}$ is consistent. Let (S',B') be a μ^+-saturated model of $K' \cup D^+_{(S,B)} \cup D^-_{(S,B)}$ of power $\leq \mu^+$. By 3.3, (S',B') is permeable to (S,B), and hence, K' is permeable to (S,B).

Let K and K' each be consistent theories. K' is permeable to K if for each μ^+-saturated model (S,B) of K of power $\leq \mu^+$, K' is permeable to (S,B). Otherwise K' is impermeable to K.

Theorem 3.5: Suppose that the consistent theory K has no trivial models. The consistent theory K' is impermeable to K if and only if there is a pair (θ,φ) of positive L-sentences such that $K \cup \{\theta \,\&\, \sim\varphi\}$ is consistent, but $K' \vdash \theta \supset \varphi$.

Proof: First suppose that there is a pair (θ,φ) of positive L-sentences such that $K \cup \{\theta \,\&\, \sim\varphi\}$ is consistent but $K' \vdash \theta \supset \varphi$. Let (S,B) be a μ^+-saturated model of $K \cup \{\theta \,\&\, \sim\varphi\}$ of power $\leq \mu^+$. Then, by the previous theorem, K' is impermeable to (S',B'). Hence K' is impermeable to K.

Conversely, suppose that for every pair (θ,φ) of positive L-sentences such that $K \cup \{\theta \,\&\, \sim\varphi\}$ is consistent, $K' \cup \{\theta \,\&\, \sim\varphi\}$ is also consistent. If (S',B') is an arbitrary μ^+-saturated model of K of power $\leq \mu^+$, and if (θ,φ) is an arbitrary pair of positive L-sentences defined in (S,B) such that $(S,B) \models \theta \,\&\, \sim\varphi$, then $K' \cup \{\theta \,\&\, \sim\varphi\}$ is consistent. Thus, by the previous theorem, K' is permeable to (S,B). Hence K' is permeable to K.

§4 Weak Projective Limits

Let $\{\alpha_i : (S_{i+1},B_{i+1}) \longrightarrow (S_i,B_i)\}$ be a sequence of surjections. Let C be $\bigcup C_{B_i}$. To each element $c \in C$ we associate an element $b_c \in \prod S_i$ as follows: if $c \in C_{B_i}$, let the ith component of b_c, $(b_c)_i$ be the corresponding element of B_i. Otherwise, choose $j \geq i$ such that $c \in C_{B_j}$, and let $(b_c)_i = \alpha_i\alpha_{i+1},\ldots,\alpha_{j-1}((b_c)_j)$. Let $B = \{b_c | c \in C\}$ and let S be the substructure of $\prod S_i$ generated by B. We denote this weak projective limit construction by $(S,B) = \overleftarrow{\lim}^*(S_i,B_i)$. Since $\overleftarrow{\lim}(S_i,B_i)$ consists of all sequences (a_i) when $\alpha_i(a_{i+1}) = a_i$, $\overleftarrow{\lim}^*(S_i,B_i) \subset \overleftarrow{\lim}(S_i,B_i)$. For each i let $\pi_i : (S,B) \longrightarrow (S_i,B_i)$ be the surjection given by projection onto the ith component. A sentence φ in the vocabulary of (S,B) is in the vocabulary of all but finitely many (S_i,B_i).

__Lemma__ 4.1: If each surjection $\alpha_i : (S_{i+1},B_{i+1}) \longrightarrow (S_i,B_i)$ is an elementary surjection, then the surjections $\pi_i(S,B) \longrightarrow (S_i,B_i)$ are also elementary surjections.

__Proof__. Let $\varphi(c_1,\ldots,c_n)$ be a sentence in the vocabulary of (S,B) where $c_i,\ldots,c_n \in C_{i_0}$. We prove by induction on $\sigma(\varphi)$ that $(S,B) \models \varphi$ if and only if $(S_{i_0},B_{i_0}) \models \varphi$.

If φ is atomic, the assertion follows from the definitions. If φ is $\sim\theta$, $\theta \,\&\, \psi$, or $\theta \vee \psi$, the proof is a simple application of the inductive hypothesis. Now suppose φ is $\exists x\theta(x)$.

If $(S_{i_0},B_{i_0}) \models \varphi$, $(S_{i_0},B_{i_0}) \models \theta(t)$ for some closed term t in the vocabulary of (S_{i_0},B_{i_0}). Hence, $(S,B) \models \theta(t)$, and $(S,B) \models \varphi$. Conversely, if $(S,B) \models \varphi$, $(S,B) \models \theta(t)$ where t is a closed term in the vocabulary of (S,B). Let $j \geq i_0$ be an index such that all constants in $\theta(t)$ appear in C_{B_j}. Then $(S_j,B_j) \models \theta(t)$ and $(S_j,B_j) \models \varphi$. Since $\alpha_{i_0},\alpha_{i_0+1},\ldots,\alpha_{j-1}$ is an elementary surjection and φ is in the vocabulary of (S_{i_0},B_{i_0}), $(S_{i_0},B_{i_0}) \models \varphi$.

__Theorem__ 4.2: Let K be a consistent theory with no trivial models. Then the following are equivalent:

(i) The truth of the set H of sentences is preserved under $\underset{\longleftarrow}{\lim}^*(S_i,B_i)$ for all sequences $\{\alpha_i : (S_{i+1},B_{i+1}) \longrightarrow (S_i,B_i)\}$ of models of K.

(ii) H is equivalent relative to K to a set H' of sentences $\varphi \supset \psi$ where both φ and ψ are positive L-sentences.

Proof: (i) $\Longrightarrow$ (ii). Let $H' = \{\varphi \supset \psi | \varphi$ and ψ are both positive and $K,H \vdash \varphi \supset \psi\}$, then $K,H \vdash H'$; if $K,H' \vdash H$, we are done.

Suppose that there is a sentence $\theta \in H$ such that $K_0 = K \cup H' \cup \{\sim\theta\}$ is consistent. Since $H' \subset K_0$, we have by 3.5 that if $(S_0,B_0) \models K_0$ is μ^+-saturated of cardinal $\leq \mu^+$, there exist μ^+-saturated models $(T_0,D_0) \models K \cup H$ and $(S_1,B_1) \models K_0$ and surjections $(S_1,B_1) \xrightarrow{g_0} (T_0,D_0) \xrightarrow{h_0} (S_0,B_0)$ such that $h_0 g_0$ is an elementary surjection. Continuing in this fashion one obtains

$$\cdots \longrightarrow (S_i,B_i) \longrightarrow (T_{i-1},D_{i-1}) \longrightarrow (S_{i-1},B_{i-1}) \longrightarrow \cdots \longrightarrow (S_0,B_0).$$

Let (S^*,B^*) be $\underset{\longleftarrow}{\lim}^*$ of this chain. Then we have also that (S^*,B^*) is $\underset{\longleftarrow}{\lim}^*(S_i,B_i)$, a chain of elementary surjections. Thus $(S^*,B^*) \models \sim\theta$. On the other hand, $(S^*,B^*) = \lim^*(T_i,D_i)$, each $(T_i,B_i) \models K \cup H$ and H is preserved under $\underset{\longleftarrow}{\lim}^*$. Therefore $(S^*,B^*) \models \theta$. This contradiction forces us to conclude that $K,H' \vdash H$.

One limitation of the previous theorem is that many common theories have trivial models; notably, the theory of groups has such a model. Suppose that there is a positive $\bar{L}$-sentence ψ such that for all non-trivial models (S,B) of the consistent theory K_0, $(S,B) \models K_1 = K_0 \cup \{\psi\}$. Observe that if φ is a positive L-sentence and (S,B) is a trivial L-structure, then $(S,B) \models \varphi$. Thus if θ and φ are any positive L-sentences, $(S,B) \models \theta \supset \varphi$. Consequently, for any sentence its truth is preserved under all weak projective limits of models of $K_0 \cup H$ if and only if its truth is preserved under all weak projective limits of models of $K_1 \cup H$, unless $K_1 \cup H$ is inconsistent. (If, $K_1 \cup H$ is inconsistent, then $K_0 \cup H$ has the trivial model of K_0 as its only model.) Therefore, if the truth of H is preserved under all projective limits of models of $K_0 \cup H$, and if $K_0 \cup H$ has non-trivial models, then H is equivalent relative to K_1 to a set of sentences of form $\theta \supset \varphi$, where θ and φ are positive L-sentences.

<u>Corollary</u> 4.3: Let H be a set of sentences in the language of group theory which is satisfied by some non-trivial group. Let K_2 be the axioms of group theory together with the sentence $\exists x,y[\sim x = y]$. If H is preserved under all weak projective limits of groups, H is equivalent (relative to K_2) to a set of sentences of form $\theta \supset \varphi$, where θ and φ are positive sentences of L.

A group is <u>divisible</u> if it satisfies all of the following axioms in addition to the group axioms:

$$\varphi_n : \forall x \exists y[y^n = x] \quad n = 2,3,4,\ldots .$$

The corresponding axioms of form $\theta \supset \varphi$ may be taken to be

$$\psi_n : \exists z[z = z] \supset \forall x \exists y[y^n = x] \quad n = 2,3,4,\ldots .$$

A group is infinite if it satisfies all of the following axioms:

$$\Phi_n : \exists x_1,x_2,\ldots,x_n[\sim x_1 = x_2 \;\&\; \sim x_1 = x_3 \;\&\; \ldots \;\&\; \sim x_{n-1} = x_n] \quad n = 2,3,4,\ldots$$

An equivalent set of axioms, relative to K_2 is given by

$$\psi_n : \exists x_1,\ldots,x_n \forall y[y = x_1 \vee y = x_2 \vee \ldots \vee y = x_n] \supset \forall x,z[x = z].$$

We continue this section by showing that if a theory K has a projective model completion K^*, then the class of models of K^* is closed under weak projective limits. That is, K^* is equivalent to a set of sentences of form $\theta \supset \varphi$, where θ and φ are positive.

<u>Theorem</u> 4.4: Let K^* be the projective model completion of a consistent set of sentences K. Then the class of models of K^* is closed under weak projective limits.

Proof: Let $\{\beta_i : (S_{i+1}, B_{i+1}) \longrightarrow (S_i, B_i)\}$ be a sequence of surjections of models of K^*. Let $(S^*, B^*) = \underleftarrow{\lim}^*(S_i, B_i)$. Each $(S_i, B_i) \models K^*$, and $(S_{i+1}, B_{i+1}) \models D^-_{(S_i,B_i)}$. Since $K^* \cup D^-_{(S_i,B_i)}$ is complete, each β_i is an elementary surjection onto (S_i, B_i). By 4.1, the projection $\pi_1 : (S^*, B^*) \longrightarrow (S_1, B_1)$ is an elementary surjection. Hence $(S^*, B^*) \models K^*$.

Corollary 4.5: Let ψ_0 be any negative statement in K where K is a consistent theory having a projective model completion K^*, and having no one-element models. Then K^* is equivalent relative to $\{\psi_0\}$ to a set of sentences of form $\theta \supset \varphi$, where θ and φ are positive L-sentences.

§5 Introduction to Coforcing

As we mentioned in the Introduction, we wish to formalize the idea that for non-abelian free groups all of their properties follow from the fact that they satisfy no positive L-sentence other than those sentences forced upon them by the axioms of group theory. In this section we introduce the coforcing relation between finite sets P of sentences of ρ-rank 0 (positive $\bar{L}$-sentences) and other sentences of $\bar{L}$; the coforcing relation describes the way our conditions P determine the properties of structures which satisfy P. As we shall see in §7, there exist structures (S,B) which satisfy precisely the sentences coforced by some finite set of sentences of ρ-rank 0 in their vocabulary.

Let K be a consistent set of sentences. A condition relative to K is a finite set P of positive $\bar{L}$-sentences such that $K \cup P$ is consistent and such that all relation and function symbols of P are in K. If P is a condition (relative to K) and if φ is an $\bar{L}$-sentence, we define a relation $P \dashv\!\mid \varphi$, P coforces φ, by induction on $\rho(\varphi)$: (a) If $\rho(\varphi) = 0$, $P \dashv\!\mid \varphi$ if and only if $\varphi \in P$. (b) If $\rho(\varphi) > 0$ and φ is $\theta \,\&\, \psi$, $P \dashv\!\mid \varphi$ if and only if $P \dashv\!\mid \theta$ and $P \dashv\!\mid \psi$. (c) If $\rho(\varphi) > 0$ and φ is $\theta \vee \psi$, $P \dashv\!\mid \varphi$ if and only if $P \dashv\!\mid \theta$ or $P \dashv\!\mid \psi$ or both. (d) If $\rho(\varphi) > 0$ and φ is $\exists x \theta(x)$, $P \dashv\!\mid \varphi$ if and only if $P \dashv\!\mid \theta(t)$ for some closed term t. (e) If $\rho(\varphi) > 0$ and φ is $\forall x \theta(x)$,

$P \dashv\| \varphi$ if and only if $P \dashv\| \sim \exists x \sim \theta(x)$. (Observe that $\rho(\varphi) = 6\,\rho(\theta) > \rho(\sim \exists x \sim \theta(x)) = 2[\rho(\theta) + 1] + 1$.) (f) If $\rho(\varphi) > 0$ and φ is $\sim \theta$, $P \dashv\| \varphi$ if and only if for all conditions $Q \supseteq P$, not $Q \dashv\| \theta$.

We say that a condition P <u>weakly</u> <u>coforces</u> φ, $P \dashv\|^* \varphi$, if $P \dashv\| \sim\sim \varphi$. $P \dashv\|^* \varphi$ if and only if for each condition $Q_1 \supseteq P$ there is a condition $Q_0 \supseteq Q_1$ such that $Q_0 \dashv\| \varphi$.

<u>Lemma</u> 5.1: For no $\bar{L}$-sentence φ and for no condition P do we have $P \dashv\| \varphi$ and $P \dashv\| \sim \varphi$.

<u>Proof</u>: If $P \dashv\| \varphi$, then $P \supseteq P$ implies not $P \dashv\| \sim \varphi$.

<u>Lemma</u> 5.2: If $P \dashv\| \varphi$ and $Q \supseteq P$ is a condition, then $Q \dashv\| \varphi$.

<u>Proof</u>: If $\rho(\varphi) = 0$, the statement is trivial. Now suppose $\rho(\varphi) > 0$. We proceed by induction. If φ is $\theta \,\&\, \psi$, $\theta \vee \psi$, $\exists x\theta(x)$ or $\forall x\theta(x)$, the argument is clear. If φ is $\sim \theta$ and $P \dashv\| \varphi$, then no condition $P' \supseteq P$ satisfies $P' \dashv\| \theta$. In particular, for no condition $P'' \supseteq Q$ does $P'' \dashv\| \theta$. Thus $Q \dashv\| \varphi$.

<u>Lemma</u> 5.3: If $\varphi \in P$, $P \dashv\| \varphi$.

<u>Proof</u>: If $\varphi \in P$, $\rho(\varphi) = 0$. Hence $P \dashv\| \varphi$.

<u>Lemma</u> 5.4: If $\rho(\varphi) = 0$ and $P \dashv\| \varphi$, then $\varphi \in P$.

<u>Proof</u>: This is definition (a) above.

<u>Lemma</u> 5.5: Let $\bar{R}$ be an n-ary relation symbol not appearing in K. Let $t_1,\ldots,t_n$ be any terms. Then $\emptyset \dashv\| \sim \bar{R}(t_1,\ldots,t_n)$.

<u>Proof</u>: If P is a condition $\bar{R}(t_1,\ldots,t_n) \notin P$ since all relation symbols in P must appear in K.

<u>Lemma</u> 5.6: (i) If $P \dashv\| \varphi$, then $P \dashv\|^* \varphi$.

(ii) If $P \dashv\|^* \sim \varphi$, then $P \dashv\| \sim \varphi$.

Proof: (i) If $P \dashv\Vert \varphi$, then for no conditions $Q_0 \supseteq Q_1 \supseteq P$ can $Q_0 \dashv\Vert \sim\varphi$. Hence $P \dashv\Vert^* \varphi$.

(ii) If this were false, then for some condition $Q_1 \supseteq P$, $Q_1 \dashv\Vert \sim\sim\varphi$ and $Q_1 \dashv\Vert \sim\sim\sim\varphi$. This is impossible.

Let A be a set of individual constants which includes the set of constants in K and such that $\{a \mid a \in A,\ a$ does not appear in $K\}$ is non-empty. Let H be the set of $\bar{L}$-sentences all of whose relation and function symbols appear in K and all of whose individual constants appear in A. Let N be the set of positive $\bar{L}$-sentences in H. For any condition $P \subseteq N$ and $\varphi \in H$ we define a relativized coforcing as follows: (i) If $\rho(\varphi) = 0$, $P \dashv\Vert_A \varphi$ if and only if $\varphi \in P$. (ii) If $\rho(\varphi) > 0$ and φ is $\theta \,\&\, \psi$, $P \dashv\Vert_A \varphi$ if and only if $P \dashv\Vert_A \theta$ and $P \dashv\Vert_A \psi$. (iii) If $\rho(\varphi) > 0$ and φ is $\theta \vee \psi$, $P \dashv\Vert_A \varphi$ if and only if $P \dashv\Vert_A \theta$ or $P \dashv\Vert_A \psi$ or both. (iv) If $\rho(\varphi) > 0$ and φ is $\exists x\theta(x)$, $P \dashv\Vert_A \varphi$ if and only if there exists a closed term t all of whose individual constants appear in A such that $P \dashv\Vert_A \theta(t)$. (v) If $\rho(\varphi) > 0$ and φ is $\forall x\theta(x)$, $P \dashv\Vert_A \varphi$ if and only if $P \dashv\Vert_A \sim\exists x \sim\theta(x)$. (vi) If $\rho(\varphi) > 0$ and φ is $\sim\theta$, $P \dashv\Vert_A \varphi$ if and only if for no condition Q with $N \supseteq Q \supseteq P$, do we have $Q \dashv\Vert_A \theta$.

Let P be a condition and φ a sentence such that $P \dashv\Vert \varphi$. Let $a_1,\ldots,a_n$ be the individual constant symbols in P and φ which do not appear in K. Let $b_1,\ldots,b_n$ be new constant symbols which also do not appear in K. Let Q and ψ be obtained from P and φ by replacing each a_i by b_i. Then Q is a condition and $Q \dashv\Vert \psi$.

Lemma 5.7: Suppose that $\{a \in A \mid$ does not appear in $K\}$ is infinite. Let $\varphi \in H$ and let $P \subseteq N$ be a condition. Then $P \dashv\Vert_A \varphi$ if and only if $P \dashv\Vert \varphi$.

Proof: If $\rho(\varphi) = 0$, the assertion follows at once from the definitions. Also, if $\rho(\varphi) > 0$ and φ is of form $\theta \,\&\, \psi$, $\theta \vee \varphi$, or $\forall x\theta(x)$, the result is easy to obtain by an inductive argument.

Suppose φ is $\exists x\theta(x)$. If $P \dashv\|_A \varphi$, then $P \dashv\| \varphi$. Conversely, suppose $P \dashv\| \varphi$. Then there is a closed term t such that $P \dashv\| \theta(t)$. If t involves only individuals from A, then $P \dashv\|_A \theta(t)$ and $P \dashv\|_A \varphi$. We prove that t can be so chosen: Let $a_1,\ldots,a_n$ be the constants in t which lie outside of A. Let $a'_1,\ldots,a'_n$ be constants in A which do not appear in $K \cup P \cup \{\varphi\}$. Then by the discussion preceding this lemma, $P \dashv\| \theta(t(a'_1,\ldots,a'_n))$. If t' is $t(a'_1,\ldots,a'_n)$, $P \dashv\|_A \theta(t')$. Hence $P \dashv\|_A \varphi$.

Now suppose φ is $\sim\theta$. If $P \dashv\| \varphi$, then $P \dashv\|_A \varphi$. Conversely, suppose that $P \dashv\|_A \varphi$. If Q is any condition such that $Q \supseteq P$ and $Q \dashv\| \theta$, we may, by the substitution trick of the preceding paragraph, find a condition $Q' \subseteq N$, such that $Q' \supseteq P$ and $Q' \dashv\| \theta$. Then $Q' \dashv\|_A \theta$, and hence not $P \dashv\|_A \varphi$. Therefore, no such Q can exist and $P \dashv\| \varphi$.

§6 Cogeneric Structures and Complete Sequences of Conditions

Let K be a consistent theory which is projectively model consistent with a structure (S,B). We say that (S,B) is K-<u>cogeneric</u> if for all sentences φ defined in (S,B), $(S,B) \models \varphi$ if and only if there is a condition P defined and true in (S,B) such that $P \dashv\| \varphi$.

<u>Theorem</u> 6.1: If K is projectively model consistent with (S,B), (S,B) is K-cogeneric if and only if for every sentence φ in the vocabulary of (S,B) there is a condition P defined and true in (S,B) such that $P \dashv\| \varphi$ or $P \dashv\| \sim\varphi$.

<u>Proof</u>: If (S,B) is K-cogeneric, the fact that for all φ defined in (S,B), $(S,B) \models \varphi$ or $(S,B) \models \sim\varphi$ entails that for all φ, there is a condition P defined and true in (S,B) such that $P \dashv\| \varphi$ or $P \dashv\| \sim\varphi$.

Conversely, suppose that for any sentence φ defined in (S,B), there is a condition P defined and true in (S,B) such that $P \dashv\| \varphi$ or $P \dashv\| \sim\varphi$. By induction on $\rho(\varphi)$ we prove that $(S,B) \models \varphi$ if and only if $P \dashv\| \varphi$. If $\rho(\varphi) = 0$, and $(S,B) \models \varphi$, $\{\varphi\} \dashv\| \varphi$. Conversely, if

$P \dashv\!| \varphi$, $\varphi \in P$, and $(S,B) \models \varphi$ since P is true in (S,B). Now suppose that $\rho(\varphi) > 0$. The cases in which φ is $\exists x\theta(x)$, $\forall x\theta(x)$, $\theta \,\&\, \psi$, or $\theta \vee \psi$ are straightforward since the inductive definitions of $\dashv\!|$ and $\models$ are parallel. If φ is $\sim\theta$, and $(S,B) \models \varphi$, then not $(S,B) \models \theta$; hence no condition P defined and true in (S,B) can coforce θ. Hence by hypothesis, some condition $P' \dashv\!| \sim\theta$. Conversely, suppose that there is a condition P defined and true in (S,B) such that $P \dashv\!| \varphi$; if $(S,B) \models \theta$, then by induction hypothesis, there is a condition P' defined and true in (S,B) such that $P' \dashv\!| \theta$. Since K is projectively model consistent with (S,B), $K \cup \bar{D}_{(S,B)}$ is consistent; that is $P \cup P'$ is a condition. Unfortunately, $P \cup P' \dashv\!| \theta$ and $P \cup P' \dashv\!| \sim\theta$, violating 5.1. Consequently, not $(S,B) \models \theta$; in other words $(S,B) \models \varphi$.

<u>Corollary</u> 6.2: Let K be projectively model consistent with (S,B). If (S,B) is K-cogeneric and φ is defined in (S,B), then $(S,B) \models \varphi$ if and only if there is a condition P defined and true in (S,B) such that $P \dashv\!|^{*} \varphi$.

<u>Proof</u>: ($\Longrightarrow$) If $P \dashv\!| \varphi$, $P \dashv\!|^{*} \varphi$. If $(S,B) \models \varphi$, there is P such that $P \dashv\!| \varphi$.

($\Longleftarrow$) If $P \dashv\!|^{*} \varphi$, $P \dashv\!| \sim\sim\varphi$. Thus not $P \dashv\!| \sim\varphi$. On the other hand, if there is a P' such that $P' \dashv\!| \sim\varphi$, then $P' \cup P'$ is a condition since $(S,B) \models P \cup P'$. But $P \cup P' \dashv\!| \sim\varphi$ and $P \cup P' \dashv\!| \sim\sim\varphi$.

We turn next to the question of the existence of cogeneric structures. Suppose that K and A are at most countable. Let $\mathcal{S} = \{P_n\}$ for $n = 0,1,2,\ldots,$ be a sequence of conditions such that for each n, $P_n \subseteq P_{n+1}$. $\mathcal{S}$ is <u>complete</u> if the following three conditions are met:

(i) for all $\varphi \in H$, there is $P \in \mathcal{S}$ such that $P \dashv\!| \varphi$ or $P \dashv\!| \sim\varphi$;

(ii) for all $\varphi \in N$ of form $\exists x\theta(x)$, if for some $P \in \mathcal{S}$, $P \dashv\!| \varphi$, then for some $Q \in \mathcal{S}$, and some closed term t, $Q \dashv\!| \theta(t)$;

(iii) for all $\varphi \in N$ of form $\forall x\theta(x)$, if for some $P \in \mathcal{S}$, $P \dashv\!| \varphi$, then for all closed terms t there is $Q \in \mathcal{S}$ such that $Q \dashv\!| \theta(t)$.

Theorem 6.3: Suppose that A contains infinitely many constants not in K. Let $P \subseteq N$ be any condition. Then there exists a complete sequence $\mathcal{S}$ whose initial element P_0 is P.

Proof: Let $\varphi_0, \varphi_1, \ldots,$ be an enumeration of all $\bar{L}$-sentences involving relations, functions and constants from $K \cup A$. We define $\mathcal{S}$ inductively. Suppose that P_{3i} has been defined.

Stage 3i+1: If $P_{3i} \dashv\!\vdash \varphi_i$ as $P_{3i} \dashv\!\vdash \sim\varphi_i$, let P_{3i+1} be P_{3i}. Otherwise, let P_{3i+1} be a condition extending P_{3i} such that $P_{3i+1} \dashv\!\vdash \varphi_i$.

Stage 3i+2: For each sentence in $P_{3i+1} - P_{3i-2}$ of form $\exists x\theta(x)$, choose a new constant a not appearing in $K \cup P_{3i+1}$. Obtain P_{3i+2} from P_{3i+1} by adjoining these instances $\theta(a)$. This will be a condition; for if $\exists x\theta(x)$ is satisfiable in a model of K, $\theta(a)$ is also.

Stage 3i+3: For all closed terms t of length at most i involving only constants and function symbols in P_{3i+2}, and for all sentences $\varphi \in P_{3i+2}$ of form $\forall x\theta(x)$, form the instances $\theta(x)$. Obtain P_{3i+3} from P_{3i+2} by adding these instances.

Given a complete sequence $\mathcal{S}$ we may define a structure $M_{\mathcal{S}}$ as follows: the underlying set of $M_{\mathcal{S}}$ is the set of closed terms involving constants in $K \cup A$. The constants are interpreted as themselves. A relation $\bar{R}(t_1,\ldots,t_n)$ holds in $M_{\mathcal{S}}$ if and only if $P \dashv\!\vdash \bar{R}(t_1,\ldots,t_n)$ for some $P \in \mathcal{S}$.

Theorem 6.4: Let K be a consistent set of sentences in Skolem normal form and let $\mathcal{S}$ be a complete sequence. Then $M_{\mathcal{S}}$ is a K-cogeneric model of K.

Proof: Since $\mathcal{S}$ is a complete sequence, every sentence or its negation is coforced by a condition $P \in \mathcal{S}$. Parts (ii) and (iii) of the definition guarantee that each $P \in \mathcal{S}$ is defined and true in $M_{\mathcal{S}}$. To show $M_{\mathcal{S}}$ is K-cogeneric, it remains to show that $M_{\mathcal{S}}$ is projectively model consistent with K. In fact, we show $M_{\mathcal{S}} \models K$.

Let $\mathcal{S}_{At}$ be the set of atomic $\bar{L}$-sentences in $\bigcup\mathcal{S}$. Since $K \cup \bigcup\mathcal{S}$ is consistent, $K \cup \mathcal{S}_{At}$ is consistent; let (S,B) be a model of this theory. Then $A \subseteq C_B$, and if we let (S',B') be the substructure of (S,B) generated by the elements of B corresponding to constants in A, $M_{\mathcal{S}} \cong (S',B')$. Also, since K is universal, $(S',B') \models K$.

Remark 6.5: The hypothesis that K be in Skolem normal form is needed to guarantee the existence of enough functions for $M_{\mathcal{S}}$ to be a model of K. Otherwise, the best that we can say is that $M_{\mathcal{S}}$ is embeddable in a model of K. For example if K_0 is the following axiomatization of divisible abelian group then every countable such group which is torsion-free and of rank ≥ 1 can be seen to be $M_{\mathcal{S}}$ for some complete sequence $\mathcal{S}$.

(DA1) $\forall xy(x+y = y+x)$

(DA2) $\forall xyz(x + (y+z) = x + (y+z))$

(DA3) $\forall x(x + (-x) = 0)$

(DA4) $\forall x(x+0 = x)$

(DA5)$_n$ $\forall x(\underbrace{f_{1/n}(x) + \cdots + f_{1/n}(x)}_{n \text{ times}} = x)$

If we obtain a different axiomatization K, by replacing (DA5)$_n$ by

(DA6)$_n$ $\forall x\exists y(\underbrace{y + \cdots + y}_{n \text{ times}} = x)$,

one can see that every $M_{\mathcal{S}}$ is not divisible. Namely, for each closed term t and each constant a, $t^n \neq a$ $(n > 1)$ is consistent with any condition P. Thus $t^n \neq a$ is satisfied by $M_{\mathcal{S}}$ since $\sim t^n \neq a$ can never be coforced for any t. Indeed, one can verify that $M_{\mathcal{S}}$ is free abelian of infinite rank, for any complete sequence $\mathcal{S}$.

For the rest of this paper we assume that K is in Skolem normal form.

Remark 6.6: If K is in Skolem normal form, then every K-cogeneric structure is a model of K. This follows from the fact that every structure projectively model consistent with K is in fact a surjective image of a model of K when K is a universal theory. Let (S,B) be K-cogeneric and let (S',B') be a pre-image of (S,B) which is a model of K; we may suppose $C_{B'} = C_B$ since K is universal. For any atomic formula $\overline{R}(t_1,\ldots,t_n)$, if $(S',B') \models \overline{R}(t_1,\ldots,t_n)$, then $(S,B) \models \overline{R}(t_1,\ldots,t_n)$ for no condition P defined and true in (S,B) can coforce $\sim\overline{R}(t_1,\ldots,t_n)$ because if P is such a condition, $P \cup \{\overline{R}(t_1,\ldots,t_n)\}$ is an extension of P which coforces $\overline{R}(t_1,\ldots,t_n)$. Consequently, $(S',B') \cong (S,B)$ and $(S,B) \models K$.

Corollary 6.7: Every condition $P \subseteq N$ is satisfied in some K-cogeneric structure.

Proof: This follows at once from 6.3 and 6.4.

Corollary 6.8: Let $P \subseteq N$ be a condition. If $P \vdash \varphi$, then $P \dashv\Vdash^* \varphi$.

Proof: If $P \vdash \varphi$, then φ holds in every K-cogeneric structure which satisfies P. If P had an extension Q such that $Q \dashv\Vdash \sim\varphi$, then there is a complete sequence $\mathcal{S}$ whose initial element is Q. $M_{\mathcal{S}} \models P$, hence $M_{\mathcal{S}} \models \varphi$. On the other hand, $M_{\mathcal{S}} \models Q$, and so, $M_{\mathcal{S}} \models \sim\varphi$. Consequently no such Q exists and $P \dashv\Vdash \sim\sim\varphi$.

Corollary 6.9: Suppose that $P \dashv\Vdash \varphi$ and $K \cup P \vdash \varphi \supset \psi$. Then $P \dashv\Vdash^* \psi$.

Proof: If $P \dashv\Vdash \varphi$, then every K-cogeneric structure which satisfies P satisfies φ and ψ. P has no extension Q such that $Q \dashv\Vdash \sim\psi$, by an argument similar to that of the preceding corollary.

Corollary 6.10: Let P be any condition. Let $K^f(P) = \{\varphi \mid \varphi$ is in the vocabulary of $P \cup K$ and $P \dashv\Vdash^* \varphi\}$. Then $\varphi \in K^f(P)$ if and only if φ holds in all K-cogeneric models of P.

Proof: If $P \dashv\Vdash^* \varphi$, then for all $Q \supseteq P$, not $Q \dashv\Vdash \sim\varphi$. Hence if (S,B) is a K-cogeneric model of P, $(S,B) \models \varphi$. Conversely, suppose it is not the case that $P \dashv\|^* \varphi$. Then there exists $Q \supseteq P$ such that $Q \dashv\| \sim\varphi$. Let $\mathcal{S}$ be a complete sequence with initial condition $P_0 = Q$. Then $M_{\mathcal{S}}$ is a K-cogeneric model of P satisfying $\sim\varphi$.

<u>Theorem</u> 6.11: Let (S,B) be projectively model consistent with K, where K is in Skolem normal form. (S,B) is K-cogeneric if and only if for all φ defined in (S,B), if $(S,B) \models \varphi$ then there is a condition P defined and true in (S,B) such that $P \dashv\|^* \varphi$.

<u>Proof</u>: If (S,B) is K-cogeneric then the desired conclusion is trivial. Conversely, suppose that any φ satisfied by (S,B) is weakly coforced by a condition defined and true in (S,B), but that (S,B) is not K-cogeneric. Choose a sentence φ of minimal ρ-rank such that neither φ nor $\sim\varphi$ is coforced by a condition in (S,B). Then certainly, $(S,B) \models \varphi$ for if $(S,B) \models \sim\varphi$, there is P such that $P \dashv\| \sim\sim\sim\varphi$; consequently, $P \dashv\| \sim\varphi$. Furthermore, $\rho(\varphi) > 0$, for $\{\varphi\} \dashv\| \varphi$, if $\rho(\varphi) = 0$, and φ is not $\theta \,\&\, \psi$, $\theta \vee \psi$, $\forall x\theta(x)$, or $\sim\theta$, by minimality of $\rho(\varphi)$. Moreover, if φ is $\exists x(\theta(x))$ and $(S,B) \models \varphi$, $(S,B) \models \theta(t)$ for some closed term t. Choosing P such that $P \dashv\|^* \theta(t)$, we see that no P' defined and true in (S,B) can coforce $\sim\theta(t)$. By minimality of $\rho(\varphi)$, there is P" defined and true in (S,B) such that $P'' \dashv\| \theta(t)$, whence $P'' \dashv\| \varphi$. This contradicts the choice of φ, from which we conclude that (S,B) is indeed K-cogeneric.

§7 Coforcing and Projective Model Completion

<u>Theorem</u> 7.1: Let (S,B) be K-cogeneric. Let θ_1, θ_2 be positive $\bar{L}$-sentences in the vocabulary (S,B) such that $K \cup D^-_{(S,B)} \cup \{\theta_1 \supset \theta_2\}$ is consistent. Then $(S,B) \models \theta_1 \supset \theta_2$.

<u>Proof</u>: Suppose $(S,B) \models \theta_1 \,\&\sim\theta_2$. Therefore, for some condition P defined and true in (S,B), $P \dashv\| \sim\theta_2$. That is $K \cup P \cup \{\theta_2\}$ is inconsistent. Then, $K \cup D^-_{(S,B)} \cup \{\theta_1 \supset \theta_2\}$ is inconsistent.

Corollary 7.2: If φ and ψ are positive L-sentences and $K \vdash \varphi \supset \psi$, then $(S,B) \models \varphi \supset \psi$.

Proof: Let θ_1 be $\sim \psi$ and θ_2 be $\sim \varphi$, and apply 6.11.

Theorem 7.3: Let K be a countable consistent theory in Skolem normal form with a projective model completion K^*. Suppose that K has no 1-element models. Let (S,B) be a K-cogeneric structure. Then $(S,B) \models K^*$.

Proof: Since K has no 1-element models, we may assume that K contains a pair of constants C and C' such that $K \models C \neq C'$. By 4.5, K^* is equivalent relative to $C \neq C'$ to a set of sentences of form $\varphi \supset \psi$ where φ and ψ are positive L-sentences. Also $K^* \supseteq K$. Thus K^* may be obtained from K by the adjunction of the sentences $\varphi \supset \psi$. By 7.2 $(S,B) \models \varphi \supset \psi$ since $\varphi \supset \psi$ is consistent with K.

Theorem 7.4: Let K be a countable consistent theory in Skolem normal form with no one-element models, having a projective model completion K^*. Suppose A contains infinitely many constants not in K. Then every model (S,B) of K^* is K-cogeneric.

Proof: $K^* \cup D^-_{(S,B)}$ is a consistent complete theory. Without loss of generality we may suppose that K^* consists of $C \neq C'$ and a set of axioms of form $\varphi \supset \psi$ where φ and ψ are $\bar{L}$-sentences of ρ-rank 0. Obviously (S,B) is projectively model consistent with K^*, so that we need only check that every sentence in the vocabulary of (S,B) or its negation is coforced by a condition defined and true in (S,B).

First we observe that every sentence φ in $K^* \cup D^-_{(S,B)}$ is coforced by a finite condition P defined and true in (S,B). If $\varphi \in D^-_{(S,B)}$, then $\{\varphi\} \dashv\!| \ \varphi$. If φ is $\theta \supset \psi$ where $\rho(\theta) = \rho(\psi) = 0$, then either $(S,B) \models \psi$, or $(S,B) \models \sim \theta$. In the first instance, $\{\psi\} \dashv\!| \ \theta \supset \psi$. Now suppose $(S,B) \models \sim \theta$; we claim that there is a condition P defined and true in (S,B) such that $P \dashv\!| \sim \theta$. Suppose not; then, by compactness $K^* \cup D^-_{(S,B)} \cup \{\theta\}$ is consistent. Hence, there is an elementary extension

(S',B') of (S,B) which is a surjective image of a model (S'',B'') of $K^* \cup D^-_{(S,B)} \cup \{\theta\}$. But then the surjection from (S'',B'') onto (S',B') is not an elementary surjection since $(S'',B'') \models \theta$ and $(S',B') \models \sim\theta$. Consequently, if K is projectively model complete, there is a condition P defined and true in (S,B) such that $P \dashv\!| \sim\theta$.

Now since $D^-_{(S,B)} \cup K^*$ is complete, and for all $\varphi \in K^*$ there is $P \subseteq D^-_{(S,B)}$ such that $P \dashv\!| \varphi$ we have that every sentence satisfied by (S,B) is weakly coforced by a condition defined and true in (S,B) by 6.9. Then, by 6.11 we may conclude that (S,B) is K-cogeneric.

<u>Corollary</u> 7.5: If K is a countable consistent theory in Skolem normal form, with a projective model completion K^* then

(i) the models of K^* are precisely the K-cogeneric structures

(ii) $K^* = K^f[\emptyset]$.

§8 Coforcing and Completeness

Let K be a consistent theory. K is a <u>turreted</u> theory if for any two models M_1 and M_2 of K, there exist models N_1, N_2 and N^* of K such that $M_1 \prec N_1$, $M_2 \prec N_2$, and N_1 and N_2 are surjective images of N^*.

If α and β are the surjections from N^* to N_1 and N_2, we call the following diagram a <u>turret</u> over M_1 and M_2:

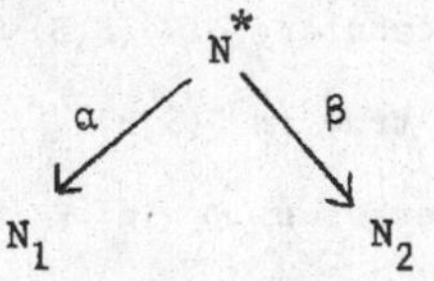

N^* is called the <u>top of the turret</u>.

<u>Lemma</u> 8.1: K is a turreted theory if and only if for any pair of positive $\overline{L}$-sentences, $\{\varphi,\psi\}$, if $K \cup \{\varphi\}$ and $K \cup \{\psi\}$ are consistent, then so is $K \cup \{\varphi,\psi\}$.

Proof: First suppose K is turreted. Let $M_1 \models K \cup \{\varphi\}$ and $M_2 \models K \cup \{\psi\}$. Then, if N^* is the top of a turret over M_1 and M_2, $N^* \models \varphi$, $N^* \models \psi$, and $N^* \models K$. Thus $K \cup \{\varphi, \psi\}$ is consistent.

Conversely, let M_1 and M_2 be any two models of K. Let D be $D^-_{M_1} \cup D^-_{M_2}$. By the hypothesis and the compactness theorem $K \cup D$ is consistent. If $N^* \models K \cup D$, $N^* \models D^-_{M_1}$ and $N^* \models D^-_{M_2}$. Therefore there exist elementary extensions $N_1 \succ M_1$ and $N_2 \succ M_2$, and surjections $\alpha : N^* \longrightarrow N_1$ and $\beta : N^* \longrightarrow N_2$. Thus K is turreted.

Let $K^f = K^f[\emptyset]$.

Theorem 8.2: K^f is complete if and only if K is a turreted theory.

Proof: First suppose that K is a turreted theory. Choose a sentence φ of $\overline{L}$ in the vocabulary of K. If neither $\emptyset \dashv\!| \sim\varphi$ nor $\emptyset \dashv\!| \sim\sim\varphi$, there exist conditions P and Q such that $P \dashv\!| \varphi$ and $Q \dashv\!| \sim\varphi$. Since K is a turreted theory, $P \cup Q$ is a condition. Since this is impossible, we must conclude either $\emptyset \dashv\!|^* \varphi$ or $\emptyset \dashv\!| \sim\varphi$.

Conversely, suppose K^f is complete. Let φ and ψ be two positive $\overline{L}$-sentences each of which is consistent with K. Then $\{\varphi\}$ and $\{\psi\}$ are conditions. Since $\{\varphi\} \dashv\!| \varphi$ and $\{\psi\} \dashv\!| \psi$, $\emptyset \dashv\!|^* \varphi$ and $\emptyset \dashv\!|^* \psi$. That is for any condition P there is a condition $Q \supset P$ such that $Q \dashv\!| \varphi$ and $Q \dashv\!| \psi$. Since $K \cup Q$ is consistent, there is a model (S,B) of $K \cup Q$. Since $\varphi \in Q$ and $\psi \in Q$, $K \cup \{\varphi, \psi\}$ is consistent. Therefore K is a turreted theory.

Corollary 8.3: If K is a turreted theory, then all K-cogeneric structures are elementarily equivalent.

Proof: K^f is the set of sentences true of all K-cogeneric structures.

If K is in Skolem normal form, then K is turreted if and only if K has the joint pre-image property, that is, if and only if for any pair of models of K there is a model of K which is a surjective pre-image of each of them. In general, the joint pre-image properties fails for turreted theories.

§9 Some Remarks on the Theory of Free Groups

Tarski has conjectured that all of the non-abelian free groups are elementarily equivalent. In this section we apply some of the techniques of projective model theory in order to obtain some reductions of this problem. Let ETF be the set of sentences true in all non-abelian free groups, and for $k = 2,3,4,\ldots,\omega$, let ETF_k be the elementary theory of the free group of rank k. Also, let the free group of rank k have the presentation $< \xi_1,\xi_2,\xi_3,\ldots,\xi_k >$. Thus in our earlier notation we would describe F_k by the notation $(F_k,\{\xi_1,\ldots,\xi_k\})$.

<u>Theorem</u> 9.1: Let $k \geq k' \geq 2$. Let φ be a positive sentence defined in $F_{k'}$. $F_{k'} \models \varphi$ if and only if $F_k \models \varphi$.

<u>Proof</u>: This is Theorem C of [16].

Note that if $\{F_k | \text{cardinals } k \geq 2\}$ were an elementary class, then ETF would be projectively model complete. Since this class has the joint pre-image property, ETF would be complete.

<u>Lemma</u> 9.2: The free group F_ω is $\mathcal{G}$-cogeneric when $\mathcal{G}$ is the theory of groups.

<u>Proof</u>: Let P_i be the set of sentences of $D^-_{F_\omega}$ which involve at most i symbols. Clearly each P_i is a condition relative to $\mathcal{G}$ and $P_{i+1} > P_i$. Moreover $F_\omega \models \mathcal{G}$, so that it suffices to prove that for all φ of $\bar{L}$, there is i such that $P_i \dashv\!| \varphi$ or $P_i \dashv\!| \sim\varphi$. We proceed by induction on $\rho(\varphi)$.

First suppose $\rho(\varphi) = 0$. In this case there is an i such that $P_i \dashv\!| \varphi$ if and only if there is an i such that $\varphi \in P_i$. If $\mathcal{G} \cup \{\varphi\}$ is consistent, then there is a group G such that $G \models \varphi$. Since φ is positive in $\bar{L}$, we may suppose that G is a free group of infinite rank. By the Skolem-Löwenheim Theorem, we may suppose that G is countable; that is, we may suppose that $G \simeq F_\omega$. Thus $\mathcal{G} \cup \{\varphi\}$ is consistent if and only if $\varphi \in P_j$ where j is the number of symbols in φ. Consequently, either $\mathcal{G} \cup \{\varphi\}$ is inconsistent; and hence $\emptyset \dashv\!| \sim\varphi$ or else $\mathcal{G} \cup \{\varphi\}$

is consistent and some $P_j \dashv\| \varphi$.

Now suppose $\rho(\varphi) > 0$. If φ is $\theta \,\&\, \psi$, then there is i such that $P_i \dashv\| \varphi$ if and only if there exist j and k such that $P_j \dashv\| \theta$ and $P_k \dashv\| \psi$. By induction hypothesis there is j such that $P_j \dashv\| \theta$ or $P_j \dashv\| \sim\theta$ and there is k such that $P_k \dashv\| \psi$ or $P_k \dashv\| \sim\psi$. If $P_j \dashv\| \theta$ and $P_k \dashv\| \psi$ then $P_{j+k} \dashv\| \varphi$. In any other combination of circumstances, $P_{j+k} \dashv\| \sim\varphi$. A similar argument applies in φ is $\theta \vee \psi$, $\sim\theta$, or $\forall x\theta(x)$.

Finally suppose φ is $\exists x\theta(x)$. There is a condition P_i such that $P_i \dashv\| \varphi$ if and only if there is a closed term t all of whose constants are in A such that $P_i \dashv\| \theta(t)$. By induction hypothesis, for each closed term t there is an index j such that $P_j \dashv\| \theta(t)$ or $P_j \dashv\| \sim\theta(t)$. If for some term t on A the first case holds, then $P_j \dashv\| \varphi$ and we are done. On the other hand, if for each term t on A there is a j such that $P_j \dashv\| \sim\theta(t)$, we claim that for some k, $P_k \dashv\| \sim\varphi$.

For the sake of a contradiction, suppose that there is no such k, that is, suppose that for each k there is a condition (relative to $\mathcal{G}$) $Q_k \supset P_k$ such that $Q_k \dashv\| \exists x\varphi(x)$. Then, for some closed term s_k, $Q_k \dashv\| \theta(s_k)$. Let $b_1,\ldots,b_n$ be the constants in s_k or Q_k which lie outside of A. For each $j \geq k$, $P_j \cup Q_k$ must be a condition since the free group $\widetilde{F} = \langle b_1,\ldots,b_n, \xi_1,\xi_2,\ldots \rangle$ is a model of $\mathcal{G} \cup P_j \cup Q_k$. Since $\widetilde{F}$ and F_ω are isomorphic, let α be an isomorphism from $\widetilde{F}$ onto F_ω; we may suppose that for each b_i, $\alpha(b_i) = \xi_{m+i}$ where m is the largest subscript appearing on a constant $\xi_q \in P_j$. Then, the application of the isomorphism α to the constants in Q_k and s_k, converts Q_k into a condition $P_{k'}$ and s_k into a term all of whose constants lie in A and $P_{k'} \dashv\| \theta(\alpha(s_k))$. On the other hand, $P_{k'}$ has an extension $P_{k''}$ such that $P_{k''} \dashv\| \sim\theta(\alpha(s_k))$. This contradiction forces us to conclude that Q_k and s_k as described do not exist. That is, some $P_k \dashv\| \sim[\exists x\theta(x)]$. Thus the sequence $\{P_i\}$ coforces every sentence or its negation.

To complete the proof that $\{P_i\}$ is a complete sequence we must verify (i) if $\exists x\varphi(x) \in \bigcup P_i$, then $P_i(t) \in \bigcup P_i$ for some closed t and

(ii) if $\forall x\varphi(x) \in \bigcup P_i$, then $\varphi(t) \in \bigcup P_i$ for all closed t. In re (i), if $\varphi(t) \notin \bigcup P_i$ for all t, $F_\omega \models \varphi_i(t)$ for all t and $F_\omega \models \sim \exists x\varphi(x)$ and so $\exists x\varphi(x) \notin \bigcup P_i$. In re (ii), if $\forall x\varphi(x) \in \bigcup P_i$, $\varphi(t) \in P_i$ for all t because $F_\omega \models \varphi(t)$.

<u>Theorem</u> 9.3: If k is infinite, F_k is $\mathcal{G}$-cogeneric.

<u>Proof</u>: Let φ be a sentence defined in F_k. Let $\xi_{i_1},\ldots,\xi_{i_n}$ be the constants which appear in φ. We may extend $\xi_{i_1},\ldots,\xi_{i_n}$ to a countably infinite list of constants and let $\overline{F}$ be the subgroup of F_k generated by this set. Let P be a condition defined and true in $\overline{F}$ which coforces φ or $\sim\varphi$. P is defined and true in F_k also. Moreover, $F_k \models \mathcal{G}$. Hence, by 6.1, F_k is K-cogeneric.

<u>Theorem</u> 9.4: If (S,B) is a group which is $\mathcal{G}$-cogeneric, then S is free on B and B has at least two elements. That is, every $\mathcal{G}$-cogeneric group is free of rank ≥ 2.

<u>Proof</u>: First, since no condition defined and true in an abelian group can coforce $\exists xy(xy \neq yx)$, we see $|B| \geq 2$. Now suppose (S,B) satisfies a non-trivial relation $t = 1$ where t is a closed term which cannot be reduced to the empty word by elimination of inverse pairs. (S,B) has a pre-image (S',B') which satisfies $t \neq 1$. Thus from any condition P defined and true in (S,B) we may obtain $P \cup \{t \neq 1\}$ as a condition; consequently, not $P \dashv\!\mid\!\mid \sim(t \neq 1)$. On the other hand, certainly not $P \dashv\!\mid\!\mid t \neq 1$. Thus (S,B) is not $\mathcal{G}$-cogeneric.

<u>Theorem</u> 9.5: The theory of the $\mathcal{G}$-cogeneric groups is a complete theory.

<u>Proof</u>: $\mathcal{G}$ has the joint pre-image property; for any pair of groups, their direct product is a pre-image of both elements of the pair. Hence, by 8.2 $\mathcal{G}^f$ = theory of the $\mathcal{G}$-cogenerics is complete.

<u>Corollary</u> 9.6: Any two finitely generated free groups which are $\mathcal{G}$-cogeneric are elementarily equivalent.

Theorem 9.7: ETF is not projectively model complete.

Proof: If ETF were projectively model complete, then the $\mathcal{G}$-cogenerics would be the models of ETF. All $\mathcal{G}$-cogenerics are free groups and the class of free groups is not closed under ultrapower. Consequently, there are models of ETF which are not $\mathcal{G}$-cogeneric.

We remark that in order to show that a finitely generated free group F_k is $\mathcal{G}$-cogeneric all that is required is that for all φ defined in F_k, $\emptyset \dashv\Vdash^* \varphi$ if $F_k \models \varphi$.

Bibliography

[1] K. J. Barwise and A. Robinson, Completing Theories by Forcing, Ann. of Math. Logic, 2, 119-143.

[2] J. L. Bell and A. B. Slomson, Models and Ultraproducts, North-Holland, Amsterdam, 1971.

[3] N. Bourbaki, Éléments de Mathématiques, Algèbre, Livre II, Hermann, Paris, 1962.

[4] C. C. Chang, On Unions of Chains of Models, Proc. of the A.M.S., 10, 120-127.

[5] P. J. Cohen, Set Theory and the Continuum Hypothesis, Benjamin, New York, 1966.

[6] S. Feferman, Some Applications of the Notions of Forcing and Generic Sets, Fund. Math., 56, 325-346.

[7] G. Grätzer, Universal Algebra, Van Nostrand, Princeton, 1968.

[8] J. Łoś and R. Suszko, On the Extending of Models (IV), Fund. Math., 44, 52-60.

[9] R. Lyndon, An Interpolation Theorem in the Predicate Calculus, Pac. J. of Math., 9, 129-142.

[10] ————, Properties Preserved Under Homomorphism, Pac. J. of Math., 9, 143-154.

[11] A. Robinson, Obstructions to Arithmetical Extension and the Theorem of Łoś and Suszko, Proc. Roy. Acad. of Sci. of Amsterdam, sec. A.62, 489-495.

[12] ————, Introduction to Model Theory and the Metamathematics of Algebra, North-Holland, Amsterdam, 1965.

[13] ————, Forcing in Model Theory, Istituto Nazionale di Alta Matematica, Symposia Mathematica, V, 69-82.

[14] ——————, Infinite Forcing in Model Theory, Proceedings of the Second Scandinavian Logic Symposium (J. E. Fenstad, editor), North-Holland, Amsterdam, 1971.

[15] ——————, Generic Categories (Mimeographed).

[16] G. S. Sacerdote, Elementary Properties of Free Groups, Trans. of the A.M.S., 178, 127-138.

[17] ——————, Projective Model Completeness, J. of Symbolic Logic, to appear.

[18] A. Tarski and R. L. Vaught, Arithmetical Extensions of Relational Systems, Compositio Mathematica, 13, 81-102.

[19] H. J. Keisler, Finite Approximations of Infinitely Long Formulas in Theory of Models, (Addison, Henkin, and Tarski, editors), North-Holland, Amsterdam, 1965.

[20] ——————, Some Applications of Infinitely Long Formulas, J. of Symbolic Logic, 30, 339-349.

ON ALGEBRAIC CURVES OVER COMMUTATIVE REGULAR RINGS

D. Saracino and V. Weispfenning

Colgate University and University of Heidelberg

It has been known for some time that the theory of fields has a model completion, namely the theory of algebraically closed fields. It has recently been shown [10] that the theory K of commutative (von Neumann —) regular rings has a model completion K'; the models of K' are just the commutative "monically closed" regular rings with no minimal idempotents. In light of the evidence provided by the study of classical algebraic geometry over fields, it would seem reasonable to suppose that the models of K' should provide a natural setting for discussing algebraic geometry over commutative regular rings. It is our purpose in this paper to provide some justification for this supposition by examining some properties of algebraic curves over commutative regular rings. Actually for the present purposes it turns out that it suffices to consider a weakened version of K', namely the theory of K* of commutative monically closed regular rings.

The outline of the paper is as follows: In part I we study the concept of a *-ring. A *-ring consists of a ring with an additional operation satisfying certain conditions. Section 1 deals with elementary properties of *-rings and regular rings, and with a canonical representation of such rings as subdirect products of integral domains. Section 2 deals with locally constant integer-valued functions on an appropriate spectrum of such rings. In particular, the local constancy of such functions is approached through the metamathematical notion of definability in a certain first-order language. Section

3 deals with further general theory of *-rings, the central theme being the extension theory for such rings and the compatibility of certain extensions with the canonical representation. Section 4 is concerned with an analogue of Hilbert's Nullstellensatz for finitely generated polynomial ideals over models of K*.

Part II deals with those special cases of *-rings which arise naturally in an attempt to generalize in as direct a way as possible the theory of algebraic curves over fields. Sections 1 and 2 introduce the appropriate *-rings in affine and projective settings, respectively. The remaining sections are concerned with the interpretation of various geometric concepts as locally constant functions on Spec(R), where R is a model of K*. Section 3 deals with multiplicities and intersection numbers, section 4 with the resolution of singularities, and section 5 with divisors.

We wish to acknowledge the particular debt we owe Abraham Robinson with respect to this paper. Without his encouragement it would not have been written.

I. *-rings

I.1. Regular rings and *-rings.

All rings in this paper will be commutative rings with identity. For any ring R the set B(R) of all idempotents in R forms a Boolean algebra under the operations $a \cap b = a \cdot b$, $a \cup b = a + b - ab$, $\sim a = 1 - a$. Thus $a \leq b$ for $a, b \in B(R)$ if and only if $ab = a$. If R is an integral domain, $B(R) = \{0,1\}$. A ring R is called <u>regular</u> (in the sense of von Neumann), if for every $a \in R$ there exists $b \in R$ such that $a^2b = a$. We recall some elementary properties of regular rings R (see [7]). For all $a,b \in R$ such that $a^2b = a$ the element $a^{-1} = ab^2$ is uniquely determined by a and called <u>the quasi-inverse of a</u>. The map $a \longrightarrow a^{-1}$ is uniquely determined by the properties

I.1.1. $$a^2a^{-1} = a\ ,\ (a^{-1})^2\, a = a^{-1}\ .$$

Other properties include:

I.1.2 (i) $(ab)^{-1} = a^{-1}b^{-1}$

(ii) $(a^{-1})^{-1} = a$

(iii) $(-a)^{-1} = -(a^{-1})$

(iv) $ab^{-1} = 0 \longleftrightarrow ab = 0$

(v) $ab = 0 \longrightarrow (a+b)^{-1} = a^{-1} + b^{-1}$.

For every $a \in R$ the element $a^* = aa^{-1}$ is an idempotent, called <u>the idempotent of a</u>. The map $a \longrightarrow a^*$ satisfies

I.1.3 (i) $aa^* = a$, (ii) $\forall x(ax = a \longrightarrow a^*x = a^*)$.

Notice that both the quasi-inverse and the operation $*$ are defined by equations in R. So any ring homomorphism $\varphi : R \longrightarrow R'$ of R into a regular ring R' is compatible with $^{-1}$ and $*$. In particular for any regular ring $R' \supset R$ $\langle R', ^{-1}, * \rangle$ extends $\langle R, ^{-1}, * \rangle$.

We say that a regular ring R is <u>monically closed</u> if every monic polynomial with coefficients in R has a root in R.

For the purpose of this paper the class of regular rings is somewhat too narrow. Remark, e.g., that the polynomial ring R[X] over a regular ring R is not regular. So we are forced to consider a wider class of rings, namely the class of all p.p. rings. A ring R is called a <u>p.p. ring</u> (see [1]), if every principal ideal in R is projective. An equivalent definition is the following: R is a p.p. ring if for every $a \in R$ there exists $b \in R$ such that

I.1.4 (i) $ab = a$

(ii) $\forall x(ax = a \longrightarrow bx = b)$.

Notice that every regular ring R is a p.p. ring, since $b = aa^{-1}$ satisfies I.1.4. For any p.p. ring R the element b satisfying I.1.4 is obviously uniquely determined by a. So there exists a unique map

$*: R \longrightarrow R$ satisfying I.1.3. Unfortunately, however, the definition of $*$ in a p.p. ring R is in general not compatible with extensions, i.e. there exist p.p. rings $R \subset R'$ such that the operation $*$ in R' does not extend the operation $*$ in R. Take, e.g., $R = \mathbb{Z}$, $R' = \prod_{p \text{ prime}} \mathbb{Z}/_p$, and consider R as a subring of R' under the embedding $a \longmapsto \langle a \bmod p : p \text{ prime}\rangle$. The operation $*$ in R is given by $0^* = 0$ and $a^* = 1$ for $a \neq 0$, in R' by $a^*(p) = 0$ if $a(p) = 0$ and $a^*(p) = 1$ otherwise for all primes p. So $2^* = 1$ in R but $2^* \neq 1$ in R'. To overcome this difficulty we shall treat the operation $*$ not as defined, but as a primitive operation (cf. [2], ex. 2'). This leads to the following definition:

A *-ring is a structure $\langle R,*\rangle$, where R is a ring and $*$ a map from R into R satisfying I.1.3.

Lemma I.1.5. Let $\langle R,*\rangle$ be a *-ring and $a, b \in R$. Then (i) $(a^*)^2 = a^*$ (ii) $(-a)^* = a^*$ (iii) $a^{**} = a^*$ (iv) $(ab)^* = a^*b^*$ (v) $0^* = 0$, $1^* = 1$ (vi) $a^2 = a \longleftrightarrow a^* = a$ (vii) $ab = 0 \longrightarrow a^*b = 0$ (viii) $ab = 0 \longrightarrow (a+b)^* = a^* + b^*$.

All these properties are easily derived from I.1.3.

Let $\langle R,*\rangle \subset \langle R',*\rangle$ be *-rings. We say $\langle R',*\rangle$ is a regular closure of $\langle R,*\rangle$ if R' is regular and for every regular *-ring $\langle R'',*\rangle \supset \langle R,*\rangle$ there exists an embedding $\varphi : \langle R',*\rangle \longrightarrow \langle R'',*\rangle$ such that $\varphi \restriction \langle R,*\rangle$ is the identity.

Theorem I.1.6 (compare [1], lemma 3.1).

Let $\langle R,*\rangle$ be a *-ring, S the multiplicative semigroup of non zero-divisors in R, R' the quotient ring of R with respect to S, and define the operation $*$ on R' by $(a/b)^* = a^*$ for $a \in R$, $b \in S$. Then $\langle R',*\rangle$ is a regular closure of $\langle R,*\rangle$ and any other regular closure of $\langle R,*\rangle$ is isomorphic to $\langle R',*\rangle$ over $\langle R,*\rangle$.

<u>Proof</u>. Notice that S consists exactly of all elements $b \in R$ such that $b^* = 1$. By the definition of $*$ in R', $\langle R',*\rangle$ extends $\langle R,*\rangle$. To show that R' is regular, we define the quasi-inverse by $(a/b)^{-1} = ba^*/(a+1-a^*)$ for $a \in R$, $b \in S$. By I.1.5. $(a+1-a^*)^* = a^* + (1-a^*) = 1$, so $(a/b)^{-1}$ is well-defined. By I.1.5 we have moreover for $a \in R$, $b \in S$
$(a/b)^2 \ (a/b)^{-1} = a^2b/b^2(a+1-a^*) = a(a+1-a^*)/b(a+1-a^*) = (a/b)$, and
$((a/b)^{-1})^2 \ (a/b) = ab^2/b(a+1-a^*)^2 = ba^*(a+1-a^*)/(a+1-a^*)^2 = (a/b)^{-1}$, and
$(a/b)(a/b)^{-1} = ab/b(a+1-a^*) = a^*(a+1-a^*)/(a+1-a^*) = a^* = (a/b)^*$. So I.1.1 is satisfied and $cc^{-1} = c^*$ with c^* defined as above for all $c \in R'$. Thus $\langle R',*\rangle$ is a regular *-ring extending $\langle R,*\rangle$. $\langle R',*\rangle$ is also minimal with this property, since $a/b = ab^{-1}$ for all $a \in R$, $b \in S$. Next suppose $\langle R'',*\rangle$ is another regular *-ring extending $\langle R,*\rangle$ and define $\varphi : R' \longrightarrow R''$ by $\varphi(a/b) = ab^{-1}$ for $a \in R$, $b \in S$. Since $\langle R'',*\rangle$ is a *-extension of $\langle R,*\rangle$, $b^* = 1$ in R" for all $b \in S$, so by a well-known property of total quotient rings, φ is a well-defined embedding of R' into R". Since R' and R" are regular, φ is moreover a *-embedding. Finally assume $\langle R'',*\rangle$ is a regular closure of $\langle R,*\rangle$ and let $\psi : \langle R'',*\rangle \longrightarrow \langle R',*\rangle$ be an embedding over $\langle R,*\rangle$. Then ψ must be onto $\langle R',*\rangle$ by the minimality of $\langle R',*\rangle$.

Let $\langle R,*\rangle$ be a *-ring and I an ideal in R. We call I a <u>*-ideal</u> in $\langle R,*\rangle$ if I is closed under the operation $*$, i.e. $a \in I$ implies $a^* \in I$. The following theorem establishes a 1-1 correspondence between *-ideals in $\langle R,*\rangle$ and Boolean ideals in B(R).

<u>Theorem</u> I.1.7. Let $\langle R,*\rangle$ be a *-ring.

(i) For any *-ideal I in $\langle R,*\rangle$, $I_B = I \cap B(R)$ is a Boolean ideal in B(R). If I is prime, then I_B is prime.

(ii) For any Boolean ideal J in B(R),
$J_R = R \cdot J = \{a \cdot b : a \in R, b \in J\}$ is a *-ideal in $\langle R,*\rangle$.
If J is prime, then J_R is prime.

(iii) $I = (I_B)_R$ and $J = (J_R)_B$.

(iv) For all *-ideals $I \subsetneq I'$ in $\langle R,*\rangle$, $I_B \subsetneq I'_B$; and for all Boolean ideals $J \subsetneq J'$ in $B(R)$, $J_R \subsetneq J'_R$.

<u>Proof</u>. (i) Put $B = B(R)$ and assume $a, b \in I \cap B$, $c \in B$. Then $0 \in I \cap B$, $a \cup b = a + b - ab \in I \cap B$, and $c \cap a = ca \in I \cap B$. If I is prime, let $d \in B$ be such that $cd = c \cap d \in I \cap B$. Then $c \in I \cap B$ or $d \in I \cap B$.

(ii) Suppose $a,b \in R$, $i,j \in J$. Then $b(ai) = (ba)i \in RJ$, $ai + bj = (ai+bj)(i \cup j) \in RJ$, and $(ai)^* = a^* i \in RJ$. If J is prime and $ab = ci \in RJ$ for some $c \in R$, then $a^*b^* = (ab)^* = c^*i \leq i \in J$. So $a^* \in J$ or $b^* \in J$ and hence $a = aa^* \in RJ$ or $b = bb^* \in RJ$.

(iii) $R(I \cap B) \subset RI \subset I$ and conversely $a \in I$ implies $a = aa^* \in R(I \cap B)$. $J \subset (J \cap B) \subset (RJ \cap B)$ and conversely $a \in R$, $i \in J$, $ai \in B$ implies by (ii) $ai = (ai)^* = a^*i \leq i \in J$, and so $ai \in J$. (iv) follows immediately from (iii).

We call a *-ideal I in a *-ring $\langle R,*\rangle$ <u>*-maximal</u> if I is maximal among all the proper *-ideals of $\langle R,*\rangle$. (<u>Warning</u>: A *-ideal I can be *-maximal but not maximal). Notice that a Boolean ideal is prime if and only if it is maximal. So theorem I.1.7 implies that a *-ideal I in a *-ring $\langle R,*\rangle$ is prime if and only if I is *-maximal. For a <u>regular</u> ring R any ideal I in R is a *-ideal in $\langle R,*\rangle$, since $a \in I$ implies $a^* = aa^{-1} \in I$. So we arrive by a somewhat circuitous route at the well-known result that I is prime if and only if it is maximal. As another consequence of I.1.7 we obtain

<u>Corollary</u> I.1.8. Let $\langle R,*\rangle \subset \langle S,*\rangle$ be *-rings with the same idempotents, i.e. $B(R) = B(S)$. Then any *-ideal I in $\langle R,*\rangle$ has a unique extension to a *-ideal J in $\langle S,*\rangle$ such that $J \cap R = I$. J is given by $J = S \cdot I_B = S \cdot I = \{a \cdot b : a \in S, b \in I\}$.

Proof. Let $B = B(R) = B(S)$. By I.1.7 $J = S\cdot I_B$ is a *-ideal extending $I = R\cdot I_B$ and $J \cap R = R\cdot I_B = I$; also $SI = SRI_B = SI_B$. For any *-ideal J' in $\langle S,*\rangle$ such that $J' \cap R = I$ we have $J'_B = J' \cap B = I \cap B = I_B$ and so by I.1.7 $J' = S\cdot J'_B = S\cdot I_B = J$.

For a ring R the spectrum of R, Spec(R), is defined as the set of all prime ideals of R; by analogy we define the *-spectrum of a *-ring $\langle R,*\rangle$, $\mathrm{Spec}^*(R)$, as the set of all prime *-ideals of $\langle R,*\rangle$. Thus $\mathrm{Spec}(R) = \mathrm{Spec}^*(R)$ for a regular ring R. For any *-ring $\langle R,*\rangle$ the intersection of all prime *-ideals is $\{0\}$. For if a is in this intersection, then a^* is by I.1.7 in the intersection of all Boolean prime ideals in $B(R)$, and so $a = 0$.

After these preparations we are now in a position to prove a representation theorem for *-rings. The easiest examples of *-rings are direct products $R = \prod_{i\in I} R_i$ of integral domains $\{R_i\}_{i\in I}$ with the operation $*$ defined by $a^*(i) = 0$ if $a(i) = 0$, and $a^*(i) = 1$ if $a(i) \neq 0$, for all $i \in I$. For any subring R' of R closed under $*$, $\langle R',*{\restriction}R'\rangle$ is obviously also a *-ring. The following theorem shows that every *-ring is of this form (compare Bergman [1], lemma 3.1).

Theorem I.1.9. Let $\langle R,*\rangle$ be a *-ring, $R_p = R/p$ for $p \in \mathrm{Spec}^*(R)$, $\langle R',*\rangle$ the direct product $\prod_{p\in \mathrm{Spec}^*(R)} R_p$ together with the canonical operation $*$ as defined above, and let $\xi_R : \langle R,*\rangle \longrightarrow \langle R',*\rangle$ be defined by $\xi_R(a) = \langle a \bmod p : p \in \mathrm{Spec}^*(R)\rangle$ for $a \in R$. Then ξ_R is an embedding and $\xi_R(R)$ is a subdirect product of $\{R_p\}_{p\in \mathrm{Spec}^*(R)}$. We refer to ξ_R as the canonical representation of $\langle R,*\rangle$ as a subdirect product of integral domains.

Proof. Evidently ξ_R is a ring homomorphism. Since $\ker(\xi_R)$ is the intersection of all prime *-ideals in $\langle R,*\rangle$, ξ_R is in fact a ring embedding. It is also obvious from the definition of ξ_R that $\xi_R(R)$ is a subdirect product of $\{R_p\}_{p \in \mathrm{Spec}^*(R)}$. So it remains to show that $(\xi_R(a))^* = \xi_R(a^*)$ for all $a \in R$. Let $a \in R$ and $p \in \mathrm{Spec}^*(R)$. Since $a^* \bmod p$ is an idempotent in the integral domain R_p, $a^* \equiv 0 \bmod p$ or $a^* \equiv 1 \bmod p$. If $a \equiv 0 \bmod p$, then $a^* \equiv 0 \bmod p$, since p is a $*$-ideal. If $a \not\equiv 0 \bmod p$, $a^* \not\equiv 0 \bmod p$, since $aa^* = a$, and so $a^* \equiv 1 \bmod p$. This finishes the proof.

Recall that for a regular $*$-ring $\langle R,*\rangle$ every prime ideal is a $*$-ideal and maximal. So R_p is a field for each $p \in \mathrm{Spec}^*(R) = \mathrm{Spec}(R)$. In this case our representation coincides with the usual representation of regular rings as subdirect products of fields (see [11]).

<u>Theorem</u> I.1.10. Let $\langle R,*\rangle \subset \langle R',*\rangle$ be $*$-rings with the same idempotents. For each $p \in \mathrm{Spec}^*(R)$ let $p' \in \mathrm{Spec}^*(R')$ be the unique extension of p such that $p' \cap R = p$ and $\xi_p : R_p \longrightarrow R'_{p'}$ the natural embedding. Let

$$\varphi : \prod_{p \in \mathrm{Spec}^*(R)} R_p \longrightarrow \prod_{p' \in \mathrm{Spec}^*(R')} R'_{p'}$$

be defined by $\varphi(a)(p') = \xi_p(a(p))$ for $p \in \mathrm{Spec}^*(R)$. Then the diagram

$$\begin{array}{ccc} R & \xrightarrow{\ \xi_R\ } & \prod_{p \in \mathrm{Spec}^*(R)} R_p \\ \cap\downarrow & & \downarrow\varphi \\ R' & \xrightarrow{\ \xi_{R'}\ } & \prod_{p' \in \mathrm{Spec}^*(R')} R'_{p'} \end{array}$$

commutes.

We shall frequently replace the factors $R_p = R/p$ in the canonical representation of a $*$-ring $\langle R,*\rangle$ by isomorphic integral domains R'_p for every $p \in \mathrm{Spec}^*(R)$. Let $\{\psi_p\}$ be isomorphisms between R_p and R'_p and define $\varphi: \langle R,*\rangle \longrightarrow \langle \prod_{p \in \mathrm{Spec}^*(R)} R'_p, *\rangle$ by $\varphi(a)(p) = \psi_p(\xi_R(a)(p))$ for $a \in R$, $p \in \mathrm{Spec}^*(R)$. Then φ is evidently an embedding of $*$-rings and for each $p \in \mathrm{Spec}^*(R)$ φ composed with the canonical projection

$\pi_p : \prod_{q \in \mathrm{Spec}^*(R)} R'_q \longrightarrow R'_p$ is onto R'_p. Any map φ obtained in this way will be called a <u>representation</u> of $\langle R,*\rangle$. We have obviously the following criterion for a map φ to be a representation of $\langle R,*\rangle$.

<u>Proposition</u> I.1.11. Let $\langle R,*\rangle$ be a *-ring and let $\{R'_p\}_{p \in \mathrm{Spec}^*(R)}$ be a family of integral domains. Then an embedding of *-rings $\varphi: \langle R,*\rangle \longrightarrow \langle \prod_{p \in \mathrm{Spec}^*(R)} R'_p, *\rangle$ is a representation of $\langle R,*\rangle$ iff φ together with the canonical projection π_p is onto R'_p and $\ker(\pi_p \circ \varphi) = p$ for all $p \in \mathrm{Spec}^*(R)$.

Whenever we deal in the following either with an arbitrary representation φ of a *-ring $\langle R,*\rangle$ or with a representation φ which is fixed and clear from the context, we denote $\varphi(a)(p)$ simply by a_p for $a \in R$ and $\varphi(M)(p)$ by M_p for a subset M of R, $p \in \mathrm{Spec}^*(R)$.

I.2 <u>Locally constant functions and properties.</u>

For every *-ring $\langle R,*\rangle$ I.1.7 provides a 1-1 correspondence between prime *-ideals in $\langle R,*\rangle$ and Boolean prime ideals in $B(R)$. So we can identify $\mathrm{Spec}^*(R)$ with the Stone space $S(B(R))$ of the Boolean algebra $B(R)$. $\mathrm{Spec}^*(R)$ with the topology induced by this identification forms a Boolean space, i.e. a compact Hausdorff space with a base of clopen sets. For every representation $\varphi : \langle R,*\rangle \longrightarrow \langle \prod_{p \in \mathrm{Spec}^*(R)} R_p, *\rangle$, $\varphi \restriction B(R)$ is just the Stone representation of $B(R)$; so we can talk about the idempotent e in $\langle R,*\rangle$ corresponding to a given clopen subset U of $\mathrm{Spec}^*(R)$; e is the unique element of $B(R)$ for which $\varphi(e) = \chi_U$, the characteristic function of U.

In this section we shall combine the topological properties of $\mathrm{Spec}^*(R)$ (in particular the compactness of $\mathrm{Spec}^*(R)$) with a metamathematical property of algebraically closed fields — namely the existence of an elimination of quantifiers — to prove a compactness principle for monically closed regular rings. This principle and related results of this section will be basic for the further development of the theory of *-rings in I.3.

We will first be concerned with functions on the *-spectrum of a *-ring $\langle R,*\rangle$ taking values in $\mathbb{Z} \cup \{\infty\} = \mathbb{Z}'$. The set $\mathbb{Z}'^{\mathrm{Spec}^*(R)}$ of all these functions is obviously partially ordered by the relation $f \leq g$ iff $f(p) \leq g(p)$ for all $p \in \mathrm{Spec}^*(R)$. We say f is <u>constant on a subset</u> U of $\mathrm{Spec}^*(R)$ if $f(p) = f(q)$ for all $p,q \in U$. We call f <u>locally constant</u> if for every $p \in \mathrm{Spec}^*(R)$ there exists a neighborhood U of p such that f is constant on U. Let $U_1,\ldots, U_n$ be a finite system of clopen subsets of $\mathrm{Spec}^*(R)$. We say $U_1,\ldots, U_n$ is a <u>clopen partition</u> of $\mathrm{Spec}^*(R)$ if the U_i's are pairwise disjoint and their union is $\mathrm{Spec}^*(R)$. (If this is the case, we say the corresponding idempotents $e_1,\ldots, e_n$ form a partition of 1.) If $k_1,\ldots, k_n \in \mathbb{Z}'$ and $U_1,\ldots, U_n$ is a clopen partition of $\mathrm{Spec}^*(R)$ then the function $f : \mathrm{Spec}^*(R) \longrightarrow \mathbb{Z}'$ defined by $f(p) = k_i$ for $P \in U_i$, $1 \leq i \leq n$, is obviously locally constant. The following lemma shows that every locally constant function can be obtained in this way.

<u>Lemma</u> I.2.1 For any locally constant function $f : \mathrm{Spec}^*(R) \longrightarrow \mathbb{Z}'$ there exists a unique partition $V_1,\ldots, V_k$ of $\mathrm{Spec}^*(R)$ such that f is constant on each V_i and assumes different values on different V_i's.

<u>Proof</u>. For every $p \in \mathrm{Spec}^*(R)$ pick a clopen neighborhood $U^{(p)}$ of p such that f is constant on $U^{(p)}$. Then $\bigcup_{p \in \mathrm{Spec}^*(R)} U^{(p)} = \mathrm{Spec}^*(R)$; so by the compactness of $\mathrm{Spec}^*(R)$ there exist finitely many $U_i = U^{(p_i)}$, $1 \leq i \leq n$, such that $\bigcup_{i=1}^{n} U_i = \mathrm{Spec}^*(R)$ and f is constant on each U_i. Uniting all the sets U_i on which f has the same value, we arrive at a clopen partition $V_1,\ldots, V_k$ of $\mathrm{Spec}^*(R)$ such that f is constant on each V_i and assumes different values on different V_i's. Moreover the

partition $V_1,\ldots,V_k$ is obviously uniquely determined by this property.

Besides locally constant functions on $\mathrm{Spec}^*(R)$ we will study in the sequel mainly <u>locally constant properties of</u> $\langle R,*\rangle$. By this we mean the following: Let $\varphi(x_1,\ldots,x_n)$ be an $L_{\infty\omega}$-formula in the language of rings, i.e. a formula built up from polynomial equations by means of negation, quantification, and finite or infinite conjunction and disjunction. Let $\varphi : \langle R,*\rangle \longrightarrow \langle \prod_{p\in \mathrm{Spec}^*(R)} R_p, *\rangle$ be a fixed representation of $\langle R,*\rangle$, $p \in \mathrm{Spec}^*(R)$, $U \subset \mathrm{Spec}^*(R)$, and $a_1,\ldots,a_n \in R$. Then we say $\varphi(a_1,\ldots,a_n)$ <u>holds at p</u> if $R_p \models \varphi((a_1)_p,\ldots,(a_n)_p)$, and we say $\varphi(a_1,\ldots,a_n)$ <u>holds on U</u> if φ holds at every point $p \in U$. Finally, we say $\varphi(x_1,\ldots,x_n)$ is a <u>locally constant property</u> for $\langle R,*\rangle$ if for every $p \in \mathrm{Spec}^*(R)$ and all $a_1,\ldots,a_n \in R$, whenever $\varphi(a_1,\ldots,a_n)$ holds at p, then there exists a neighborhood U of p such that $\varphi(a_1,\ldots,a_n)$ holds on U. Notice that whether or not $\varphi(a_1,\ldots,a_n)$ holds at p and whether or not $\varphi(x_1,\ldots,x_n)$ is a locally constant property for $\langle R,*\rangle$ does not depend on the particular representation used in the definition. The following is the basic result on locally constant properties.

<u>Compactness principle</u> I.2.2. Let $\langle R,*\rangle$ be a $*$-ring, S an additive subgroup of R closed under multiplication with idempotents from R, $c_1,\ldots,c_m \in R$, and $\varphi(x_1,\ldots,x_n,y_1,\ldots,y_m)$ a locally constant property for $\langle R,*\rangle$. If for every $p \in \mathrm{Spec}^*(R)$ there exist $a_1^{(p)},\ldots,a_n^{(p)} \in S_p$ such that $R_p \models \varphi(a_1^{(p)},\ldots,a_n^{(p)},(c_1)_p\ldots,(c_m)_p)$, then there exist $b_1,\ldots b_n \in S$ such that $\varphi(b_1,\ldots,b_n,c_1,\ldots,c_m)$ holds on $\mathrm{Spec}^*(R)$.

<u>Proof</u>. For every $p \in \mathrm{Spec}^*(R)$ pick $b_i^{(p)} \in S$ such that $(b_i^{(p)})_p = a_i^{(p)}$, $1 \leq i \leq n$, and let $U^{(p)}$ be a neighborhood of p such that $\varphi(b_1^{(p)},\ldots,b_n^{(p)},c_1,\ldots,c_m)$ holds on $U^{(p)}$. By the compactness of $\mathrm{Spec}^*(R)$ finitely many of these neighborhoods, say $U^{(p_1)},\ldots,U^{(p_k)}$

cover $\mathrm{Spec}^*(R)$. Thus the sets $V_i = U^{(p_i)} - \bigcup_{j<i} U^{(p_j)}$, $1 \leq i \leq k$, form a clopen partition of $\mathrm{Spec}^*(R)$. Now let e_j, $1 \leq j \leq k$, be the idempotent in $\langle R,*\rangle$ corresponding to V_j, and let $b_i = \sum_{j=1}^{k} e_j b_i^{(p_j)}$, $1 \leq i \leq n$. Then $(b_i)_p = (b_i^{(p_j)})_p = a_i^{(p)}$ for $p \in V_j$, $1 \leq i \leq n$, and so $\varphi(b_1,\ldots, b_n, c_1,\ldots,c_m)$ holds at all $p \in \mathrm{Spec}^*(R)$.

To find locally constant properties for $\langle R,*\rangle$, we need

<u>Lemma</u> I.2.3 (compare [10]). Let $\varphi(x_1,\ldots, x_n)$ be a finite quantifier-free formula in the language of rings, $a_1,\ldots, a_n \in R$ and $\varphi' = \varphi(a_1,\ldots, a_n)$. Then the set $U_\varphi = \{p \in \mathrm{Spec}^*(R) : \varphi' \text{ holds at } p\}$ is a clopen subset of $\mathrm{Spec}^*(R)$.

<u>Proof</u> by induction on the complexity of φ'. If φ' is an equation $t_1 = t_2$, $U_{\varphi'} = \{p:((t_1-t_2)^*)_p = 0\}$, so $U_{\varphi'}$ is clopen. For $\varphi' = \neg\psi'$ we have $U_{\varphi'} = \sim U_{\psi'}$, for $\varphi' = \psi_1' \wedge \psi_2'$, $U_{\varphi'} = U_{\psi_1'} \cap U_{\psi_2'}$, and for $\varphi' = \psi_2' \vee \psi_2'$, $U_{\varphi'} = U_{\psi_1'} \cup U_{\psi_2'}$; so in all these cases $U_{\varphi'}$ is clopen.

<u>Corollary</u> I.2.4. Let $\langle R,*\rangle$ be a $*$-ring, $\{\varphi_i(x_1,\ldots,x_n)\}_{i \in I}$ a family of finite formulas in the ring language having no free variables besides $x_1,\ldots,x_n$, and let $\varphi(x_1,\ldots,x_n) = \bigvee_{i \in I} \varphi_i(x_1,\ldots,x_n)$.

(i) If all the formulas φ_i are quantifier-free, then φ is a locally constant property for $\langle R,*\rangle$.

(ii) If R is monically closed and regular, and φ_i are arbitrary, then φ is a locally constant property for $\langle R,*\rangle$.

<u>Proof</u>. (i) follows immediately from I.2.3. For (ii) we remark that R_p is an algebraically closed field for every $p \in \mathrm{Spec}^*(R)$. Since the theory K_{AF} of algebraically closed fields allows an elimination of quantifiers (see [13]), each φ_i is in K_{AF} equivalent to a finite quantifier-free formula ψ_i in the ring language. Thus for $a_1,\ldots, a_n \in R$, $\varphi(a_1,\ldots,a_n)$ holds at a point $p \in \mathrm{Spec}^*(R)$ if and only if

$\psi(a_1,\ldots,a_n) = \bigvee_{i \in I} \psi_i(a_1,\ldots,a_n)$ holds at p. Hence $\varphi(x_1,\ldots,x_n)$ is a locally constant property for $\langle R,*\rangle$ by (i). The following lemma illustrates an easy application of the compactness principle.

Lemma I.2.5. Let $\langle R,*\rangle$ be a $*$-ring and $S,S' \subset R$ additive subgroups of R which are closed under multiplication with idempotents from R.

(i) For any $a \in R$, $a_p \in S'_p$ for all $p \in \mathrm{Spec}^*(R)$ implies $a \in S'$.

(ii) $S_p \subset (=) S'_p$ for all $p \in \mathrm{Spec}^*(R)$ implies $S \subset (=) S'$.

Proof. (ii) follows immediately from (i). To prove (i) we observe that the formula $x = y$ is locally constant and that for every $p \in \mathrm{Spec}^*(R)$ there exists $b^{(p)} \in S'_p$ such that $R_p \models a_p = b^{(p)}$. So by the compactness principle there exists $b \in S'$ such that $R_p \models a_p = b_p$ for all $p \in \mathrm{Spec}^*(R)$ and hence $a = b$.

We conclude this section with two examples of locally constant functions.

(i) Let $\langle R,*\rangle$ be an arbitrary $*$-ring, $f \in R[X_1,\ldots,X_n]$, and $p \in \mathrm{Spec}^*(R)$. Let $f_p \in R_p[X_1,\ldots,X_n]$ be the polynomial obtained from f by replacing all coefficients a of f by a_p. Define $\deg(f) : \mathrm{Spec}^*(R) \longrightarrow \mathbb{Z}$ by $\deg(f)(p) = \deg(f_p)$. Then $\deg(f)$ is a locally constant function. To see this assume $\deg(f_p) = r$, let $a_1,\ldots,a_n$ be all the coefficients of terms in f with formal degree $> r$, and let a be the coefficient of a term in f of formal degree r which does not vanish at p. Then $\bigwedge_{i=1}^{m} a_i = 0 \wedge a \neq 0$ holds at p, and so by I.2.4 on a clopen neighborhood U of p.

(ii) let $\langle R,*\rangle$ be a model of K^* and (a_{ij}) an $n \times m$ matrix with entries from R. Define rank $(a_{ij}) : \mathrm{Spec}^*(R) \longrightarrow \mathbb{Z}$ by rank $(a_{ij})(p) = \mathrm{rank}\ ((a_{ij})_p)$. Then rank is locally constant. For it is an easy exercise in linear algebra to write down a first order formula

$\rho_r(a_{ij})$ in the ring language such that for all $p \in \mathrm{Spec}^*(R)$ $\rho_r(a_{ij})$ holds at p iff rank $((a_{ij})_p) = r$. From this we conclude by I.2.4 that rank (a_{ij}) is locally constant.

I.3. Further theory of *-rings.

In this section we will use the results of section I.2 to derive further results on *-rings. We will first be concerned with isomorphisms between *-rings, next with the extension theory of *-rings, and finally with modules over *-rings.

Let $\langle R,*\rangle \subset \langle S,*\rangle$, $\langle S',*\rangle$ be *-rings with the same idempotents and $\varphi : \langle S,*\rangle \longrightarrow \langle S',*\rangle$ an isomorphism over R. Then φ induces isomorphisms $\varphi_p : S \longrightarrow S'_p$ over R_p for every $p \in \mathrm{Spec}^*(R)$. φ_p can by I.1.8 and I.1.10 be defined by $\varphi_p(a_p) = \varphi(a)_p$ for $a \in S$. In order to characterize the systems $\{\psi_p\}_{p \in \mathrm{Spec}^*(R)}$ of isomorphisms between S_p and S'_p over R_p that are induced in this way by an isomorphism between $\langle S,*\rangle$ and $\langle S',*\rangle$ over R, we introduce the following definitions. Let $U \subset \mathrm{Spec}^*(R)$ and $\Psi = \{\psi_p : S_p \longrightarrow S'_p\}_{p \in U}$ be a system of isomorphisms over R_p on U. We say Ψ is a <u>compatible system of isomorphisms over R on U</u>, if for every $a \in S$ there exists $b \in S'$ such that $\psi_p((a)_p) = (b)_p$ for all $p \in U$. We call a system Ψ of isomorphisms over R on $\mathrm{Spec}^*(R)$ <u>locally compatible</u> if every $p \in \mathrm{Spec}^*(R)$ has a clopen neighborhood $U^{(p)}$ such that Ψ restricted to $U^{(p)}$ is a compatible system of isomorphisms over R on U_p.

<u>Theorem I.3.1.</u> (compare Pierce [12], lemma 7.2) Let $\langle R,*\rangle \subset \langle S,*\rangle$, $\langle S',*\rangle$ be *-rings with the same idempotents. Then for any isomorphism $\varphi : \langle S,*\rangle \longrightarrow \langle S',*\rangle$ over R the induced system $\Phi = \{\varphi_p : S_p \longrightarrow S'_p\}_{p \in \mathrm{Spec}^*(R)}$ of isomorphisms is locally compatible. Conversely for any locally compatible system $\Psi = \{\psi_p : S_p \longrightarrow S'_p\}_{p \in \mathrm{Spec}^*(R)}$ of isomorphisms over R on $\mathrm{Spec}^*(R)$ there exists an isomorphism $\varphi : \langle S,*\rangle \longrightarrow \langle S',*\rangle$ over R such that

$\varphi_p = \psi_p$ for all $p \in \mathrm{Spec}^*(R)$.

Proof. The first part of the theorem is obvious from the definition of Φ. To prove the second part, define

$\psi : \langle \prod_{p \in \mathrm{Spec}^*(R)} S_p, *\rangle \longrightarrow \langle \prod_{p \in \mathrm{Spec}^*(R)} S'_p, *\rangle$ by

$\psi(a)(p) = \psi_p(a(p))$ for $a \in \prod_{p \in \mathrm{Spec}^*(R)} S_p$. ψ is evidently an isomorphism of $*$-rings. Let

$\xi_S : \langle S,*\rangle \longrightarrow \langle \prod_{p \in \mathrm{Spec}^*(R)} S_p, *\rangle$,

$\xi_{S'} : \langle S',*\rangle \longrightarrow \langle \prod_{p \in \mathrm{Spec}^*(R)} S'_p, *\rangle$ be the canonical representations of $\langle S,*\rangle$ and $\langle S',*\rangle$. We claim that ψ maps $\xi_S(S)$ into $\xi_{S'}(S')$. To see this let $U^{(p)}$ be a clopen neighborhood of $p \in \mathrm{Spec}^*(R)$ as in the definition of local compatibility. As usual we find a finite subset M of $\mathrm{Spec}^*(R)$ such that $\bigcup_{p\in M} U^{(p)} = \mathrm{Spec}^*(R)$; we may also assume that the $U^{(p)}$'s are pairwise disjoint for $p \in M$. For any $p \in M$ and $a \in S$ pick $b^{(p)} \in S'$ such that $\psi_q(a_q) = (b^{(p)})_q$ for all $q \in U^{(p)}$. Let $e^{(p)}$ be the idempotent in R corresponding to $U^{(p)}$. Then we have $\psi(\xi_S(e^{(p)}a)) = \xi_{S'}(e^{(p)}b^{(p)})$ and so $\psi(\xi_S(a)) =$

$$= \psi\Big(\sum_{p \in M} \xi_S(e^{(p)}a)\Big) = \sum_{p \in M} \xi_{S'}(e^{(p)}b^{(p)}) \in \xi_{S'}(S').$$

A similar argument shows that ψ maps $\xi_S(S)$ onto $\xi_{S'}(S')$. Hence the map $\varphi = \xi_{S'}^{-1} \circ \psi \circ \xi_S : \langle S,*\rangle \longrightarrow \langle S',*\rangle$ is an isomorphism over R and by definition $\varphi_p = \psi_p$ for all $p \in \mathrm{Spec}^*(R)$.

Next we study polynomial rings over a $*$-ring $\langle R,*\rangle$.

Theorem I.3.2. Let $\langle R,*\rangle$ be a $*$-ring, $S = R[X_1,\dots,X_n]$, and let the operation $* : S \longrightarrow S$ be defined by $f^* = \bigcup_{i=0}^{k} a_i^*$ for a polynomial

$f \in S$ with coefficients $a_0, \ldots, a_k \in R$. Then $\langle S, * \rangle$ is a $*$-ring extending $\langle R, * \rangle$ and $B(S) = B(R)$.

Proof. The theorem follows by induction from the case $n = 1$. So we assume that $S = R[X]$ and $f(X) = \sum_{i=0}^{k} a_i X^i \in S$. Then

$$f \cdot f^* = \sum_{i=0}^{k} a_i \Big(\bigcup_{j=0}^{k} a_j^* \Big) X^i = \sum_{i=0}^{k} a_i X^i = f,$$ and so I.1.3.(i)

is satisfied in $\langle S, * \rangle$. To show I.1.3.(ii) assume $fg = f$ for some $g \in R[X]$. For any $h = \sum_{i=0}^{m} b_i X^i \in R[X]$ and $p \in \mathrm{Spec}^*(R)$ define

$h_p = \sum_{i=0}^{m} (b_i)_p X^i \in R_p[X]$. Then the map $h \longmapsto h_p$ is obviously a homomorphism from $R[X]$ into $R_p[X]$. So we have $f(g-1) = 0$ and therefore $f_p(g-1)_p = 0$ for all $p \in \mathrm{Spec}^*(R)$. Since $R_p[X]$ is an integral domain this implies that $f_p = 0$ or $(g-1)_p = 0$, and hence

$\Big(\bigcup_{i=0}^{k} a_i^* \Big)_p = 0$ or $(g-1)_p = 0$ for all $p \in \mathrm{Spec}^*(R)$. Consequently $f^*(g-1) = 0$, and so $f^* g = f^*$. This shows that $\langle S, * \rangle$ is a $*$-ring extending $\langle R, * \rangle$. Moreover for any idempotent $f \in S$, $f = f^* \in R$.

Corollary I.3.3. Let $\langle R, * \rangle \subset \langle S, * \rangle$ be as in theorem I.3.2 and $S_p = R_p[X_1, \ldots, X_n]$ for $p \in \mathrm{Spec}^*(R)$. Then there exists a representation φ such that the diagram

$$\begin{array}{ccc}
\langle R,*\rangle & \xrightarrow{\ \xi_R\ } & \langle \prod_{p\in \mathrm{Spec}^*(R)} R_p, *\rangle \\
\cap & & \cap \\
\langle S,*\rangle & \xrightarrow{\ \varphi\ } & \langle \prod_{p\in \mathrm{Spec}^*(R)} S_p, *\rangle
\end{array}$$

commutes.

Proof. For any $f \in S$, $p \in \mathrm{Spec}^*(R)$, let $f_p \in S_p$ be the polynomial obtained from f by taking the coefficients of f modulo p. Then the map $f \longmapsto f_p$ is as above an epimorphism from S onto S_p, and so the map $\varphi: S \longrightarrow \prod_{p\in \mathrm{Spec}^*(R)} S_p$ given by $\varphi(f) = \langle f_p : p \in \mathrm{Spec}^*(R)\rangle$ for $f \in S$ is a homomorphism. Moreover $\varphi(f) = 0$ means that $f_p = 0$ for all $p \in \mathrm{Spec}^*(R)$, and so that $(a_0)_p = 0,\dots,(a_k)_p = 0$ for all $p \in \mathrm{Spec}^*(R)$, where $a_0,\dots,a_k$ are the coefficients of f. But this can only be the case if all the coefficients of f vanish, i.e. $f = 0$. So we have shown that φ is an embedding. A similar argument shows that φ is compatible with the operation $*$, i.e. $\varphi(f^*) = (\varphi(f))^*$.

We consider now the following question. Let $\langle R,*\rangle$ be a $*$-ring and I an ideal in $S = R[X_1,\dots,X_n]$, $I \cap R = \{0\}$. What is a necessary and sufficient condition on I in order that $S' = S/I$ can be turned into a $*$-ring $\langle S',*\rangle$ extending $\langle R,*\rangle$ with $B(S') = B(R)$?

Definition. Let $\langle S,*\rangle$ be a $*$-ring and I an ideal in S. Then we say I is everywhere prime if I_p is a prime ideal in S_p for every $p \in \mathrm{Spec}^*(S)$.[1] We say I is smooth if for every $a \in S$ the set $\{p \in \mathrm{Spec}^*(S) : a_p \in I_p\}$ is clopen.

Examples. Let $S = R[X_1,\dots,X_n]$ as above. Then the ideals $I_0 = \{0\}$, $I_j = (X_j)$, $1 \le j \le n$, are everywhere prime and smooth. (To see that I_j is smooth for $j > 0$, let $f \in S$ and write f in the form $f = g + X_j h$, where g does not contain X_j. Then the set

1) Equivalently, $ab \in I$ implies that there exists $e \in B(R)$ such that $ae \in I$ and $b(1-e) \in I$.

$\{p \in \mathrm{Spec}^*(S) : f_p \in (I_j)_p\} = \{p : g_p = 0\}$ is clopen by I.2.3.

We now have the following theorem.

Theorem I.3.4. Let $\langle R,*\rangle$ be a *-ring, I an ideal in $S = R[X_1,\ldots,X_n]$ such that $I \cap R = \{0\}$. Then $S' = S/I$ can be turned into a *-ring $\langle S',*\rangle$ extending $\langle R,*\rangle$ with $B(S') = B(R)$ iff I is a smooth everywhere prime ideal.

Proof. Let $\varphi : S \longrightarrow S'$ be the canonical epimorphism. Assume first that $\langle S',*\rangle$ is a *-ring extending $\langle R,*\rangle$ and that $B(S') = B(R)$. For every $p \in \mathrm{Spec}^*(R)$ let $\varphi_p : S_{pS} \longrightarrow S'_{pS'}$ be defined by $\varphi_p(f_{pS}) = (\varphi(f))_{pS'}$. Then φ_p is a well-defined homomorphism with kernel I_{pS}. For, $f_{pS} \in I_{pS}$ iff there exists $g \in I$ such that $f-g \in pS$, iff $\varphi(f) - \varphi(g) \in pS'$, iff $(\varphi(f))_{pS'} = 0$. Since $S'_{pS'}$ is an integral domain by I.1.9, I_{pS} is prime for all $p \in \mathrm{Spec}^*(R)$, and so I is everywhere prime. To see that I is smooth we remark that for every $f \in S$, $p \in \mathrm{Spec}^*(R)$, $f_{pS} \in I_{pS}$ iff $\varphi_p(f_{pS}) = (\varphi(f))_{pS'} = 0$. Since $\langle S',*\rangle$ is by assumption a *-ring with the same idempotents as R, $\{p \in \mathrm{Spec}^*(R) : (\varphi(f))_{pS'} = 0\}$ is a clopen subset of $\mathrm{Spec}^*(R)$, and so I is indeed smooth.

Conversely assume I is everywhere prime and smooth. We define a homomorphism $\psi : S \longrightarrow \prod_{p \in \mathrm{Spec}^*(S)} S_p/I_p$ by

$\psi(f) = \langle f_p \bmod I_p : p \in \mathrm{Spec}^*(S)\rangle$. Then $\psi(f) = 0$ iff for all $p \in \mathrm{Spec}^*(S)$ $f_p \in I_p$ iff $f \in I$ by I.2.5. So $\ker(\psi) = I$, and therefore the map $\varphi : S' \longrightarrow \prod_{p \in \mathrm{Spec}^*(S)} S_p/I_p$ given by

$\varphi(f \bmod I) = \psi(f)$ is an embedding. For any $f \in S$ $(\psi(f))^*$ is by definition the characteristic function χ_U of the clopen set $\{p \in \mathrm{Spec}^*(S) : f_p \notin I_p\}$. So we can define the operation $*$ on S' by $(f \bmod I)^* = \xi_R^{-1}((\psi(f))^*)$ and obtain $\langle S',*\rangle$ as a *-ring extending

$\langle R,*\rangle$ with $B(S') = B(R)$.

Let $\langle R,*\rangle$ be a regular $*$-ring and $\langle S,*\rangle$ a $*$-ring extending $\langle R,*\rangle$ with $B(R) = B(S)$. Then we define the transcendence degree of S over R as the function $\mathrm{trdeg}(S:R)\colon \mathrm{Spec}^*(R) \longrightarrow \mathbb{Z}'$ whose value at $p \in \mathrm{Spec}^*(R)$ is the transcendence degree of S (as an integral domain) over R. We now modify our question as follows: Let $\langle R,*\rangle$ be a regular $*$-ring and I an ideal in $S = R[X_1,\ldots, X_n]$, $I \cap R = \{0\}$. What is a necessary and sufficient condition on I in order that $S' = S/I$ can be turned into a $*$-ring $\langle S',*\rangle$ extending $\langle R,*\rangle$ such that $B(S') = B(R)$ and $\mathrm{trdeg}\ (S' :R)$ is locally constant? The following theorem provides a complete answer to this problem in case $\langle R,*\rangle$ is a monically closed regular $*$-ring.

<u>Theorem I.3.5</u>. Let $\langle R,*\rangle$ be a monically closed regular $*$-ring and I an ideal in $S = R[X_1,\ldots, X_n]$ such that $I \cap R = \{0\}$. Then $S' = S/I$ can be turned into a $*$-ring $\langle S',*\rangle$ extending $\langle R,*\rangle$ such that $B(S') = B(R)$ and $\mathrm{trdeg}(S' :R)$ is locally constant iff I is a finitely generated everywhere prime ideal.

The proof of this theorem requires some more preparation. We will make use of a series of definability results in the theory of polynomial ideals over fields.

<u>Theorem I.3.6</u> (Greta Hermann [6]). For all $n, k, d \in \mathbb{N}$ there exists an existential formula $\varphi_{n,k,d}\left(\vec{x}_0,\ldots, \vec{x}_k\right)$ in the ring language such that for all fields K and all polynomials $f_0,\ldots, f_k \in K[X_1,\ldots, X_n]$ of degree $\leq d$ with coefficients $\vec{a}_0,\ldots, \vec{a}_k$, $f_0 \in (f_1,\ldots, f_k)$ iff $K \models \varphi_{n,k,d}(\vec{a}_0,\ldots, \vec{a}_k)$.

<u>Theorem I.3.7</u>. (Lambert [8], compare also P. Eklof [4]). For all $n, k, d \in \mathbb{N}$ there exists a formula $\pi_{n,k,d}\left(\vec{x}_1,\ldots, \vec{x}_k\right)$ in the ring language such that for all fields K and all polynomials

$f_1, \ldots, f_k \in K[X_1, \ldots, X_n]$ of degree $\leq d$ with coefficients $\vec{a_1}, \ldots, \vec{a_k}$, $(f_1, \ldots, f_k)$ is a prime ideal iff $K \models \pi_{n,k,d}(\vec{a_1}, \ldots, \vec{a_k})$.

Theorem I.3.8. (P. Eklof [4]). For all $n, k, d \in \mathbb{N}$, $0 \leq r \leq n$, there exists a formula $\delta_{n,k,d,r}(\vec{x_1}, \ldots, \vec{x_k})$ in the ring language such that for all fields K and all polynomials $f_1, \ldots, f_k \in K[X_1, \ldots, X_n]$ of degree $\leq d$ with coefficients $\vec{a_1}, \ldots, \vec{a_k}$, $(f_1, \ldots, f_k)$ is a prime ideal of dimension r iff $K \models \delta_{n,k,d,r}(\vec{a_1}, \ldots, \vec{a_k})$.

A proof of the last theorem can be outlined as follows: Let I be a prime ideal in $K[X_1, \ldots, X_n]$ generated by the polynomials $f_1, \ldots, f_k$ of degree $\leq d$ with coefficients $\vec{a_1}, \ldots, \vec{a_k}$. By Zariski-Samuel [17], p.193, thm. 20, $\dim(I) = r$ iff there exist $n-r$ prime ideals $I_1, \ldots, I_{n-r}$ in $K[X_1, \ldots, X_n]$ such that $\{0\} \neq I_1 \subsetneq I_2 \subsetneq \ldots \subsetneq I_{n-r} = I$ and $r+1$ prime ideals $J_0, \ldots, J_r$ such that $I = J_0 \subsetneq \ldots \subsetneq J_r \subsetneq K[X_1, \ldots, X_n]$. With the help of I.3.6 and I.3.7 this statement can be expressed as an infinite disjunction of formulas $\bigvee_{i=1}^{\infty} \psi_{n,k,d,r,i}(\vec{a_1}, \ldots, \vec{a_k})$ in the ring language. Thus the theory of fields entails $\pi_{n,k,d}(\vec{x_1}, \ldots, \vec{x_k})$

$$\longrightarrow \bigvee_{r=0}^{n} \bigvee_{i=1}^{\infty} \psi_{n,k,d,r,i}(\vec{x_1}, \ldots, \vec{x_k}).$$

Hence by the compactness theorem for first order theories the infinite disjunctions $\bigvee_{i=1}^{\infty} \psi_{n,k,d,r,i}(\vec{a_1}, \ldots, \vec{a_k})$ can be replaced by finite subdisjunctions in the definition of $\dim(I) = r$.

With the help of these results we can now give an alternative characterization of finitely generated everywhere prime ideals.

Lemma I.3.9. Let $\langle R,*\rangle$ be a monically closed regular *-ring, $S = R[X_1,\dots, X_n]$, and I a finitely generated ideal in S. Then I is smooth.

Proof. Let $f_1,\dots, f_k$ be generators of I with coefficients $\vec{a}_1,\dots, \vec{a}_k$ and $f_0 \in S$ with coefficients $\vec{a}_0$ and assume that d is an integer such that the maximum value of $\deg(f_i)$ is $\leq d$ for $0 \leq i \leq k$. Then by I.3.6 $(f_0)_{pS} \in I_{pS}$ iff $R_p \models \varphi_{n,k,d}(\vec{a}_0, \vec{a}_1,\dots, \vec{a}_k)$. So by I.2.3 the set $\{pS : p \in \mathrm{Spec}^*(R),\ (f_0)_{pS} \in I_{pS}\}$ is clopen.

Let $\langle R,*\rangle$ be a regular *-ring, $\langle S,*\rangle$ a *-ring extending $\langle R,*\rangle$ with $B(S) = B(R)$, and I an everywhere prime ideal in $\langle S,*\rangle$ such that $I \cap R = \{0\}$. Then we define the dimension of I as the function $\dim(I) : \mathrm{Spec}^*(R) \longrightarrow \mathbb{Z}$ with value $\dim(I_{pS})$ at $p \in \mathrm{Spec}^*(R)$.

Theorem I.3.10. Let $\langle R,*\rangle$ be a monically closed regular *-ring, $S = R[X_1,\dots, X_n]$, and I an everywhere prime ideal in S such that $I \cap R = \{0\}$. Then I is finitely generated iff $\dim(I)$ is locally constant.

Proof. Assume to begin with that I is finitely generated, say $I = (f_1,\dots, f_k)$, where f_i are polynomials of formal degree $\leq d$ with coefficients $\vec{a}_i$, $1 \leq i \leq k$. Let $p \in \mathrm{Spec}^*(R)$ and assume $\dim(I)(pS) = r$. Then by I.3.8 $R_p \models \delta_{n,k,d,r}((\vec{a}_1)_p,\dots, (\vec{a}_k)_p)$, and so by I.2.3 there exists a clopen neighborhood U of p such that $R_q \models \delta_{n,k,d,r}((\vec{a}_1)_q,\dots, (\vec{a}_k)_q)$, i.e. $\dim(I)(qS) = r$, for all $q \in U$.

To show the converse assume $\dim(I)$ is locally constant. For every $p \in \mathrm{Spec}^*(R)$, S_{pS} is Noetherian; so we can take

$f_1^{(p)}, \ldots, f_{k_p}^{(p)} \in I$ such that $\left(f_1^{(p)}\right)_{pS}, \ldots, \left(f_{k_p}^{(p)}\right)_{pS}$ generate I_{pS}. Let d_p be the maximum of the formal degrees of $f_i^{(p)}$, let $\vec{a_i}^{(p)}$ be the coefficients of $f_i^{(p)}$, $1 \leq i \leq k_p$, and let $r_p = \dim(I)(pS)$. Then by I.3.8 $\delta_{n,k_p,d_p,r}\left((\vec{a_i})_p, \ldots, (\vec{a_k})_p\right)$ holds in R_p; so by I.2.3 there exists a clopen neighborhood $U^{(p)}$ of p such that $\delta_{n,k_p,d_p,r}\left((\vec{a_1})_q, \ldots, (\vec{a_k})_q\right)$ holds in R_q for every $q \in U^{(p)}$. Let $V^{(p)} \subset U^{(p)}$ be a clopen neighborhood of p on which $\dim(I)$ has constant value r_p. Then for every $q \in V^{(p)}$ the ideal generated by $\left(f_1^{(p)}\right)_{qS}, \ldots, \left(f_{k_p}^{(p)}\right)_{qS}$ in S_{qS} is a prime ideal of dimension r_p. Since $\dim(I)(qS)$ is also r_p this means that $\left(f_1^{(p)}\right)_{qS}, \ldots, \left(f_{k_p}^{(p)}\right)_{qS}$ generate I_{qS}. Let M be a finite subset of $\mathrm{Spec}^*(R)$ such that

$$\bigcup_{p \in M} V^{(p)} = \mathrm{Spec}^*(R).$$

Then $\{f_1^{(p)}, \ldots, f_{k_p}^{(p)} : p \in M\}$ is by I.2.5 a system of generators for I.

We now return to the proof of theorem I.3.5.: Assume first that I is finitely generated and everywhere prime, and $I \cap R = \{0\}$. Then I is smooth by I.3.9, and so $S' = S/I$ can be turned into a $*$-ring $\langle S', *\rangle \supset \langle R, *\rangle$ with $B(S') = B(R)$. Moreover $\mathrm{trdeg}(S' : R)(p) = \dim(I)(pS)$ for every $p \in \mathrm{Spec}^*(R)$; so $\mathrm{trdeg}(S' : R)$ is locally constant by I.3.10. Conversely assume that $\langle S', *\rangle \supset \langle R, *\rangle$, $B(S') = B(R)$, and that $\mathrm{trdeg}\,(S' : R)$ is locally constant. Then I is everywhere prime by I.3.4. We conclude as above that $\dim(I)$ is locally constant. So I is finitely generated by I.3.10.

We will now consider the problems solved in I.3.4 and I.3.5 in greater generality by allowing $\langle S, *\rangle$ to be an arbitrary $*$-ring extending $\langle R,*\rangle$ which is a ring-finite extension of R such that $B(S) = B(R)$ and $trdeg(S:R)$ is locally constant. We then have the following companion theorems to I.3.4. and I.3.5.

Theorem I.3.11. Let $\langle R,*\rangle \subset \langle S,*\rangle$ be *-rings with $B(R) = B(S)$ and I an ideal in S, $I \cap R = \{0\}$. Then $S' = S/I$ can be turned into a *-ring $\langle S',*\rangle$ extending $\langle R,*\rangle$ with $B(S') = B(R)$ iff I is a smooth everywhere prime ideal.

Theorem I.3.12. Let $\langle R,*\rangle$ be a monically closed regular *-ring, $\langle S,*\rangle$ a *-ring extending $\langle R,*\rangle$ which is ring-finite over R and such that $B(S) = B(R)$ and $trdeg(S:R)$ is locally constant, and let I be an ideal in S such that $I \cap R = \{0\}$. Then $S' = S/I$ can be turned into a *-ring $\langle S',*\rangle$ extending $\langle R,*\rangle$ such that $B(S') = B(R)$ and $trdeg(S':R)$ is locally constant iff I is a finitely generated everywhere prime ideal.

The proof of theorem I.3.4 can be carried over verbatim to prove I.3.11. Theorem I.3.12 is proved by replacing I.3.4, I.3.9, I.3.10 in the proof of theorem I.3.5 by their counterparts I.3.11, I.3.13, and I.3.14. So the proof of theorem I.3.12 will be completed after we have stated and proved I.3.13 and I.3.14.

Lemma I.3.13. Let $\langle R,*\rangle$ be a monically closed regular *-ring, $\langle S,*\rangle$ a *-ring extending $\langle R,*\rangle$ which is ring-finite over R and such that $B(S) = B(R)$ and $trdeg(S:R)$ is locally constant, and let I be a finitely generated ideal in S. Then I is smooth.

Proof. By reduction to I.3.9. Let $x_1,\dots, x_n$ be generators of S over R, $R' = R[X_1,\dots,X_n]$, $\varphi : R' \longrightarrow S$ the canonical epimorphism over R mapping X_i onto x_i, $1 \leq i \leq n$, and let $J = \ker(\varphi)$. Then J is by I.3.5 a finitely generated everywhere prime ideal, say $J = (g_1,\dots, g_m)$. Let $I = (\varphi(f_1),\dots, \varphi(f_k))$, $f_i \in R'$, and let $I' = \varphi^{-1}(I)$. Then it is not difficult to see that I' is generated by

$f_1,\ldots,f_k, g_1,\ldots,g_m$. So I' is smooth by I.3.9. To show that I is smooth let $f \in R'$. Then $(\varphi f)_{pS} \in I_{pS}$ iff there exists $g \in I'$ such that $\varphi(f) - \varphi(g) \in pS$, iff $f-g \in pR'$ iff $f_{pR'} \in (I')_{pR'}$, for all $p \in \mathrm{Spec}^*(R)$. So $\{pS : (\varphi(f))_{pS} \in I_{pS}\}$ is clopen in $\mathrm{Spec}^*(S)$, since $\{pR' : f_{pR'} \in (I')_{pR'}\}$ is clopen in $\mathrm{Spec}^*(R)$.

Theorem I.3.14. Let $\langle R,*\rangle$ be a monically closed regular $*$-ring, $\langle S,*\rangle$ a $*$-extension of $\langle R,*\rangle$ which is ring-finite over R such that $B(S) = B(R)$ and $\mathrm{trdeg}(S{:}R)$ is locally constant, and let I be an everywhere prime ideal in S. Then I is finitely generated iff $\dim(I)$ is locally constant.

Proof. Let R', φ, J, I', $g_1,\ldots,g_m$, be as in the preceding proof. If I is finitely generated, say $I = (f_1,\ldots,f_k)$, then $I' = (f_1,\ldots, f_k, g_1,\ldots, g_m)$, and so by I.3.10, if $S' = S/I$, $\dim(I') = \mathrm{trdeg}(S'{:}R) = \dim(I)$ is locally constant. If $\dim(I)$ is locally constant, then $\dim(I')$ is locally constant by the same argument, and so by I.3.10 I' is finitely generated, say $I' = (h_1,\ldots, h_r)$. Then I is generated by $(\varphi(h_1),\ldots, \varphi(h_r))$.

Next we shall be concerned with the quotient ring of a $*$-ring $\langle S,*\rangle$ with respect to a smooth everywhere prime ideal.

Lemma I.3.15. Let $\langle S,*\rangle$ be a $*$-ring, I an everywhere prime ideal in S, and $M = \{a \in S : \text{for all } p \in \mathrm{Spec}^*(S),\ a_p \notin I_p\}$. Then M is a multiplicatively closed subset of S containing no zero-divisors.

Proof. For any $a \in M$, $p \in \mathrm{Spec}^*(S)$, $a_p \neq 0$ and so $a^* = 1$, i.e. a is not a zero-divisor. Next let $a, b \in M$; then for any $p \in \mathrm{Spec}^*(S)$, $a_p \notin I_p$ and $b_p \notin I_p$, so $(ab)_p \notin I_p$ since I_p is a prime ideal. Thus $ab \in M$.

Let $\langle S,*\rangle$ be a $*$-ring, $\langle T,*\rangle$ its regular closure, and let I be a smooth everywhere prime ideal in S. Then we can form the quotient ring $S_I = \{ab^{-1} : a,b \in S,\ b_p \notin I_p$ for all $p \in \mathrm{Spec}^*(S)\}$ of S in T. It is well-known that S_I is a subring of T containing S. Since $B(S) = B(S_I) = B(T)$, S_I is closed under the operation $*$ of $\langle T,*\rangle$ and hence $\langle S_I,*\rangle$ is a $*$-ring sitting between $\langle S,*\rangle$ and $\langle T,*\rangle$.

The following theorem provides representations for various $*$-rings defined in this section which will be useful in the geometric context of part II.

<u>Theorem I.3.16</u>. Let $\langle R,*\rangle$ be a monically closed regular $*$-ring, $R' = R[X_1,\ldots, X_n]$, I a finitely generated everywhere prime ideal in R', $\langle S,*\rangle = \langle R'/I,*\rangle$ as defined in I.3.5, $\langle T,*\rangle$ the regular closure of $\langle S,*\rangle$, J a smooth everywhere prime ideal in S. Let $\varphi : \langle R',*\rangle \longrightarrow \prod_{p \in \mathrm{Spec}^*(R)} R'_p$ be the representation defined in I.3.3, and let $S_p = R'_p/I_p$ for every $p \in \mathrm{Spec}^*(R)$. Then there exist representations σ, ρ_J, ρ of $\langle S,*\rangle$, $\langle S_J,*\rangle$, and $\langle T,*\rangle$, respectively, such that the following diagram commutes.

$$\begin{array}{ccc}
\langle R,*\rangle & \xrightarrow{\ \xi_R\ } & \langle \prod_{p \in \mathrm{Spec}^*(R)} R_p, *\rangle \\
\cap\downarrow & & \downarrow\cap \\
\langle S,*\rangle & \xrightarrow{\ \sigma\ } & \langle \prod_{pS \in \mathrm{Spec}^*(S)} S_p, *\rangle \\
\cap\downarrow & & \downarrow\cap \\
\langle S_J,*\rangle & \xrightarrow{\ \rho_J\ } & \langle \prod_{pS_J \in \mathrm{Spec}^*(S_J)} (S_p)_{J_p}, *\rangle \\
\cap\downarrow & & \downarrow\cap \\
\langle T,*\rangle & \xrightarrow{\ \rho\ } & \langle \prod_{pT \in \mathrm{Spec}^*(T)} T_p, *\rangle
\end{array}$$

In this diagram $J_p = \sigma(J)(pS)$, $(Sp)_{J_p}$ is the localization of S_p at the prime ideal J_p, and T_p is the quotient field of S_p.

Proof. Define an epimorphism $\psi_p : S \longrightarrow R'_p/I_p$ by $\psi_p(f \bmod I) = f_p \bmod I_p$ for $p \in \mathrm{Spec}^*(R)$. ψ_p is well-defined and has kernel pS. For, $f_p \bmod I_p = 0$ iff $f_p \in I_p$ iff there exists $g \in I$ such that $(f-g)_p = 0$ or equivalently $f-g \in pR'$, and this is so iff $f \bmod I \in pS$. Thus ψ_p induces an isomorphism $\psi_p' : S/pS \longrightarrow R'_p/I_p = S_p$ with $\psi_p' \upharpoonright R_p = \mathrm{id}$. Consequently, the map $\sigma : \langle S,*\rangle \longrightarrow \langle \prod_{pS \in \mathrm{Spec}^*(S)} S_p, *\rangle$ given by $\sigma(a)(pS) = \psi_p'(\xi_S(a)(pS))$

is a representation of $\langle S,*\rangle$ and we have established the commutativity of the upper part of the diagram. In the following we denote $\sigma(a)(p)$ by a_p for elements a of S.

Next let $\rho : T \longrightarrow \prod_{pT \in \mathrm{Spec}^*(T)} T_p$ be defined by

$\rho(ab^{-1}) = \sigma(a) \cdot (\sigma(b))^{-1}$ for $a, b \in S$, $b^* = 1$. Thus ρ is the canonical extension of the isomorphism $\sigma : S \longrightarrow \sigma(S)$ to the total ring T of quotients of S. Moreover, ρ is compatible with the operation $*$: $\rho((ab^{-1})^*) = \rho(a^*) = \sigma(a^*) = (\sigma(a))^* = \sigma(a)^* ((\sigma b)^{-1})^* = (\rho(ab^{-1}))^*$. So $\rho : \langle T,*\rangle \longrightarrow \langle \prod_{pT \in \mathrm{Spec}^*(T)} T_p, *\rangle$ is indeed a

representation of $\langle T,*\rangle$, and we know that the diagram with the third row deleted commutes.

Finally, we define $\rho_J : S_J \longrightarrow \prod_{pS_J \in \mathrm{Spec}^*(S_J)} (S_p)_{J_p}$ by

$\rho_J(c)(pS_J) = \rho(c)(pT)$ for $c \in T$, $p \in \mathrm{Spec}^*(R)$. (Notice $\rho_J(c)(pS_J) \in (S_p)_{J_p}$ by definition of S_J.) Then ρ_J is clearly

compatible with the operation $*$. To show that $\rho_J(c)(pS_J)$ ranges over the whole ring $(S_p)_{J_p}$ for any $p \in \mathrm{Spec}^*(R)$, pick elements $a, b \in S$ such that $a_p \cdot (b^p)^{-1} \in (S_p)_{J_p}$. Then $b_p \notin J_p$, and so there exists a clopen neighborhood U of p such that $b_q \notin J_q$ for $q \in U$, since is smooth by assumption. Let e be the idempotent in R corresponding to J and put $c = eb + (1-e)$. Then $ac^{-1} \in S_J$ and $\rho_J(ac^{-1})(pS_J) = a_p(b_p)^{-1}$. Thus $\rho_J : \langle S_J, * \rangle \longrightarrow \langle \prod_{pS_J \in \mathrm{Spec}^*(S_J)} (S_p)_{J_P}, * \rangle$ is a representation of $\langle S_J, * \rangle$ and the proof is complete.

It might be of interest to note at this point that in any polynomial ring $R[X_1,\ldots,X_n]$, $n \geq 2$, over a monically closed regular ring R there are finitely generated everywhere prime ideals I besides $\{0\}$ and (X_i), $1 \leq i \leq n$. Actually there is an abundance of them.

<u>Proposition I.3.17</u> Let $\langle R, * \rangle$ be a monically closed regular *-ring, $S = R[X_1,\ldots,X_n]$, and let $I_1,\ldots,I_k$ be finitely generated everywhere prime ideals in S. Pick $p_1,\ldots,p_m \in \mathrm{Spec}^*(S)$ and proper prime ideals J_i in S_{p_i} different from $(I_1)_{p_i},\ldots,(I_k)_{p_i}$, $1 \leq i \leq m$. Then there exists a finitely generated everywhere prime ideal I such that $I_q \neq (I_1)_q,\ldots,(I_k)_q$ for all $q \in \mathrm{Spec}^*(S)$ and $I_{p_i} = J_i$ for $1 \leq i \leq m$.

The proposition is proved by a "piecing together" argument similar to the one used in the proof of I.3.10; we leave the details to the reader. Using a lifting argument as in the proof of I.3.13 a corresponding result can be shown for arbitrary *-extensions $\langle S, * \rangle$ of $\langle R, * \rangle$ which are ring-finite over R and such that $B(S) = B(R)$ and $\mathrm{trdeg}(S:R)$ is locally constant.

We conclude this section with some remarks on modules over *-rings based on results of Pierce [12]. Let $\langle R, * \rangle$ be a *-ring, M a (unitary) R-module, and M_p an R_p-module for every $p \in \mathrm{Spec}^*(R)$. M is called a <u>subdirect product</u> of $\{M_p\}_{p \in \mathrm{Spec}^*(R)}$ if M is a subdirect product of $\{M_p\}_{p \in \mathrm{Spec}^*(R)}$ as a group and $(\lambda a)(p) = \lambda_p \cdot a(p)$ for all $\lambda \in R$, $a \in M$, $p \in \mathrm{Spec}^*(R)$.

<u>Theorem I.3.18</u> (compare [12], 1.3 and 1.7). Let $\langle R,*\rangle$ be a *-ring and M an R-module. Then M is isomorphic to a subdirect product of R_p-modules M_p, $p \in \mathrm{Spec}^*(R)$.

<u>Proof</u>. We set $M_p = M/p\cdot M$ for $p \in \mathrm{Spec}^*(R)$. Then the map $\varphi : M \longrightarrow \prod_{p \in \mathrm{Spec}^*(R)} M_p$ defined by

$\varphi(a) = \langle a \bmod p\cdot M : p \in \mathrm{Spec}^*(R)\rangle$ for $a \in M$ is a homomorphism of R-modules and the image of M under φ is a subdirect product of $\{M_p\}_{p \in \mathrm{Spec}^*(R)}$ as a group. Each M_p becomes an R_p-module under the definition $\lambda_p \cdot (a \bmod pM) = \lambda a \bmod pM$. To show that φ is injective, assume $a \in M$, $a \neq 0$. Then $\mathrm{Ann}_B(a) = \{\lambda \in B(R) : \lambda a = 0\}$ is a proper Boolean ideal in $B(R)$. Let I be a proper Boolean prime ideal extending $\mathrm{Ann}_B(a)$ and $p = I\cdot R$ the corresponding prime *-ideal. We claim that $a \notin p\cdot M$ and so $\varphi(a) \neq 0$: Assume for a contradiction that $a = \mu\lambda b$ with $\mu \in I$, $\lambda \in R$, $b \in M$. Then $(1-\mu)\, a = (1-\mu)\mu\lambda b = 0$, i.e. $(1-\mu) \in \mathrm{Ann}_B(a) \subset I$ contradicting the fact that $\mu \in I$.

If $\varphi : M \longrightarrow \prod_{p \in \mathrm{Spec}^*(R)} M_p$ is the representation of M given in I.3.17, we will in the following write a_p for $\varphi(a)(p)$, $a \in M$, $p \in \mathrm{Spec}^*(R)$.

<u>Corollary I.3.19</u>. Let $\langle R,*\rangle$ be a *-ring, M an R-module, $p \in \mathrm{Spec}^*(R)$, and $a, b \in M$ such that $a_p = b_p$. Then there exists a neighborhood U of p such that $a_q = b_q$ for all $q \in U$.

<u>Proof</u>. $a_p = b_p$ implies that there exists $\lambda \in p \cap B(R)$ and $c \in M$ such that $a-b = \lambda\cdot c$. Let U be the clopen neighborhood of p corresponding to $(1-\lambda)$. Then $a-b = \lambda c$, $\lambda \in q \cap B(R)$, and hence $a_q = b_q$ for all $q \in U$.

I.4 A Nullstellensatz for regular rings.

Hilbert's Nullstellensatz for fields may be formulated as follows:

Theorem I.4.1. Let k be a field, K an algebraically closed extension of k, $f \in k[X_1, \ldots, X_n]$ and I and ideal in $k[X_1, \ldots, X_n]$. Then f vanishes at every zero of I in K^n iff f is in the radical of I.

Since every ideal in $k[X_1, \ldots, X_n]$ is finitely generated, this theorem is equivalent to the following.

Theorem II.4.2. Let k be a field, K an algebraically closed extension of k, and $f, g_1, \ldots, g_m \in k[X_1, \ldots, X_n]$. Then f vanishes at every common zero of $g_1, \ldots, g_m$ in K^n iff there exists $r \in \mathbb{N}$ and $h_1, \ldots, h_m \in k[X_1, \ldots, X_n]$ such that $f^r = \sum_{i=1}^{m} h_i g_i$.

We will now prove an analogue of theorem II.4.2 for regular rings. The following remarks will be helpful. We recall that a first-order existential formula is said to be primitive if its matrix consists of a conjunction of atomic and negated atomic formulas. If $A \subset B$ are structures for a first-order language, then let us say that they satisfy Robinson's condition if, for any sentence ψ obtained by substituting constants denoting elements of A into a primitive formula of the language, $B \models \psi$ implies $A \models \psi$. Robinson's test says that in order for a theory T to be model-complete it is necessary and sufficient that when $A \subset B$ are models of T, then the extension satisfies Robinson's condition.

K' is the model-completion of K because K has the amalgamation property, K' is model-consistent with K, and K' is model-complete (see [10]). The heart of the proof that K' is model-complete is the demonstration that if $R \subset S$ are two models of K' then the extension satisfies Robinson's condition. An examination of the proof reveals that if $R \subset S$ are models of K^* then Robinson's condition holds when restricted to primitive formulas whose matrix contains at most one negation.

Theorem I.4.3. Let R be a regular ring, R' a monically closed regular ring extending R, $f, g_1, \ldots, g_m \in R[X_1, \ldots, X_n]$ and assume $(g_1, \ldots, g_m) \cap R = \{0\}$. Then f vanishes at every common zero of $g_1, \ldots, g_m$ in R'^n iff there exist $r \in \mathbb{N}$ and $h_1, \ldots, h_m \in R[X_1, \ldots, X_n]$ such that $f^r = \sum_{i=1}^{m} h_i g_i$.

Proof. Assume first that $f^r = \sum_{i=1}^{m} h_i g_i$ and that $\langle a_1, \ldots, a_n \rangle \in R'^n$ is a common zero of $g_1, \ldots, g_m$. Then $f^r(a_1, \ldots, a_n) = 0$ and so $f(a_1, \ldots, a_n) = 0$, since R' has no non-zero nilpotent elements. Next assume f is not in the radical J of $(g_1, \ldots, g_m)$. Notice that $R \cap J = \{0\}$; for, $b \in R \cap J$ implies $b^k \in R \cap (g_1, \ldots, g_m)$ for some $k \in \mathbb{N}$, hence by assumption $b^k = 0$, and so $b = 0$. Consequently $R[X_1, \ldots, X_n]/J$ is a ring without non-zero nilpotent elements extending R. Let R'' be a regular ring extending $R[X_1, \ldots, X_n]/J$ and amalgamate R' and R'' over R into a monically closed regular ring R'''. Let $x_1, \ldots, x_n$ denote the images of $X_1, \ldots, X_n$ under the canonical homomorphism from $R[X_1, \ldots, X_n]$ onto $R[X_1, \ldots, X_n]/J$. Then $\langle x_1, \ldots, x_n \rangle$ is a common zero of $g_1, \ldots, g_m$ in $R[X_1, \ldots, X_n]/J$ and a non-zero of f. So the statement

$$\varphi = \exists x_1 \ldots \exists x_n \left(\bigwedge_{i=1}^{m} g_i(x_1, \ldots, x_n) = 0 \wedge f(x_1, \ldots, x_n) \neq 0 \right)$$

holds in $R[X_1, \ldots, X_n]/J$, hence in R''', and so by the remark above in R'. Thus we have shown that f does not vanish at every common zero of $g_1, \ldots, g_m$ in R'^n.

Corollary I.4.4. For all positive integers n,d,m there exist positive integers $r(n,d,m)$ and $s(n,d,m)$ such that for all models R of K^* and all $f,g_1,\ldots,g_m \in R[X_1,\ldots,X_n]$ of degrees $\leq d$, f is in the radical of $(g_1,\ldots,g_m)$ iff there exist $h_1,\ldots,h_m \in R[X_1,\ldots,X_n]$ of degrees $\leq s(n,d,m)$ such that $f^{r(n,d,m)} = \sum_{i=1}^{m} h_i g_i$.

Proof. By theorem II.4.3 the following statements are equivalent for every model R of K^*.

(i) There exist positive integers r,s, and $h_1,\ldots,h_m \in R[X_1,\ldots,X_n]$ of degrees $\leq s$ such that $f^r = \sum_{i=1}^{m} h_i g_i$.

(ii) $R \models \varphi$, where φ is defined as in the proof of the theorem. The first statement can be written as an infinite disjunction $\bigvee_{r,s=1}^{\infty} \psi_{r,s}$ of first-order sentences about the coefficients of $f,g_1,\ldots,g_m, h_1,\ldots,h_m$. So the statement $\bigvee_{r,s=1}^{\infty} \psi_{r,s} \longleftrightarrow \varphi$ holds in every model of K^* and hence is a consequence of K^*. Now the compactness theorem for first order theories provides bounds $r(n,d,m)$, $s(n,d,m)$ such that

$$\bigvee_{\substack{r \leq r(n,d,m) \\ s \leq s(n,d,m)}} \psi_{r,s} \longleftrightarrow \varphi$$

is a consequence of K^* and hence holds in every monically closed regular ring R.

Remark: The condition that $R \cap (g_1,\ldots,g_m) = \{0\}$ is necessary for theorem I.4.3: Consider, e.g. the case where $n=1$, $m=1$, $g = e \in B(R)$, $e \neq 0,1$, and $f = X$. Then (g) has no zero in R'; so f vanishes at every zero of (g) in R'. However, no power of f can be a multiple of e.

The Nullstellensatz I.4.1 for arbitrary ideals I (even with the side condition $I \cap R = \{0\}$) can not be transferred to regular rings. Consider the following counterexample (compare [3], Section 4). Let $R = R'$ be a model of K', p a fixed point in $\mathrm{Spec}^*(R)$, $f = X$, and let $I = \{gX : g \in R[X], (g)_p = 0\}$. Then it is easy to see that I is a radical ideal in $R[X]$ and that $I \cap R = \{0\}$. We claim that the only zero of I in R' is 0. To prove this let a be a zero of I and $q \in \mathrm{Spec}^*(R)$, $q \neq p$. Pick $e \in B(R)$ such that $e_q = 1$ and $e_p = 0$. Then $eX \in I$ and so $(ea)_q = 0$ which shows that $a_q = 0$. To show that $a_p = 0$, assume for a contradiction that $a_p \neq 0$. Then there exists $q \in \mathrm{Spec}^*(R)$, $q \neq p$ such that $a_q \neq 0$, a contradiction. We conclude from this claim that every zero of I is a zero of f. On the other hand $f \notin I$.

II. Algebraic Curves

II. 1. Affine Varieties over Monically Closed Commutative Regular Rings.

In this section we shall give some definitions which are straightforward generalizations of some familiar definitions from the theory over fields.

To begin, let $R \models K^*$. Affine n-space $A^n(R)$ over R is the set of all n-tuples $(a_1, \ldots, a_n)$ of elements of R. An algebraic set S in $A^n(R)$ is the set of all zeroes of a finitely generated ideal in $R[X_1, \ldots, X_n]$. A variety V in $A^n(R)$ is the set of all zeros of a finitely generated everywhere prime ideal in $R[X_1, \ldots, X_n]$. If $p \in \mathrm{Spec}^*(R)$, then S_p (resp. V_p) denotes the canonical image of S (resp. V) in $A^n(R_p)$.

We point out that every nonempty algebraic set can be decomposed into finitely many varieties, in the sense of the following theorem.

(Of course the empty set is itself already a variety.)

Theorem II.1.1. For any algebraic set S in $A^n(R)$ there exist finitely many varieties $V_1, \dots, V_r$, each contained in S, such that for any $s \in S$ there exist $s_i \in V_i$, $1 \le i \le r$, and a partition $e_1, \dots, e_r$ of 1 such that $s = \sum_{i=1}^{r} s_i e_i$. ($\Sigma$ denotes addition of n-tuples.)

Proof: If S is empty the result is clear, so assume $S \neq \emptyset$. Let $p \in \mathrm{Spec}^*(R)$. S_p is, by II.1.2(i) below, a finite union of varieties (in the sense of the field theory), $V_{p,1}, \dots, V_{p,k_p}$. For each i, $1 \le i \le k_p$, $I(V_{p,i})$ is generated by polynomials $f_{p,i,1}, \dots, f_{p,i,m_i}$ over R/p, and $I(V_{p,i})$ is a prime ideal in $R/p[X_1, \dots, X_n]$. For each i, $1 \le i \le k_p$, let $\psi_i\,(x_1, \dots, x_{t_i})$ — where $t_i = \sum_{s=1}^{m_i} (\deg f_{p,i,s} + 1)$ — be a predicate which expresses the fact that $I(V_{p,i})$ is a prime ideal (I.3.7).

For each i, $1 \le i \le k_p$, and each s, $1 \le s \le m_i$, let $F_{p,i,s}$ be a polynomial in $R[X_1, \dots, X_n]$ whose image modulo p is $f_{p,i,s}$; let $a_{i,1}, \dots, a_{i,t_i}$ be the coefficients of the $F_{p,i}$. Then if S is the set of zeros of $F_1, \dots, F_v$, the formula

$$\varphi_p = \forall x_1 \dots \forall x_n \left[\bigwedge_{j=1}^{v} F_j(x_1, \dots, x_n) = 0 \longleftrightarrow \bigvee_{i=1}^{k_p} \left(\bigwedge_{s=1}^{m_i} F_{p,i,s}(x_1, \dots, x_n) = 0 \right) \right] \wedge \psi_1(a_{1,1}, \dots, a_{1,t_1}) \wedge \dots \wedge \psi_{k_p}(a_{k_p,1}, \dots, a_{k_p,t_{k_p}})$$

holds at p and hence at each point of some idempotent $e_p \in R$ which contains p, since $R \models K^*$. The e_p, $p \in \mathrm{Spec}^*(R)$, cover $\mathrm{Spec}^*(R)$, so there exist $e_{p_1}, \dots, e_{p_w}$ which cover $\mathrm{Spec}^*(R)$; we can assume that

$e_{p_1}, \dots, e_{p_w}$ are disjoint.

Consider all sets of the form $T = \{T_{p_1}, \dots, T_{p_w}\}$, where each T_{p_j} is one of $V_{p_j,1}, \dots, V_{p_j,k_{p_j}}$ for $1 \le j \le w$. If $r = \max_{1 \le j \le w} k_{p_j}$ then it is clear that there exist r such sets $T^1, \dots, T^r$ such that any $V_{p_j,i}$ appears in at least one of $T^1, \dots, T^r$. $T^1, \dots, T^r$ will give us the varieties we seek, in the following way. Consider T^q, $1 \le q \le r$. Suppose $T^q_{p_j}$ is V_{p_j, i_j} for each j, where $1 \le i_j \le k_{p_j}$. Consider the polynomials $F_{p_j,i_j,1}, \dots, F_{p_j,i_j,m_{i_j}}$ for each j. If $N_q = \max_{1 \le j \le w} m_{i_j}$ then for each j define $F_{p_j,i_j,t} = F_{p_j,i_j,m_{i_j}}$ for $m_{i_j} \le t \le N_q$. Define $F_{q,1} = \sum_{j=1}^{w} e_j F_{p_j,i_j,1}, \dots, F_{q,N_q} = \sum_{j=1}^{w} e_j F_{p_j,i_j,N_q}$. Then $F_{q,1}, \dots, F_{q,N_q}$ determine a variety $V_q = V(F_{q,1}, \dots, F_{q,N_q})$. We claim that $V_1, \dots, V_r$ satisfy the requirements of the theorem.

First notice that by construction $S_p = (V_1)_p \cup \dots \cup (V_r)_p$ for any $p \in \mathrm{Spec}^*(R)$. Thus each $V_q \subseteq S$, because if a point $x \in A^N(R)$ satisfies $F_1, \dots, F_v$ at every $p \in \mathrm{Spec}^*(R)$ then x satisfies $F_1, \dots, F_v$ in R. On the other hand if $s = (a_1, \dots, a_n) \in S$ then for any $p \in \mathrm{Spec}^*(R)$, p is in some e_{p_j}, so since φ_j holds on e_{p_j} we have $\bigwedge_{s=1}^{m_i} F_{p_j,i,s}(a_1, \dots, a_n) = 0$ at p for some i, $1 \le i \le k_{p_j}$.

Hence by the definition of the T^q, $\bigwedge_{z=1}^{N_q} F_{q,z}(a_1,\ldots, a_n) = 0$ at p for some q, $1 \leq q \leq r$. Hence $\bigwedge_{z=1}^{N_q} F_{q,z}(a_1,\ldots, a_n) = 0$ on some idempotent c_p containing p. As usual we see that finitely many c_p's cover $\mathrm{Spec}^*(R)$, say $c_{p_1},\ldots, c_{p_n}$, and we can assume these are disjoint. (These p_j's aren't necessarily the same as the p_j's we had before, of course.) Let d_1 be the union of those c_p's on which $\bigwedge_{z=1}^{N_1} F_{1,z}(a_1,\ldots, a_n) = 0$ holds identically, let $\left(b_1^1,\ldots,b_n^1\right)$ be an arbitrary element of V_1, and let $g_i^1 = d_1 a_i + (1-d_1)\, b_i^1$ for $1 \leq i \leq n$. Then $\left(g_1^1,\ldots, g_n^1\right) \in V_1$ and $\left(g_1^1,\ldots, g_n^1\right) \cdot d_1 = (a_1,\ldots, a_n) \cdot d_1$. Next let d_2 be the union of the remaining c_p's (i.e. those not used in d_1) on which $\bigwedge_{z=1}^{N_2} F_{2,z}(a_1,\ldots, a_n) = 0$ holds identically, let $\left(b_1^2,\ldots,b_n^2\right)$ be an arbitrary element of V_2 and let $g_i^2 = d_2 a_i + (1-d_2)\, b_i^2$ for $1 \leq i \leq n$. Then $\left(g_1^2,\ldots,g_n^2\right) \in V_2$ and $\left(g_1^2,\ldots,g_n^2\right) \cdot d_2 = (a_1,\ldots,a_n)\cdot d_2$. Continuing in this way define d_i, $1 \leq i \leq r$ and $\left(g_1^i,\ldots,g_n^i\right) \in V_i$ such that $\left(g_1^i,\ldots,g_n^i\right) \cdot d_i = (a_1,\ldots,a_n) \cdot d_i$ and $\bigcup_{i=1}^{r} d_i = 1$, $d_i \cap d_j = 0$ if $i \neq j$. (The d_i may be 0 from some point on). Then

$$(a_1,\ldots,a_n) = \sum_{i=1}^{r} d_i \cdot (a_1,\ldots,a_n) = \sum_{i=1}^{r} d_i \cdot \left(g_1^i,\ldots,g_n^i\right),$$ as desired.

Remark. The number of V_i's needed is bounded by the maximum number of components of the V_p's, $p \in \mathrm{Spec}^*(R)$; that this maximum exists follows from the proof.

If S is an algebraic set the ideal $I(S)$ of S is the set of all elements f of $R[X_1,\dots,X_n]$ such that $f(a_1,\dots,a_n) = 0$ for all $(a_1,\dots,a_n) \in S$. The following lemma will be useful in dealing with varieties.

Lemma II.1.2. (i) If I is a finitely generated ideal and $S(I)$ is nonempty then $S(I_p) = (S(I))_p$ for all $p \in \mathrm{Spec}^*(R)$.

(ii) If S is an algebraic set then $I(S_p) = (I(S))_p$ for all $p \in \mathrm{Spec}^*(R)$.

(iii) If I is a finitely generated everywhere prime ideal such that $I \cap R = \{0\}$, and $V = V(I)$ is its variety, then $I(V(I)) = I$.

Proof: (i) It is clear that $(S(I))_p \subseteq S(I_p)$. For the opposite inclusion suppose $I = (F_1,\dots,F_k)$ and $b_1,\dots,b_n \in R/p$ satisfy $\bigwedge_{i=1}^{k} F_{i_p}(b_1,\dots,b_n) = 0$. Let $a_1,\dots,a_n \in R$ be such that

$(a_i)_p = b_i$, $1 \leq i \leq n$; then $\bigwedge_{i=1}^{k} F_i(a_1,\dots,a_n) = 0$ holds on an

idempotent e containing p. Since $S(I)$ is nonempty we can let $(c_1,\dots,c_n) \in S(I)$; then $(e(a_1,\dots,a_n) + (1-e)(c_1,\dots,c_n)) \in S(I)$ and $[e(a_1,\dots,a_n) + (1-e)(c_1,\dots,c_n)]_p = (b_1,\dots,b_n)$. So $S(I_p) \subseteq (S(I))_p$.

(ii) If $S = \emptyset$ the result is clear, so suppose S is nonempty. If F is a polynomial which vanishes on S then F_p vanishes on S_p for each p, so $(I(S))_p \subseteq I(S_p)$. For the other inclusion, let S be the set of zeros of the ideal $I = (F_1,\dots,F_k)$. If $f \in R/p[X_1,\dots,X_n]$ vanishes on S_p, then by (i) f vanishes on all zeroes of I_p.

Therefore if we let $F \in R[X_1,\ldots,X_n]$ be such that $F_p = f$, then

$$\forall x_1 \ldots \forall x_n \left(\bigwedge_{i=1}^{k} F_i(x_1,\ldots,x_n) = 0 \longrightarrow F(x_1,\ldots,x_n) = 0 \right)$$

holds at p and hence on an idempotent e including p. Now $eF + (1-e)F_1 \in I(S)$ and $(eF + (1-e)F_1)_p = f$, so $I(S_p) \subseteq (I(S))_p$.

(iii) Suppose I is finitely generated and everywhere prime. It suffices, to prove (iii), to show that I is a radical ideal, for then since $I \cap \{R\} = 0$, the Nullstellensatz of section I.4 guarantees that $I(V(I)) = I$. Suppose $F \in R[X_1,\ldots,X_n]$ and $F^m \in I$ for some positive integer m; then $F_p^m \in I_p$ for all $p \in \mathrm{Spec}^*(R[X_1,\ldots,X_n])$, so since I is everywhere prime, $F_p \in I_p$ for all p. Thus $F \in I$ by I.2.5.

<u>Remark</u>. (iii) could have been proved by using (i) and (ii) to conclude $I(V(I))_p = I_p$ for all p and then using I.2.5.

If V is a variety the coordinate ring $\Gamma(V)$ of V is defined to be $R[X_1,\ldots,X_n] / I(V)$.

<u>Theorem II.1.3</u>. Assume V is a nonempty variety. Then $\Gamma(V)$ is a commutative *-ring extending R, with the same idempotents as R. We have a representation σ of $\Gamma(V)$ (as a *-ring) such that the diagram

$$\begin{array}{ccc} R & \xrightarrow{\xi_R} & \prod R_p \\ \downarrow & & \downarrow \\ \Gamma(V) & \xrightarrow{\sigma} & \prod \Gamma(V_p) \end{array}$$

commutes, where $\Gamma(V_p)$ denotes the coordinate ring of V_p as a variety over R_p, i.e. $R_p[X_1,\ldots,X_n] / I_p$.

<u>Proof</u>. Direct application of I.3.16.

We remark that $\Gamma(V)$ is isomorphic to the ring of polynomial functions on V, by II.1.2(iii). Thus it is clear what $b(Q)$ means, for $b \in \Gamma(V)$ and $Q \in V$. We define $k(V)$ the <u>regular ring of rational functions</u> on V, to be the regular closure of $\Gamma(V)$.

<u>Theorem II.1.4</u>. Let V be a nonempty variety. Then $k(V)$ is a regular extension of $\Gamma(V)$, with the same idempotents. We have a representation ρ of $k(V)$ such that the diagram

$$\begin{array}{ccc}
R & \xrightarrow{\xi_R} & \prod R_p \\
\downarrow & & \downarrow \\
\Gamma(V) & \xrightarrow{\sigma} & \prod \Gamma(V_p) \\
\downarrow & & \downarrow \\
k(V) & \xrightarrow{\rho} & \prod k(V_p)
\end{array}$$

commutes, where $k(V_p)$ denotes the usual rational function field of V_p as a variety over R_p, i.e. the quotient field of $\Gamma(V_p)$.

<u>Proof</u>. By I.3.16.

Now if $Q \in V$ define $\mathcal{O}_Q(V)$ as the subset of $k(V)$ consisting of all elements which have a representation ab^{-1}, where $a, b \in \Gamma(V)$ and $b(Q)$ is a unit in R.

<u>Theorem II.1.5</u>. $\Gamma(V) \subseteq \mathcal{O}_Q(V)$ and $\mathcal{O}_Q(V)$ is a *-subring of $k(V)$ with the same idempotents as $k(V)$. We have a representation ρ_Q of $\mathcal{O}_Q(V)$ (as a *-ring) such that the diagram

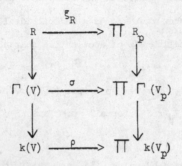

$$\begin{array}{ccc}
R & \xrightarrow{\ \xi_R\ } & \prod R_p \\
\downarrow & & \downarrow \\
\Gamma(V) & \xrightarrow{\ \sigma\ } & \prod \Gamma(V_p) \\
\downarrow & & \downarrow \\
\mathcal{O}_Q(V) & \xrightarrow{\ \rho_Q\ } & \prod \mathcal{O}_{Q_p}(V_p) \\
\downarrow & & \downarrow \\
k(V) & \xrightarrow{\ \rho\ } & \prod k(V_p)
\end{array}$$

commutes, where $\mathcal{O}_{Q_p}(V_p)$ denotes the local ring of V_p at Q_p.

<u>Proof</u>. This will follow from I.3.16 as soon as we know that the ideal I in $\Gamma(V)$ consisting of all b such that $b(Q) = 0$ is a finitely generated everywhere prime ideal. (Notice that in the notation of I.3.16, $\mathcal{O}_Q(V)$ is $\Gamma(V)_{I^*}$.) It is easy to see that I is everywhere prime. And I is finitely generated because if $Q = (a_1,\ldots,a_n)$ then the residues in $\Gamma(V)$ of $X_1-a_1,\ldots,X_n-a_n$ generate I.

With this result as justification, we call $\mathcal{O}_Q(V)$ the <u>subdirectly</u> <u>local</u> (<u>sublocal</u>, for short) ring of V at Q.

II.2. <u>Projective Varieties</u>

We now consider the concepts of Section 1 in a projective setting.

Let R be a commutative regular ring. Let X be the set of all $(n+1)$-tuples $(a_0,\ldots,a_n)$ of elements of R such that $\bigcup_{i=0}^{n} a_i^* = 1$. We define projective n-space $P^n(R)$ over R to be the set of equivalence classes of elements of X under the equivalence relation $(a_0,\ldots,a_n) \sim (b_0,\ldots,b_n)$ iff there exists a unit $u \in R$ such that

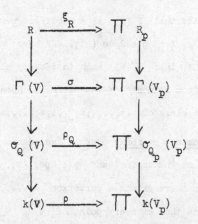

$u \cdot a_i = b_i$, $0 \leq i \leq n$. Notice that $P^n(R)$ is a subdirect product of the projective n-spaces $P^n(R_p)$, for $p \in \mathrm{Spec}^*(R)$. We have $n+1$ embeddings $\varphi_0, \ldots, \varphi_n$ of $A^n(R)$ into $P^n(R)$ such that every element of $P^n(R)$ can be pieced together from elements of the $\varphi_i(A^n)$. Specifically, $\varphi_i(a_1, \ldots, a_n)$ = the equivalence class of $(a_1, \ldots, a_{i-1}, 1, a_i, \ldots, a_n)$.

By a homogeneous polynomial (or form) over R we mean a polynomial F such that F_p is homogeneous for each $p \in \mathrm{Spec}^*(R)$. Observe that if F is homogeneous then there exists a partition $e_1, \ldots, e_m$ of 1 and integers $n_1, \ldots, n_m$ such that at every point of e_i, F_p is homogeneous of degree n_i (see example (i) in I.2).

Proposition II.2.1. Any non-zero polynomial F can be written uniquely as a sum of forms $F = \sum_{i=1}^{m} F_i$ where (i) $\deg F_m = \deg F$;

(ii) if $i_1 \geq i_2$ then $\deg(F_{i_1}) \geq \deg(F_{i_2})$, and the inequality is strict except at points p where F_{i_1} is the zero polynomial;

(iii) no F_i is the zero polynomial.

Proof. Given F, we claim that there is a unique polynomial G over R such that G_p is the sum of the terms of highest degree in F_p, for each $p \in \mathrm{Spec}^*(R)$. For consider $p \in \mathrm{Spec}^*(R)$. Let H be the sum of those terms of F which give the terms of highest degree in F_p. Let e be the idempotent in R on which the coefficients of the terms in H do not vanish and the coefficients of the terms of F with formal degree $\geq$ the formal degree of H do vanish. Then $p \in e$ and on e H_p gives the sum of the highest terms of F_p. As usual we can argue that we have a partition $e_1, \ldots, e_k$ of 1 and polynomials $H_1, \ldots, H_k \in R[X_1, \ldots, X_n]$ such that H_{i_p} gives the terms of highest degree of F_p on e_i. Now $G = \sum_{i=1}^{k} e_i H_i$ satisfies our claim; it is clear that there is at most one such G.

Now let $G = G_1$ and form G_2 from $F-G_1$ in the same way that G was formed from F. Continuing in this way, define G_3, G_4, etc. After a finite number m of steps we arrive at $F-G_1-G_2-\ldots G_m = 0$, (with G_m not the zero polynomial) since the degree of

$\left(F-\sum_{i=1}^{k+1} G_i\right)_p$ is less than that of $\left(F-\sum_{i=1}^{k} G_i\right)_p$ at every p (unless the latter is already the zero polynomial) and all degrees in sight are bounded by the formal degree of F.

Let $F_i = G_{m+1-i}$, $1 \leq i \leq m$; then $F = \sum_{i=1}^{m} F_i$.

As for uniqueness, if $F = \sum_{i=1}^{r} H_i$ as in the statement of the proposition then H^r satisfies the condition in the definition of G_1, so by the uniqueness of G_1, $H^r = G_1 = F^m$. Inductively, we see that if $1 \leq j \leq m$ then $H_{r+1-j} = F_{m+1-j}$, and in particular $r \geq m$, since no F_i is the zero polynomial. Since $\sum_{i=r-m+1}^{r} H_i = \sum_{i=1}^{m} F_i = H = \sum_{i=1}^{r} H_i$, it follows from the assumed properties of the H_i that in fact $r = m$, and this finishes the proof.

We remark that, at each p, $F_p = \sum_{i=1}^{m} F_{i_p}$ is the unique expression of F_p as a sum of forms, with (possibly) some zero terms tacked on.

We say that a point $\{(a_o,\ldots,a_n)\} \in P^n(R)$ is a zero of a form $F \in R[X_o,\ldots,X_n]$ iff for all units $u \in R$, $F(ua_o,\ldots,ua_n) = 0$. This is the case iff $F(a_o,\ldots,a_n) = 0$. We call an ideal I in $R[X_o,\ldots,X_n]$ <u>homogeneous</u> if whenever $F \in I$ and $F = \sum_{i=1}^{m} F_i$ as in lemma II.2.1, we have

$F_i \in I$, $1 \leq i \leq m$.

Lemma II.2.2. I is homogeneous $\Longleftrightarrow$ I is generated by forms $\Longleftrightarrow$ each I_p is homogeneous in the usual sense (from the field theory).

Proof. If I is homogeneous and generated by $\{F_\alpha\}$ then the forms associated to the F_α's by Proposition II.2.1 generate I. Conversely if I is generated by forms then so is each I_p, so each I_p is homogeneous in the usual sense. Thus if $F = \sum_{i=1}^{m} F_i$ as in II.2.1 , then by the remark following the proof of that Proposition, $(F_i)_p \in I_p$ for any i and p. Thus by I.2.5 $F_i \in I$ for each i. The proof of the second equivalence is contained in what we have already said.

Now assume that R is a model of K^*.

By a projective algebraic set on $P^n(R)$ we mean the set of all zeroes in $P^n(R)$ of a finitely generated homogeneous ideal I in $R[X_o,\ldots,X_n]$. If I is in addition everywhere prime, the algebraic set is called a projective variety. We have a decomposition of algebraic sets into varieties, as in the affine case. We also have the analogue of lemma II.1.2 for projective varieties; in particular, if I is a finitely generated homogeneous everywhere prime ideal in $R[X_o,\ldots,X_n]$ which determines a nonempty variety in $P^n(R)$, then $I(V(I)) = I$. (The remark following the proof of II.1.2 is useful in this connection.)

If now $V = V(I)$ is a projective variety, we define $\Gamma_h(V)$ as the *-ring $\langle R[X_o,\ldots,X_n] / I, * \rangle$, and $K_h(V)$ as the regular closure of $\Gamma_h(V)$, just as in the affine case. As in the affine case, we have representations σ of $\Gamma_h(V)$ and ρ of $K_h(V)$ such that the diagram

$$
\begin{array}{ccc}
R & \xrightarrow{\ \xi_R\ } & \prod R/p \\
\downarrow & & \downarrow \\
\Gamma_h(V) & \xrightarrow{\ \sigma\ } & \prod \Gamma_h(V_p) \\
\downarrow & & \downarrow \\
K_h(V) & \xrightarrow{\ \rho\ } & \prod K_h(V_p),
\end{array}
$$

commutes, where $\Gamma_h(V_p)$ and $K_h(V_p)$ denote the homogeneous coordinate ring and homogeneous function field of V_p.

As in the field theory, elements of $K_h(V)$ need not determine functions on V, so we restrict ourselves to a subring $k(V)$, defined as follows. Recall that every element of $K_h(V)$ has a representation of the form ab^{-1} where $a, b \in \Gamma_h(V)$ and b is a unit in $K_h(V)$. $k(V)$ consists of all elements of $K_h(V)$ which have such a representation in which $a, b \in \Gamma_h(V)$ have homogeneous polynomial representatives $A, B, \in R[X_o, \ldots, X_n]$ such that A and BA^* have the same degree. Before verifying that $k(V)$ is a subring of $K_h(V)$ we define $\mathcal{O}_Q(V)$, for $Q \in V$, as the subset of $k(V)$ consisting of all elements which have such a representation in which $B(Q)$ is a unit in R. (Notice that although B might assume different values for different homogeneous coordinates of Q, the value for one set of coordinates is a unit iff the value for every set of coordinates is a unit, since B is homogeneous.) We again call $\mathcal{O}_Q(V)$ the <u>sublocal ring</u> of V at Q. Now we have

<u>Theorem II.2.3</u>. $k(V)$ is a regular subring of $K_h(V)$ extending R. $\mathcal{O}_Q(V)$ is a *-subring of $k(V)$ with the same idempotents. We have a commutative diagram

$$\begin{array}{ccc}
R & \longrightarrow & \prod R/p \\
\downarrow & & \downarrow \\
\sigma_Q(V) & \longrightarrow & \prod \sigma_{Q_p}(V_p) \\
\downarrow & & \downarrow \\
k(V) & \longrightarrow & \prod k(V_p) \\
\downarrow & & \downarrow \\
K_h(V) & \longrightarrow & \prod K_h(V_p)
\end{array}$$

Proof. We claim first that $k(V)$ is a subring of $K_h(V)$. For let ab^{-1}, cd^{-1} be elements of $k(V)$, and let A, B, C, D be forms representing a, b, c, d, as in the definition of $k(V)$. Then $(ab^{-1}) \cdot (cd^{-1}) = (ac) \cdot (bd)^{-1}$, and $ac, bd \in \Gamma_h(V)$ are represented by the forms AC, BD. Now on $(AC)^* = A^* \cap C^*$, A and B have the same degree and C and D have the same degree, so AC and BD have the same degree. For the sum, $ab^{-1} + cd^{-1} = (ad+bc) \cdot (bd)^{-1}$, and $ad + bc, bd \in \Gamma_h(V)$ are represented by $AD + BC$, BD respectively. Notice that $AD + BC$ is a form, since at any p where neither $(AD)_p$ nor $(BC)_p$ is the zero polynomial, A_p and B_p have the same degree and C_p and D_p have the same degree. In fact at any such point $\deg(AD)_p = \deg(BC)_p = \deg(BD)_p$; thus at a point p of $(AD+BC)^*$ where neither $(AD)_p$ nor $(BC)_p$ is the zero polynomial, $(AD+BC)_p$ and $(BD)_p$ have the same degree. At a point of $(AD+BC)^*$ where one of $(AD)_p$, $(BC)_p$ is the zero polynomial, the other is not, and, as above, has the same degree as $(BD)_p$. Thus $(AD+BC)$ and $(BD)(AD+BC)^*$ have the same degree. Now it is easy to see that $k(V)$ is a subring of $K_h(V)$.

That $k(V)$ is regular follows from the fact that it is closed under the formation of quasi-inverses in $K_h(V)$. To see that $k(V)$ does in fact have this closure property suppose $ab^{-1} \in k(V)$, and A,B

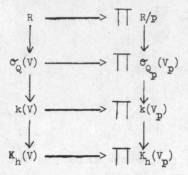

are forms representing a,b such that A and A^*B have the same degree. Notice that $ab^{-1} = [a+(1-a^*)] \cdot (ba^*)^{-1}$ and $[a+(1-a^*)]$ is a unit in $K_h(V)$, so the quasi-inverse of ab^{-1} in $K_h(V)$ is $(ba^*) \cdot [a+(1-a^*)]^{-1}$, and to finish the proof it suffices to find forms F, G which represent ba^*, $a+(1-a^*)$ respectively, such that F and F^*G have the same degree. We claim that $F = a^*B$, $G = a^*A + (1-a^*)$ will do the job. Clearly these are forms and represent ba^*, $a+(1-a^*)$ respectively. Now $F^*G = (a^*B^*)(a^*A + (1-a^*)) = a^*AB^*$. Since b is a unit in $K_h(V)$, $b^* = 1$; thus since $b^* \leq B^*$, $B^* = 1$, so $F^*G = a^*A$. Thus to say that F^*G and F have the same degree is to say that a^*A and a^*B have the same degree. But we know that A and A^*B have the same degree, so a^*A and a^*A^*B have the same degree; and $a^*A^*B = a^*B$ since $a^* \leq A^*$.

The proof that $\mathcal{O}_Q(V)$ is a subring of $k(V)$ is easy and we omit it. Since $R \subseteq \mathcal{O}_Q(V)$, $\mathcal{O}_Q(V)$ has the same idempotents as $k(V)$; and therefore $\mathcal{O}_Q(V)$ (with the restriction of the *-function of $k(V)$) is a *-subring of $k(V)$.

To finish the proof it suffices to show that $k(V) / p \cdot k(V) \cong k(V_p)$ and $\mathcal{O}_Q(V) / p \cdot \mathcal{O}_Q(V) \cong \mathcal{O}_{Q_p}(V_p)$. Recall from the proof of theorem II.1.4 that we have maps $\rho_p : K_h(V) \longrightarrow K_h(V_p)$ with kernel $pK_h(V)$ for every $p \in \mathrm{Spec}^*(R)$. A glance at the definition of ρ_p reveals that $\rho_p(k(V)) \subseteq k(V_p)$, so $\ker(\rho_p|k(V))$ is a prime ideal in $k(V)$, and this kernel clearly includes $p \cdot k(V)$. Since I_p is proper, $\ker(\rho_p|k(V))$ is a proper ideal in $k(V)$. Thus $\ker(\rho_p|k(V)) = p \cdot k(V)$, so we have an injection $k(V)/p \cdot k(V) \longrightarrow k(V_p)$. We claim that this map is surjective. For suppose $f, g \in R_p[X_o, \ldots, X_n]$ are forms of the same degree and $g \notin I_p$. Then there are polynomials $F, G \in R[X_o, \ldots, X_n]$ such that $F_p = f$ and $G_p = g$. By I.3.9 there is a neighborhood e_1 of p such that $G \in I$ on e_1. Let e_2 be a neighborhood of p on which F, G have the same degree, and let

$e_3 = e_1 \cap e_2$. Let $H = e_3G + (1-e_3)$, $K = e_3 F + (1-e_3)$. Then $H_p = f$, $K_p = g$, H and H^*K have the same degree, and the image of H in $\Gamma_h(V)$ is not a zero-divisor (consider the representation σ of $\Gamma_h(V)$).

To show that $\mathcal{O}_Q(V) / p\cdot\mathcal{O}_Q(V) \cong \mathcal{O}_{Q_p}(V_p)$, look at $\rho_p | \mathcal{O}_Q(V)$. As above the kernel of this map must be $p\cdot\mathcal{O}_Q(V)$. That its image is contained in $\mathcal{O}_{Q_p}(V_p)$ is easy. That its image is all of $\mathcal{O}_{Q_p}(V_p)$ follows by an argument similar to the one we just gave for $k(V) / p\cdot k(V) \cong k(V_p)$, if we add to the requirements on the idempotent e of that proof the requirement that $G(Q)$ be nonzero on e.

This completes our list of definitions. We now sketch the correspondence between affine and projective varieties. (cf. [5], pp. 96-97.) We consider $A^n(R) \subseteq P^n(R)$ via φ_o.

If F is a polynomial in $R[X_1,\dots,X_n]$ we define the homogenization $F^{\#}$ of F to be the homogeneous polynomial in $R[X_o,\dots,X_n]$ obtained by homogenizing F_p at every point p. (It is easy to see that this polynomial <u>is</u> in $R[X_o,\dots,X_n]$.) If $F \in R[X_o,\dots,X_n]$ is a form we define the dehomogenization $F_{\#}$ of F to be the polynomial in $R[X_1,\dots,X_n]$ obtained by dehomogenizing F at every point (i.e. by substituting 1 for X_o).

If I is an ideal in $R[X_1,\dots,X_n]$ we define $I^{\#}$ to be the ideal in $R[X_o,\dots,X_n]$ generated by $\{f^{\#} : f \in I\}$. If I is a homogeneous ideal in $R[X_o,\dots,X_n]$ we define $I_{\#}$ to be ideal $\{f_{\#} : f \in I\}$. It is easy to see that for any $p \in \mathrm{Spec}^*(R)$ we have $(I^{\#})_p = (I_p)^{\#}$ for any ideal I in $R[X_1,\dots,X_n]$ and $(I_{\#})_p = (I_p)_{\#}$ for any ideal I in $R[X_o,\dots,X_n]$.

<u>Proposition II.2.4</u>. If I is a finitely generated everywhere prime ideal in $R[X_1,\dots,X_n]$ and $I \cap R = \{0\}$ then $I^{\#}$ is a finitely

generated everywhere prime ideal.

Proof. $(I^{\#})_p = (I_p)^{\#}$ for any p, so I^* is everywhere prime. And since I is finitely generated, I.3.10 implies that Dim I is locally constant, so Dim $I^{\#}$ = Dim I + 1 is locally constant, so $I^{\#}$ is finitely generated by I.3.10.

If V is a variety in A^n we define $V^{\#} = V((I(V))^{\#})$. $V^{\#}$ is a projective variety. For if V is empty this is obvious. If V is not empty then by II.1.2 and II.2.4 $(I(V))^{\#}$ is finitely generated and everywhere prime, so $V^{\#}$ is a variety. $V^{\#}$ is called the projective closure of V.

If V is a projective variety we define $V_{\#} = V((I(V))_{\#})$. $V_{\#}$ is an affine variety. If $V = \emptyset$ then $V_{\#} = \emptyset$. If V is not empty then I(V) is finitely generated, by the projective analogue of II.1.2. Thus $V_{\#}$ is an affine algebraic set, and since $(I_{\#})_p = (I_p)_{\#}$ is prime for every $p \in \mathrm{Spec}^*(R)$, $V_{\#}$ is a variety.

If V is an affine variety, then $(V^{\#})_{\#} = V$. For if $p \in \mathrm{Spec}^*(R)$, then by lemma II.1.2(ii), $(I((V^{\#})_{\#}))_p = I(((V^{\#})_{\#})_p) = I((V_p{}^{\#})_{\#}) = I(V_p) = (I(V))_p$; thus $I(V^{\#})_{\#} = I(V)$, so $(V^{\#})_{\#} = V$. Similarly, if V is a projective variety such that for each $p \in \mathrm{Spec}^*(R)$ we have $(X_o)_p \notin I(V)_p$ then $(V_{\#})^{\#} = V$. Thus we have a one-one correspondence between affine varieties and projective varieties which are nowhere contained in the hyperplane at infinity.

If V is a nonempty affine variety and $f \in \Gamma_h(V^{\#})$ we define $f_{\#} \in \Gamma(V)$ as follows: Pick $F \in R[X_o,\ldots,X_n]$ which represents f, and let $f_{\#}$ be the I(V)-residue of $F_{\#}$, the polynomial obtained by substituting 1 for X_o in F. It is not difficult to see that this definition is independent of the choice of F, and that the map $f \longmapsto f_{\#}$ is a homomorphism from $\Gamma_h(V^{\#})$ into $\Gamma(V)$. We define a map α: $k(V^{\#}) \longrightarrow k(V)$ by sending fg^{-1} to $f_{\#}(g_{\#})^{-1}$. Observe that this makes sense because if g is not a zero-divisor in $\Gamma_h(V^{\#})$ then $g_{\#}$ is not a zero divisor in $\Gamma(V)$ (and hence is a unit in $k(V)$). For suppose $G,H \in R[X_o,\ldots,X_n]$ are forms such that $G_{\#}$ represents $g_{\#}$ in $R[X_1,\ldots,X_n]$, $G_{\#} H_{\#} \in I(V)$, $G_{\#} \notin I(V)$, $H_{\#} \notin I(V)$. Then $G, H \notin (I(V))^{\#}$; and $V \neq \emptyset \Longrightarrow V^{\#} \neq \emptyset$, so by the projective analogue of II.1.2, $I(V^{\#}) = (I(V))^{\#}$, so $G, H \notin I(V^{\#})$. Also it is easy to see that $(GH)_{\#} \in I(V)$ implies $GH \in (I(V))^{\#}$, since there is a partition $e_1,\ldots,e_n$ of 1 (corresponding to the idempotents on which various powers of X_o divide GH) such that on each e_i, GH is $((GH)_{\#})^{\#}$ multiplied by some power of X_o. Thus g is a zero-divisor in $\Gamma_h(V^{\#})$. We note the following properties of α:

<u>Lemma II.2.5</u>. (i) $\alpha : k(V^{\#}) \longrightarrow k(V)$ is an isomorphism.

(ii) For any $Q \in V$, $\alpha(\mathcal{O}_{\varphi_o(Q)}(V^{\#})) = \mathcal{O}_Q(V)$.

<u>Proof</u>: (i) It is easy to check that α is a homomorphism. α is injective because if $F \in R[X_o,\ldots,X_n]$ is a form such that $F_{\#} \in I(V)$, then $F \in I(V^{\#})$ (as we have just seen). α is surjective because given $ab^{-1} \in k(V)$, if $A,B \in R[X_1,\ldots,X_n]$ represent a,b, respectively, then by multiplying $A^{\#}$ and/or $B^{\#}$ by appropriate powers of X_o on suitable idempotents we get forms A^+, $B^+ \in R[X_o,\ldots,X_n]$ such that A^+ and $(A^+)^* B^+$ have the same degree and $(A^+)_{\#} = A$, $(B^+)_{\#} = B$, so if a^+, b^+ are the images of A^+, B^+ in $\Gamma_h(V^{\#})$, $a^+(b^+)^{-1} \in k(V^{\#})$ and $\alpha(a^+(b^+)^{-1}) = ab^{-1}$. (Notice that b^+ is not a zero-divisor in $\Gamma_h(V^{\#})$ because $(B^+)_{\#} = B$ and b is not a zero-divisor in $\Gamma(V)$.)

(ii) If $F \in F[X_o,\ldots,X_n]$ and $Q \in V$ then $F(\varphi_o(Q)) = F_{\#}(Q)$, so if $F(\varphi_o(Q))$ is a unit then $F_{\#}(Q)$ is a unit. This implies that $\alpha(\mathcal{O}_{\varphi_o(Q)}(V^{\#})) \subseteq \mathcal{O}_Q(V)$. To see that $\alpha(\mathcal{O}_{\varphi_o(Q)}(V^{\#})) = \mathcal{O}_Q(V^{\#})$, observe that, in the notation of the proof of part (i), if $B(Q)$ is a unit then so is $B^+(\varphi_o(Q))$; thus if $ab^{-1} \in \mathcal{O}_Q(V)$ then $\alpha^+(b^+)^{-1} \in \mathcal{O}_{\varphi_o(Q)}(V^{\#})$, finishing the proof.

3. Multiplicities and Intersection Numbers

Let R be a commutative regular monically closed ring. If F is a polynomial over R, we call F nowhere constant if F_p is a nonconstant polynomial, for each $p \in \mathrm{Spec}(R)^*$. If F and G are two polynomials over R, we say that F and G are equivalent if $G = u \cdot F$ for some unit $u \in R$.

By an affine plane curve F over R we mean an equivalence class of nowhere constant polynomials in $R[X,Y]$. A projective plane curve F over R is an equivalence class of nowhere constant forms in $R[X,Y,Z]$. The degree of a curve is the degree of a representative polynomial (form, in the projective case) for the curve.

A projective curve with representative F is irreducible if there do not exist nowhere constant forms G,H such that F = GH. Since irreducibility is first-order, it is easy to see that F is irreducible iff F_p is irreducible for all $p \in \mathrm{Spec}^*(R)$. A similar remark applies to affine curves.

If F is an affine (resp. projective) plane curve and $P \in A^2(R)$ (resp. $P \in P^2(R)$) we define the multiplicity $m_P(F)$ of F at P to be the function on $\mathrm{Spec}^*(R)$ whose value at $p \in \mathrm{Spec}^*(R)$ is $m_{P_p}(F_p)$, the multiplicity of F_p at P_p. (Here F_p is the curve over R_p given by the canonical images at p of all the representatives for F.)

Proposition II.3.1. Consider plane curves f over fields. For any integer m, the notion "f has multiplicity m at P" is definable in field theory by a formula involving the coefficients of a representative for f and a set of coordinates for P (in fact the formula is in the ring language).

Proof. We have only to say that all the derivatives of (a representative for) f of order less than m vanish at P and one of the m-th order derivatives of f does not vanish at P.

Remark. Obviously whether or not the formula holds is independent of which representative of f we pick. In the projective case it is also independent of the choice of projective coordinates for P.

Corollary II.3.2. For any affine (resp. projective) plane curve over R and $P \in A^2(R)$ (resp. $P \in P^2(R)$), the function $m_P(F)$ is locally constant.

Proof. Apply the proposition and I.2.4(ii).

Now let $P \in P^2(R)$. It is easy to see that there is a line $L : aX + bY + cZ = 0$, $a,b,c \in R$ such that if $P = (p_1,p_2,p_3)$ then $ap_1 + bp_2 + cp_3$ is a unit in R, i.e. P is nowhere on L. If now $F \in R[X,Y,Z]$ is a form of degree d (recall that d is a locally constant function on $\mathrm{Spec}^*(R)$) then there are a partition $e_1,\dots,e_n$ of 1 and integers $d_1,\dots,d_n$ such that on e_i, F has degree d_i. Thus $L^d = \sum_{i=1}^{n} e_i L^{d_i}$ is a form in $R[X,Y,Z]$, $L^d(P)$ is a unit in R, and F and $F^* L^d$ have the same degree, so $F/L^d \in \mathcal{O}_P(P^2)$. We denote F/L^d by F_*. Notice that if we choose a different L, say L', then $F/(L')^d = (L/L')^d F_*$ and $(L/L')^d$ is a unit in $\mathcal{O}_P(P^2)$. It is not very hard to see that we can perform a projective change of coordinates so that L becomes the line X at infinity; then F_* as we have just defined it corresponds to our original $F_{\#} = F(1,Y,Z)$ under our natural isomorphism of $k(A^2)$ with $k(P^2)$.

Let F and G be projective plane curves, and let $Q \in P^2(R)$. In analogy with the theory over fields, we seek to define the intersection multiplicity $I(Q,F \cap G)$ of F and G at Q to be the "dimension" of the R-module $\mathcal{O}_Q(P^2)/(F_*,G_*)$, where (F_*,G_*) denotes the ideal in $\mathcal{O}_Q(P^2)$ generated by F_* and G_*. (Here we have replaced F and G by representatives, and it doesn't matter which ones.) To define this dimension, we will use the following:

Lemma II.3.3. Ther is a natural R-module embedding

$$\mathcal{O}_Q(P^2)/(F_*,G_*) \longrightarrow \prod_{p \in \operatorname{Spec}^*(R)} \mathcal{O}_{Q_p}(P^2)/(F_{p_*},G_{p_*})$$

as a subdirect sum.

Proof. The general consideration of section I.3 give us an embedding

$$\mathcal{O}_Q(P^2)/(F_*,G_*) \longrightarrow \prod_{p \in \operatorname{Spec}^*(R)} [\mathcal{O}_Q(P^2)/(F_*,G_*)]/p[\mathcal{O}_Q(P^2)/(F_*,G_*)]$$

as a subdirect sum. We claim that for each $p \in \operatorname{Spec}^*(R)$, $[\mathcal{O}_Q(P^2)/(F_*,G_*)]/p[\mathcal{O}_Q(P^2)/(F_*,G_*)]$ is naturally isomorphic to $\mathcal{O}_{Q_p}(P^2)/(F_{p_*},G_{p_*})$. To see this we resurrect the maps ρ_p of section II.2, and consider $\rho_p : \mathcal{O}_Q(P^2) \longrightarrow \mathcal{O}_{Q_p}(P^2)$. Composing this with natural map $\mathcal{O}_{Q_p}(P^2) \longrightarrow \mathcal{O}_{Q_p}(P^2)/(F_{p_*},G_{p_*})$ we get a map whose kernel obviously includes (F_*,G_*). (If L is used in forming F_* and G_*, then $\rho_p(L)(Q_p) \neq 0$ since $L(Q)$ is a unit, so $\rho_p(F_*)$ and $\rho_p(G_*)$ generate (F_{p_*},G_{p_*}).) Thus we have a map

$$\mathcal{O}_Q(P^2)/(F_*,G_*) \longrightarrow \mathcal{O}_{Q_p}(P^2)/(F_{p_*},G_{p_*}),$$

and the kernel of this map clearly includes $p[\mathcal{O}_Q(P^2)/(F_*,G_*)]$. But suppose $x \in \mathcal{O}_Q(P^2)$ represents an element $\bar{x}$ of the kernel. Then $\rho_p(x) \in (F_{p_*},G_{p_*})$, so there is $y \in (F_*,G_*)$ such that $\rho_p(y) = \rho_p(x)$ (recall that ρ_p is surjective, and, again, $\rho_p(F_*)$ and $\rho_p(G_*)$

generate (F_{p_*}, G_{p_*})). Thus $x = y + z$, where $z \in \ker(\rho_p)$. But $\ker(\rho_p | \mathcal{O}_Q(P^2)) = p\,\mathcal{O}_Q(P^2)$ by Theorem II.2.3, so, taking residues modulo (F_*, G_*), $\bar{x} \in p[\mathcal{O}_Q(P^2)/(F_*, G_*)]$. This finishes the proof.

We now define the dimension of the R-module $\mathcal{O}_Q(P^2)/(F_*, G_*)$ — and hence the intersection number $I(P, F \cap G)$ of F and G at P — to be the function on $\mathrm{Spec}^*(R)$ whose value at p is $\dim_{R/p} \mathcal{O}_{Q_p}(P^2)/(F_{p_*}, G_{p_*})$. The values of this function are either nonnegative integers or ∞.

__Theorem II.3.4.__ $I(P, F \cap G)$ is a locally constant function on $\mathrm{Spec}^*(R)$.

To prove this it suffices to prove

__Theorem II.3.5.__ If m is a nonnegative integer or ∞, there is a formula $\varphi(\vec{x}, \vec{y}, \vec{v})$ of ring language such that for any algebraically closed field k, any projective plane curves F,G over k, and any $P \in P^2(k)$, $k \models \varphi$ (coefficients of F, coefficients of G, coordinates of P) iff $\dim_k \mathcal{O}_P(P^2)/(F_*, G_*) = m$.

__Remark.__ Again it will be irrelevant which homogeneous coordinates we use for P and which representatives we pick for F and G.

__Proof of the theorem.__ If $m = \infty$ let φ say that F and G have a common component through P. Now assume $m \neq \infty$.

The statement that $\dim_k \mathcal{O}_P(P^2)/(F_*, G_*) = m$, i.e. the intersection number of F and G at P is m, is ([14], p. 44) equivalent to the assertion that there is a projective change of coordinates T such that, if F^T and G^T denote the transforms of F and G by T, (i) at least one of F^T and G^T does not pass through (0,0,1), (ii) no two points of intersection of F^T and G^T are collinear with (0,0,1) and (iii) the resultant $R_Z(F^T, G^T)$ of F^T and G^T with respect to Z has multiplicity m at the point (a,b) whose coordinates are the first two coordinates of $T^{-1}(P)$. (Recall that F^T is obtained by simultaneously substituting $a_{11}X + a_{12}Y + a_{13}Z$ for X, $a_{21}X + a_{22}Y + a_{23}Z$ for Y, and $a_{31}X + a_{32}Y + a_{33}Z$ for Z, in F.) We claim that this assertion is first-order. For we have only to say that there exist elements $a_{11}, a_{12}, a_{13}, a_{21}, a_{22}, a_{23}, a_{31}, a_{32}, a_{33}$ of k such that $\det(a_{ij}) \neq 0$ and (i) $F(a_{13}, a_{23}, a_{33}) \neq 0$ or $G(a_{13}, a_{23}, a_{33}) \neq 0$,

(ii) $\forall a,b,c(a\neq 0 \vee b\neq 0 \vee c\neq 0 \longrightarrow \forall x,y,z,u,v,w\ \{[(x\neq 0 \vee y\neq 0 \vee z\neq 0) \wedge$
$(u\neq 0 \vee v\neq 0 \vee w\neq 0) \wedge F^T(x,y,z) = 0 \wedge G^T(u,v,w) = 0 \wedge ax + by + cz = 0$
$\wedge\ au + bv + cw = 0] \longrightarrow c \neq 0\})$, and (iii) $m_{(a,b)}R_Z(F^T, G^T) = m$.

Notice that (iii) is first-order by II.3.1. This concludes the proof.

Remark. We have employed the equivalence between the modern and classical definitions of intersection number. Using Beth's definability theorem, one can argue directly that the modern definition is first-order, by making use of the axioms for intersection numbers ([5], pp. 74-75). The resulting proof is much longer and much less clear than the above.

We can now state an analogue of Bezout's Theorem for the theory over fields. Recall that Bezout's Theorem states that if F and G are projective plane curves of degrees m and n respectively over an algebraically closed field k, and F and G have no common components, then the sum of all the intersection numbers $I(P, F\cap G)$ for all P in $P^2(k)$ is mn. In our setting the number of points of intersection in $P^2(R)$ of two plane curves F and G may of course be infinite. However, we have

Theorem II.3.6. Let F and G be projective plane curves over R, of degrees m and n respectively, such that for every $p \in \mathrm{Spec}^*(R)$, F_p and G_p have no common component over R_p. Then

(i) there exists a finite set $\{P_1,\ldots,P_M\}$ of intersection points of F and G in $P^2(R)$ such that for any $P \in F \cap G$ there exists a partition $e_1, \ldots, e_{j_n}$ of 1 such that $P = \sum_{i=1}^{M} e_i P_i$, and

(ii) for any such finite set $\{P_1, \ldots, P_{j_n}\}$,

$$\sum_{i=1}^{M} \left(\prod_{j<i} (P_i - P_j)^* \; I(P_i, F \cap G) \right) = mn,$$

an equality between locally constant functions on Spec(R).

Remarks. (i) The equation $P = \sum_{i=1}^{M} e_i P_i$ is strictly speaking an abuse of notation, but it should be clear what is meant: if we replace the P_i's by representatives then the indicated sum represents an element of $P^2(R)$, which does not depend on the choice of representatives for the P_i's, and this element is in fact P.

(2) $(P_i - P_j)^*$ denotes the idempotent in R on which P_i and P_j are projectively distinct. Namely, if $(P_{i,1}, P_{i,2}, P_{i,3})$ and $(P_{j,1}, P_{j,2}, P_{j,3})$ are chosen coordinates of P_i and P_j, then $(P_i - P_j)^*$ is the idempotent on which

$\forall x (x \neq 0 \longrightarrow \bigvee_{\alpha=1}^{3} (x P_{i,\alpha} \neq P_{j,\alpha}))$ holds.

Proof of II.3.6. Denote by M the maximum value attained by mn on $\text{Spec}^*(R)$. We have a partition $f_1, \ldots, f_M$ of 1 such that f_i is the idempotent on which F and G have precisely i projectively distinct points of intersection. (We have used Bezout's Theorem.) An easy piecing-together argument shows that for $1 \leq i \leq M$ there are points $P_{i,1}, \ldots, P_{i,i}$ in $P^2(R)$ such that for any $p \in f_i$, $(P_{i,1})_p, \ldots, (P_{i,i})_p$ are precisely the distinct intersection points of F_p and G_p in $P^2(R_p)$.

Now for $1 \leq i \leq M$ define

$$P_i = \sum_{j \leq i} f_j P_{j,j} + \sum_{i < j \leq M} f_j P_{j,i} .$$

Then for any $P \in F \cap G$ and any $p \in \mathrm{Spec}^*(R)$, P_p is one of $(P_1)_p, \ldots, (P_M)_p$. From this it is easy to see that $\{P_1, \ldots, P_M\}$ satisfy (i).

It is immediate from Bezout's Theorem and our definition of intersection numbers that (ii) holds.

II.4 Resolution of Singularities

In this section we shall use the resolution of singularities for projective curves over fields in order to reduce any projective curve over a model R of K^* to a non-singular curve. This result provides the basis for the discussion in the next section.

Let $\langle R, * \rangle$ be a monically closed regular *-ring and V an affine variety in $A^n(R)$. Then we define $\dim(V) = \dim(I(V))$; if W is a projective variety in $P^n(R)$ we define the (projective) dimension of W by $\dim(W) = \dim(I(W))-1$. Thus for every $p \in \mathrm{Spec}^*(R)$ $\dim(V)(p) = \dim(V_p)$ and $\dim(W)(p) = \dim(W_p)$, where $\dim(V_p)$ and $\dim(W_p)$ are defined as in the field theory (see Zariski-Samuel [17] for the projective case). We call $V(W)$ a <u>curve</u> if $\dim(V) \equiv 1$ $(\dim(W) \equiv 1)$.

<u>Lemma II.4.1</u>. Let $V = V(I)$ be a projective variety in $P^2(R)$. Then V is an irreducible projective curve as defined in section II.2 iff V is a projective curve as defined above.

<u>Proof</u>. If V is a plane projective curve, then V_p is a plane projective curve over R_p and so by Fulton [5], a projective variety of dimension 1 for every $p \in \mathrm{Spec}^*(R)$. Conversely, if V is a projective variety of

dimension 1, let $f_1,\dots,f_k$ be forms with coefficients $\vec{a}_1,\dots,\vec{a}_k$ generating $I(V)$. By Fulton [5], V_p is a plane projective curve over R_p for each $p \in \mathrm{Spec}^*(R)$, and so there exist forms $g^{(p)} \in R[X_1,X_2,X_3]$ of degree d_p such that $(g^{(p)})_p$ generates $I(V_p)$. Let $\vec{b}^{(p)}$ be the coefficients of $g^{(p)}$ and $\psi_{d_p}(\vec{b}^{(p)};\vec{a}_1,\dots,\vec{a}_k)$ a first-order formula such that ψ_{d_p} holds at $q \in \mathrm{Spec}^*(R)$ iff $(g^{(p)})_q$ is irreducible and $V((g^{(p)})_q) = V(I_q)$. Then ψ_{d_p} holds at $p \in \mathrm{Spec}^*(R)$, and hence by I.2.3 on a clopen neighborhood $U^{(p)}$ of p. We conclude as in the proof of the Compactness principle I.2.2 that there exists a form $g \in R[X_1,X_2,X_3]$ such that $V(g) = V(I)$ and g is everywhere irreducible.

A corresponding lemma holds for affine curves.

Next we consider the singularities of affine and projective varieties. Let $V \subseteq A^n(R)$ be an affine variety, $Q \in A^n(R)$, and let $F_1,\dots,F_k$ be generators for the ideal $I(V)$. Then we define the <u>multiplicity</u> of V at Q, $m_Q(V) : \mathrm{Spec}^*(R) \longrightarrow \mathbb{Z}$ as follows:

$m_Q(V) \equiv -1$, if $V = \emptyset$

$m_Q(V)(p) = 0$, if $Q_p \notin V_p$;

$m_Q(V)(p) = n + 1 - \dim(V)(p) - \mathrm{rank}\,(\partial F_i/\partial X_j(Q)(p)$, otherwise.

Let $\vec{c}$ be the coordinates of Q and $\vec{a}_i$ the coefficients of F_i, $1 \le i \le k$. Using the results of I.2 and I.3 we can write down for every $r \in \mathbb{Z}$ a first order formula $\mu_r(\vec{c},\vec{a}_1,\dots,\vec{a}_k)$ in the ring language such that μ_r holds at $p \in \mathrm{Spec}^*(R)$ iff $m_Q(V)(p) = r$. Consequently $m_Q(V)$ is a locally constant function. Notice also that $m_Q(V)(p)$ coincides with the multiplicity $m_{Q_p}(V_p)$ as defined in the field theory (see Weil [16]). For a projective variety $W \subseteq P^n(R)$ and

$Q \in P^n(R)$ we define the multiplicity of W at Q, $m_Q(W)$; $\mathrm{Spec}^*(R) \longrightarrow \mathbb{Z}$ as follows: For $0 \leq j \leq n$ let $\varphi_j : A^n(R) \longrightarrow P^n(R)$ be the embedding defined in II.2 and let $W_{\#j}$, $Q_{\#j}$ be the dehomogenizations of W,Q with respect to X_j. Then we set

$m_Q(W) \equiv -1$, if $W = \emptyset$;

$m_Q(W)(p) = m_{Q_{\#0}}(W_{\#0})(p)$ for $Q_p \in (\varphi_0(A^n))_p$,

.

.

.

$m_Q(W)(p) = m_{Q_{\#n}}(W_{\#n})(p)$ for $Q_p \in (\varphi_n(A^n))_p$.

Let $\vec{c}$ be projective coordinates for Q, $F_1,\dots,F_k$ forms generating $I(W)$, and let $\vec{a}_i$ be the coefficients of F_i , $1 \leq i \leq k$. Using the formulas μ_r defined above it is now easy to write down first-order formulas $\mu_r'(\vec{c},\vec{a}_1,\dots,\vec{a}_k)$ in the ring language such that

$\mu_r'(\vec{c},\vec{a}_1,\dots,\vec{a}_k)$ holds at $p \in \mathrm{Spec}^*(R)$ iff $m_Q(W)(p) = r$. Again $m_Q(W)(p)$ coincides with the multiplicity $m_{Q_p}(W_p)$ as defined in the field theory. We call Q a <u>simple</u> point of W if $m_Q(W) \equiv 1$. Every point in W which is not simple will be called a <u>singularity</u> of W. We say "W is non-singular" if every point in W is simple.

<u>Lemma II.4.2</u>. Let $W \subseteq P^n(R)$ be a projective curve, $Q_1,\dots, Q_m \in P^n(R)$, $p \in \mathrm{Spec}^*(R)$ such that $(Q_i)_p$ are the only multiple points of W_p . Then there exists a clopen neighborhood U of p such that for all $q \in U$, $(Q_i)_q$ are the only multiple points of W_q .

<u>Proof</u>. Let $\vec{c}_1,\dots,\vec{c}_m$ be projective coordinates for $Q_1,\dots, Q_m$, $F_1,\dots, F_k$ generators of $I(W)$, $\vec{a}_i$ the coefficients of F_i, $1 \leq i \leq k$, and assume $m_{Q_i}(W)(p) = r_i > 1$. Then the formula

$$\bigwedge_{i=1}^{m} \mu'_{r_i}(\vec{c_i}\ ;\ \vec{a_1},\ldots,\vec{a_k}) \wedge \forall\vec{x} \Big(\Big(\bigwedge_{i=1}^{m} \neg\, \exists y\Big(y \neq 0 \wedge \bigwedge_{i=0}^{n} yx_j = (c_i)_j\Big) \wedge$$

$$\bigvee_{j=0}^{n} x_j \neq 0 \wedge \bigwedge_{i=1}^{k} F_i(\vec{x}) = 0\Big) \longrightarrow \mu'_1(\vec{x}\ ;\ \vec{a_1},\ldots,\vec{a_k})$$

holds at p and hence on a clopen neighborhood U of p.

Corollary II.4.3. Let $W \subseteq P^n(R)$ be a projective curve. Then there exists an integer B such that for all $p \in \mathrm{Spec}^*(R)$ W_p has at most B multiple points.

Proof. By II.4.2 the number of multiple points in W_q is a locally constant function on $\mathrm{Spec}^*(R)$. So by I.2.1 this function assumes only finitely many values.

Assume next that $W = W(F) \subseteq P^2(R)$ is a plane projective curve and $Q \in P^2(R)$. Then we say Q is an ordinary multiple point of W at $p \in \mathrm{Spec}^*(R)$ if Q_p is an ordinary multiple point of W_p in the sense of the field theory (see Fulton [5]).

Lemma II.4.4 Let W be as above and assume that W has only ordinary multiple points at some $p \in \mathrm{Spec}^*(R)$. Then there exists a clopen neighborhood U of P on which W has only ordinary multiple points.

Proof. Recall that Q_p is an ordinary multiple point of W_p if $Q_p \in W_p$ and there exist $m_{Q_p}(W_p)$ many different tangents to W_p at Q_p. Let L be a projective line in $P^2(R)$. Then L_p is a tangent to W_p at Q_p if and only if L_p is not degenerate and $I(Q, L \cap W)(p) > m_Q(W)(p)$. Using the results of II.3 and this section we can write down a first-order formula $o\mu(\vec{a})$ (where $\vec{a}$ are the coefficients of F) such that $o\mu(\vec{a})$ holds at $q \in \mathrm{Spec}^*(R)$ iff W_q has only ordinary multiple points. Since $o\mu(\vec{a})$ holds at $p \in \mathrm{Spec}^*(R)$, it holds on a clopen neighborhood U of p.

We recall that for the theory over fields two projective varieties are birationally equivalent if and only if their function fields are isomorphic. We shall adopt a global form of the latter condition as our definition of birational equivalence.

Let $V \subseteq P^n(R)$, $W \subseteq P^n(R)$ be projective curves. We say V and W are birationally equivalent, if $k(V)$ and $k(W)$ are isomorphic over R.

Theorem II.4.5. V and W are birationally equivalent iff there is a locally compatible system of isomorphisms over R_p between $k(V_p)$ and $k(W_p)$, $p \in \mathrm{Spec}^*(R)$.

The theorem follows immediately from I.3.1.

After these preparations we are now in a position to state the main theorems on resolution of singularities, which are the exact analogues of the corresponding theorems in field theory (see Fulton [5]).

Theorem II.4.6. Every projective curve $W \subseteq P^n(R)$ is birationally equivalent to an irreducible plane projective curve $V \subseteq P^2(R)$.

Theorem II.4.7. Every irreducible plane projective curve $W \subseteq P^2(R)$ is birationally equivalent to an irreducible plane projective curve $V \subseteq P^2(R)$ with only ordinary multiple points for singularities.

Theorem II.4.8. Every irreducible plane projective curve $W \subseteq P^2(R)$ with only ordinary multiple points is birationally equivalent to a non-singular projective curve $V \subseteq P^n(R)$.

The crux of the proof of these theorems is the following lemma.

Lemma II.4.9. Let $W \subseteq P^n(R)$ be a projective curve over a model R of K^*, $p \in \mathrm{Spec}^*(R)$, and $W' \subseteq P^n(R_p)$ a projective curve over R_p which is birationally equivalent to W_p. Then there exists a finitely generated homogeneous ideal $J \subseteq R[X_o,\ldots,X_m]$ and a clopen neighborhood U of p such that $V(J_p) = W'$ and such that for all $q \in U$

(i) J_q is a homogeneous prime ideal in $R_q[X_o,\ldots,X_n]$,

(ii) There exists a compatible system of isomorphisms

$\{\varphi_q : W_q \longrightarrow V(J_q)\}_{q \in U}$ over R on U.

<u>Proof</u>. Let $I = I(W)$, $I' = I(W')$. We may assume modulo a change of variables that $(X_o)_p \notin I_p$ and $(X_o)_p \notin I'$. We denote the dehomogenization with respect to X_o by $(\ )_\#$. Let $F_1,\dots,F_h$ be forms in $R[X_o,\dots,X_m]$ such that $(F_1)_p,\dots,(F_h)_p$ generate I', let $J = (F_1,\dots,F_h)$, $W_{\#p} = (V(I_\#))_p = V(I_{\#p})$, $W'_\# = V(I'_\#)$ and φ: $K(W'_\#) \longrightarrow K(W_{\#p})$ the isomorphism existing by assumption. Denote the image of X_i mod $J_\#$ by $\bar{X}_i \in k(V(J_\#))$ and the image of X_i mod $I_\#$ by $x_i \in k(W_\#)$. Pick $a_1,\dots,a_m \in k(W_\#)$ such that $\varphi((\bar{X}_i)_p) = (a_i)_p$. Then $(a_1)_p,\dots,(a_m)_p$ generate $k(W_{\#p})$ as a field, and so there exist B_j, $C_j \in R[X_1,\dots,X_m]$ such that $(x_j)_p = (B_j(a_1,\dots,a_m)\ C_j^{-1}(a_1,\dots,a_m))_p$ for $1 \leq j \leq n$. Moreover we have $(F_{i\#}(a_1,\dots,a_m))_p = 0$ for $1 \leq i \leq h$, since φ is an isomorphism. So by the results of I.3 and II.2 there exists a clopen neighborhood U of p such that for all $q \in U$:

(o) $(X_o)_q \notin I_q$, $(X_o)_q \notin J_q$

(i) $(x_j)_q = (B_j(a_1,\dots,a_m)\ C_j^{-1}(a_1,\dots,a_m))_q$, $1 \leq j \leq n$,

(ii) $(F_{i\#}(a_1,\dots,a_m))_q = 0$, $1 \leq i \leq h$,

(iii) $\dim(W_\#)(q) = \dim(W_\#)(p)$

(iv) $J_{\#q}$ is a prime ideal, $J_{\#q} \cap R_q = \{0\}$ and $\dim(J_\#)(q) = \dim(J_\#)(p)$.

For $q \in U$ let $\psi_q : R_q[X_1,\dots,X_m] \longrightarrow k(W_{\#q})$ be the homomorphism over R_q mapping X_i onto $(a_i)_q$, $1 \leq i \leq m$. By (i) $k(W_{\#q})$ is the quotient field of $\mathrm{Im}(\psi_q)$, and by (ii) $\ker(\psi_q)$ is a prime ideal containing $J_{\#q}$. So $\dim(\ker(\psi_q)) = \mathrm{tr\,deg}_{R_q}(k(W_{\#q}) = \dim(I_{\#q}) =$

$\dim(I_{\#p}) = \operatorname{tr}\deg_{R_p}(k(W_{\#p})) = \dim(\ker(\psi_p)) = \dim(J_{\#p}) = \dim(J_{\#_q})$, and

hence $\ker(\psi_q) = J_{\#q}$. Thus $\{\psi_q\}_{q \in U}$ induces a compatible system of

isomorphisms $\{\varphi_q : k(V(J_{\#q})) \longrightarrow k(W_{\#q})\}$ over R_q on U with

$\varphi_p = \varphi$. By II.2.5 we have moreover a compatible system of isomorphisms $k(V(J_q)) \cong k(V(J_{q\#})) = k(V(J_{\#q}))$ and $k(W_q) \cong k(W_{q\#}) = k(W_{\#q})$ over R on U, and so there is a compatible system of isomorphisms from $k(W_q)$ onto $k(V(J_q))$ over R on U. This completes the proof of the lemma.

Theorem II.4.6 is now proved as follows.

Let $W \subseteq P^n(R)$. For every $p \in \operatorname{Spec}^*(R)$ there exists an irreducible projective plane curve $W' \subseteq P^2(R_p)$ which is birationally equivalent to W_p (see Fulton [5], Cor. to Prop. 12, p. 155). Applying the lemma we find a clopen neighborhood $U^{(p)}$ of p and a homogeneous finitely generated ideal $J^{(p)}$ as specified in the lemma. The proof of the lemma shows moreover that we can take $J^{(p)}$ as a principal ideal generated by a form $F^{(p)}$. Then there exists a finite subset M of $\operatorname{Spec}^*(R)$ such that $\bigcup_{p \in M} U^{(p)} = \operatorname{Spec}^*(R)$; we may also assume without restriction that the sets $U^{(p)}$ are pairwise disjoint for $p \in M$. We define

$F = \sum_{p \in M} e^{(p)} F^{(p)}$, and $J = (F)$, where $e^{(p)}$ is the idempotent in R

corresponding to $U^{(p)}$. Then J is everywhere prime and by the lemma there is a locally compatible system of isomorphisms between $V(J_q)$ and W_q. Thus $V = V(J)$ is birationally equivalent to W by theorem II.4.5.

The proofs of theorem II.4.7 and theorem II.4.8 are very similar , using theorem 2, p. 177 together with the remark on p. 170 and proposition 1, p. 170, in Fulton [5], respectively. The essential difference is in the selection of the neighborhoods $U^{(p)}$ of $p \in \operatorname{Spec}^*(R)$.

To show theorem II.4.8, for example, we remark first that by II.4.3 there is a uniform bound B on the number of non-simple points of W at $p \in \mathrm{Spec}^*(R)$. The proof of Prop 1, p. 170 in [5], together with the Segre embedding, p. 102 in [5], provides us now with a nonsingular curve $W'^{(p)} \subseteq P^{n(B)}(R)$ birationally equivalent to W_p for every $p \in \mathrm{Spec}^*(R)$, where $n(B)$ is a number independent from p which is determined by the Segre embedding. Using the lemma we find clopen neighborhoods $U^{(p)}$ of $p \in \mathrm{Spec}^*(R)$ and homogeneous ideals $J^{(p)} = \left(F_1^{(p)},\dots,F_{k(p)}^{(p)}\right) \subseteq R[X_o,\dots,X_{n(B)}]$ as before. But this time we cut down each $U^{(p)}$ to a possibly smaller clopen neighborhood $U'^{(p)}$ of p in such a way that $V(J^{(p)})_q$ is a nonsingular projective curve for all $q \in U'^{(p)}$. Replacing the $U^{(p)}$'s by the $U'^{(p)}$'s we find M and $e^{(p)}$ as above and define J as the ideal generated by

$$\{ \sum_{p \in M} F_{i_p}^{(p)} e^{(p)} , 1 \le i_p \le h(p)\}.$$ We conclude as before that $V = V(J)$ satisfies the statement of theorem II.4.8.

The proof of theorem II.4.7 is now left to the reader.

5. Concepts Related to the Riemann-Roch Theorem.

In this section we shall investigate those concepts which enter into a discussion of the Riemann-Roch Theorem.

To begin with, let C be a projective curve over R, $R \models K^*$. Let $X \subseteq P^n(R)$ be a nonsingular model of C, as in section II.4. We wish to define the notion of a *divisor* on X.

Definition. A *divisor* on X is a function which assigns to each $P \in X$ an integer-valued function $D(P)$ on $\mathrm{Spec}^*(R)$, in such a way that

(i) for every $p \in \mathrm{Spec}^*(R)$, if P,Q are two points on X such that $P_p = Q_p$, then $D(P)(p) = D(Q)(p)$,

(ii) each $D(P)$ is locally constant, and

(iii) there exist finitely many points $P_1,\dots,P_k$ in X such that for all $P \in X$, $D(P)(p) = 0$ for any p such that $P_p \neq (P_1)_p,\dots,(P_k)_p$.

Notice that by conditions (i) and (iii), D provides us with a divisor D_p for each $p \in \mathrm{Spec}^*(R)$; $D_p(P_p) = D(P)(p)$ for each $P \in X$. Notice also that, by a piecing-together argument, (iii) is equivalent to its local counterpart (iii)' : for any $p \in \mathrm{Spec}^*(R)$ there exist a finite set $P_1,\dots,P_s$ of elements of X and a neighborhood e_p of p such that for all $P \in X$, $D(P)(q) = 0$ for any $q \in e_p$ such that $P_q \neq (P_1)_q,\dots,(P_s)_q$.

We define the degree $\mathrm{Deg}(D)$ of a divisor D to be the function on $\mathrm{Spec}^*(R)$ whose value at any $p \in \mathrm{Spec}^*(R)$ is $\deg(D_p)$, in the usual sense (i.e. the sum of the coefficients of D_p). It follows from clauses (ii) and (iii) of the definition of a divisor that $\mathrm{Deg}(D)$ is a locally constant function. For, given $p \in \mathrm{Spec}^*(R)$, let $P_1,\dots,P_k$ be as in (ii) and choose a neighborhood e of p such that each of $D(P_1),\dots,D(P_k)$ is constant on e. It is immediate the $\mathrm{Deg}(D)$ is constant on e.

If D_1 and D_2 are divisors then $D_1 \succ D_2$ means $D_1(P) \geq D_2(P)$ for every $P \in X$.

Our first goal is to show that to each unit $z \in k(X)$ (with $k(X)$ as in section II.2) we may associate a divisor $\mathrm{Div}\,(z)$. To do so we define, for a unit $z \in k(X)$ and $P \in X$, the order of z at P to be the function on $\mathrm{Spec}^*(R)$ whose value at p is $\mathrm{ord}_{P_p}(z_p)$, the order of z_p in the discrete valuation ring $\mathcal{O}_{P_p}(X_p)$. (Here we use the results of section II.2). We will denote the order of z at P by $\mathrm{Ord}_P\,(z)$. With this definition we can define the divisor of z

by setting

$$\mathrm{Div}\ (z)\ (P) = \mathrm{Ord}_P\ (z).$$

Notice that $(\mathrm{Div}\ (z))_p = \mathrm{Div}\ (z_p)$ for each $p \in \mathrm{Spec}^*(R)$.

Theorem II.5.1. For any unit $z \in k(X)$, $\mathrm{Div}\ (z)$ is in fact a divisor on X.

Proof. We verify conditions (i)-(iii) of the definition. (i) is trivially satisfied. Next we verify (ii) by proving

Lemma II.5.2. If z is a unit in $k(X)$, $P \in X$, and $p \in \mathrm{Spec}^*(R)$, then there is a neighborhood of p on which $\mathrm{Ord}_P\ (z)$ is constant.

Proof. Let $(P_0, \ldots, P_n)$ be a set of coordinates for P such that $(P_j)_p = 1$ for some j, $0 \leq j \leq n$. Then $P_j = 1$ on a neighborhood e_1 of p. For every $q \in e_1$, $\{(x_i-(P_i)_q\, x_j)/x_j \mid 0 \leq i \leq n,\ i \neq j\}$ generates the maximal ideal $M_{P_q}(X_q)$ of $\mathcal{O}_{P_q}(X_q)$, where $x_i - (P_i)_q x_j$ and x_j are the residues of $X_i - (P_i)_q\, X_j$ and X_j in $\Gamma_h\,(X_q)$. If t is a uniformizing parameter in $\mathcal{O}_{P_p}(X_p)$ then there are elements $a_i \in \mathcal{O}_{P_p}(X_p)$, $0 \leq i \leq n$, $i \neq j$ such that

$$(*) \qquad (x_i-(P_i)_p\, x_j)\ /\ x_j = a_i\ t.$$

Let T, A_i be elements of $\mathcal{O}_P(X)$ such that $T_p = t$ and $(A_i)_p = a_i$ (by II.2), and let L be a line such that P is nowhere on L and L is just X_j on e_1. If $\overline{X_i-P_iX_j}$ and $\bar{L}$ are the residues of $X\text{-}P_iX_j$ and L in $\Gamma_h\,(X)$ then $\overline{X_i-P_iX_j}\ /\ L \in \mathcal{O}_P(X)$ and

$$(**) \qquad (\overline{X_i-P_i\, X_j}\ /\ \bar{L})_q = (x_i-(P_i)_q\, x_j)\ /\ x_j$$

for each $q \in e_1$. Let e_2 be an idempotent in the regular ring $k(X)$ which contains p and on which

$$\overline{X_i - P_i X_j} / \bar{L} = A_i T, \quad 0 \leq i \leq n, \ i \neq j$$

hold identically (by (*) and (**)).

Let $e_3 = e_1 \cap e_2$; then $e_3 \in R$ since R and $k(X)$ have the same idempotents (section II.2). At each $q \in e_3$,

$$(x_i - (P_i)_q x_j) / x_j = (A_i)_q T_q,$$

the $(x_i - (P_i)_q x_j) / x_j$ generate $M_{P_q}(X_q)$ (as remarked above), and each $(A_i)_q \in \mathcal{O}_{P_q}(X_q)$ (by section II.2). Thus at each $q \in e_3$, T_q is a uniformizing parameter in $\mathcal{O}_{P_q}(X_q)$.

Now suppose $\mathrm{ord}_{P_p}(z_p) = n$. Then there is a unit $u \in \mathcal{O}_{P_p}(X_p)$ such that

$$z_p = u (T_p)^n \quad (\text{if } n \geq 0) \quad \text{or} \quad z_p(T_p)^{-n} = u \quad (\text{if } n < 0).$$

There is a unit $U \in \mathcal{O}_P(X)$ such that $U_p = u$. Let e_4 be a neighborhood of p on which one of the equations $z = U T^n$ or $z(T)^{-n} = U$, whichever is appropriate, holds **identically**. Let $e = e_3 \cap e_4$. Then $\mathrm{ord}_{P_q}(z_q) = n$ for each $q \in e$. This proves the Lemma.

To verify (iii) we will actually check (iii)'. Fix $p \in \mathrm{Spec}^*(R)$.

Let C' be a plane curve with only ordinary multiple points which is birationally equivalent to X. Then $k(X)$ and $k(C')$ are isomorphic over R, and z corresponds to an element $g/h \in k(C')$, where g,h are the residues in $\Gamma(C')$ of forms $G(X,Y,Z)$, $H(X,Y,Z)$. By the results of section II.4, $\mathrm{Ord}_P(z) = \mathrm{Ord}_P(G) - \mathrm{Ord}_P(H)$ for every P, so it suffices to find a set $P_1, \ldots, P_k$ and a neighborhood e of p

such that for any $q \in e$ $\mathrm{ord}_{P_q}(G_q) = 0$ unless P_q is one of $(P_1)_q, \ldots, (P_k)_q$; for we then find similarly $Q_1, \ldots, Q_{k'}$ and e^1 for H and consider the neighborhood $e \cap e'$ and the set $\{P_1, \ldots, P_k, Q_1, \ldots, Q_{k'}\}$. (For $P \in X$, $\mathrm{Ord}_P(G)$ is the function $\mathrm{Ord}_P(g/\ell^d)$, where $d = \deg G$ and ℓ is the residue in $\Gamma(C')$ of a suitably chosen line (see section II.3), similarly for $\mathrm{Ord}_P(H)$.) Observe that $\mathrm{Ord}_P(G)$ and $\mathrm{Ord}_P(H)$ assume only nonnegative values.

Now let $P_1, \ldots, P_k \in X$ be points such that $(P_1)_p, \ldots, (P_k)_p$ are the distinct points of X_p where $\mathrm{ord}\,(G_p) > 0$. By lemma II.5.2 let e_1 be a neighborhood of p on which $\mathrm{Ord}_{P_1}(G), \ldots, \mathrm{Ord}_{P_k}(G)$ are all constant; we can assume the P_j's are distinct on e_1 . If $n = \deg(C')$, $d = \deg(G)$, then for every $q \in \mathrm{Spec}^*(R)$ we have

$$\sum_{P_q \in X_q} \mathrm{ord}_{P_q}(G_q) = \sum_{Q_q \in C'_q} I(Q_q, C'_q \cap G_q) = nd,$$

by Prop. 2, page 182 of [5] and Bezout's Theorem. Let e_2 be a neighborhood of p on which n and d have constant values n_o and d_o. Let $e = e_1 \cap e_2$. Then for every $q \in e$,

$$\sum_{P_q \in X_q} \mathrm{ord}_{P_q}(G_q) = n_o d_o \quad \text{and}$$

$$\sum_{i=1,\ldots,k} \mathrm{ord}_{(P_i)_q}(G)_q = n_o d_o ,$$

so since for any P $\mathrm{ord}_{(P_q)}(G_q) \geq 0$, $\mathrm{ord}_{(P_q)}(G_q) > 0$ implies that P_q must be one of $(P_1)_q, \ldots, (P_k)_q$. This concludes the proof.

We introduce the notion of a <u>canonical</u> <u>divisor</u> by considering the module $\Omega_R(k(X))$ of differentials on $k(X)$. $\Omega_R(k(X))$ is obtained by taking the free $k(X)$-module F on $\{[x] : x \in k(X)\}$ (where $[x]$ is a symbol corresponding to x) and factoring by the submodule generated by the set

$$\{[x+y] - [x] - [y] \mid x,y \in k(X)\} \cup$$

$$\{[\lambda x] - \lambda[x] \mid x \in k(X), \lambda \in R\} \cup$$

$$\{[xy] - x[y] - y[x] \mid x,y \in k(X)\}$$

(compare [5], p. 204.) We denote the image of $[x]$ in $\Omega_R(k(X))$ by dx.

We recall (I.3.18) that there is a representation of $\Omega_R\, k(X)$ as a subdirect product of $k(X_p)$-modules.

In particular let $\omega \in \Omega_R(k(X))$ be nowhere zero (for example take $\omega = dx$ where $x \in k(X)$ is an element which is everywhere a uniformizing parameter at some $P \in X$ — the existence of such an x follows from the first part of the proof of II.5.2). If ω' is also nowhere zero then there exists a unit $f \in k(X)$ such that $\omega' = f\omega$. For given $p \in \mathrm{Spec}^*(R)$ there exists $f \in k(X)$, $f_p \neq 0$, such that $\omega' = f\omega$ at p and hence (I.3.19) on a neighborhood of p on which we can assume $f \neq 0$; thus a piecing-together argument does the job.

To define the divisor Div (ω) of ω take $P \in X$ and let $T \in k(X)$ be everywhere a uniformizing parameter at P; then $\omega = f\, dT$ for some unit f in $k(X)$; we define $\mathrm{Ord}_P(\omega) = \mathrm{Ord}_P(T)$. It is easy to see that this doesn't depend on the choice of T, using Prop. 7, page 205 of [5]. We let Div (ω) have the value $\mathrm{Ord}_P(\omega)$ at P. Notice that if $\omega = g\, dx$ then $(\mathrm{Ord}_P(\omega))_q = (\mathrm{Ord}_P\, g)_q$ for every q such that $(dx)_q$ is the differential of a uniformizing parameter in $\mathcal{O}_{P_q}(X_q)$.

Div (ω) is called a <u>canonical divisor</u> on X. If ω' is also nowhere zero then as above $\omega' = f\omega$ for a unit $f \in k(X)$, so $\mathrm{Div}(\omega') = \mathrm{Div}(f) + \mathrm{Div}(\omega)$, i.e. $\mathrm{Div}(\omega')$ and $\mathrm{Div}(\omega)$ differ by the divisor of a unit in $k(X)$, i.e. are "linearly equivalent".

Observe that $(\mathrm{Div}(\omega))_p$ is a canonical divisor in the usual sense, for any $p \in \mathrm{Spec}^*(R)$.

<u>Theorem II.5.3</u>. If ω is nowhere zero then $\mathrm{Div}(\omega)$ is a divisor on X.

<u>Proof</u>. We again refer to clauses (i)-(iii) in the definition of "divisor". It is easy to see that condition (i) is satisfied. That (ii) is satisfied follows from the fact that for any $P \in X$, $\mathrm{Ord}_P(\omega)$ is just $\mathrm{Ord}_P(f)$ for some unit f in $k(X)$, so II.5.2 applies.

Now we check condition (iii)'. Fix $p \in \mathrm{Spec}^*(R)$. We claim that there is a neighborhood e of p and a finite set S of points in X, and a unit f in $k(X)$ such that for $q \in e$, $(\mathrm{Ord}_P(\omega))_q = (\mathrm{Ord}_P(f))_q$ for all P such that $P_q \notin \{Q_q : Q \in S\}$. Then the union of S and a set $\{P_1,\ldots,P_k\}$ which satisfies (iii) for $\mathrm{Div}(f)$ provides us with a finite set of P's which satisfies (iii)' for $\mathrm{Div}(\omega)$ on e.

To verify the claim we will for convenience work in the affine setting; it is not hard to transfer the result to the projective situation. So we assume we have an affine variety $V = V(F_1,\ldots,F_m)$ in $A^n(R)$.

Now we can assume that for all except finitely many $\vec{a} \in V_p$, the first $n-1$ columns of the matrix

$$\begin{pmatrix} \frac{\partial F_1}{\partial X_1}(\vec{a}) & \cdots & \frac{\partial F_1}{\partial X_n}(\vec{a}) \\ \cdot & & \cdot \\ \cdot & & \cdot \\ \cdot & & \cdot \\ \frac{\partial F_m}{\partial X_1}(\vec{a}) & \cdots & \frac{\partial F_m}{\partial X_n}(\vec{a}) \end{pmatrix}$$

are linearly independent. For the dependence of any set of n-1 columns of $\left\{\frac{\partial F_i}{\partial X_j}\right\}$ at a point $\vec{a}$ is expressed by the vanishing of all the determinants of $(n-1)x(n-1)$ submatrices formed using these n-1 columns. This vanishing occurs either at finitely many points or on all of V_p. If for each choice of the n-1 columns the vanishing occurred on all of V_p then every point $\vec{a}$ of V_p would be a multiple point, since the matrix $\left\{\frac{\partial F_i}{\partial X_j}(\vec{a})\right\}$ would have rank $\leq n-2$ for each $\vec{a} \in V_p$; but this cannot happen. So for convenience we assume that except at $\vec{a_1},\ldots,\vec{a_r} \in V_p$, the first n-1 columns of $\left\{\frac{\partial F_i}{\partial X_j}(\vec{a})\right\}$ are independent.

Let $A_1,\ldots, A_r \in V$ be such that $(A_i)_p = \vec{a_i}$, $1 \leq i \leq r$.

Let c be a neighborhood of p on which

$$\forall \vec{x} \left(\left(\bigwedge_{i-1}^{m} F_i(\vec{x}) \wedge \bigwedge_{i=1}^{r} \vec{x} \neq A_i\right) \longrightarrow \varphi(\vec{x})\right)$$

holds identically, where φ says that one of the $(n-1)x(n-1)$ submatrices of the first n-1 columns of $\left\{\frac{\partial F_i}{\partial X_j}(\vec{x})\right\}$ has nonzero determinant.

Now we assert that for any $q \in e$ and $A = (b_1,\dots,b_n)$ in V, the residue $x_n-(b_n)_q$ of $X_n-(b_n)_q$ in $\Gamma(V_q)$ is a uniformizing parameter in $\mathcal{O}_{A_q}(V_q)$ provided that A_q is none of $(A_1)_q,\dots,(A_r)_q$.

Assuming this for the moment, we can conclude the proof. For if $\bar{X}_n$ is the residue of X_n in $\Gamma(V)$ then $d\bar{X}_n$ is nowhere zero on e. (To see that $(d\bar{X}_n)_q \neq 0$, pick A such that $A_q \neq (A_1)_q,\dots,(A_1)_q$. Then $d(x_n-(b_n)_q) \neq 0$ since $x_n-(b_n)_q$ is a uniformizing parameter in $\mathcal{O}_{A_q}(V_q)$, so $(d\bar{X}_n)_q = dx_n = d(x_n-(b_n)_q) \neq 0$.) Thus we have $\omega = f\, d\bar{X}_n$ on e for some unit $f \in \mathbf{k}(V)$. Thus for any P, $\mathrm{ord}_{P_q}(\omega) = \mathrm{ord}_{P_q}(f)$ for any $q \in e$ such that $P_q \neq (A_i)_q$, $1 \leq i \leq r$.

To verify the assertion let $q \in e$, $A \in V$, and suppose $A_q \neq (A_1)_q,\dots,(A_r)_q$. Then by definition of e the first $(n-1)$ columns of $\left\{\frac{\partial F_i}{\partial X_j}\ (A_q)\right\}$ are linearly independent.

Say A_q is $(c_1,\dots,c_n)$. Then at q we can write

$$
\begin{aligned}
F_1 &= F_1(c_1,\dots,c_n) + \sum_{j=1}^{n} (X_j-c_j)\, H_{1_j}\\
&\ \vdots\\
F_m &= F_m(c_1,\dots,c_n) + \sum_{j=1}^{n} (X_j-c_j)\, H_{m_j}
\end{aligned}
$$

for polynomials H_{i_j} with $H_{i_j}(\vec{c}) = \partial F_i/\partial X_j(\vec{c})$. In $\Gamma(V_q)$ this becomes

$$
(*)\qquad
\begin{aligned}
0 &= \sum_{j=1}^{n} (x_j-c_j)\, H_{1_j}(\vec{x})\\
&\ \vdots\\
0 &= \sum_{j=1}^{n} (x_j-c_j)\, H_{m_j}(\vec{x})\,.
\end{aligned}
$$

We claim that this system can be transformed into a system

$$(**)\qquad \begin{aligned} 0 &= (x_1-c_1) - d_1(x_n-c_n) \\ &\vdots \\ 0 &= (x_{n-1}-c_{n-1}) - d_{n-1}(x_n-c_n) \end{aligned}$$

where $d_1,\ldots,d_{n-1} \in \mathcal{O}_{A_q}(V_q)$, so that x_n-c_n is in fact a uniformizing parameter in $\mathcal{O}_{A_q}(V_q)$. The transformation of the system is possible because the independence of the first $n-1$ columns of

$\left\{\frac{\partial F_i}{\partial X_j}(\vec{c})\right\} = \left\{H_{i_j}(\vec{c})\right\}$ guarantees that by a sequence of

elementary row operations we can transform $\{H_{i_j}(\vec{c})\}$ into

$$\begin{pmatrix} 1 & & 0 & \alpha_1 \\ & \ddots & & \vdots \\ 0 & & 1 & \alpha_{n-1} \\ 0 & \cdots & & 0 \\ \vdots & & & \vdots \\ 0 & \cdots & & 0 \end{pmatrix}$$

Since a rational function having nonzero value at $\vec{c}$ corresponds to its image in $k(V_q)$ being a unit in $\mathcal{O}_{\vec{c}}(V_q)$, a corresponding sequence of operations transforms $(*)$ to $(**)$.

Now for any divisor D on X we introduce the R-module $L(D) = \{z \in k(X) : \mathrm{Div}(z+(1-z^*)) \succ -D \text{ on } z^*\}$. At any $p \in \mathrm{Spec}^*(R)$, $L(D)$ gives us $L(D_p) = \{z \in k(X_p) : z = 0 \text{ or } \mathrm{div}(z) \succ -D_p\}$, the usual vector space over R_p from the theory over fields. We define <u>dimension</u> $\ell(D)$ of $L(D)$ to be the function on $\mathrm{Spec}^*(R)$ such that $\ell(D)(p) = \ell(D_p)$, the dimension of $L(D_p)$, for each $p \in \mathrm{Spec}^*(R)$.

We will show that $\ell(D)$ is locally constant; but first we introduce the notion of the genus of X. The definition should by this time be no surprise: the genus of X is the function on $\mathrm{Spec}^*(R)$ whose value at p is the genus of X_p.

Theorem II.5.4. The genus g of X is a locally constant function.

Proof. Let C' be a plane model of $k(X) = k(C)$ with only ordinary multiple points, as in section II.4. Then by Prop. 5, p. 199 of [5], we have for every $q \in \mathrm{Spec}^*(R)$

$$g(q) = \frac{(n(q)-1)(n(q)-2)}{2} - \sum_{Q \in C'_q} \frac{r_Q(r_Q-1)}{2} ,$$

where $n(q) = \deg(C'_q)$ and $r_Q = m_Q(C'_q)$.

Let $p \in \mathrm{Spec}^*(R)$ and suppose $P^1,\ldots,P^m$ are points on C' such that $P^1_p,\ldots,P^m_p$ are the nonsimple points on C'_p. By II.3.1, let φ be a formula which says "for each i, $1 \leq i \leq m$, the multiplicity of C' at P^i is the integer $m_{P^i_p}(C'_p)$; for all points P other than $P^1,\ldots,P^m$, the multiplicity of C' at P is 1; and $P^1,\ldots,P^m$ are distinct". Let e_1 be a neighborhood of p on which φ holds identically. Then if $q \in e_1$, $P^1_q,\ldots,P^m_q$ are precisely the nonsimple points of C'_q. Thus if $q \in e_1$

$$g(q) = \frac{(n(q)-1)(n(q)-2)}{2} - \sum_{Q = (P^1)_q,\ldots,(P^m)_q} \frac{r_Q(r_Q-1)}{2} =$$

$$= \frac{(n(q)-1)(n(q)-2)}{2} - \sum_{i=1,\ldots,m} \frac{r_{P^i}(q)\left(r_{P^i}(q)-1\right)}{2} ,$$

where $r_{p^i} = m_{p^i}(C')$. Let e_2 be a neighborhood of p on which n is constant, and let $e = e_1 \cap e_2$. Then g is constant on e.

The following lemma will be useful in showing that $\ell(D)$ is locally constant.

Lemma II.5.5. Suppose $z \in k(X)$, D is a divisor, and $p \in \mathrm{Spec}^*(R)$. If $z_p \in L(D_p)$ there is a neighborhood e of p such that $z_q \in L(D_q)$ for all $q \in e$. If $z_p \notin L(D_p)$ there is a neighborhood e of p such that $z_q \notin L(D_q)$ for all $q \in e$.

Proof. First suppose $z_p \in L(D_p)$. Then if $z_p = 0$, choose e as a neighborhood of p on which $z = 0$. If $z_p \neq 0$, then $z' = z + (1-z^*)$ is a unit in $k(X)$ and $z = z'$ on a neighborhood e_1 of p. Then $\mathrm{Div}(z')_q = \mathrm{div}(z_q)$ for all $q \in e_1$. Since D and $\mathrm{Div}(z')$ are divisors we can find a neighborhood e_2 of p and points $P_1,\ldots,P_s \in X$ such that for every $q \in e_2$, all the nonzero values of D_q and $\mathrm{Div}(z')_q$ are assumed at one of $(P_1)_q,\ldots,(P_s)_q$. We know that $\mathrm{Div}(z')_p \succ -D_p$; now let e_3 be a neighborhood of p on which $\mathrm{Div}(z')(P_1),\ldots,\mathrm{Div}(z')(P_s),D(P_1),\ldots,D(P_s)$ are all constant. Let $e = e_1 \cap e_2 \cap e_3$. Then on e $\mathrm{Div}(z')(P_i) \succ -D(P_i)$ for $1 \leq i \leq s$, and by the definition of e_2 this implies that on e $\mathrm{Div}(z')(P) \succ -D(P)$ for all P. Hence for all $q \in e$, $z_q \in L(D_q)$, since $\mathrm{div}(z_q) = \mathrm{Div}(z')_q$.

Now suppose $z_p \notin L(D_p)$. Then $z_p \neq 0$; so again the unit $z' = z + (1-z^*)$ equals z on a neighborhood e_1 of p, and $\mathrm{Div}(z')_q = \mathrm{div}(z_q)$ for all $q \in e_1$. Since $z_p \notin L(D_p)$ there is $P \in X$ such that $\mathrm{Div}(z')(P)(p) < -D(P)(p)$. Let e_2 be a neighborhood of p on which $\mathrm{Div}(z')(P)$ and $D(P)$ are constant. Then if $e = e_1 \cap e_2$, $\mathrm{div}(z_q) = \mathrm{Div}(z')_q \notin L(D_q)$ for $q \in e$. This concludes the proof.

Theorem II.5.6. $\ell(D)$ is locally constant for any divisor D on X.

Proof. We shall use the Riemann-Roch Theorem ([5], p. 210). Take $p \in \mathrm{Spec}^*(R)$. Let e_1 be a neighborhood of p on which both $\mathrm{Deg}(D)$ and the genus g of X are constant. Suppose for the moment that $\mathrm{Deg}(D) \geq 2g - 1$ on e_1. Then at every point in e_1,

$$\ell(D) = \mathrm{Deg}(D) + 1 - g$$

([5], Cor. 2, p. 212). Thus $\ell(D)$ is constant on e_1.

Now drop the assumption that $\mathrm{Deg}(D) \geq 2g - 1$ on e_1. Let $P \in X$. Since g and $\mathrm{Deg}(D)$ are both constant on e_1, there is an integer m such that $\mathrm{Deg}(D+mP) \geq 2g - 1$ on e_1. Then by the above $\ell(D+mP)$ is constant on e_1. Since we may obtain D from $D + mP$ by successively subtracting P, it now suffices to prove that if $\ell(D+P)$ is constant on a neighborhood of p then $\ell(D)$ is constant on a neighborhood of p.

So let e be a neighborhood of p on which g, $\mathrm{Deg}(D)$, and $\ell(D+P)$ are all constant. At each point q of e, $\mathrm{Deg}(D) = \mathrm{Deg}(D+P) - 1$, $\ell(D)$ is either $\ell(D+P)$ or $\ell(D+P)-1$, and if W denotes a canonical divisor then $\ell(W-D)$ is either $\ell(W-D-P)+1$ or $\ell(W-D-P)$. The Riemann-Roch Theorem tells us that

(1) $\ell(D) = \mathrm{Deg}(D) + 1 - g + \ell(W-D)$ and

(2) $\ell(D+P) = \mathrm{Deg}(D+P) + 1 - g + \ell(W-D-P)$.

In particular, (2) implies that

(3) $\ell(W-D-P)$ is constant on e.

Now suppose that $\ell(D_p) = \ell(D_p+P_p) - 1$. Then there is $z \in k(X)$ such that $z_p \in L(D_p+P_p)$ and $z_p \notin L(D_p)$. By II.5.5 there is a neighborhood e' of p such that for all $q \in e'$, $z_q \in L(D_q+P_q)$ and

$z_q \notin L(D_q)$. Hence $\ell(D) = \ell(D + P) - 1$ on e', so $\ell(D)$ is constant on $e \cap e'$.

If on the other hand $\ell(D_p) = \ell(D_p + P_p)$, then by Riemann-Roch $\ell(W_p - D_p) = \ell(W_p - D_p - P_p) + 1$ and by a similar argument to that in the last paragraph $\ell(W - D) = \ell(W - D - P) + 1$ on a neighborhood e" of p. Hence by (3) $\ell(W - D)$ is constant on $e \cap e''$, so by (1) $\ell(D)$ is constant on $e \cap e''$. This finishes the proof.

We observe that, as already noted above, the statement of the Riemann-Roch Theorem applies in our setting, if all the terms involved are interpreted according to our definitions, i.e. as locally constant functions on Spec (R).

6. Concluding Remarks

As we indicated in section II.3, the representation of modules which we have employed was already studied by Pierce [12] for the case of a regular ring, but from a viewpoint slightly different from ours. Pierce considers the regular ring R as a sheaf of fields over a Boolean space. R-modules are then considered as sheaves of modules over this sheaf of fields. Pierce defines a dimension function for such sheaves of modules; the function assigns to each X in the Boolean space the dimension of the stalk at X as a vector space over the corresponding field. Thus this notion corresponds precisely to our notion of dimension for the R-modules we have considered.

Local constancy of the dimension function is equivalent to its continuity as a function from the Boolean space to the integers (with the integers given the discrete topology). Pierce ([12], 15.3) proves that his dimension function is continuous iff the sheaf of modules under consideration is coherent. Thus our "local constancy results" for modules may be viewed as proofs that certain sheaves are coherent. As far as we are aware, the sheaf machinery does not, however, provide easier proofs that our sheaves are in fact coherent.

Similarly, all the *-rings we have considered can be viewed in an appropriate sheaf-theoretic setting. It does not appear that any simplification of the associated local constancy proofs results. And as far as our approach via locally constant properties is concerned, the subdirect product set-up is more appropriate.

References

1. G. M. Bergman, Hereditary Commutative Rings and Centres of Hereditary Rings, Proc. London Math. Soc., 3rd series, vol. 23, 1971. pp. 214-236.

2. __________, Sulle Classi Filtrali di Algebre, Annali universitata di Ferrarra, sezione 7, scienze matematiche, vol. 17, 1971. pp. 35-42.

3. G. L. Cherlin, Algebraically Closed Commutative Rings, J.S.L. vol. 38, 1973. pp. 493-499.

4. P. C. Eklof, Resolutions of Singularities in Prime Characteristic for almost all Primes, AMS Transactions, vol. 146, 1969. pp. 429-438.

5. W. Fulton, Algebraic Curves, W. A. Benjamin, New York, 1969.

6. G. Hermann, Die Frage der endlich vielen Schritte in der Theorie der Polynomideale, Math. Ann. 95, 1926, pp. 736-788.

7. J. Lambek, Lectures on Rings and Modules, Blaisdell, Waltham, Mass., 1966.

8. W. M. Lambert, A Notion of Effectiveness in Algebraic Structures, J.S.L. vol. 33, 1967. pp. 577-602.

9. S. Lang, Introduction to Algebraic Geometry, Interscience, New York, 1958.

10. L. Lipshitz and D. Saracino, The Model Companion of the Theory of Commutative Rings Without Nilpotent Elements, AMS Proceedings, vol. 38, 1973. pp. 381-387.

11. N. H. McCoy, Rings and Ideals, Carus Mathematical Monographs, MAA, 1948.

12. R. S. Pierce, Modules Over Commutative Regular Rings, Memoires of the AMS, no. 70, Providence, R.I., 1967.

13. A. Robinson, Introduction to Model Theory and to the Metamathematics of Algebra, North Holland, Amsterdam, 1963.

14. A. Seidenberg, Elements of the Theory of Algebraic Curves, Addison-Wesley, Reading, Mass., 1968.

15. R. J. Walker, Algebraic Curves, Dover, New York, 1962.

16. A. Weil, Foundations of Algebraic Geometry, revised edition, AMS Colloquium Publications, vol. 29, Providence, R.I., 1962.

17. O. Zariski and P. Samuel, Commutative Algebra, vol. II, van Nostrand, Princeton, New Jersey, 1960.

Existence of rigid-like families of abelian p - groups

S. Shelah

Dedicated to the memory of A. Robinson

ABSTRACT: We prove that for arbitrarily large λ, there are large families of abelian groups, with only the necessary homomorphisms between them.

INTRODUCTION: Here a group means an abelian group. Improving results of Fuchs (see [Fu2], [Fu4]), Shelah [Sh] proved the existence of 2^λ rigid groups of cardinality λ; i.e. for every cardinal λ, there are groups $G_i (i < 2^\lambda)$ each of cardinality λ, such that if h is a non-zero homomorphism from G_i into G_j, then $i = j$, and $h(x) = n x$ for some integer n.

We try to generalize this theorem to separable p - groups.

We cannot have rigid systems of separable p-groups because any basic subgroup of a separable p-group G, is the image of an endomorphism of G. Also the multiplication by a p-adic integer is an endomorphism. Weakening accordingly the notion of rigid systems, we prove existence theorems in §1, §2 (for possible extensions, see a remark at the end of **section 2**).

Pierce [P] asked, and this is repeated in [Fu.2], p.55, problem 55, whether there are essentially indecomposable p-groups of arbitrarily large cardinalities (G is essentially indecomposable if $G = G_1 \oplus G_2$ implies that G_1, or G_2 is bounded). Our result implies a positive answer (because each member of a rigid-like family is essentially indecomposable).

Fuchs [Fu.2], p.55, problem 53 asked to construct large systems of p-groups such that all homomorphisms between different members are small. As a zero-like homomorphism is the same thing as a small homomorphism (as defined in [Fu.1] 46.3 p.195) theorem 5.1 answers this question. The construction in theorem 1.2 gives for $\mu = \lambda^{\aleph_0} = 2^\lambda > 2^{\aleph_0}$, a family of 2^μ separable p-groups of power μ so that any homomorphism between different members has range of power $\leq\lambda$.

We assume knowledge of naive set theory, and of separable p-groups as in [Fu. 1], VI; [Fu.2], XI.

Notation: Let λ, μ, κ denote infinite cardinals, α, β, γ, δ, i, j ordinals, δ a limit ordinal, k, ℓ, m,n, M, N natural numbers or integers, ω the first infinite ordinal. We let η, τ, ν be sequences of ordinals. Let $\ell(\eta)$ be the length of η, $\eta(i)$ its ith element. Let cf[α] be the cofinality of α.

G, H and sometimes K, I, R are groups, h, g are homomorphisms, p, q are prime natural numbers, r a rational or sometimes a p-adic integer. Here a group means a reduced separable p-group.

When notations become complex, $a_i(j)$ is written as $a[j,i]$, a_i as $a[i]$.

1 Rigid-like systems of p-groups

DEFINITION 1.1:

(1) A homomorphism $h:G \to H$ is called zero-like if there are no $m < \omega$ and $a_n \in G$ for $n < \omega$ such that a_n has exponent $n+m$ and $h(a_n)$ has exponent $\geq n+1$. We call h semi-zero-like if there are no $m < \omega$, $a_n^i \in G$ for $n < \omega$, $i < (2^{\aleph_0})^+$, such that a_n^i has exponent $n+m$, $h(a_n^i)$ has order $\geq n+1$, and $p^n h(a_n^i) \neq p^n h(a_n^j)$ for $i \neq j$.

(2) Let G be a subgroup of H, $h:G \to H$ a homomorphism. Then h is called simple if $h = h_1 + h_2$, h_1 is zero-like, h_2 is a multiplication by a p-adic integer. Similarly h is semi-simple when h_1 is semi-zero-like.

(3) A family $\{G_i : i < i_0\}$ (of separable p-groups) is called rigid-like if whenever $h: G_i \to G_j$ is a non-zero-like homomorphism then $i = j$ and h is simple. Similarly a semi-rigid-like family is defined.

DEFINITION 1.2: G is essentially indecomposable if $G = G_1 \oplus G_2$ implies G_1 or G_2 is bounded.

CLAIM 1.1:

(1) Suppose $\bigoplus_{\substack{i<\beta(n)\\ n<\omega}} \langle x_i^n \rangle$ is a basic subgroup of G, x_i^n of exponent $n+1$, $h: G \to H$ a homomorphism. Then h is zero-like iff $f(n) = \min_i\{n -$ "the exponent of $h(x_i^n)$"$\}$ goes to infinity.

(2) Suppose I is a basic subgroup of G, and $|p^m I| \geq \lambda$ for $m < \omega$. Then for any group H of cardinality $\leq \lambda$ there is a zero-like homomorphism from G onto H.

(3) If G belongs to a rigid-like family then G is essentially indecomposable.

(4) If G belongs to a semi-rigid-like family and $G = G_1 \oplus G_2$ then for some $m < \omega$, $\ell = 1$ or 2, $|p^m G_\ell| \leq 2^{\aleph_0}$.

PROOF: Immediate (Part (2) is similar to a theorem in [Fu. 1]).

THEOREM 1.2: Assume $\mu = \lambda^{\aleph_0} = 2^\lambda > 2^{\aleph_0}$.

(A) There is a semi-rigid-like family of 2^μ groups of cardinality μ, with basic subgroups of cardinality $\leq \lambda$.

(B) Moreover if G, H are members of the family, I a pure subgroup of G, closed in it, and $p^m I$ has power $\geq \lambda$ for each m, and $h: I \rightarrow H$ is a non-semi-zero-like homomorphism then $G = H$, and h is semi-simple.

REMARK: In (A) we can demand the basic subgroups have cardinality λ.

PROOF

NOTATION: W.l.o.g. $\chi < \lambda$ implies $\chi^{\aleph_0} < \mu$, hence λ has cofinality ω and $\chi < \lambda$ implies $\chi^{\aleph_0} < \lambda$. So let $\lambda = \sum_{n<\omega} \lambda_n$, $2^{\aleph_0} < \lambda_0 < \lambda_1 < \ldots$, each λ_n regular, $\lambda_n^{\aleph_0} = \lambda_n$. Let G be the group generated by x_i^n, $i < \lambda_n^+$ $0 \leq n < \omega$, $p^{n+1} x_i^n = 0$. Let H be the torsion-completion of G, so each $a \in H$ is of the form $\Sigma k_i^n x_i^n$, where $\{i: k_i^n \neq 0\}$ is finite for each n, and for some m $p^m k_i^n x_i^n = 0$, for every n, i. Let $d(a) = \{x_i^n: k_i^n x_i^n \neq 0, n < \omega, i < \lambda_n^+\}$ and $d_m(a) = \{x_i^n: p^m k_i^n x_i^n \neq 0, p^{m+1} k_i^n x_i^n = 0, n < \omega, i < \lambda_n^+\}$.

If $d(a)$ is infinite $d_*(a)$ is $d_m(a)$ for the maximal m for which $d_m(a)$ is infinite. We attribute properties to a instead of $d(a)$, sometimes. Let d_1, d_2 be almost **disjoint** if $d_1 \cap d_2$ is finite; let d_1 be almost included in d_2 if $d_1 - d_2$ is finite.

Let $X_m = \{x_i^m: i < \lambda_m^+\}$. If $A \subseteq H$, let $PC(A)$ be the smallest subgroup I of H such that $A \subseteq I$, and $\sum_{i,n} pk_i^n x_i^n \in I$, where

$k_i^n x_i^n \neq 0 \Rightarrow p k_i^n x_i^n \neq 0$, implies $\Sigma k_i^n x_i^n \in I$. Clearly I is a pure subgroup of H. Note that if I is any pure subgroup of H then any homomorphism $h: I \to H$ has an extension to a homomorphism $h: H \to H$. If I is dense (in the p-adic topology) the extension is unique. Also each closed pure subgroup I of H is determined by any basic subgroup of it (so its cardinality is either μ or $\leq\lambda$).

Hence, H has $2^\lambda = \mu$ pure closed subgroups, and there are μ homomorphisms from H into H. Let $\{ (h_i, I_i) : i < \mu\}$ be a list of all pairs of homomorphisms $h: H \to H$ and pure closed subgroups I of H, each pair appearing μ times.

We now define by induction on $i < \mu$, $a_i^*, b_i^* \in H$ which will satisfy the following induction assumptions, and then for $C \subseteq \mu$ let $G(C) = PC[G \cup \{a_i^*: i \in C\}]$; our family will be $\subseteq \{G(C): C \subseteq \mu\}$.

The induction assumptions are

(1) $d_*(a_\alpha^*)$ is not almost included in a finite union of $d(a_i^*)$, $d(b_i^*)$, $(i < \alpha)$

(2) a_α^* has exponent $m+1$ when $d_*(a_\alpha^*) = d_m(a_\alpha^*)$.

(3) If for every m $|p^m I_\alpha| \geq \lambda$ and $h_\alpha | I_\alpha$ is not semi-simple <u>then</u> $h(a_\alpha^*) = b_\alpha^*$ and $b_\alpha^* \notin PC[G \cup \{a_i^*: i \leq \alpha\}]$

(4) If for every m $|p^m I_\alpha| \geq \lambda$ and $h_\alpha | I_\alpha$ is semi-simple but not zero-like <u>then</u> $h_\alpha(a_\alpha^*) = b_\alpha^*$, $b_\alpha^* \notin PC[G \cup \{a_i^*: i < \alpha\}]$

Notice that (1), (2) implies $\{a_\alpha^* + G : \alpha < \mu\}$ is an independent family. We first prove:

<u>CLAIM 1.3</u>: Suppose μ_n are regular cardinals, $\mu_n \leq \mu_{n+1}$, and $\kappa < \mu_n \to \kappa^{\aleph_0} < \mu_n$.

(1) If K is a subgroup of H, and $|p^m K| \geq \Sigma_n \mu_n$ for any m, <u>then</u> we can find $y_i^n \in K$ $(n < \omega, i < \mu_n)$ which are pairwise

disjoint, and y_i^n has exponent $\geq n+1$. Hence we can assume the $d(y_i^n)$'s are disjoint to some prescribed set of cardinality $< \mu_n$.

(2) Suppose $Y^n \subseteq K$, and $z_1 \neq z_2 \in Y_n$ implies $p^n z_1 \neq p^n z_2$, and $|Y^n| = \mu_n$. Then we can find y_i^n which is $z_1 - z_2$ for some $z_1, z_2 \in Y^n$; for $n < \omega$, $i < \mu_n$, such that these y_i^n's satisfy the conclusion of (1).

PROOF:

(1) As $|p^m K| \geq \sum_n \mu_n$, we can find $Y^m \subseteq K$ such that $z_1, z_2 \in Y^m \to p^m z_1 \neq p^m z_2$, hence it suffices to prove (2).

(2) By Erdös and Rado [ER], (as $|Y^n| = \mu_n$, μ_n regular, $\kappa < \mu_n \to \kappa^{\aleph_0} < \kappa_n$ and as $d(z)$ is countable for $z \in Y^n$) there are $Y_n \subseteq Y^n$, $|Y_n| = \mu_n$ and a set d^n such that for $z_1 \neq z_2 \in Y_n$, $d(z_1) \cap d(z_2) = d^n$. As necessarily $\mu_n > 2^{\aleph_0}$, we can assume that if $x_j^m \in d^{n(0)}$, $z_1, z_2 \in Y_{n(0)}$, $z_\ell = \sum_{n,i} k_i^{n,\ell} x_i^n$ then $k_j^{m,1} = k_j^{m,2}$. Now notice that if z_ℓ $\ell = 1,\ldots,4$, are distinct members of Y_n, then $p^n(z_1 - z_2) \neq p^n(z_3 - z_4)$.

Define by induction on $\alpha < \sum_n \mu_n$ and on $m < \omega$, the y_α^m $(\alpha < \mu_m)$. If we have defined y_β^n for $\beta < \alpha$, $\beta < \mu_n$ then for $\beta = \alpha < \mu_n$, $n < m$ we define y_α^m as follows:

Clearly the number of $z \in Y_m$ such that $d(z) \cap d(y_\beta^n) \not\subseteq d^m$ for some n, β as above, is $\leq |\alpha| + \aleph_0 < \mu_m$, hence choose $z_1, z_2 \in Y_m$ which do not satisfy it, and let $y_\alpha^m = z_1 - z_2$.

* * * * *

Suppose we have defined a_j^*, b_j^* for $j < \alpha$, and we shall define them for α.

CASE I: $h_\alpha | I_\alpha$ is semi-zero-like or $p^n I_\alpha$ has power $< \lambda$ for some n.

If for every m, $p^m I_\alpha$ has power $\geq \lambda$ let $K = I_\alpha$, otherwise let $K = H$. Then by Claim 1.3 we can find in

K elements y^n_i, $i < \lambda^+_n$, $n < \omega$, y^n_i of exponent $n+1$, the $d(y^n_i)$'s pairwise disjoint. For any $\eta \in \prod_{n<\omega} \lambda^+_n$ let $y_\eta = \prod_{n<\omega} p^n y^n_{\eta(n)} \in K$, so we can find a_i $i < \mu$, such that $pa_i = 0$ and $d_*(a_i) = d(a_i)$, and no $d(a_i)$ is almost included in a finite union of $d(a_j)$ $j \neq i$, $j < \mu$. Hence for any set $A \subseteq H$, of power $<\mu$, some $d(a_i)$ is not almost included in a finite union of $d(a)$, $a \in A$. (Otherwise for each a_i there is a corresponding finite set $A_i \subseteq A$; so for some A^* $\{i < \mu : A_i = A^*\}$ has power $> 2^{\aleph_0}$, so a countable set has $> 2^{\aleph_0}$ distinct subsets, contradiction). So we can define $a^* \in \{a_i : i < \mu\}$ to satisfy (1), (2), and b^* so that $d_*(b^*)$ is not almost included in a finite union of $d(a^*_i), d(b^*_j)$ $(i \leq \alpha, j < \alpha)$. Clearly (3), (4) are satisfied.

<u>CASE II</u>: Not Case 1, but $p^m K_\alpha$ has power $\geq \lambda$ for every m, where K_α is the kernel of $h_\alpha | I_\alpha$.

Let the image of $h_\alpha | I_\alpha$ be R_α and as not case 1, $h_\alpha | I_\alpha$ is not semi-zero-like; hence there are $m < \omega$, $z^n_i \in K$ of exponent $n+m$ for $i < (2^{\aleph_0})^+$, such that $p^n y^n_i \neq p^n y^n_j$ for $i \neq j$ where $y^n_i = h_\alpha(z^n_i)$. Letting $Y^n = \{y^n_i : i < (2^{\aleph_0})^+\}$ we can, by Claim 1.3 and renaming assume the y^n_i are pairwise disjoint. Let $y_n = y^n_o$; as R is reduced, $h_\alpha(\Sigma k_n z^n_o) = \Sigma k_n y_n$.

So $\Sigma k_n y_n$ exists and belongs to R_α whenever for some ℓ $p^\ell k_n y_n = 0$ for every n. Hence, $\Sigma k_n p^n y_n \in R$ for every k_n, and $p^n y_n \neq 0$. Now for any sequence $\bar{k} = \langle \ldots, k_n, \ldots \rangle$, $-p < k_n < p$, let $b_{\bar{k}} = \Sigma k_n p^n y_n$ only when $d(\bar{k}) = \{n : k_n \neq 0\}$ is infinite. Notice that $d(b_{\bar{k}}) = \bigcup_{n \in d(\bar{k})} d(y_n)$. We shall find a $\bar{k}$ such that $b_{\bar{k}} \notin PC[G \cup \{a^*_i : i < \alpha\}]$. Suppose there is no such $\bar{k}$; then there are $m = m(\bar{k})$, and $\alpha > i(0,\bar{k}) > i(1,\bar{k}) > \ldots > i(m,\bar{k})$ and integers $M_\ell = M(\ell,\bar{k})$ such that:

$$(*) \quad b_{\bar{k}} + G = M_o a^*_{i(0,\bar{k})} + \ldots + M_m a^*_{i(m,\bar{k})} + G \quad (\text{where } M_\ell a^*_{i(\ell,\bar{k})} \neq 0).$$

Notice that by condition (1), this expression is determined uniquely.

Suppose $d(\bar{k}) \subseteq d(\bar{\ell})$ but $i(0,\bar{k}) > i(0,\bar{\ell})$ Then:

(A) $d_*(a^*_{i(0,\bar{k})}) \subseteq^* d(b_{\bar{k}}) \cup \bigcup_{0<n\leq m(\bar{k})} d(a^*_{i(n,\bar{k})})$ [as $pb_{\bar{k}} = 0$, $d(M_o a^*_{i(0,\bar{k})}) = d_*(a^*_{i(0,\bar{k})})$ by conditions (1), (2)]. ($\subseteq^*$- almost included).

(B) $d(b_{\bar{k}}) \subseteq d(b_{\bar{\ell}})$ (by the expression for $d(b_{\bar{k}})$).

(C) $d(b_{\bar{\ell}}) \subseteq \bigcup_{0\leq n\leq m(\bar{\ell})} d(a^*_{i(n,\bar{\ell})})$

Combining we get $d_*(a^*_{i(0,\bar{k})})$ is almost included in a finite union of $d(a^*_j)$, $j < i(0,\bar{k})$, contradicting condition (1). Hence $d(\bar{k}) \subseteq d(\bar{\ell})$ implies $i(0,\bar{k}) \leq i(0,\bar{\ell})$.

So choose $\bar{k}$ with minimal $i(0,\bar{k})$; as $d(\bar{k})$ is infinite, there are $\bar{k}^1$, $\bar{k}^2$ such that $d(\bar{k}) = d(\bar{k}^1) \cup d(\bar{k}^2)$, $d(\bar{k}^1) \cap d(\bar{k}^2) = \emptyset$, and $n \in d(\bar{k}^1) \rightarrow k^1_n = k_n$, $n \in d(\bar{k}^2) \rightarrow k^2_n = k_n$.

By the previous statement and the choice of $\bar{k}$, $i(0,\bar{k}) = i(0,\bar{k}^1) = i(0,\bar{k}^2)$. By condition (1) necessarily $M(0,\bar{k}^1) = M(0,\bar{k}) = M(0,\bar{k}^2)$. Define $\bar{k}^*$ so that $n \in d(\bar{k}^1) \rightarrow k^*_n = k_n$, $n \in d(\bar{k}^2) \rightarrow n = -k_n$, $n \notin d(\bar{k}) \rightarrow k^*_n = 0$. Then $\bar{b}_{\bar{k}^*} = b_{\bar{k}^1} - b_{\bar{k}^2}$, hence the expression for $b_{\bar{k}^*}$ in (*) can be obtained by subtracting those of $\bar{b}_{\bar{k}^1}$, $\bar{b}_{\bar{k}^2}$. Then $a^*_{i(0,k)}$ vanishes, and we get a contradiction to the choice of $\bar{k}$.

Hence for some $\bar{k}$, $b_{\bar{k}} \notin PC[G \cup \{a_i : i < \alpha\}]$, and we define $b^*_\alpha = b_{\bar{k}}$.

As h_α is into a reduced group, K_α is closed in H's topology (but is not necessarily a pure subgroup), hence K_α is closed in its (p-adic) topology hence it is torsion-complete. Remember $p^m K_\alpha$ has power $\geq \lambda$ for every m. Let $a \in I_\alpha$ be such that $h_\alpha(a) = b^*_\alpha$, and let its exponent be m. So by claim 1.3, as in Case I, we can find $a' \in K$ such that $d_*(a') = d_m(a')$ is not almost contained in a finite union of $d(a^*_i)$, $d(b^*_j)$, $d(a)$ $(i < \alpha$, $j \leq \alpha)$ and let $a^*_\alpha = a' + a$.

Case III: Neither case I, nor case II, but $|p^m K_\alpha| > 2^{\aleph_0}$ for every m

So $p^m I_\alpha$ has cardinality $\geq \lambda$ for every m, and for some $m(*)$ $p^{m(*)} K_\alpha$ has cardinality $< \lambda$, and as mentioned before it is torsion complete; hence $K_\alpha = K^1 \oplus K^2$, K^1 bounded, K^2 torsion-complete, unbounded and of power $< \lambda$, but $> 2^{\aleph_0}$. The situation is dual to that of Case II, the kernel and image interchanging roles. Using claim 1.3 twice we get $a_i \in I_\alpha$ $(i < \mu)$ such that a_i has exponent $m+1$, $h_\alpha(a_i)$ has exponent 1, and no $d_*(a_i) = d_m(a_i)$ is almost included in a finite union of $d(a_j)$ $j \neq i$; and similarly for the $d(h_\alpha(a_i)) = d_*(h_\alpha(a_i))$ and $d(a_i), d(h_\alpha(a_i))$ are disjoint to $d(b)$, $b \in K^2$. Hence there is $a \in \{a_i : i < \mu\}$ so that $d_*(a)$, $d_*(h_\alpha(a))$ are not almost included in a finite union of $d(a_i^*)$, $d(b_i^*)$ $(i < \alpha)$. We let $b_\alpha^* = h_\alpha(a)$. Suppose we let $a_\alpha^* = a$; the only thing that can go wrong is that $b_\alpha^* \in PC[G \cup \{a_i^* : i \leq \alpha\}]$, hence, as in Case II (**) $b_\alpha^* + G = M_0 a_{i(0)}^* + \ldots + k a_\alpha^* + G$ where $\alpha > i(0) > \ldots$

Multiplying by p, $pb_\alpha^* = 0$, hence $pka_\alpha^* \in G$ hence [as $d_*(a_\alpha^*) = d_m(a_\alpha^*)$ $m+1$ is the exponent of a_α^*] $pka_\alpha^* = 0$, but $ka_\alpha^* \neq 0$. So $k = p^m k_1$, k_1 not divisible by p.

As in the definition of the y_n's in Case II, we can find $w_n \in K^2$ of order $n+1$, $d(w_n)$ pairwise disjoint; $k_n, n < \omega$; and $J \subseteq \omega - \{0,\ldots,m\}$ so that $\sum_{n \in J} k_n p^n w_n \notin PC[G \cup \{a_i^* : i < \alpha\}]$ and let $a_\alpha^* = a + \sum_{n \in J} p^{n-m} w_n$. Then $h_\alpha(a_\alpha^*) = b_\alpha^*$, and suppose we get (**) again, with $M'_\ell, i'(0), \ldots, k'$. Then as $d(a)$, $d(h(a))$ are disjoint to $d(\sum_{n \in J} p^{n-m} w_n)$, $k = k'$. Subtracting the equations we get a contradiction to the definition of the w_n's. So in any case we can define a_α^*.

Case IV: Not cases I, II, III, but $h_\alpha \restriction I_\alpha$ is not semi-simple.

Let G^* be the smallest subgroup of H such that $PC(G^*) = G^*$,

$G \subseteq G^*$, a_i^*, $b_i^* \in G^*$ $(i < \alpha)$; $a_i \in G^*$, $d(b) \subseteq \bigcup_{i=1}^{n} d(a_i)$ implies $b \in G^*$ and for some m $p^m \mathrm{Ker}(h_\alpha | I_\alpha) \subseteq G^*$ (as not case III, $p^m \mathrm{Ker}(h_\alpha | I_\alpha) \leq 2^{\aleph_0}$ for some m) and $a \in G^* \implies h_\alpha(a) \in G^*$ for $a \in G^* + I_\alpha$. Clearly the power of G is $\lambda + |\alpha|$. Let $I^* = G^* + I_\alpha$; for every $a \in I^* - G^*$ there are no $a_i \in G^*$ such that $d(a) \subseteq \dot{\bigcup}_{i=1}^{n} d(a_i)$. We can find $a' \in a{+}G^*$ so that $a + G^*$, a' have the same order. Hence $d_*(a') \nsubseteq \bigcup_{i=1}^{n} d(a_i)$ for any $a_i \in G^*$. Notice that $h_\alpha(a) + G^* = h_\alpha(a') + G^*$. If for some such a we let $a_\alpha^* = a'$, $b_\alpha^* = h_\alpha(a')$, the only thing that can go wrong is that $h_\alpha(a') \in PC[G^* \cup \{a'\}]$, hence for some rational $r = r_a$, $h_\alpha(a') - ra \in G^*$ hence $h_\alpha(a) - ra \in G^*$; and for $a \in G^*$ we let $r_a = 0$; and r is a p-adic integer as the order of $h_\alpha(a') + G^*$ is $\leq$ the order of $a' + G$. For the same reason if r, r' are suitable r_a's then $r - r'$ is divisible by p^m where m is the exponent of $h(a') + G^*$ (divisibility among the p-adic integers). So if we choose the minimal integer r_a, r_a is defined uniquely.

If $b = p^\ell a$, $r_b - r_a$ is divisible by $p^{m-\ell}$. Also if $b = ra$ r, an integer, $(r,p) = 1$ then $r_a = r_b$. If also $h(b)+G^*$ has exponent $m+1$, $b \in I^*$, then $h(a-b) - (r_a a - r_b b) + G^* = 0$ and $h(a-b) - r_{a-b}(a-b) + G^* = 0$ hence $(r_a - r_{a-b})a + G^* = (r_b - r_{a-b})b + G^*$.

If we choose b so that $d(a)$, $d(b)$ are almost disjoint this implies that p^m divides $r_a - r_{a-b}$ and $r_b - r_{a-b}$; hence it divides $r_a - r_b$, hence $r_a = r_b$. As for every such a, b, there is $c \in I^*$ such that $h(c) + G^*$ has exponent m, and $d(c)$ is almost disjoint from $d(a)$ and $d(b)$, then $r_a = r_c = r_b$.

Combining we get a p-adic integer r such that $h_\alpha(a) - ra \in G^*$ for $a \in I^*$ so if $h^*(x) = rx$ then $h = (h_\alpha - h^*) | I_\alpha$ has range of power $< \mu$. By assumption h is not semi-zero-like hence as in case II, for some $b = h(a)$, $a \in I_\alpha$,

$b \notin PC[G \cup \{a^*_i: i < \alpha\}]$, $pb = 0$. Also there is $x \in I_\alpha \cap \ker(h)$, $d_*(x) = d_*(x+a)$, and x is not almost included in a finite union of $d(a^*_i)$, $d(b^*_i)$ $(i < \alpha)$ and $d(x)$ is disjoint to $d(a)$, $d(b)$. Then let $a^*_\alpha = a+x$, $b^*_\alpha = h_\alpha(a+x) = h_\alpha(a) + h_\alpha(x) = (b+h^*(a)+h^*(x) = b+ra^*_\alpha$ The only thing that can go wrong is $b^*_\alpha \in PC[G \cup \{a^*_i, i < \alpha\} \cup \{a^*_\alpha\}]$ i.e. $p^\ell b^*_\alpha + G = Ma^*_\alpha + .. + M_i a^*_{\alpha(i)} + ... + G$ $(\alpha(j) < \alpha)$ where $d(p^\ell b^*_\alpha)$, $d(b^*_\alpha)$ are equal up to a finite set. Using $d_*(x)$ we see that necessarily $p^\ell rx = Mx$; hence we get a contradiction to the definition of b, and x.

Case V Not any of the previous cases.

So $h_\alpha | I_\alpha$ is semi-simple; let $h_\alpha | I_\alpha = h^1 + h^2$, h^1 semi-zero-like, $h^2(x) = rx$, r a p-adic integer. Let $r = p^m r_1$, r_1 a p-adic unit. As in Case I, by claim 1.3, we can find $\{a_i: i < \mu\} \subseteq I_\alpha$ such that $d_*(a_i) = d_m(a_i)$ is not almost included in a finite union of $d(a_j)$ $(j \neq i)$, $d(a^*_j)$, $d(b^*_j)$, $(j < \alpha)$. The image of h^1 is of cardinality $\leq\lambda$, so w.l.o.g. $h^1(a_i)$ does not depend on i. So $a^*_\alpha = a_3 - a_2$, $b^*_\alpha = h_\alpha(a^*_\alpha)$ will satisfy our demands.

* * * * * *

We have defined, for $J \subseteq \mu$, $G(J) = PC[G \cup \{a^*_i: i \in J\}]$. It is easy to check that $b^*_i \notin G(J)$ when $h_\alpha | I_\alpha$ is not semi-simple; and $a^*_j \in G(J)$ iff $j \in J$. Suppose $J^* \subseteq J_0, J_1$ where $J^* = \{a:$ not case V$\}$ and $h: G(J_0) \to G(J_1)$ is a homomorphism, so, for some α $I_\alpha = H$, $h_\alpha | G(J_0) = h$ and so $h(a^*_\alpha) \neq b^*_\alpha$. As in cases II, III, IV equality holds, h_α is zero-like (case I), or simple (case V.) In the second case it is easy to check this implies $J_0 \subseteq J_1$. So if $\{J_i: i < 2^\mu\}$ is a family of subsets J of μ, $J^* \subseteq J$, no one included in the other, then $\{G(J_i) : i < 2^\mu\}$ is the required family in (1.2) (A).

For 1.2(B) it suffices to choose $\{J_i : i < 2^\mu\}$ so that if $I \subseteq H$ is a closed pure subgroup, $p^m I$ has power $\geq \lambda$ for every m, $h: I \to H$ is a simple homomorphism, $i \neq j < 2^\mu$, then for some α $I_\alpha = I$, $h = h_\alpha | I_\alpha$, $\alpha \in J_i$, $\alpha \notin J_j$, [so by $h(a^*_\alpha)$ we get the non-existence of the homomorphism h from $I \cap G(J_i)$ into $G(J_j)$]. This can be easily done as in the list $\{(I_\alpha, h_\alpha): \alpha < \mu\}$ each pair appears μ times.

<u>Conclusion 1.4</u>: For $\mu = \lambda^{\aleph_0} = 2^\lambda > 2^{\aleph_0}$ there is a family of 2^μ groups each of cardinality μ, such that any homomorphism from one member to another has range of cardinality $\leq \lambda$.

A complement to 1.1(3) is:

<u>Claim 1.5</u>: Suppose $|p^m G| = \lambda$ for every m, but $\lambda^{\aleph_0} > \lambda \geq 2^{\aleph_0}$. Then $G = G_1 \oplus G_2$ where for every $m < \omega$, $\ell = 1,2$ $|p^m G_\ell|^{\aleph_0} \geq \lambda$. (So G_1, G_2 are unbounded, hence G is essentially decomposable).

<u>Proof</u>: Let $\bigoplus_{\substack{i<\beta(n)\\ n<\omega}} \langle x^n_i \rangle$ be a basic subgroup of G, where x^n_i has exponent $n + 1$. Let $\lambda_0 = \min_n \sum_{m \geq n} |\beta(n)|$, so by the assumption $\lambda_0^{\aleph_0} = \lambda^{\aleph_0}$. Define for $a \in G$, d(a) as in the proof of 1.2. We can easily find $Y_j \subseteq \{x^n_i : n < \omega, i < \beta(n)\}$ for $j < \lambda_0^{\aleph_0}$ such that $\min_n \sum_{m \geq n} |Y_j \cap \{x^m_i : i < \beta(m)\}|$ is λ_0, and $\alpha \neq \beta$ implies $Y_\alpha \cap Y_\beta$ is a subset of $\{x^m_i : i < \beta(m), m < n\}$ for some n. As for each $a \in G$, the number of α's such that $Y_\alpha \cap d(a)$ is infinite, is $\leq 2^{\aleph_0}$ (by the "almost-disjointness" of the Y_α's), clearly

$$\sum_\alpha |\{a: d(a) \cap Y_\alpha \text{ is infinite}\}| \leq |G| + 2^{\aleph_0} < \lambda^{\aleph_0}$$

As the number of α's is $\lambda^{\aleph_0}$, for some α for no $a \in G$ is $d(a) \cap Y_\alpha$ infinite. So let G_1 be the closure in G of the subgroup generated by Y_α, and G_2 be the closure in G of the subgroup generated by $\{x^n_i : i < \beta(n), n < \omega, x^n_i \notin Y_\alpha\}$. Clearly $G = G_1 \oplus G_2$, $|p^m G_\ell| \geq \lambda_0$.

2. Large rigid-like systems

A group means an (abelian) reduced separable p-group.

Theorem 2.1: Suppose μ is a strong limit cardinal of cofinality $> \aleph_0$. Then there is a rigid-like system of 2^μ groups each of cardinality μ.

Remark: Here we elaborate less than §1.

Proof: Let G be $\oplus\langle x_i^n\rangle$, $i < \mu$, $n < \omega$, where x_i^n has exponent $n + 1$, and H the torsion completion of G. We use the notation of Th. 1.2, and in addition $G_\alpha = \bigoplus_{\substack{i<\alpha\\ n<\omega}} \langle x_i^n\rangle$, H_α the torsion completion of G_α, and $H^*_\delta = \bigcup_{\alpha<\delta} H_\alpha$, $X_n^\alpha = \{x_i^n: i < \alpha\}$, $X^\alpha = \bigcup_{n<\omega} X_n^\alpha$.

We define $a_\alpha^\lambda, b_\alpha^\lambda \in H$ for $\lambda < \mu$ a strong limit cardinal of cofinality $\aleph_0$, $\alpha < 2^\lambda$, so that

(1) $a_\alpha^\lambda, b_\alpha^\lambda \in H_\lambda - H^*_\lambda$

(2) $b_\alpha^\lambda \notin PC[G_\lambda \cup \{a_\beta^\kappa: (\kappa,\beta) < (\lambda,\alpha)\}]$ where $(\kappa,\beta) < (\lambda,\alpha)$ means $\kappa < \lambda$ or $\kappa = \lambda, \beta < \alpha$.

(3) The intersection of $d_*(a_\alpha^\lambda)$ with any $d(a_\beta^\kappa), d(b_\beta^\kappa)$, $(\kappa,\beta) < (\lambda,\alpha)$ is included in some X^γ, $\gamma < \lambda$.

(4) $d_*(a_\alpha^\lambda)$ is not almost included in any X_γ, $\gamma < \lambda$.

We define them by induction on λ, so suppose $a_\alpha^\kappa, b_\alpha^\kappa$ were defined for $\kappa < \lambda$. Let $\{(I_\alpha^\lambda, h_\alpha^\lambda): \alpha < 2^\lambda\}$ be a list of all pairs of closed pure subgroups I of H_λ, and a homomorphism $h : I \to H_\lambda$, each appearing 2^λ times. Now we define by induction on α, as follows. If there are $a_\alpha^\lambda \in I_\alpha^\lambda$, and $b_\alpha^\lambda = h_\alpha^\lambda(a_\alpha^\lambda)$ which satisfy conditions (1), (2), (3), (4) and $b_\alpha^\lambda \notin PC[G \cup \{a_\beta^\kappa: (\kappa,\beta) \leq (\lambda,\alpha)\}]$ we choose them in this way. If not but there are $a_\alpha^\lambda \in I_\alpha$, $b_\alpha^\lambda = h_\alpha^\lambda(a_\alpha)$ which satisfy conditions (1) - (4) we choose them in this way. Otherwise we choose $a_\alpha^\lambda, b_\alpha^\lambda$ so that $a_\alpha^\lambda, b_\alpha^\lambda$ are almost disjoint, and almost disjoint with any $a_\beta^\kappa, b_\beta^\kappa$ $(\kappa,\beta) < (\lambda,\alpha)$ and (1) - (4) are satisfied. So we have three possibilities which we denote respectively by A, B, C. Let $J_0^* = \{(\lambda,\alpha):$ possibility A or B holds$\}$ and $J^* = \{(\lambda,\alpha):$ possibility A holds$\}$.

Let $G(J) = PC[G \cup \{a_\alpha^\lambda: (\lambda,\alpha) \in J\}]$. Suppose $J^* \subseteq J_1, J_2 \subseteq J_0^*$

and , h: $G(J_1) \to G(J_2)$ a non-zero-like homomorphism. Let h^c be the unique extension of h to a homomorphism $h^c: H \to H$. We shall show that h is simple and that for arbitrarily large $\lambda < \mu$ there is α, $h^\lambda_\alpha = h^c | I^\lambda_\alpha$ and for (λ,α) possibility B holds, hence

$$(I^\lambda_\alpha, h^\lambda_\alpha) = (I^\lambda_\beta, h^\lambda_\beta), (\lambda,\beta) \in J_1 \Rightarrow (\lambda,\beta) \in J_2$$

From this it will be easy to draw our conclusion, as in 1.2.

Notice that by conditions (1) - (4), $b^\lambda_\alpha \in G(J_\ell)$, only if (λ,α) satisfies possibility B and $(\lambda,\alpha) \in J$. Also $a^\lambda_\alpha \in G(J_\ell)$ iff $(\lambda,\alpha) \in J_\ell$ (for $\ell = 1,2$).

But first we prove:

Claim 2.2: Assume $I' \subseteq H$ is a closed subgroup, $h': I' \to H$ a non-zero-like homomorphism. Then for some $a \in I'$, $h'(a) \notin PC(G \cup \{a^\kappa_\beta: (\kappa,\beta) < (\mu,\mu)\})$ and $h'(a)$ has exponent 1.

Proof: As h' is not zero-like, there are $m < \omega$ and $z_n \in I'$ such that z_n has order $n + m$, and $y_n = h'(z_n)$ has order $n + 1$. Let $\zeta(n) = \min\{\zeta: d(p^n y_n) \subseteq X^\zeta\}$; as the $\zeta(n)$ are ordinals there are $n_0 < n_1 < \ldots$ such that $\zeta(n_0) < \zeta(n_1) < \ldots$ or $\zeta(n_o) = \zeta(n_1) = \ldots$ As we can replace y_{n_ℓ} by $p^{n_\ell - \ell} y_{n_\ell}$, we can assume $n_\ell = \ell$. Let $\delta = \sup_n \zeta(n)$. As $d(p^n y_n)$ is countable each $\zeta(n)$, hence δ has cofinality ω or is a successor. Define δ_n so that if cf $\delta = \omega$ then $\delta_0 < \delta_1 < \ldots$, $\delta = \bigcup_{n<\omega} \delta_n$ and $\delta_\ell < \zeta(\ell)$; and if δ is a successor $\delta_0 = \delta_1 = \ldots$; $\delta = \delta_0 + 1$; and clearly then $\zeta(n) = \delta$, hence $\delta_n < \zeta(n)$. We now define inductively $n(\ell)$, $t_\ell \notin d[p^{n(\ell)} y_{n(\ell)}]$ such that $n(\ell) < n(\ell+1)$, and when $\ell < m$, $t_\ell \in d(p^{n(m)} y_{n(m)})$ and $t_\ell \notin X^{\delta_\ell}$. Let $n(0) = 0$, and suitable t_0 exists as $\delta_n < \zeta(n)$. If we have defined for ℓ, let $n(\ell+1)$ be greater than the orders of $t_0,\ldots,t_\ell$, and than $n(\ell)$; hence $t_0,\ldots,t_\ell \notin d(p^{n(\ell+1)} y_{n(\ell+1)})$, and we can choose $t_{\ell+1} \in d(p^{n(\ell+1)} y_{n(\ell+1)})$, $t_{\ell+1} \notin X^{\delta_{\ell+1}}$ as $\delta_{\ell+1} < \zeta(n(\ell+1))$. By renaming assume $n(\ell) = \ell$. As we can replace y_n

by $y_n + \sum_{\omega > \ell \geq n+1} k_\ell p^{\ell-n} y_\ell$, we can assume $t_m \in d(p^n y_n)$ iff $n = m$. For $J \subseteq \omega$ let $b_J = \sum_{n \in J} p^n y_n$; then $d(b_J) \subseteq \bigcup_n d(p^n y_n)$ hence $d(b_J) \subseteq X^\delta$, and $t_n \in d(b_J)$ iff $n \in J$. Also $b_J \in$ Range h' $[b_J = h'(\Sigma p^n z_n)]$. We shall show that for some J $b_J \notin PC[G \cup \{a^\kappa_\gamma : (\kappa,\gamma) < (\mu,\mu)\}]$. If δ is a successor every infinite J suffices, by the induction assumptions. If δ has cofinality ω, and $J = \omega$ is not sufficient, necessarily δ is a strong limit cardinal of cofinality ω; and for some $w \in \{a^\delta_\gamma : \gamma < 2^\delta\}$, $d = d_*(w) \cap \{t_n : n < \omega\}$ is infinite, and then any J such that $d \cap \{t_n : n \in J\}$, $d \cap \{t_n : n \notin J\}$ are infinite will suffice. As we mentioned $b_J \in$ Range h', so for some $a \in I'$, $h'(a) = b_J$. So we have proved the claim; and we continue with $h: G(J_1) \to G(J_2)$.

<u>Observation 1:</u> If h^* is a zero-like homomorphism, $h^*: H \to H$, then for every m $p^m \mathrm{Range}(h^c - h^*)$ has cardinality μ.

For suppose $m = m(*)$ is a counter example, let $h^1 = h^c - h^*$. By the claim 2.2 there is an a such that

$$h^c(a) \notin PC(G \cup \{a^\kappa_\beta : \kappa < \mu, \beta < 2^\kappa\})$$

$h^c(a)$ has exponent 1, a has exponent $n(*)$.

We define by induction, strong limit cardinals of cofinality $\aleph_0$, $\lambda_n < \lambda_{n+1} < \mu$, $a \in H_{\lambda_0}$, $\lambda_0 > |p^{m(*)} \text{ Range } h^1|$; and sets $Y_n \subseteq X^{\lambda_{n+1}}_{n+1}$, $Y_n \cap d(a) = \emptyset$ such that h^1 is constant on $p^{m(*)}$ Y_n, and $|Y_n| = (\lambda_n^{\aleph_0})^+$, and for $y \in Y_n$, $h^1(y) \in H_{\lambda_{n+1}}$. (This is easy to do.) We can also assume, as in 1.3, that for distinct $y_\ell \in Y_n$, $h^*(y_1) - h^*(y_2)$, $h^*(y_3) - h^*(y_4)$ are disjoint, and $\subseteq X^{\lambda_{n+1}}$ and disjoint to X^{λ_n}.

Let $\lambda = \sum_n \lambda_n$, so λ too is a strong limit cardinal of cofinality $\aleph_0$, Let I' be the torsion completion of $PC(\{a\} \cup \bigcup_n Y_n)$ in H and $h' = h^1 | I'$. Clearly $h': I' \to H$, and for some α, $(I', h') = (I^\lambda_\alpha, h^\lambda_\alpha)$. We can find $n(0) < \omega$

$n(0) \geq n(*) + m(*)$ such that $h^*(p^{n-n(*)}y) = 0$ for $n \geq n(0)$, $y \in Y_n$ and for $i < 2^\lambda$, $y^i = \sum_{n(0)<n<\omega} p^{n-n(*)} (y_n^i - z_n^i)$; $y_n^i \neq z_n^i \in Y_n$, and the y^i's are pairwise almost disjoint. Clearly $h^1(y^i) = 0$, and except for $< 2^\lambda$ i's, y^i is almost disjoint from a_β^κ, b_β^κ for $(\kappa,\beta) < (\lambda,\alpha)$ and to a, $h^c(a)$. Notice $h^c(a + y^i) = h^c(a) + h^1(y^i) = h^c(a)$. It is now easy to check that $h^c(a + y^i) \notin PC[G_\lambda \cup \{a_\beta^\kappa : (\kappa,\beta) < (\lambda,\alpha)\} \cup \{a + y^i\}]$ hence for (λ,α) possibility A occurs, hence we get a contradiction to $h: G(J_1) \to G(J_2)$.

Observation II: Suppose h is not simple but h_0 is simple, $h' = h - h_0$; and $y_i^n \in PC(X_n)$ for $i < \mu$ are pairwise disjoint and $\neq 0$; and the exponent of y_i^n is $n + 1$. Then there are $\ell = \ell_0$ and n_0 such that for every $n \geq n_0$, $S \subseteq \mu$ of cardinality μ; the set $\{p^{n-\ell} h'(y_i^n) : i \in S\}$ has cardinality μ.

Otherwise we can find $n(\ell)$ $\ell < \omega$ and $S_\ell \subseteq \mu$, $|S_\ell| = \mu$ such that $n(\ell) < n(\ell+1)$ and $|\{p^{n(\ell)-\ell} h'(y_i^n);\ i \in S_\ell\}| < \mu$. Let $h_0 = h_1 + h_2$, h_2 zero-like, $h_1(x) = rx$, r a p-adic integer.

The rest is like the proof of observation I noticing that for every $c \in H_\lambda$ $h^c(c) \in PC[G_\lambda \cup \{a_\beta^\kappa : (\kappa,\beta) < (\lambda,\alpha)\} \cup \{c\}]$ iff $(h^c - h_1)(c) \in PC[G_\lambda \cup \{a_\beta^\kappa : (\kappa,\beta) < (\lambda,\alpha)\} \cup \{c\}]$.

Observation III: We can find an increasing continuous sequence of strong limit cardinals $\lambda_\alpha < \mu$ $\alpha < cf\,\mu$, (say $\lambda_0 = 0$) and $y_i^n \in G$ for $i < \mu$ such that:

(i) y_i^n is $x_{j(i,1)}^n - x_{j(i,2)}^n$

(ii) the y_i^n are pairwise disjoint

(iii) if $\lambda_\alpha \leq i < \lambda_{\alpha+1}$ then $d(h(y_i^n)) \subseteq X^{\lambda_{\alpha+2}}$, but is disjoint to $X^{\lambda_{\alpha+1}}$

(iv) the $h(y_i^n)$ are pairwise disjoint.

The proof is by induction, using a theorem of Erdos-Rado [ER].

*　　*　　*　　*　　*

For each y_i^n choose a maximal $\ell = \ell(n,i)$ and a rational p-adic integer r_i^n such that $p^{n-\ell}(h(y_i^n) - r_i^n y_i^n) = 0$. (ℓ may be -1, so it is always defined). As the number of possible ℓ's is $\aleph_0$, $cf\mu > \aleph_0$, we can assume $\ell(n,i) = \ell(n)$; similarly $r_i^n = r_n$.

Observation IV: For every ℓ there is a k_ℓ such that $n, m \geq k_\ell \Rightarrow r_n - r_m$ is divisible by p^ℓ. So for some p-adic integer r, $r_n - r$ is divisible by p^ℓ for $n \geq k_\ell$. Also $\lim_{n\to\infty} \ell(n) = \infty$.

Let ℓ be given. Let $\lambda = \lambda_\omega$ and I be the torsion completion of $PC(\{y_i^n: i < \lambda_\omega, n < \omega\})$ and $\alpha < 2^\lambda$ be such that $(I_\alpha^\lambda, h_\alpha^\lambda) = (I, h^c | I)$. Clearly we can find $i(n) < \lambda$ such that $y^\ell = \sum_{\ell \leq n} p^{n-\ell} y_{i(n)}^n$ satisfies the following: if $(\kappa,\beta) < (\lambda,\alpha)$ then for some m, for every $n \geq m$, $p^{n-\ell} y_{i(n)}^n$, $p^{n-\ell} h(y_{i(n)}^n)$ are disjoint to a_β^κ. As (λ,α) is not in possibility A , $h^c(y^\ell) \in PC[G_\lambda \cup \{a_\beta^\kappa: (\kappa,\beta) < (\lambda,\alpha) \cup \{y^\ell\}]$ hence there is an m such that $d(h^c(y^\ell)) = d(p^m h^c(y^\ell))$ and

$$p^m h^c(y^\ell) = My^\ell + M_{i(0)}^{\kappa(0)} a_{i(0)}^{\kappa(0)} + \ldots + M_{i(n)}^{\kappa(n)} a_{i(n)}^{\kappa(n)} + c$$

$c \in G_\lambda$, $(\kappa(0), i(0)), \ldots,$ are $< (\lambda,\alpha)$. Then clearly $M' = M/p^m$ is an integer, and for some k_ℓ for every $k \geq k_\ell$ $h(p^{k-\ell} y_{i(k)}^k) = M' p^{k-\ell} y_{i(k)}^k$ hence $\ell(k) \geq \ell$ and $M' p^{k-\ell} y_{i(k)}^k = p^{k-\ell} r_k y_{i(k)}^k$ hence $(M' - r_k)$ is divisible by p^ℓ. Clearly we have proved the observation.

* * *

So for every ℓ for every n big enough $p^{n-\ell}(h(y_i^n) - r y_i^n) = 0$, so the only way not to contradict II is to assume

Observation V: h is simple.

* * *

So let $h = h_1 + h_2$, $h_1(x) = rx$, h_2 is zero-like. Let $r = p^k r'$, r' a p-adic unit. Hence for some $n_0 < \omega$ for every $n \geq n_0$ $h_2(y_i^n)$ has order $< n-k$. The only thing left to be proved is that if $\lambda^0 < \mu$ there are λ, $\lambda^0 < \lambda < \mu$ and $\alpha < 2^\lambda$ such that $h^\lambda_\alpha = h^c | I^\lambda_\alpha$ and (λ,α) satisfies possibility B. Choose $\lambda = \lambda_\ell > \lambda^0$, δ a limit ordinal and α such that $I^\lambda_\alpha = PC(\{y^n_i : i < \lambda,\ n < \omega\}]$ $h^\lambda_\alpha = h^c | I^\lambda_\alpha$. By the assumption on h, the pair (λ,α) does not satisfy possibility A. Now we can find $y = \sum_{n_0 \le n} p^{n-k} y^n_{i(n)}$ which is almost disjoint to a^κ_β $(\kappa,\beta) < (\lambda,\alpha)$. Clearly $h_2(y) = 0$, $d(h_1(y) = d(y)$, so it proves (λ,α) satisfies possibility B.

So we have finished the proof of 2.1.

Conclusion 2.3: There are arbitrarily large essentially indecomposable groups. (This answers a question of Pierce [P], which is [Fu. 2], Pr.55, p. 55.)

Remark: We can prove 2.1 also for $\mu = \aleph_{\alpha+n}$ where $\aleph_\alpha$ is a strong limit cardinal of cofinality $> \aleph_0$ and $0 \le n < \omega$ or $\aleph_\alpha = \lambda^{\aleph_0} = 2^\lambda$, $0 < n < \omega$.

Moreover, if G, H are members of the family, I a closed subgroup of G, $|p^k I| \ge \aleph_{\alpha+n}$ for every k, $h: I \to H$ a homomorphism then $G = H$, and h is simple. (The purity of I is not needed).

Question: Can we prove 2.1 when $\mu = \lambda^{\aleph_0} = 2^\lambda$? (assuming G.C.H this is the only open case). Can we prove 1.2 when $\mu = 2^{\aleph_0}$?

REFERENCES

[ER] P. Erdös and R. Rado, Intersection theorems for systems of sets, J. London Math. Soc. 44 (1969), 467-479.

[Fo] G. Fodor, Eine Bemerkung zur Theorie der regressiven Funktionen, Acta. Sci. Math. 17 (1956), 139-142.

[Fu.1] L. Fuchs, Infinite abelian groups, Vol. I, Academic Press, N. Y. & London 1970.

[Fu.2] L. Fuchs, Infinite abelian groups, Vol. II, Academic Press, N. Y. & London, 1973.

[Fu.3] L. Fuchs, Abelian groups, Publishing house of the Hungarian Academy of Sciences, Budepest, 1958.

[Fu.4] L. Fuchs, Indecomposable abelian groups of measurable cardinals, dedicated to R. Baer, to appear.

[P] R. S. Pierce, Homomorphism of primary abelian groups, topics in abelian groups, (Chicago, Illinois 1963), 215-310.

[Sh] S. Shelah, Infinite abelian groups, Whitehead problem and some contradiction, Israel Journal of Mathematics, 1974 to appear.

Institute of Mathematics
The Hebrew University of Jerusalem
Jerusalem, Israel

THE COMPLEXITY OF T^f and OMITTING TYPES IN F_T

H. Simmons

University of Aberdeen

Let L be some fixed countable first order language. Let $\forall_1, \forall_2, \forall_3, \ldots$ be the usual sets of L-formulas logically equivalent to formulas in prenex normal form with the indicated prenex. Let T be some fixed L-theory (where a theory is a deductively closed consistent set of sentences).

In this note I will compute an upper bound for the complexity of the set $T^f \cap \forall_{n+1}$ (where T^f is the finite forcing companion of T) and then use this bound to obtain an omitting types result for the class F_T of finite generic structures for T .

The theory T^f is completely determined by the set

$$A = T \cap \forall_1$$

so it is natural that the complexity of T^f must be computed relative to the complexity of A . We use the arithmetical hierachy relativized to A . For each $n \in \omega$ we write $\Sigma_n[A]$, $\Pi_n[A]$ instead of the Σ_n^A, Π_n^A of [1; p. 3o4]. We prove the following theorem.

Theorem 1. For each $n \in \omega$, $T^f \cap \forall_{n+1} \in \Pi_n[A]$.

Notice that this theorem implies the well known result that T^f is hyperarithmetical in A . M.Boffa also has obtained a result similar to Theorem 1.

To prove the omitting types theorem we will use a slightly weaker version of Theorem 1. Let B be any set (of sentences) such that $A \in \Sigma_1[B]$ (so typically B is a set of axioms for T). In particular if A is r.e. then we can put $B = \emptyset$.

COROLLARY 2. $T^f \cap \forall_1 \in \Sigma_1[B]$ and for each $n \in \omega$,

$$T^f \cap \forall_{n+2} \in \Pi_{n+2}[B] .$$

Proof. Since $\Pi_{n+1}[\Sigma_1[B]] = \Pi_{n+2}[B]$.

Before we prove Theorem 1 let us look at some examples.

First let T be full number theory N . Here $T^f = T$ and $A \in \Pi_1$. Theorem 1 gives $T^f \cap \forall_{n+1} \in \Pi_{n+1}$. But $T \cap \forall_{n+1}$ is a complete Π_{n+1} set, so the computed bounds for this case ($A \in \Pi_1$) are optimal.

Second let T be peano number theory P , or group theory, or the theory of division rings. Here $A \in \Sigma_1$ so Corollary 2 gives $T^f \cap \forall_{n+2} \in \Pi_{n+2}$. But in each of these three cases there is a translation of full number theory N into T^f . These translations show that each Π_{n+1} set is one-one reducible to $T^f \cap \forall_{n+2}$ so, with the obvious abuse of notation, we have

$$\Pi_{n+1} \leq T^f \cap \forall_{n+2} \leq \Pi_{n+2} .$$

I do not know the actual complexity of $T^f \cap \forall_{n+2}$ or whether for this case ($A \in \Sigma_1$) the bounds given in Theorem 1 are optimal.

To prove Theorem 1 we need some information about T^f . We use the equality

$$T^f \cap \forall_1 = T \cap \forall_1 \qquad (*)$$

together with the following lemma. (For each formula ϕ, $fv(\phi)$ is the set of free variables of ϕ .)

LEMMA 3. For each formula ϕ the following are equivalent.

(i) ϕ is consistent with T^f .

(ii) There is some $\exists_1$-formula θ , consistent with T^f , such that $fv(\theta) \subseteq fv(\phi)$ and $T^f \vdash \theta \rightarrow \phi$.

It is well known that equality (*) and Lemma 3 uniquely determine T^f .

Proof of Theorem 1. The proof is by induction on n . Since the initial case $(n = 0)$ follows trivially from (*) it is sufficient to prove the induction step (from n to $n + 1$).

Let σ be any $\forall_{n+2}$-sentence and let $\sigma = (\forall \underset{\sim}{v})\psi(\underset{\sim}{v})$ where $\psi(\underset{\sim}{v})$ is an $\exists_{n+1}$-formula and $\underset{\sim}{v}$ is the sequence of free variables of ψ . For each $\exists_1$-formula $\theta(\underset{\sim}{v})$ let $\sigma(\theta)$ be $(\forall \underset{\sim}{v})[\theta \rightarrow \neg\psi]$, so that $\sigma(\theta)$ is an $\forall_{n+1}$-sentence.

By Lemma 3 we have

$$\sigma \notin T^f \cap \forall_{n+2}$$

if and only if

there is some $\exists_1$-formula $\theta(\underset{\sim}{v})$ consistent with T^f such that $\sigma(\theta) \in T^f \cap \forall_{n+1}$.

The required result now follows.

We now turn to the omitting types theorem.

The following is the classical omitting types theorem stated for T^f .

THEOREM 4. Let Γ be a countable collection of types each non-principal over T^f . Then there is some model of T^f which omits each member of Γ .

We will improve this theorem and at the same time generalize [2; Theorem1]. Notice that our proof does not use any forcing machinery.

A type is a set Γ of formulas such that $fv(\Gamma)$ is finite (where $fv(\Gamma)$ is the set of free variables of Γ). We do not require any consistency or maximality conditions on Γ . A type Γ is principal over T^f if there is some formula ϕ , consistent with T^f , such that $fv(\phi) \subseteq fv(\Gamma)$ and for each $\gamma \in \Gamma$, $T^f \vdash \phi \rightarrow \gamma$. By lemma 3 we may assume that such a formula ϕ is an $\exists_1$-formula θ .

Let F_T be the class of finite generic structures for T , so F_T is exactly the class of completing models of T^f . We use the following description of F_T .

<u>LEMMA 5.</u> There is a countable collection F of types, each non-principal over T^f, such that for each model $\underline{A}$ of T^f, $\underline{A} \in F_T$ if and only if $\underline{A}$ omits each member of F .

<u>COROLLARY 6.</u> Let $\boldsymbol{\Gamma}$ be a countable collection of types each non-principal over T^f . Then there is some $\underline{A} \in F_T$ which omits each member of $\boldsymbol{\Gamma}$.

<u>Proof.</u> Apply Theorem 4 to the countable collection $\boldsymbol{\Gamma}' = \boldsymbol{\Gamma} \cup F$.

For each $n \in \omega$ let $B^{(n)}$ be the nth jump of B (where B is the set of Corollary 2), and for each set X let $d(X)$ be the Turing-degree of X . Thus, by [1; p. 314, Theorem VIII]

$$d(X) \leq d(B^{(n)}) \leftrightarrow X \in \Sigma_{n+1}[B] \cap \Pi_{n+1}[B] .$$

For each type Γ let $\Gamma^* = \Gamma \cup \{\neg\gamma : \gamma \in \Gamma\}$. We can now prove our omitting types theorem.

<u>THEOREM 7.</u> Let $\boldsymbol{\Gamma}$ be a countable collection of types such that for each $\Gamma \in \boldsymbol{\Gamma}$ there is some $n \in \omega$ such that the following hold.

(i) $\Gamma \subseteq \forall_n \cup \exists_n$.

(ii) $d(\Gamma^*) \leq d(B^{(n)})$.

(iii) $d(\Gamma) \not\leq d(B^{(n)})$.

Then there is some $\underline{A} \in F_T$ which omits each member of $\boldsymbol{\Gamma}$.

<u>Proof.</u> By Corollary 6 it is sufficient to show that each such type Γ is non-principal over T^f .

Let Γ be a type which satisfies (i), (ii) and is principal over T^f . We show that $d(\Gamma) \leq d(B^{(n)})$ i.e. Γ does not satisfy (iii).

Let θ be the $\exists_1$-formula controlling Γ. For each $\phi \in \forall_n \cup \exists_n$ let $s(\phi)$ be the universal closure of $\theta \rightarrow \phi$. Thus $s(\phi)$ is an $\forall_{n+1}$-sentence. We have

$$\phi \in \Gamma \Rightarrow s(\phi) \in T^f$$

$$\phi \in \Gamma^* - \Gamma \Rightarrow s(\neg \phi) \in T^f .$$

Now, since θ is consistent with T^f, at most one of $s(\phi) \in T^f$, $s(\neg\phi) \in T^f$ can hold. Thus for each $\phi \in \Gamma^*$ the above implications are equivalences. This with (ii) and Corollary 2 gives $d(\Gamma) \leq d(B^{(n)})$, as required.

References

[1] H. Rogers, Jr., Theory of Recursive Functions and Effective Computability, McGraw-Hill, 1967.

[2] A. Macintyre, Omitting quantifier-free types in generic structures, J.S.L. 37 (1972), 512-52o.

MODEL-COMPLETENESS
and
SKOLEM EXPANSIONS

Peter M. Winkler

Yale University

Any first-order theory T , given by a set of axioms, may be expanded to a universal theory in a standard fashion (see, for example, Shoenfield [25]) by introducing Skolem functions for each axiom which is not already a universal sentence. For example, the theory of algebraically closed fields is "universalized" by introducing root functions f_n which, given the coefficients of an nth-degree polynomial, choose a root.

In the Fall of 1973 Abraham Robinson presented the following problem to students and faculty at Yale:

> Let T be either the theory of algebraically closed fields or the theory of real-closed fields, with the usual axiomatizations. Then the "universalization" T^+ of T is no longer model-complete. What is the (finite or infinite) forcing companion of T^+ ?

* This work constitutes part of the authors doctoral dissertation, Yale U. 1975; an NSF Graduate Fellowship provided part of the support. The author is deeply indebted to all the logicians at Yale from 1971 to 1974, but especially to Dan Saracino and to the author's advisors Abraham Robinson and Angus Macintyre.

In what follows we show that in either case T^+ actually possesses a model-completion; and that this fact is traceable to a simple property shared by many algebraic theories. The methods employed are then used to obtain results for general Skolem expansions, and finally applied to a problem of H. Jerome Keisler's.

Notation and preliminaries are established in §0. In §1 the notion of algebraic boundedness is defined, and the key theorems are proved using a sort of model-theoretic "surgery". These results are then used in §2 to examine a strengthening of the property of possessing a model-companion.

In §3 we attempt to establish the reasonableness of the property of algebraic boundedness by example; then, in §4, we attack the problem of when a model of T can be expanded to a model of the model-completion of T^+. Finally in §5, results from §1, §3, and §4 are used to attain some limited progress on Keisler's problem.

§0

For the most part standard notation for first-order model theory will be employed; similar, for example, to the notation used in Sacks [20]. A language L will consist of function, relation, and constant symbols (the "non-logical" part of L); logical symbols from the set $\{()\neg\wedge\vee\exists\forall\to=\}$; and variables from the set $V = \{v_i : i \in \omega\}$. The equality symbol will be assumed always to represent the identity relation. The symbols x , y , and z , with or without subscripts, will stand for members of V. Two languages will be called <u>disjoint</u> if their non-logical parts have no symbols in common.

If $\mathcal{A}$ is an L-structure, $|\mathcal{A}|$ denotes its underlying set and $L(\mathfrak{A})$ is the language $L \cup \{\underline{a} : a \in |\mathfrak{A}|\}$. A bar over a letter indicates an n-tuple subscripted by 1 through n . If $A(x_1 \ldots x_n)$ is an L-formula, and $a_1, \ldots, a_n \in |\mathfrak{A}|$, then the following are all equivalent: "$\mathfrak{A} \models A[a_1 \ldots a_n]$", "$\mathfrak{A} \models A[\bar{a}]$", "$\mathfrak{A} \models A(\bar{\underline{a}})$", $\bar{a} \in A^{\mathfrak{A}}(\bar{x})$", and "$\bar{a}$ is a solution to $A(\bar{x})$ in $\mathfrak{A}$."

A theory T consists of a set of sentences of L(T) which is closed under logical deduction. T is "universal" or "$\forall\exists$" if it has a set of axioms each member of which is a universal or $\forall\exists$ sentence respectively. A basic formula is an atomic or negated atomic formula.

The inclusion symbol $\subset$ will be used here to allow for the possibility of equality, whether used set-theoretically or model-theoretically. If $\mathfrak{A}$ and $\mathfrak{B}$ are L-structures, the expression "$\mathfrak{A} \prec_1 \mathfrak{B}$" means that $\mathfrak{A} \subset \mathfrak{B}$ and any existential sentence of $L(\mathfrak{A})$ which holds in $\mathfrak{B}$ already holds in $\mathfrak{A}$.

The theory of model-companions is expounded in Barwise and Robinson [3] , Eklof and Sabbagh [8], and in other papers on model companions and/or model-theoretic forcing. The basic definitions and results needed here are reviewed below:

A theory T is <u>model-consistent</u> with a theory S if any model of S can be embedded in a model of T . T* is the (unique) <u>model-companion</u> of T if T and T* are mutually model-consistent and T* is model-complete. (In this case T* coincides with the forcing companions of T .)

An L(T)-structure $\mathfrak{A}$ is <u>existentially closed</u> with respect to T if $\mathfrak{A} \subset \mathfrak{B} \models T$ implies $\mathfrak{A} \prec_1 \mathfrak{B}$.

A theory is <u>inductive</u> if its class of models is closed under unions of chains; equivalently, if it is an $\forall\exists$ theory. Any model-complete theory is inductive.

A theory T has the <u>amalgamation</u> property if whenever $\mathcal{B} \supset \mathcal{A} \subset \mathcal{C}$ are models of T , the diagram

$$\begin{array}{ccccc} & & \mathcal{D} & & \\ & \subset & & \supset & \\ \mathcal{B} & & & & \mathcal{C} \\ & \supset & & \subset & \\ & & \mathcal{A} & & \end{array}$$

can be completed with a model $\mathcal{D}$ of T . If in addition we can insure that $|\mathcal{B}| \cap |\mathcal{C}| = |\mathcal{A}|$ inside $\mathcal{D}$, then T is said to have the <u>strong</u> amalgamation property.

If T has a model-completion T* then T* is also its model-companion. The converse holds if T is inductive and has the amalgamation property, that is, the model-companion is also the model-completion.

The following characterization of the model-companion will be very useful: let T be an inductive theory. Then T has a model-companion if and only if the class of models of T which are existentially closed (with respect to T) is axiomatizable, i.e. it is the class of models of some theory S ; and in that case S is, in fact, the model-companion.

§ 1

The following notation is <u>not</u> standard in the literature.

Two n-tuples $\bar{a} = \langle a_1 \ldots a_n \rangle$ and $\bar{b} = \langle b_1 \ldots b_n \rangle$ are said to be <u>dissimilar</u> if $a_i \neq b_i$ for every i , $1 \leq i \leq n$.

If $A(x_1 \ldots x_n)$ is any formula, $1 \leq k \leq n$, and $N > 0$, we define the multi-variable quantifiers $\exists$, $\exists^{<N}$, $\exists^{>N}$, $\exists^{=N}$, and $\exists^{\infty}$ as follows:

$$\exists x_1 \ldots x_k A(\bar{x}) \leftrightarrow \exists x_1 \exists x_2 \ldots \exists x_k A(\bar{x})$$

$$\exists^{<N} x_1 \ldots x_k A(\bar{x}) \leftrightarrow \neg \exists x_1^1 \ldots x_k^1 \; . \; . \; . \; x_1^N \ldots x_k^N$$

$$\Big(\bigwedge_{1 \le j \le N} A(x_1^j \ldots x_k^j x_{k+1} \ldots x_n)$$

$$\wedge \bigwedge_{\substack{1 \le i < j \le N \\ 1 \le m \le k}} x_m^i \ne x_m^j \Big)$$

(i.e. $A(\bar{x})$ has no *pairwise dissimilar* set of N solutions.)

$$\exists^{>N} x_1 \ldots x_k A(\bar{x}) \leftrightarrow \neg \exists^{<N+1} x_1 \ldots x_k A(\bar{x})$$

$$\exists^{=N} x_1 \ldots x_k A(\bar{x}) \leftrightarrow \exists^{>N-1} x_1 \ldots x_k A(\bar{x}) \wedge \exists^{<N+1} x_1 \ldots x_k A(\bar{x})$$

$$\exists^{\infty} x_1 \ldots x_k A(\bar{x}) \leftrightarrow \bigwedge_{N \in \omega} \exists^{>N} x_1 \ldots x_k A(\bar{x})$$

(to be regarded as a formula of an infinitary language.)

Let T be a fixed first-order theory, $A(x_1 \ldots x_n)$ a formula of $L(T)$, $1 \le k \le n$, and $N \in \omega$. We say that N is an *algebraic bound* for $A(\bar{x})$ *in* $x_1 \ldots x_k$ if

$$T \vdash \neg \exists x_{k+1} \ldots x_n (\exists^{=M} x_1 \ldots x_k A(\bar{x}))$$

for all $M > N$; equivalently, for any model $\mathfrak{a}$ of T and elements $a_{k+1} \ldots a_n \in |\mathfrak{a}|$, either $A^{\mathfrak{a}}(x_1 \ldots x_k \underline{a}_{k+1} \ldots \underline{a}_n)$ contains an infinite pairwise dissimilar set of k-tuples or it contains no pairwise dissimilar set of cardinality greater than N.

If there exists an N which is an algebraic bound for $A(\bar{x})$ in $x_1 \ldots x_k$, then $A(\bar{x})$ is *algebraically bounded* in $x_1 \ldots x_k$ (with respect to the theory T).

T itself will be said to be *algebraically bounded* if every formula $A(\bar{x})$ of $L(T)$ is algebraically bounded in x_1, or equivalently if each formula has an algebraic bound in any *one* variable. It is easily seen that T is algebraically bounded iff T admits elimination of the quantifier "there exist infinitely many", since if N is an algebraic bound for $A(\bar{x})$ in x_1, then "there exist infinitely many x_1 such that $A(\bar{x})$" is equivalent to the ordinary first-order formula "$\exists^{>N} x_1 A(\bar{x})$". Somewhat less obvious is the fact, which follows from Lemma 2 below, that an algebraically bounded theory admits elimination of our multi-variable quantifier $\exists^{\infty}$.

LEMMA 1: Let $\mathcal{A}$ be an L-structure, and $A(\bar{x})$ a formula of $L(\mathcal{A})$. Then $\mathcal{A} \models \exists^{\infty}\bar{x}A(\bar{x})$ if and only if there exist an L-structure $\mathcal{B} \succ \mathcal{A}$ and elements $b_1 \dots b_n \in |\mathcal{B}| - |\mathcal{A}|$ such that $\mathcal{B} \models A[\bar{b}]$.

Proof: Compactness. Note that dissimilarity enables us to get all of the b_i's outside $\mathcal{A}$. If, say, $\mathcal{A}$ were an infinite field, then the formula "$x_1 = x_2$" would have infinitely many dissimilar solutions, but "$x_1 \cdot x_2 = 0$" would have only two.

The proofs of the next two lemmas are technical, and the reader will miss little by skipping them. Lemma 2 is crucial in what follows; Lemma 3 is merely handy.

LEMMA 2: If T is algebraically bounded then any formula of L(T) is algebraically bounded in any number of its variables.

Proof: By induction; assume that any formula of L(T) is algebraically bounded in any number less than k of its variables. Let $A(x_1 \dots x_k \dots x_n)$ be an arbitrary formula of L(T), with the object of showing that $A(\bar{x})$ is algebraically bounded in $x_1 \dots x_k$. (The lemma would then follow by re-subscripting.)

Let M be an algebraic bound for $A(\bar{x})$ in $x_1 \dots x_{k-1}$; let N be an algebraic bound for $\exists^{>M} x_1 \dots x_{k-1} A(\bar{x})$ in x_k; and let P be an algebraic bound for

$$\exists x_k (A(\bar{x}) \wedge \bigwedge_{1 \le i \le N} x_k \ne y_i)$$

in $x_1 \dots x_{k-1}$. We wish to show that $N+P$ will serve as the desired algebraic bound; so let $\mathcal{A} \models T$, $a_{k+1} \dots a_n \in |\mathcal{A}|$, and $Q \in \omega$ be such that

(1) $$\mathcal{A} \models \exists^{=Q} x_1 \dots x_k A(x_1 \dots x_k \underline{a}_{k+1} \dots \underline{a}_n)$$

with the intent of showing that $Q \le N+P$.

Let $\langle b_1^1 \dots b_k^1 \rangle, \dots, \langle b_1^Q \dots b_k^Q \rangle$ witness (1) and suppose

(2) $$\mathcal{A} \models \exists^{\infty} x_k (\exists^{>M} x_1 \dots x_{k-1} A(x_1 \dots x_{k-1} x_k \underline{a}_{k+1} \dots \underline{a}_n)).$$

Then we can find an element $c_k \in |\mathcal{A}|$ not among $b_k^1 \dots b_k^Q$ such that

$$\mathcal{A} \models \exists^{>M} x_1 \dots x_{k-1} A(x_1 \dots x_{k-1} \underline{c}_k \underline{a}_{k+1} \dots \underline{a}_n)$$

so that by choice of M,

$$\mathcal{A} \models \exists^{\infty} x_1 \dots x_{k-1} A(x_1 \dots x_{k-1} \underline{c}_k \underline{a}_{k+1} \dots \underline{a}_n).$$

Thus the set $A^{\mathcal{A}}(x_1 \dots x_{k-1} \underline{c}_k \underline{a}_{k+1} \dots \underline{a}_n)$ contains infinitely many pairwise dissimilar k-1-tuples; one of these k-1-tuples, call it $\langle c_1 \dots c_{k-1} \rangle$, must differ from the b's in every coordinate; i.e. for each $i \le k-1$ and each $j \le Q$, $c_i \ne b_i^j$. But then $\langle c_1 \dots c_k \rangle$, $\langle b_1^1 \dots b_k^1 \rangle, \dots, \langle b_1^Q \dots b_k^Q \rangle$ constitute a set of $Q+1$ pairwise dissimilar solutions to $A(x_1 \dots x_k \underline{a}_{k+1} \dots \underline{a}_n)$, contradicting (1); it follows that our assumption (2) was false. Thus, by choice of N,

$$\mathcal{A} \models \exists^{<N+1} x_k (\exists^{>M} x_1 \dots x_{k-1} A(x_1 \dots x_k \underline{a}_{k+1} \dots \underline{a}_n)).$$

It follows that there are elements $d_k^1 \dots d_k^N \in |\mathfrak{A}|$ such that

$$(3) \qquad \mathfrak{A} \models \neg \exists x_k (\exists^{>M} x_1 \dots x_{k-1} A(x_1 \dots x_k \underline{a}_{k+1} \dots \underline{a}_n) \wedge \bigwedge_{1 \le i \le N} x_k \ne \underline{d}_k^i) \ .$$

Next, suppose

$$(4) \qquad \mathfrak{A} \models \exists^{\infty} x_1 \dots x_{k-1} (\exists x_k (A(x_1 \dots x_k \underline{a}_{k+1} \dots \underline{a}_n) \wedge \bigwedge_{1 \le i \le N} x_k \ne \underline{d}_k^i)) \ .$$

Then there is a set $\{<e_1^j \dots e_{k-1}^j>: 1 \le j \le M \cdot Q+1\}$ of pairwise dissimilar k-1-tuples, and elements $e_k^1 \dots e_k^{M \cdot Q+1}$ (not necessarily distinct) such that for each j ,

$$\mathfrak{A} \models A(\underline{e}_1^j \dots \underline{e}_{k-1}^j \underline{e}_k^j \underline{a}_{k+1} \dots \underline{a}_n) \wedge \bigwedge_{1 \le i \le N} \underline{e}_k^j \ne \underline{d}_k^i \ .$$

Now in view of (3), no single element of $|\mathfrak{A}|$ can appear more than M times in the list $e_k^1 \dots e_k^{M \cdot Q+1}$; hence e_k^j must take on at least $Q+1$ different values as j varies. It follows that $\{<e_1 \dots e_k>: 1 \le j \le M \cdot Q+1\}$ has a pairwise dissimilar subset of cardinality $Q+1$ whose members are solutions to the formula $A(x_1 \dots x_k \underline{a}_{k+1} \dots \underline{a}_n)$, again contradicting (1).

So, our assumption (4) is false; hence by choice of P ,

$$\mathfrak{A} \models \exists^{<P+1} x_1 \dots x_{k-1} (\exists x_k (A(x_1 \dots x_k \underline{a}_{k+1} \dots \underline{a}_n) \wedge \bigwedge_{1 \le i \le N} x_k \ne \underline{d}_k^i)) \ .$$

It follows that the set $\{<b_1^j \dots b_k^j>: j \le Q \wedge \forall i (i \le N \rightarrow b_k^j \ne d_k^i)\}$ has cardinality at most P ; and clearly $\{<b_1^j \dots b_k^j>: j \le Q \wedge \exists i (i \le N \wedge b_k^j = d_k^i)\}$ cannot have more than N elements. Thus $Q = \text{card}\{<b_1^j \dots b_k^j>: j \le Q\} \le N+P$, and the lemma is proved.

LEMMA 3: Let T be a fixed theory and suppose that any quantifier-free formula of $L(T)$ is algebraically bounded in any of its variables. Then any existential formula of $L(T)$ is algebraically bounded in any of its (free) variables.

Proof: By induction on the number of bound variables. Assume that any existential formula with fewer than m bound variables is algebraically bounded in any of its free variables, and let $A(x_1 \ldots x_k \ldots x_n y_1 \ldots y_m)$ be an arbitrary quantifier-free formula with the object of showing that $\exists \bar{y} A(\bar{x}\bar{y})$ is algebraically bounded in $x_1 \ldots x_k$.

Let M and N be algebraic bounds for $\exists y_1 \ldots y_{m-1} A(\bar{x}\bar{y})$ in $x_1 \ldots x_k$ and in $x_1 \ldots x_k y_m$ respectively. Let $\mathcal{A} \models T$, $b_{k+1} \ldots b_n \in |\mathcal{A}|$, and $Q \in \omega$ be such that

$$(5) \qquad \mathcal{A} \models \exists^{=Q} x_1 \ldots x_k (\exists y_1 \ldots y_m A(x_1 \ldots x_k \underline{b}_{k+1} \ldots \underline{b}_n y_1 \ldots y_m))$$

and assume that $Q > M \cdot N$ with the intent of deriving a contradiction.

Let $\bar{a}^1, \ldots, \bar{a}^Q$ be pairwise dissimilar k-tuples witnessing (5), and choose m-tuples $\bar{c}^1, \ldots, \bar{c}^Q$ so that for any $i \leq Q$,

$$\mathcal{A} \models A(\underline{a}^i_1 \ldots \underline{a}^i_k \underline{b}_{k+1} \ldots \underline{b}_n \underline{c}^i_1 \ldots \underline{c}^i_m) \ .$$

Now there cannot be infinitely many pairwise dissimilar k-tuples $x_1 \ldots x_k$ satisfying, for fixed $i \leq Q$,

$$\exists y_1 \ldots y_{m-1} A(x_1 \ldots x_k \underline{b}_{k+1} \ldots \underline{b}_n y_1 \ldots y_{m-1} \underline{c}^i_m)$$

because each would also satisfy

$$\exists y_1 \ldots y_m A(x_1 \ldots x_k \underline{b}_{k+1} \ldots \underline{b}_n y_1 \ldots y_m) \ ,$$

contradicting (5). Thus by the definition of M ,

$$\mathcal{A} \models \exists^{\leq M} x_1 \ldots x_k (\exists y_1 \ldots y_{m-1} A(x_1 \ldots x_k \underline{b}_{k+1} \ldots \underline{b}_n y_1 \ldots y_{m-1} \underline{c}^i_m)) .$$

It follows that at most M of the m-tuples $\bar{c}^1 , \ldots , \bar{c}^Q$ can have the same m^{th} coordinate; hence c^i_m must take on at least N+1 different values as i ranges from 1 to Q . We may assume (by re-superscripting) that $c^i_m \neq c^j_m$ for $i < j \leq N+1$. But then the k+1-tuples $\langle a^1_1 \ldots a^1_k c^1_m \rangle , \ldots , \langle a^{N+1}_1 \ldots a^{N+1}_k c^{N+1}_m \rangle$ are pairwise dissimilar, so

$$\mathcal{A} \models \exists^{\geq N+1} x_1 \ldots x_k y_m (\exists y_1 \ldots y_{m-1} A(x_1 \ldots x_k \underline{b}_{k+1} \ldots \underline{b}_n y_1 \ldots y_m)) .$$

Then by definition of N ,

$$\mathcal{A} \models \exists^{\infty} x_1 \ldots x_k y_m (\exists y_1 \ldots y_{m-1} A(x_1 \ldots x_k \underline{b}_{k+1} \ldots \underline{b}_n y_1 \ldots y_m)) ,$$

which again contradicts (5), proving the lemma.

The following definitions will be useful in proving the main theorems.

Let $\mathcal{A}$ and $\mathcal{B}$ be fixed L-structures with $\mathcal{A} \subset \mathcal{B}$. A formula $A(v_1 \ldots v_n)$ of L , together with witnesses $a_1 \ldots a_k$ and $b_{k+1} \ldots b_n$ in $|\mathcal{B}|$, will be said to constitute an <u>obstruction</u> (to $\mathcal{A} \prec_1 \mathcal{B}$) if:

(i) $A(\bar{v})$ is quantifier-free;

(ii) $\mathcal{B} \models A(\underline{a}_1 \ldots \underline{a}_k \underline{b}_{k+1} \ldots \underline{b}_n)$;

(iii) $a_1 \ldots a_k \in |\mathcal{A}|$; and

(iv) $\mathcal{A} \models \neg \exists v_{k+1} \ldots v_n A(\underline{a}_1 \ldots \underline{a}_k v_{k+1} \ldots v_n)$.

Clearly, if it is not the case that $\mathcal{A} \prec_1 \mathcal{B}$, then there is an obstruction to $\mathcal{A} \prec_1 \mathcal{B}$. We would like to specify the form of that obstruction.

A formula $A(v_1 \ldots v_n)$ is said to be in <u>splintered form</u> if it is a conjunction of basic formulas, each of which is in one of the following forms:

$$\begin{array}{ll} (a) & x_1 = x_2 \\ (b) & \neg x_1 = x_2 \\ (c) & R(x_1 \ldots x_m) \\ (d) & \neg R(x_1 \ldots x_m) \\ (e) & x_0 = f(x_1 \ldots x_m) \quad . \end{array}$$

Here f may be any function symbol, R any relation symbol, m any number and $x_0 \ldots x_m$ any members (not necessarily distinct) of $\{v_1 \ldots v_n\}$.

If $\langle A(\bar{v}), a_1 \ldots a_k, b_{k+1} \ldots b_n \rangle$ is an obstruction, we can "splinter" it (replace it by an obstruction in splintered form) as follows:

First, write $A(\bar{v})$ in disjunctive form $A_1(v) \vee \ldots \vee A_t(v)$; if i is such that $\mathcal{B} \models A_i(\underline{a}_1 \ldots \underline{a}_k \underline{b}_{k+1} \ldots \underline{b}_n)$, then $\langle A_i(\bar{v}), \ldots \rangle$ is again an obstruction. Replace $A(\bar{v})$ by $A_i(\bar{v})$.

Second, if there are any constant symbols in $A(\bar{v})$, replace each one by a new variable and augment the witness list accordingly, changing k and n.

Third, let $f(x_1 \ldots x_m)$ be the first (i.e. leftmost) simple function term appearing in $A(\bar{v})$, and let $A'(v_1 \ldots v_{n+1})$ be the result of replacing every occurence of $f(x_1 \ldots x_m)$ in $A(\bar{v})$ by the single variable v_{n+1}. Now replace $\langle A(\bar{v}), \ldots \rangle$ by $\langle A'(v_1 \ldots v_{n+1}) \wedge v_{n+1} = f(x_1 \ldots x_m), a_1 \ldots a_k, b_{k+1} \ldots b_{n+1} \rangle$ with the appropriate value for b_{n+1}. Lastly add 1 to n.

Repeat the third step until $A(\bar{v})$ is in splintered from; the result is still an obstruction.

An obstruction $<A(\bar{v}), a_1 \ldots a_k b_{k+1} \ldots b_n>$ will be called <u>clean</u> if all of the witnesses $a_1 \ldots b_n$ are distinct, and none of the b_i's lie in $|\mathfrak{A}|$. Any obstruction can be "cleaned up" by combining variables, juggling subscripts and changing some b's to a's . For example, if $a_p = a_q$, where $1 \leq p < q \leq k$, set $A'(v_1 \ldots v_{q-1} v_{q+1} \ldots v_n) = A(v_1 \ldots v_{p-1} v_p v_{p+1} \ldots v_{q-1} v_p v_{q+1} \ldots v_n)$, replace $A(\bar{v})$ by $A'(v_1 \ldots v_{n-1})$, and decrease n by 1 .

Finally, for convenience, we will say that a set of n-tuples is <u>d-infinite</u> if it contains an infinite subset whose elements are pairwise dissimilar.

<u>THEOREM 1</u>: Let T and S be model-complete, algebraically bounded theories whose languages are disjoint. Then $T \cup S$ has a model-completion.

<u>Proof</u>: Recall that two languages are disjoint if they have no function, relation or constant symbols in common except the equality symbol. If T or S has only finite models, then $T \cup S$ is already model-complete; for, if there <u>are</u> any models of $T \cup S$, they cannot be properly embedded in one another. Otherwise the model-completion of $T \cup S$ will say, roughly, the following: any two d-infinite sets of n-tuples, one definable in L(T) and the other in L(S) , must intersect.

We begin by showing that $U = T \cup S$ has a model-companion; for this it suffices to show that the class <u>E</u> of existentially closed models of the inductive theory U is axiomatizable.

Let U^* be the theory of <u>E</u> , and suppose $\mathfrak{A} \models U$ but $\mathfrak{A} \notin$ <u>E</u> , with the intent of showing that $\mathfrak{A} \not\models U^*$. Choose $\mathfrak{B} \models U$ such that $\mathfrak{A} \subset \mathfrak{B}$ but not $\mathfrak{A} \prec_1 \mathfrak{B}$; then there is a clean obstruction $<C(\bar{v}), a_1 \ldots a_k, b_{k+1} \ldots b_n>$ to $\mathfrak{A} \prec_1 \mathfrak{B}$ where $C(\bar{v})$

is in splintered form. We may then write $C(\bar{v}) = A(\bar{v}) \wedge B(\bar{v})$, where $A(\bar{v})$ is a formula of $L(T)$ and $B(\bar{v})$ is a formula of $L(S)$.

We denote by $\mathcal{A}_T$ and $\mathcal{A}_S$ the reducts of $\mathcal{A}$ to $L(T)$ and to $L(S)$ respectively, and similarly for other models of U. By the model-completeness of T, $\mathcal{A}_T \prec \mathcal{B}_T$; since $b_{k+1} \dots b_n$ are all in $|\mathcal{B}| - |\mathcal{A}|$, it follows from Lemma 1 that $\mathcal{A}_T \models \exists^\infty v_{k+1} \dots v_n A(\underline{a}_1 \dots \underline{a}_k v_{k+1} \dots v_n)$; by the same argument for S, $\mathcal{A}_S \models \exists^\infty v_{k+1} \dots v_n B(\underline{a}_1 \dots \underline{a}_k v_{k+1} \dots v_n)$. Thus $\mathcal{A} \models X$, where X is the sentence

$$\begin{aligned}\exists v_1 \dots v_k (&\exists^\infty v_{k+1} \dots v_n A(v_1 \dots v_n) \wedge \\ &\exists^\infty v_{k+1} \dots v_n B(v_1 \dots v_n) \wedge \\ &\neg \exists v_{k+1} \dots v_n (A(v_1 \dots v_n) \wedge B(v_1 \dots v_n))) .\end{aligned}$$

Now X is (equivalent in U to) an ordinary first-order sentence since by Lemma 2, $A(\bar{v})$ and $B(\bar{v})$ are each algebraically bounded in $v_{k+1} \dots v_n$. But we claim X is inconsistent with U^*.

To see this suppose there is a model $\mathcal{C} \in \underline{E}$, with $\mathcal{C} \models X$; let $c_1 \dots c_k \in |\mathcal{C}|$ be witnesses to $v_1 \dots v_k$ in X. Then in particular $\mathcal{C}_T \models \exists^\infty v_{k+1} \dots v_n A(\underline{c}_1 \dots \underline{c}_k v_{k+1} \dots v_n)$ so by Lemma 1 there is an elementary extension $\mathcal{D}$ of $\mathcal{C}_T$ and elements $d_{k+1} \dots d_n \in |\mathcal{D}| - |\mathcal{C}|$ such that $\mathcal{D} \models A(\underline{c}_1 \dots \underline{c}_k \underline{d}_{k+1} \dots \underline{d}_n)$. Furthermore, since the original b_i's in $\mathcal{B}$ were distinct, we may assume without loss of generality that for each i, j such that $k+1 \leq i < j \leq n$, the conjunct "$v_i \neq v_j$" is in $A(\bar{v})$; hence the d_i's are also distinct.

Similarly, we obtain an extension $\mathcal{E} \models S$ of $\mathcal{C}_S$, with distinct elements $e_{k+1} \dots e_n \in |\mathcal{E}| - |\mathcal{C}|$ such that $\mathcal{E} \models B(\underline{c}_1 \dots \underline{c}_k \underline{e}_{k+1} \dots \underline{e}_n)$.

Now $\mathcal{C}$, $\mathcal{D}$ and $\mathcal{E}$ are all infinite (since $\mathcal{C}$ has infinitely many solutions to a formula) and thus by Lowenheim-Skolem arguments we may insure that $\text{card}|\mathcal{C}| < \text{card}|\mathcal{D}| = \text{card}|\mathcal{E}|$. We may then find a bijection $h: |\mathcal{D}| \to |\mathcal{E}|$ such that h is the identity on $|\mathcal{C}|$, and for each i between $k+1$ and n , $h(d_i)=e_i$. Identifying $\mathcal{D}$ and $\mathcal{E}$ via h yields a model $\mathcal{F}$ of U satisfying

$$\exists v_{k+1}\ldots v_n(A(\underline{c}_1\ldots\underline{c}_k v_{k+1}\ldots v_n) \wedge B(\underline{c}_1\ldots\underline{c}_k v_{k+1}\ldots v_n)) ;$$

but then $\mathcal{C}$, being existentially closed, satisfies the same sentence, contradicting the choice of $c_1 \ldots c_k$. It follows that U* indeed axiomatizes $\underline{E}$ and is thus the model-companion of U .

To establish that in fact U* is the model-completion of U we need only show that U has the amalgamation property. Thus let $\mathcal{B} \supset \mathcal{A} \subset \mathcal{C}$ be models of U with the intent of finding a model $\mathcal{D}$ of U which completes the diagram

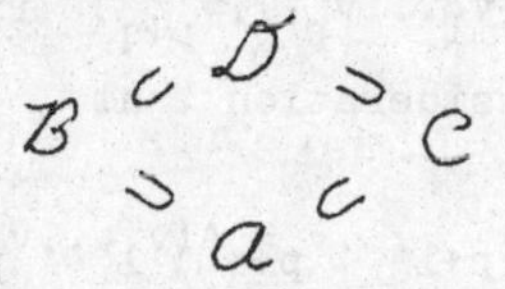

Now T and S , being model complete, each have the strong amalgamation property. (This is easily checked; it follows also from a theorem of Bacsich [1] and Eklof [7] to the effect that a theory has strong amalgamation iff it has amalgamation and every model is algebraically closed.) This means that we can complete the two diagrams

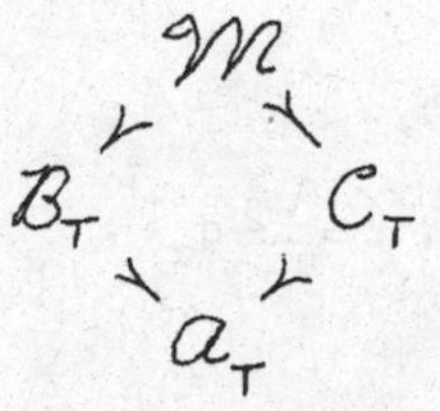

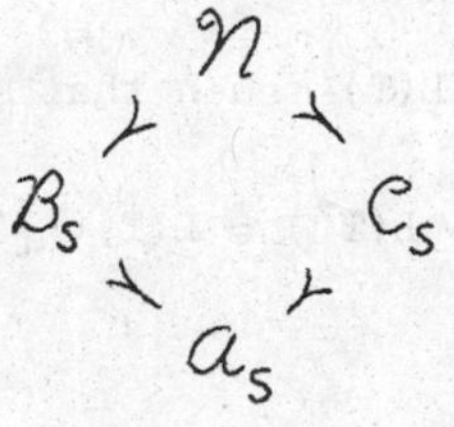

with models $\mathcal{M}$ of T and $\mathcal{N}$ of S such that in them, $|\mathcal{A}_T| = |\mathcal{B}_T| \cap |\mathcal{C}_T|$ and $|\mathcal{A}_S| = |\mathcal{B}_S| \cap |\mathcal{C}_S|$. If any of these models is finite, they are all isomorphic and the problem is trivial; otherwise another Lowenheim-Skolem argument allows us to insure $\mathrm{card}|\mathcal{B}| < \mathrm{card}|\mathcal{M}| = \mathrm{card}|\mathcal{N}| > \mathrm{card}|\mathcal{C}|$. We can then find a bijection $g: |\mathcal{M}| \to |\mathcal{N}|$ which is the identity on $|\mathcal{B}|$ and on $|\mathcal{C}|$; identifying $\mathcal{M}$ and $\mathcal{N}$ via g yields the desired model $\mathcal{D}$, completing the proof of the theorem.

<u>COROLLARY</u>: If $\{T_\alpha : \alpha < \beta\}$ is any set of model-complete, algebraically bounded theories in pairwise disjoint languages, then $\bigcup_{\alpha<\beta} T_\alpha$ has a model-completion.

<u>Proof</u>: An automatic extrapolation of the proof of Theorem 1.

A set $\{f_1,\dots,f_q\}$ of p-ary functions is a set of <u>Skolem functions for the formula</u> $A(v_1\dots v_p\dots v_{p+q})$ if it is true in the theory or structure under consideration that

$$\forall v_1\dots v_p(\exists v_{p+1}\dots v_{p+q}A(v_1\dots v_{p+q}) \to A(v_1\dots v_p f_1(v_1\dots v_p)\dots f_q(v_1\dots v_p))) .$$

A Skolem expansion of a theory is a conservative extension by Skolem functions; more precisely, we say that T^+ is a <u>Skolem expansion</u> of T if there is a set

$$\{D_\alpha(v_1\dots v_{p_\alpha}\dots v_{p_\alpha+q_\alpha}) : \alpha < \beta\}$$

of formulas of L(T) such that

$$L(T^+) = L(T) \cup \{"f_i^\alpha" : \alpha < \beta \ , \ 1 \le i \le q_\alpha\}$$

and

$$T^+ = T \cup \{\forall v_1 \dots v_{p_\alpha} (\exists v_{p_\alpha+1} \dots v_{p_\alpha+q_\alpha} A(v_1 \dots v_{p_\alpha+q_\alpha})$$
$$\rightarrow A(v_1 \dots v_{p_\alpha} f_1^\alpha(v_1 \dots v_{p_\alpha}) \dots f_q^\alpha(v_1 \dots v_{p_\alpha}))) : \alpha < \beta\} .$$

We allow the case $p_\alpha=0$, where the Skolem "functions" become merely constants.

THEOREM 2: If T is a model-complete, algebraically bounded theory then any Skolem expansion T^+ of T has a model-completion.

Proof: We proceed roughly as in the proof of Theorem 1, but a somewhat more delicate type of manipulation will be required here.

The resulting model-completion T^* of T^+ can be described roughly as follows: call a formula $B(v_1 \dots v_n)$ a configuration if it is a conjunction of equalities involving Skolem functions, and call an n-tuple $\langle a_1 \dots a_n \rangle$ of points in a model $\mathcal{A}$ of T eligible for that configuration if $\mathcal{A}$ has an expansion to T^+ satisfying $B(\underline{a}_1 \dots \underline{a}_n)$. A particular expansion $\mathcal{A}^+$ of $\mathcal{A}$ will then be a model of T^* iff for any n , any configuration $B(v_1 \dots v_n)$, and any d-infinite set $X \subset |\mathcal{A}|^n$ of eligible n-tuples which is definable with parameters in $\mathcal{A}$, some element of X actually is a solution to $B(v_1 \dots v_n)$ in $\mathcal{A}^+$. Thus T^* assures, in a sense, that the Skolem functions have been assigned in every possible way.

Now suppose that T^+ is the extension of T by Skolem functions for $\{D_\alpha(v_1 \dots v_{p_\alpha} \dots v_{p_\alpha+q_\alpha}) : \alpha < \beta\}$. For convenience we "Morleyize" the Skolem formulas as follows: introduce new relation symbols R_α , $\alpha<\beta$ and let

$$S = T\cup\{\forall v_1\ldots v_{p_\alpha+q_\alpha}(R_\alpha(v_1\ldots v_{p_\alpha+q_\alpha})\leftrightarrow(D_\alpha(v_1\ldots v_{p_\alpha+q_\alpha})\vee$$
$$\neg\exists v_{p_\alpha+q_\alpha+1}\ldots v_{p_\alpha+2q_\alpha}D_\alpha(v_1\ldots v_{p_\alpha}v_{p_\alpha+q_\alpha+1}\ldots v_{p_\alpha+2q_\alpha}))): \alpha<\beta\} .$$

Then S^+ would be simply

$$S\cup\{\forall v_1\ldots v_{p_\alpha}R_\alpha(v_1\ldots v_{p_\alpha}f_1^\alpha(v_1\ldots v_{p_\alpha})\ldots f_{q_\alpha}^\alpha(v_1\ldots v_{p_\alpha})): \alpha<\beta\}.$$

Since T is model-complete it has, essentially, the same models and embeddings as S ; it is thus easily seen that if S^+ has a model-completion S^* , then S^* restricted to $L(T^+)$ is the model-completion of T^+ . Therefore we may assume from the start that T has already been Morleyized in this fashion.

As in the proof of Theorem 1, let $\underline{E}$ be the class of all existentially closed models of T^+ and set $T^* = Th(\underline{E})$. Let $\mathcal{A}^+$ be a model of T^+ which is not in $\underline{E}$, with the intent of showing that $\mathcal{A}^+\not\models T^*$. We denote the reduct of $\mathcal{A}^+$ to L(T) by $\mathcal{A}$.

Let $\mathcal{B}^+\models T^+$ be such that $\mathcal{A}^+\subset\mathcal{B}^+$ but not $\mathcal{A}^+\prec_1\mathcal{B}^+$; then we can find a clean, splintered obstruction

$$\langle C_0(v_1\ldots v_{n_0}),\ a_1\ldots a_{k_0},\ b_{k_0+1}\ldots b_{n_0}\rangle$$

to $\mathcal{A}^+\prec_1\mathcal{B}^+$; we may then write

$$C_0(v_1\ldots v_{n_0})\equiv A_0(v_1\ldots v_{n_0})\wedge B_0(v_1\ldots v_{n_0})$$

where A_0 is a formula of L(T) and each conjunct of B_0 is of the form

$$x_0=f_j^\alpha(x_1\ldots x_{p_\alpha})$$

for some $\alpha<\beta$, some $j\leq q_\alpha$, and some list $x_0\ldots x_{p_\alpha}$ of variables, not necessarily distinct, from V .

C_0 requires some preparation before surgery; the first goal will be to remove all variable-free conjuncts from $B_0(\underline{a}_1 \dots \underline{a}_{k_0} v_{k_0+1} \dots v_{n_0})$. This is done by letting $B_1(v_1 \dots v_{n_0})$ be the result of deleting from $B_0(v_1 \dots v_{n_0})$ all conjuncts of the form $x_0 = f_j^{\alpha}(x_1 \dots x_{p_\alpha})$ in which all of the x_i's are among $\{v_1 \dots v_{k_0}\}$, and setting $C_1 = A_0 \wedge B_1$. Since the deleted conjuncts have nothing to do with the b_i's , it is clear that $\langle C_1(\bar{v}), \text{etc}\rangle$ is still an obstruction.

Second, we wish to insure that if the term $f_j^{\alpha}(x_1 \dots x_{p_\alpha})$ appears in the configuration $B_1(v_1 \dots v_{n_0})$, then its "brothers" $f_i^{\alpha}(x_1 \dots x_{p_\alpha})$, $i \neq j$, $i,j \leq q_\alpha$ are also there. To this end we say that a pair $\langle \alpha, \langle x_1 \dots x_{p_\alpha} \rangle\rangle$ is <u>represented</u> in a given formula if for some $j \leq q_\alpha$, the term $f_j^{\alpha}(x_1 \dots x_{p_\alpha})$ appears in the formula. Let $B_2(v_1 \dots v_{n_2})$ be the result of the following operations on $B_1(v_1 \dots v_{n_0})$: for each pair $\langle \alpha, \langle x_1 \dots x_{p_\alpha} \rangle\rangle$ represented in B_1 , and for each $j \leq q_\alpha$ such that the term $f_j^{\alpha}(x_1 \dots x_{p_\alpha})$ is <u>not</u> present in B_1 , we find a variable $y \in V$ not previously used and add the new conjunct

$$y = f_j^{\alpha}(x_1 \dots x_{p_\alpha})$$

to B_1 . Since $\langle C_1(\bar{v}), \text{etc}\rangle$ is clean, we may assume that no term in B_1 (or in B_2) appears more than once; B_2 is thus in a sense a <u>uniform</u> configuration. Furthermore, if we let $C_2 = A_0 \wedge B_2$ and choose suitable witnesses $b_{n_0+1} \dots b_{n_2}$ for the y's , then

$\langle C_2(v_1 \ldots v_{n_2}),\ a_1 \ldots a_{k_0},\ b_{k_0+1} \ldots b_{n_2} \rangle$ is still an obstruction; however, it may not be clean. We let

$$\langle C_3(v_1 \ldots v_n),\ a_1 \ldots a_k,\ b_{k+1} \ldots b_n \rangle$$

be the result of cleaning it up; we can still write $C_3 = A_3 \wedge B$, where $A_3(v_1 \ldots v_n)$ is a formula of $L(T)$ and B is a uniform configuration.

Incidentally, in all of the above we never had to deal with the case $p_\alpha = 0$ since the original obstruction was clean and hence had no constant symbols in it. In fact, we could assume if we wish that all new constant symbols in T^+ are already in T, since the addition of new constant symbols destroys neither model completeness nor algebraic boundedness.

Our third goal is to assure that the (n-k)-tuples which satisfy $A_3(\underline{a}_1 \ldots \underline{a}_k v_{k+1} \ldots v_n)$ are eligible for the configuration $B(\underline{a}_1 \ldots \underline{a}_k v_{k+1} \ldots v_n)$. Let

$$A_4(v_1 \ldots v_n) = \bigwedge \{ R_\alpha(x_1 \ldots x_{p_\alpha + q_\alpha}) :$$
$$\langle \alpha, \langle x_1 \ldots x_{p_\alpha} \rangle\rangle \text{ is represented in } B(\bar{v})$$
$$\text{and the variables } x_{p_\alpha+1} \ldots x_{p_\alpha+q_\alpha} \text{ are}$$
$$\text{such that for every } j \le q_\alpha \text{, the conjunct}$$
$$x_{p_\alpha+j} = f_j^\alpha(x_1 \ldots x_{p_\alpha}) \text{ appears in } B(\bar{v}) \} .$$

It is plain that $T^+ \models B(\bar{v}) \rightarrow A_4(\bar{v})$ and thus $T^+ \models C_3(\bar{v}) \equiv C_3(\bar{v}) \wedge A_4(\bar{v})$; if we let $A(\bar{v}) = A_3(\bar{v}) \wedge A_4(\bar{v}) \wedge \bigwedge_{i>j>k} v_i \neq v_j$ and $C(\bar{v}) = A(\bar{v}) \wedge B(\bar{v})$, then $\langle C(\bar{v}),\ a_1 \ldots a_k, b_{k+1} \ldots b_n \rangle$ is still an obstruction and it has all of the desired properties.

The preparation phase is now complete.

Now, taking reducts to $L(T)$, we have $\mathfrak{A} \prec \mathfrak{B}$ since T is model complete. Thus, by Lemma 1, $\mathfrak{A}$ satisfies

$$\exists^\infty v_{k+1} \ldots v_n A(\underline{a}_1 \ldots \underline{a}_k v_{k+1} \ldots v_n)$$

which is, of course, equivalent in T to an ordinary first-order sentence. Thus $\mathfrak{A}^+ \models Y$, where Y is the sentence

$$\exists v_1 \ldots v_k (\exists^\infty v_{k+1} \ldots v_n A(v_1 \ldots v_n) \wedge \neg \exists v_{k+1} \ldots v_n C(v_1 \ldots v_n)) .$$

We claim that Y is inconsistent with T^*.

To see this, suppose there is an existentially closed model $\mathfrak{C}^+$ of T^+ such that $\mathfrak{C}^+ \models Y$; let $c_1 \ldots c_k$ be witnesses to $v_1 \ldots v_k$ in Y. Then in particular $\mathfrak{C}^+$ satisfies

$$\exists^\infty v_{k+1} \ldots v_n A(\underline{c}_1 \ldots \underline{c}_k v_{k+1} \ldots v_n)$$

so by Lemma 1 there is an elementary extension $\mathfrak{D}^+$ of $\mathfrak{C}^+$ and elements $c_{k+1} \ldots c_n$ in $|\mathfrak{D}^+| - |\mathfrak{C}^+|$ such that $\mathfrak{D}^+ \models A(\underline{c}_1 \ldots \underline{c}_n)$.

We now modify the structure $\mathfrak{D}^+$ to produce a new structure $\mathfrak{E}^+$, as follows: let $\mathfrak{E}^+$ have the same underlying set and the same $L(T)$-structure as $\mathfrak{D}^+$, i.e. $\mathfrak{E} = \mathfrak{D}$. Then, if f_j^α is one of the Skolem functions and $e_1 \ldots e_{p_\alpha}$ are elements of $|\mathfrak{E}|$, we set

$$(f_j^\alpha)^{\mathfrak{E}^+}(e_1 \ldots e_{p_\alpha}) = (f_j^\alpha)^{\mathfrak{D}^+}(e_1 \ldots e_{p_\alpha})$$

unless it happens that $e_1 \ldots e_{p_\alpha}$ are all among $c_1 \ldots c_n$ and the pair $\langle \alpha, \langle \underline{e}_1 \ldots \underline{e}_{p_\alpha} \rangle \rangle$ is represented in $B(\underline{c}_1 \ldots \underline{c}_n)$. In that case, there is a unique $m \leq n$ such that the conjunct "$\underline{c}_m = f_j^\alpha(\underline{e}_1 \ldots \underline{e}_{p_\alpha})$" appears in $B(\underline{c}_1 \ldots \underline{c}_n)$, and we set

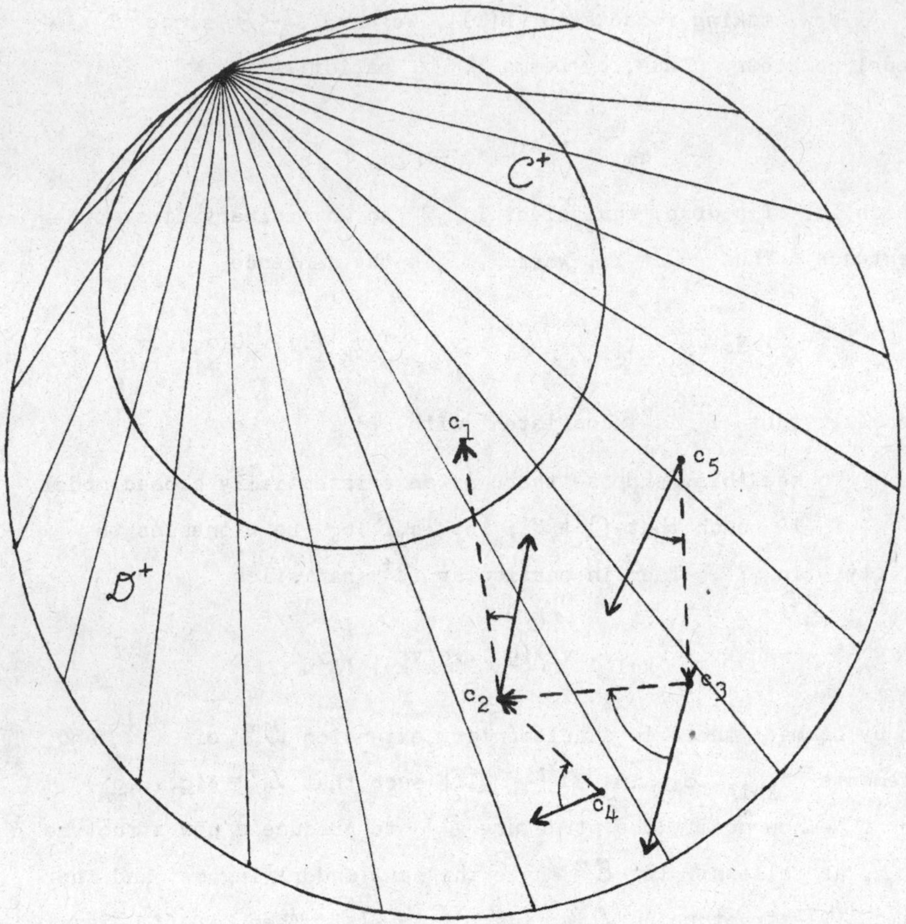

Surgery on $\mathcal{D}^+$ in the proof of Theorem 2

Here $n=5$ and $k=1$; there is just one unary Skolem function f . The configuration $B(v_1 \ldots v_5)$ is $v_1 = f(v_2) \wedge v_2 = f(v_3) \wedge v_2 = f(v_4) \wedge v_3 = f(v_5)$. Solid arrows represent the Skolem assignments on $\mathcal{D}^+$, dotted arrows the new Skolem assignments on $\mathcal{E}^+$. Rays delineate infinite definable sets; $c_1 \ldots c_5$ must be "eligible" for the configuration since the solid arrows already point into the correct sections.

$$(f_j^{\alpha})^{\mathcal{E}^+}(e_1 \ldots e_{p_\alpha}) = c_m \ .$$

Since $\mathcal{E} = \mathcal{D}$ we have that $\mathcal{E} \models T$ and $\mathcal{E}^+ \models A(\underline{c}_1 \ldots \underline{c}_n)$; by construction, $\mathcal{E}^+ \models B(\underline{c}_1 \ldots \underline{c}_n)$. From the first step in the preparation phase it follows that if f_j^{α} is a Skolem function and $d_1 \ldots d_{p_\alpha}$ are among $c_1 \ldots c_k$, and thus in particular are members of $|\mathcal{C}|$, then the term $f_j^{\alpha}(\underline{d}_1 \ldots \underline{d}_{p_\alpha})$ cannot appear in $B(\underline{c}_1 \ldots \underline{c}_n)$; so the above construction fails to alter the value of $f_j^{\alpha}(d_1 \ldots d_{p_\alpha})$. Consequently $\mathcal{E}^+$ is an extension of $\mathcal{C}^+$.

To show that $\mathcal{E}^+ \models T^+$ it suffices to check that for any $\alpha < \beta$ and any $e_1 \ldots e_{p_\alpha}$ in $|\mathcal{E}^+|$,

$$\mathcal{E}^+ \models R_\alpha(\underline{e}_1 \ldots \underline{e}_{p_\alpha} f_1^{\alpha}(\underline{e}_1 \ldots \underline{e}_{p_\alpha}) \ldots f_{q_\alpha}^{\alpha}(\underline{e}_1 \ldots \underline{e}_{p_\alpha})) \ .$$

If the pair $\langle \alpha, \langle \underline{e}_1 \ldots \underline{e}_{p_\alpha} \rangle \rangle$ is not represented in $B(\underline{c}_1 \ldots \underline{c}_n)$ then there is no problem, since $\mathcal{D}^+ \models T^+$. Otherwise, if we let d_j be the value assigned to $(f_j^{\alpha})^{\mathcal{E}^+}(e_1 \ldots e_{p_\alpha})$, the construction of the formula A_4 assures us that

$$\vdash A(\underline{c}_1 \ldots \underline{c}_n) \rightarrow R_\alpha(\underline{e}_1 \ldots \underline{e}_{p_\alpha} \underline{d}_1 \ldots \underline{d}_{q_\alpha})$$

and therefore $\mathcal{E}^+ \models R_\alpha(\underline{e}_1 \ldots \underline{e}_{p_\alpha} \underline{d}_1 \ldots \underline{d}_{q_\alpha})$.

Putting all this together, we have that $\mathcal{E}^+$ is a model of T^+ which extends $\mathcal{C}^+$, and

$$\mathcal{E}^+ \models \exists v_{k+1} \ldots v_n C(\underline{c}_1 \ldots \underline{c}_k v_{k+1} \ldots v_n) \ .$$

But by our original choice of $c_1 \ldots c_k$,

$$\mathcal{C}^+ \models \neg \exists v_{k+1} \ldots v_n C(\underline{c}_1 \ldots \underline{c}_k v_{k+1} \ldots v_n) \ ;$$

this contradicts our original assumption that $\mathcal{C}^+$ is existentially closed. It follows that T^* does axiomatize $\underline{E}$, and is therefore the model-companion of T^+.

To prove amalgamation for T^+ we again use the strong amalgamation property of T. Let $\mathcal{B}^+ \supset \mathcal{A}^+ \subset \mathcal{C}^+$ be models of T^+, and let $\mathcal{D}$ strongly amalgamate $\mathcal{B} \supset \mathcal{A} \subset \mathcal{C}$. Expand $\mathcal{D}$ to an $L(T^+)$-structure $\mathcal{D}^+$ by setting

$$(f_j^\alpha)^{\mathcal{D}^+}(d_1 \dots d_{p_\alpha}) = (f_j^\alpha)^{\mathcal{B}^+}(d_1 \dots d_{p_\alpha})$$

whenever $d_1 \dots d_{p_\alpha}$ all belong to $|\mathcal{B}|$; similarly for $d_1 \dots d_{p_\alpha} \in |\mathcal{C}|$. No conflict can result from this, since $|\mathcal{B}| \cap |\mathcal{C}| = |\mathcal{A}|$. For other $d_1 \dots d_{p_\alpha}$ in $|\mathcal{D}|$, $(f_j^\alpha)^{\mathcal{D}^+}$ is determined by choosing arbitrary witnesses to

$$\exists x_1 \dots x_{q_\alpha} R_\alpha(\underline{d}_1 \dots \underline{d}_{p_\alpha} x_1 \dots x_{q_\alpha})$$

as in the normal manner for Skolemizing a structure. $\mathcal{D}^+$ then amalgamates $\mathcal{B}^+ \supset \mathcal{A}^+ \subset \mathcal{C}^+$; consequently T^* is the model-completion of T^+ and the proof is complete.

One well-known use for a Skolem expansion of a theory is to "universalize" the theory, i.e. to provide it with a set of open axioms. A model-complete theory, being inductive, already has a set of $\forall\exists$ axioms; a given axiom, say

$$\forall v_1 \dots v_p \exists v_{p+1} \dots v_{p+q} A(v_1 \dots v_{p+q})$$

is replaced in the universalization by

$$\forall v_1 \dots v_p A(v_1 \dots v_p f_1(v_1 \dots v_p) \dots f_q(v_1 \dots v_p)) .$$

We thus have:

COROLLARY: If T is a model-complete, algebraically bounded theory, then its universalization (with respect to a given set of $\forall\exists$ axioms) has a model-completion.

Let T be a theory and suppose L' is a language containing $L(T)$, but possibly having new function, relation or constant symbols. We define the expansion of T to L' as the deductive closure of T in the language L', and denote it by $T(L')$.

THEOREM 3: Let T be model-complete and algebraically bounded, and let $L' \supset L(T)$. Then $T(L')$ has a model-completion.

Proof: Similar to the proof of Theorem 2, but much easier since there is no longer any question of "eligibility." New relation symbols are handled during the surgery on $\mathcal{D}^+$ in much the same way as new function symbols.

An alternative proof of Theorem 3 is suggested at the end of §2.

Theorem 3 has a strong converse:

THEOREM 4: Let T be a theory which is model-complete but not algebraically bounded, and let L' be any extension of $L(T)$ containing a new function or relation symbol. Then $T(L')$ fails to have even a model-companion.

Proof: Suppose first that L' has a new p-ary function symbol f, and let $A(xy_1 \dots y_n)$ be some formula of $L(T)$ which is not algebraically bounded in x (with respect to T). If $\mathcal{A}'$ is an existentially closed model of $T(L')$ with elements $a_1 \dots a_n$ such that

$$\mathcal{A}' \models \exists^\infty x A(x\bar{a})$$

then, for any $b \in |\mathcal{A}'|$,

$$\mathfrak{A}' \models \exists x(A(x\underline{\bar{a}}) \wedge f(xxx\ldots x) = \underline{b})$$

since this sentence can be realized in an extension of $\mathfrak{A}'$. Thus, if $T(L')$ has a model-companion T^* , then

$$T^* \cup \{\exists^{>M} xA(x\bar{c}) : M \in \omega\} \vdash \exists x(A(x\bar{c}) \wedge f(xxx\ldots x) = d)$$

where $c_1 \ldots c_n$ and d are new constant symbols. By compactness, there is a number $P \geq 1$ such that

$$T^* \cup \{\exists^{>P} xA(x\bar{c})\} \vdash \exists x(A(x\bar{c}) \wedge f(xxx\ldots x) = d) .$$

Now since $A(x\bar{y})$ is not algebraically bounded, there is a model $\mathfrak{B}$ of T containing elements $a_1 \ldots a_n$ such that

$$\mathfrak{B} \models \exists^{=Q} xA(x\underline{a}_1 \ldots \underline{a}_n)$$

for some $Q > P$. Choose any element e of $|\mathfrak{B}|$ and expand $\mathfrak{B}$ to a model $\mathfrak{B}'$ of $T(L')$ satisfying $\forall x(f(xxx\ldots x) = \underline{e})$. Embed $\mathfrak{B}'$ in a model $\mathfrak{C}'$ of T^* . Since T is model-complete, $\mathfrak{C}'$ can have no additional solutions to $A(x\underline{\bar{a}})$; thus, if b is an element of $|\mathfrak{B}'|$ other than e , then perforce

$$\mathfrak{C}' \models \exists^{>P} xA(x\underline{\bar{a}}) \wedge \neg\exists x(A(x\underline{\bar{a}}) \wedge f(xxx\ldots x) = \underline{b}) ,$$

a contradiction.

In the case that L' has a new relation symbol R , we have instead that

$$T^* \cup \{\exists^{>M} xA(x\bar{c}) : M \in \omega\} \vdash \exists x(A(x\bar{c}) \wedge R(xx\ldots x))$$

and a contradiction is derived in the same fashion.

<u>COROLLARY 1</u>: If T is a model-complete, algebraically bounded theory and $L' \supset L(T)$, then the model-completion of $T(L')$

is itself algebraically bounded.

Proof: Let f be a new unary function symbol, set $L'' = L' \cup \{f\}$, and let T^* be the model-completion of $T(L')$. By Theorem 3, $T(L'')$ has a model-completion, but it is immediate that this also serves as the model-companion of $T^*(L'')$. Thus, by Theorem 4, T^* must be algebraically bounded.

COROLLARY 2: If every Skolem expansion of a model-complete theory T has a model-companion, then T is algebraically bounded.

Proof: Let $T^+ = T \cup \{\forall x(f(x)=f(x))\}$. We cannot, of course, guarantee that an arbitrary Skolem expansion T^+ of a model-complete but not algebraically bounded theory has no model-companion, because for one thing T^+ might still be model-complete; this would occur for example if the Skolem function assignments were uniquely determined, or determined up to isomorphism.

COROLLARY 3: If T^+ is a Skolem expansion of a model-complete, algebraically bounded theory then the model-completion of T^+ is algebraically bounded.

Proof: Similar to the proof of Corollary 1.

COROLLARY 4: Let T be model-complete but not algebraically bounded, and suppose also that T does not have arbitrarily large finite models. Then there is a model-complete, algebraically bounded theory S in a language disjoint from T's such that $T \cup S$ has no model-companion.

Proof: Choose a unary relation symbol R not in $L(T)$ and let S be the deductive closure of $\{\exists^{>M} xy(R(x) \wedge \neg R(y)) : M \in \omega\}$. The rest of the proof proceeds as in the proof of Theorem 4, with one hitch: the model $\mathcal{B}$ of T cannot be expanded to a model of $T \cup S$ unless $\mathcal{B}$ is infinite. Hence, the extra condition on T.

§2

The purpose of this short section is to bring the algebraic boundedness condition of §1 down from model-complete theories to theories which merely possess a model-companion.

Hoping that someone will find a better term, we call a theory <u>companionable</u> if it has a model-companion. Companionability, or its absence, has been established for a fair number of theories in recent years, but in almost every case by algebraic methods related to the particular theory under consideration. In part this is because companionability is very badly behaved with respect to operations on theories----for example, expanding the language of a companionable theory could destroy its companionability. The following property at least is preserved by the operations discussed in §1.

A theory T will be said to be <u>strongly companionable</u> if it is companionable, and: for any existential formula $A(x\bar{y})$ of $L(T)$ there is a number N, such that for any model $\mathcal{A}$ of T with elements $a_1 \ldots a_n$ for which $\mathcal{A} \models \exists^{>N} x A(x\underline{\bar{a}})$, there is a model $\mathcal{B}$ of T extending $\mathcal{A}$ and satisfying $\exists^{\infty} x A(x\underline{\bar{a}})$.

<u>LEMMA 4</u>: Let T and S be mutually model-consistent theories (in the same language). Then T is strongly companionable iff S is.

<u>Proof</u>: Trivially T is companionable iff S is; suppose that T is strongly companionable. Let $A(x\bar{y})$ be an existential formula and choose N as above; we show that N also works for S. Accordingly, let $\mathcal{A}$ be an arbitrary model of S and let $a_1 \ldots a_n \in |\mathcal{A}|$ be such that $\mathcal{A} \models \exists^{>N} x A(x\underline{\bar{a}})$. Embed $\mathcal{A}$ in a model $\mathcal{B}$ of T; since existential formulas persist upward, $\mathcal{B}$ also

satisfies $\exists^{>N} xA(x\bar{a})$. By choice of N , we can embed $\mathcal{B}$ in another model $\mathcal{C}$ of T such that $\mathcal{C} \models \exists^{\infty} xA(x\bar{a})$; let $\mathcal{D}$ be a model of S containing $\mathcal{C}$. Then $\mathcal{D}$ extends $\mathcal{A}$ and satisfies $\exists^{\infty} xA(x\bar{a})$, proving the lemma.

In view of Lemma 4, one could speak of a class of cotheories (an equivalence class of theories w.r.t. mutual model-consistency) as being strongly companionable; see Henrard [9] .

LEMMA 5: A theory is strongly companionable if and only if it is companionable and its model-companion is algebraically bounded.

Proof: Automatic, using the fact that in a model-complete theory every formula is equivalent to some existential formula.

THEOREM 5: Let T be a companionable theory with no finite models. Then the following are equivalent:

(i) T is strongly companionable;

(ii) $T \cup S$ is companionable, for any strongly companionable theory S whose language is disjoint from T's;

(iii) Any Skolem expansion of T has a model-companion;

(iiii) T has a model-companion in any language (i.e. for any $L' \supset L(T)$, $T(L')$ is companionable.)

Proof: (i)$\longrightarrow$(ii), (i)$\longrightarrow$(iii), and (i)$\longrightarrow$(iiii) are consequences of Theorems 1, 2 and 3, and the converses follow from the corollaries to Theorem 4. The key fact, using * to denote the model-companion, is that $T \cup S$, T^+ , and $T(L')$ have the same model-companions as $T^* \cup S^*$, $(T^*)^+$, and $T^*(L')$ respectively;

the theorems of §1 are applied to the model-complete theory T^* via Lemma 4.

To show, for example, that $T \cup S$ and $T^* \cup S^*$ are mutually model-consistent, let $\mathcal{A} \models T \cup S$. Since $\mathcal{A}$ is infinite, we can embed $\mathcal{A}_T \subset \mathcal{B} \models T^*$ and $\mathcal{A}_S \subset \mathcal{C} \models S^*$ with $\text{card}|\mathcal{B}| = \text{card}|\mathcal{C}| > \text{card}|\mathcal{A}|$. Pasting $\mathcal{B}$ and $\mathcal{C}$ together yields a model of $T^* \cup S^*$ extending $\mathcal{A}$. Since T^* also has no finite models, the other direction is similar.

<u>COROLLARY</u>: If T is a theory (in any language) requiring no axioms, then T has a model-completion.

<u>Proof</u>: Apply (iiii) above to the theory of equality, which is easily seen to be strongly companionable.

This curious but not very useful result is already well-known for the case where $L(T)$ contains no function symbols. As it happens, we could have used the latter fact to prove Theorem 3 as follows:

We wish to show that a language-expansion $T(L')$ of a model complete, algebraically bounded theory T has a model-completion. Let L'' be L' minus the new relation symbols, and let S be the model-completion of the theory in $L'-L''$ having no axioms. $T(L'')$ is a Skolem expansion of T and thus by Theorem 2 has a model-completion, which, by Corollary 3 to Theorem 4, is itself algebraically bounded. If T^* is this model-completion, then $T^* \cup S$ is mutually model-consistent with $T(L')$; but $T^* \cup S$ has a model-completion by Theorem 1, and the rest is easy.

In the next section we establish that many of the theories which have been shown to be companionable are, in fact, strongly companionable.

§3

We begin with a few general sources of algebraically bounded theories.

THEOREM 6: Let T be a countable theory with no finite models. Then T is algebraically bounded if any of the following hold:

(i) T is $\aleph_0$-categorical,

(ii) T is $\aleph_1$-categorical, or

(iii) T is strongly minimal (but not necessarily complete).

Proof (i): Let $A(xy_1 \ldots y_n)$ be an arbitrary formula of $L(T)$; then the formulas $\{\exists^{=N}xA(xc_1 \ldots c_n) : N \in \omega\}$, where $c_1 \ldots c_n$ are new constant symbols, are pairwise inconsistent. By the theorem of Ryll-Nardzewski (see for example Sacks [20]) there are only finitely many n-types over T , so only finitely many of the above formulas can, individually, be consistent with T . It follows that $A(xy_1 \ldots y_n)$ is algebraically bounded in x .

(ii): Again let $A(x\bar{y})$ be an arbitrary formula of $L(T)$. Two structures $\mathcal{A} \precneqq \mathcal{B}$ are said to constitute a "Vaughtian pair" for the formula A if for some $\bar{a} \in |\mathcal{A}|^n$, $A^{\mathcal{A}}(x\bar{a})$ and $A^{\mathcal{B}}(x\bar{a})$ are infinite and equal. This cannot occur among models of an $\aleph_1$-categorical theory (see Sacks [20], Cor 22.5; or Baldwin and Lachlan [2] where an argument similar to the one below is used.)

Let $\underline{c}_1 \ldots \underline{c}_n$ be new constant symbols, and R a new relation symbol; we use C^R to denote the relativization of a formula C to R . The theory of pairs $\mathcal{A} \precneqq \mathcal{B}$, where $\mathcal{A} \models T$ and $c_1 \ldots c_n \in |\mathcal{A}|$, is coded up as follows:

$$S = T \cup \{\exists x \neg R(x)\} \cup \{R(\underline{c}_i) : 1 \le i \le n\}$$
$$\cup \{(R(x_1) \wedge \ldots \wedge R(x_m)) \rightarrow$$
$$(C(x_1 \ldots x_m) \equiv C^R(x_1 \ldots x_m)) :$$
$$m \in \omega, \text{ C a formula of } L(T)\} .$$

We then have that

$$S \cup \{\exists^{\infty} x A(x\underline{\bar{c}})\} \vdash \exists x (A(x\underline{\bar{c}}) \wedge \neg R(x)) \quad ,$$

so that by compactness,

$$S \cup \{\exists^{>N} x A(x\underline{\bar{c}})\} \vdash \exists x (A(x\underline{\bar{c}}) \wedge \neg R(x))$$

for some $N \in \omega$. But then N is an algebraic bound for $A(x\bar{y})$ in x ; for, suppose there exist $\mathcal{A} \models T$, $c_1 \ldots c_n \in |\mathcal{A}|$, and $M > N$ such that $\mathcal{A} \models \exists^{=M} x A(x\underline{\bar{c}})$. Since $\mathcal{A}$ is infinite, it has a proper elementary extension $\mathcal{B}$ which, perforce, has no new solutions to $A(x\underline{\bar{c}})$, a contradiction.

(iii): T is strongly minimal (Baldwin and Lachlan [2]) iff for any formula $A(x\bar{y})$ of $L(T)$,

$$T \cup \{\exists^{\infty} x A(x\underline{c}_1 \ldots \underline{c}_n)\} \cup \{\exists^{\infty} x \neg A(x\underline{c}_1 \ldots \underline{c}_n)\}$$

is inconsistent; by compactness,

$$T \cup \{\exists^{>N} x A(x\underline{c}_1 \ldots \underline{c}_n)\} \cup \{\exists^{>N} x \neg A(x\underline{c}_1 \ldots \underline{c}_n)\}$$

is inconsistent for some $N \in \omega$. Then, again N is an algebraic bound for $A(x\bar{y})$ in x , for suppose that there exist $\mathcal{A} \models T$, $c_1 \ldots c_n \in |\mathcal{A}|$ and $M > N$ such that $\mathcal{A} \models \exists^{=M} x A(x\underline{\bar{c}})$; then we must have $\mathcal{A} \models \exists^{\le N} x \neg A(x\underline{\bar{c}})$. But this is not possible, since $\mathcal{A}$ is infinite.

<u>COROLLARY</u>: All $\aleph_0$-categorical theories are strongly companionable.

<u>Proof</u>: By a theorem of Saracino [23], every $\aleph_0$-categorical theory is companionable; and it is immediate that any algebraically bounded, companionable theory is strongly companionable. We note here that algebraic boundedness by itself does not imply companionability; in fact an $\aleph_1$-categorical theory with no model-companion may be found in Saracino[21] or in Belegradek and Zil'ber [4].

An example is perhaps in order here. Let T_1 be the theory of the integers with successor function, i.e. the deductive closure of $\{\forall x \exists^{=1} y(x=s(y))\} \cup \{\forall x(x \neq s^n(x)) : n \in \omega\}$; and let T_2 be the theory of the integers with successor function and order relation. T_1 is model-complete, and, since it is $\aleph_1$-categorical, algebraically bounded. T_2 is also model-complete, but the formula $x<y<z$ is clearly not algebraically bounded in y; thus language expansions, Skolem expansions etc. of T_2 are generally not companionable. The Skolem expansion $T_2 \cup \{f(x)<x\}$, for instance, has no model-companion; but $T_2 \cup \{s(f(x))=x\}$ is still model-complete since the Skolem assignments are uniquely determined.

In each of the proofs of strong companionability below it will suffice to deal only with formulas having no disjunction symbols, for it is clear that a disjunction is algebraically bounded by the sum of the algebraic bounds of its disjuncts. In Theorems 7 and 8 the model-companions are model-completions of universal theories, and thus admit elimination of quantifiers; so for those theories we need only consider formulas which are conjunctions of basic formulas. This applies whether we prove that the model-companions are algebraically bounded, as in Theorem 7, or we show directly that a theory is strongly companionable, as in Theorem 8. In Theorem 9, the model-companion does not admit elimination of quantifiers, but Lemma 3 still enables us to look only at conjunctions of basic formulas; the price is that algebraic boundedness of the formula must now be established in more than one variable.

<u>THEOREM 7</u>: The following are strongly companionable:

(i) The theory of linear orderings;

(ii) The theory of fields;

(iii) The theory of ordered fields.

<u>Proof</u>: (i) is easy, as any conjunction of basic formulas in x defined in a dense linear ordering is reducible to one of the forms $\underline{a}=x$, $\underline{a}\neq x$, $\underline{a}<x$, $x<\underline{a}$, or $\underline{a}<x<\underline{b}$. Any of these has infinitely many solutions if it has more than one.

(ii) follows from Theorem 6, part (iii); it is also quite easy to show directly. Any polynomial equality in an algebraically closed field is algebraically bounded in x by its degree in x; any polynomial inequality is algebraically bounded by 0. One slight hitch: the usual language for the theory of fields, besides constant symbols for 0 and 1 and function symbols for addition, additive inverse and multiplication, includes a function symbol for multiplicative inverse (so that the theory of fields will be universal) which for convenience is given the value 0 at 0. Given a formula in which $(t)^{-1}$ occurs, where t is some term, we can rewrite it as the disjunction of two formulas: one for the case $t=0$, where we replace $(t)^{-1}$ by 0, and the other for the case $t\neq 0$, where we multiply through by t. In this manner all occurrences of the multiplicative inverse function can be eliminated, and we need consider only formulas of the form

$$A(x\bar{y}) = \bigwedge_{1\leq i\leq n} p_i(x\bar{y}) = \underline{0} \wedge \bigwedge_{1\leq j\leq m} q_j(x\bar{y}) \neq \underline{0}$$

where the p_i's and q_j's are polynomials; such a formula is algebraically bounded by the highest degree in x among the p_i's.

(iii) is obtainable from (i) and (ii). By eliminating inverses as above, replacing inequalities by disjunctions of $<$ and $>$, and multiplying by -1, we may assume that a given formula $A(x\bar{y})$ is of the form $B(x\bar{y}) \wedge C(x\bar{y})$, where

$$B(x\bar{y}) = \bigwedge_{1 \leq i \leq n} p_i(x\bar{y}) = \underline{0}$$

and

$$C(xy) = \bigwedge_{1 \leq j \leq m} q_j(xy) > \underline{0} \quad .$$

Now suppose that $\bar{y}$ has been interpreted as $\bar{a}$ in some real-closed field $\mathfrak{A}$. We know that $B(x\bar{y})$ is algebraically bounded, but, more than that, if $\mathfrak{A} \models \exists^{\infty} x B(x\bar{\underline{a}})$ then the polynomials $p_i(x\bar{\underline{a}})$ are all trivial and $\mathfrak{A} \models \forall x B(x\bar{\underline{a}})$. Hence it suffices to show that $C(x\bar{y})$ is algebraically bounded; but we can write each conjunct $q_j(x\bar{\underline{a}}) > \underline{0}$ of $C(x\bar{\underline{a}})$ in the form

$$\underline{b}_0(x-\underline{b}_1)\ldots(x-\underline{b}_r)((x-\underline{c}_1)^2-\underline{d}_1^2)\ldots((x-\underline{c}_s)^2-\underline{d}_s^2) > \underline{0}$$

for suitable $b_0 \ldots b_r, c_1 \ldots c_s, d_1 \ldots d_s \in |\mathfrak{A}|$. Equivalently,

$$\underline{b}_0(x-\underline{b}_1)(x-\underline{b}_2)\ldots(x-\underline{b}_r) > \underline{0}$$

but this is equivalent to a Boolean combination of the formulas $x < \underline{b}_1$, $x > \underline{b}_1$, $x < \underline{b}_2$, etc., and thus $C(x\bar{\underline{a}})$ is equivalent in $\mathfrak{A}$ to a formula in the language of linear orderings, which we will denote by $D(x\bar{\underline{b}})$. But $D(x\bar{z})$ is algebraically bounded in the theory of real-closed fields, which includes the theory of dense linear orderings. Although the structure of the formula $D(x\bar{z})$

depends on $\mathfrak{a}$ and $\bar{a}$, the above construction shows that for given $C(x\bar{y})$, there are only finitely many possibilities for $D(x\bar{z})$; the maximum of the algebraic bounds (in x) of these possibilites will serve as an algebraic bound for $C(x\bar{y})$.

An argument quite similar in technique to the one above could have been (and, in effect, has been, by Robinson & Tarski) used to show that the theory of real-closed fields admits no Vaughtian pairs, whereupon (iii) would follow from the proof of part (ii) of Theorem 6.

Theorem 7, together with the corollary to Theorem 2, answers Robinson's question, noted in the introduction; namely, any universalization of the theory of algebraically closed fields or the theory of real-closed fields has a model-completion. A recursive axiomatization of it is derivable without great difficulty from the ideas in the proof of Theorem 2.

Note that Robinson's question is in a sense less acute in the case of real-closed fields, since in this case the Skolem functions can be "pinned down"; e.g. one could specify the smallest root of each odd-degree polynomial be chosen by the Skolem functions. Thus at the expense of some additional complication the theory of real-closed fields can be universalized without destroying its model-completeness.

Robinson's original work on the model-completions of the theories of fields and of ordered fields may be found, e.g.,in [16].

In what follows all rings will be assumed to have a unit $1 \neq 0$; all modules will be left modules and all ideals left ideals. The theory T_R of modules over a ring R will include in its language the usual language of additive abelian groups, together with a unary function symbol r for each element r of R ; its action on an element c of a module will be denoted in a formula by $r\underline{c}$.

Eklof and Sabbagh ([8]) have shown that T_R has a model-companion (which is in fact a model-completion) if and only if R is coherent, i.e. any finitely generated (left) ideal of R is finitely presented.

<u>THEOREM 8</u>: The theory of modules over a coherent ring R is strongly companionable.

<u>Proof</u>: Since T_R is universal, its model-completion admits elimination of quantifiers. Let $\mathfrak{A}$ be an arbitrary model of T_R, and $A(x\underline{\bar{c}})$ a formula defined in $\mathfrak{A}$ of the form $B(x\underline{\bar{c}}) \wedge C(x\underline{\bar{c}})$, where

$$B(x\underline{\bar{c}}) = \bigwedge_{1 \leq i \leq m} r_i x = \underline{c}_i \quad \text{and}$$

$$C(x\underline{\bar{c}}) = \bigwedge_{m < i \leq n} r_i x \neq \underline{c}_i \quad .$$

To show that T_R is strongly companionable it suffices to show that if $A(x\underline{\bar{c}})$ has more than one solution in $\mathfrak{A}$, then there is a model of T_R extending $\mathfrak{A}$ which has a new solution. (An extension satisfying $\exists^{\infty} x A(x\underline{\bar{c}})$ would then be obtainable as the union of a chain.)

Accordingly, let a be a solution to $A(x\underline{\bar{c}})$, i.e. $\mathfrak{A} \models A(\underline{a}\underline{\bar{c}})$. Since $x \longmapsto x-a$ is a bijection of $|\mathfrak{A}|$, we may, by changing variables, assume that a=0 ; thus, in the expression above for $A(x\underline{\bar{c}})$, we have $c_i = 0$ for $1 \leq i \leq m$ and $c_i \neq 0$ for $m < i \leq n$.

Let I be the (left) ideal generated by $r_1 \ldots r_m$. If $I = R$, then for suitable $s_1 \ldots s_m \in R$,

$$\sum_{1\le i\le m} s_i r_i = 1$$

and thus if b is a solution to $A(x\underline{\bar{c}})$,

$$b = \Big(\sum_{1\le i\le m} s_i r_i\Big)b = \sum_{1\le i\le m} s_i(r_i b) = 0$$

so $A(x\underline{\bar{c}})$ has only one solution in $\mathcal{A}$.

Otherwise let J be a maximal ideal containing I , and, regarding R and J as R-modules, put $\mathcal{B} = \mathcal{A} \oplus R/J$. Let d be the pair $(0, 1+J)$ in $\mathcal{B}$; we claim $\mathcal{B} \models A(\underline{d\bar{c}})$.

Clearly, $\mathcal{B} \models B(\underline{d\bar{c}})$ since J annihilates d ; thus, if $\mathcal{B} \models \neg A(\underline{d\bar{c}})$, then $r_i d = c_i$ for some i with $m < i \le n$. Since $c_i \ne 0$, $r_i \notin J$; thus for suitable $s \in R$ and $t \in J$, $sr_i + t = 1$. But then $d = (sr_i + t)d = sr_i d = sc_i \in |\mathcal{A}|$, a contradiction which proves the theorem.

<u>COROLLARY</u>: The theory of abelian groups is strongly companionable.

The model-completion in this case is the theory of divisible abelian groups with infinitely many elements of each prime order. The difference in language between the theory of abelian groups and the theory of $\mathbb{Z}$-modules is not significant.

Companionability among classes of groups wider than the class of abelian groups is elusive; neither the theory of all groups (Eklof and Sabbagh, [8]) nor the theory of nilpotent groups of class n (Saracino [22]) nor the theory of groups solvable of rank n (Saracino[24]) possess model companions.

The theory of ordered abelian groups has a model completion (the theory of divisible ordered abelian groups--see Robinson [18]) which is easily seen to be algebraically bounded.

Cherlin [6] used nilpotent elements to show that the theory of commutative rings has no model-companion (his proof extends to the theory of rings). Subsequently, Lipshitz and Saracino [13] showed that the theory of commutative rings <u>without</u> nilpotent elements does have a model companion, which we will denote by T^*. An equivalent result (that T^* is the model completion of the theory of commutative regular rings) was obtained independently by Carson [5].

T^* is the theory of commutative rings which are von Neumann-regular ($\forall x \exists y(x^2y = x)$); have a root for every monic polynomial; and have no minimal idempotents ($\forall x \exists y(x \neq 0 \wedge x^2 = x \rightarrow y \neq 0 \wedge y^2 = y \wedge xy = y)$) or, equivalently, the idempotents form an atomless Boolean algebra under $e \cap f = ef$, $e \cup f = e + f - ef$. In any such ring $\mathcal{A}$ the maximal ideals have intersection $\{0\}$, and we can thus embed $\mathcal{A} \subset \mathcal{B} = \prod \mathcal{A}/M$ where the product ranges over the maximal ideals M of $\mathcal{A}$. Since each $\mathcal{A}/M$ is a field, the idempotents of $\mathcal{B}$ are exactly the points f such that for any M, $f(M) = 1$ or 0 in $\mathcal{A}/M$.

We will make use of the following result of Lipshitz and Saracino, which appears as a lemma in [13]: let $\mathcal{A}$ and $\mathcal{B}$ be as above, with $a_1 \ldots a_n \in |\mathcal{A}|$. For any quantifier-free formula $A(\bar{x})$ of $L(T^*)$, define the element $[A(\bar{a})]$ in $\mathcal{B}$ by:

$$[A(\bar{a})](M) = \begin{cases} 1 & \text{if} \quad \mathcal{A}/M \models A(\underline{\bar{a}(M)}) ; \\ 0 & \text{if} \quad \mathcal{A}/M \models \neg A(\underline{\bar{a}(M)}) . \end{cases}$$

Then $[A(\bar{a})]$ is an element of $\mathcal{A}$.

<u>THEOREM 9</u>: The theory of commutative rings without nilpotent elements is strongly companionable.

<u>Proof</u>: We show that T^* is algebraically bounded.

Let $\mathcal{A}$ and $\mathcal{B}$ be as above, and suppose that $A(\bar{x})$ is a conjunction of basic formulas in $L(\mathcal{A})$; using Lemma 3, it suffices to show that if there are two dissimilar solutions $\bar{a}$ and $\bar{b}$ to $A(\bar{x})$ in $\mathcal{A}$, then there are d-infinitely many.

We may write

$$A(\bar{x}) = \bigwedge_{1\leq j\leq p} B_j(\bar{x}) \wedge \bigwedge_{1\leq j\leq q} C_j(\bar{x})$$

where each $B_j(\bar{x})$ is an equality, each $C_j(\bar{x})$ an inequality. Then, for an arbitrary n-tuple $\bar{d}$ from $|\mathcal{A}|^n$, $\mathcal{A}\models A(\underline{\bar{d}})$ iff for each $j\leq p$, $B_j(\underline{\bar{d}(M)})$ holds in every factor $\mathcal{A}/M$; and for every $j\leq q$, $C_j(\underline{\bar{d}(M)})$ holds in <u>some</u> factor $\mathcal{A}/M$.

In particular, since $\bar{a}$ is a solution, $[C_j(\bar{a})] \neq 0$ for each $j\leq q$. Since $\mathcal{A}$ has no minimal idempotents, we may then construct an idempotent e in $\mathcal{A}$ such that

(1) $\forall j\leq q$, $e\cap[C_j(\bar{a})]\neq 0$ and

(2) $\forall i\leq n$, $(1-e)\cap[a_i\neq b_i]\neq 0$.

For each $i\leq n$, set $e_i = (1-e)\cap[a_i\neq b_i]$ and choose an infinite set of distinct idempotents $\{e_i^k : k\in\omega\}$ from $\mathcal{A}$ such that for any k, $e_i^k\subset e_i$. Let $f^k = \bigcup\{e_i^k : k\in\omega\}$ and put $c_i^k = f^k b_i + (1-f^k)a_i$. Let $\bar{c}^k$ stand for the n-tuple $\langle c_1^k\ldots c_n^k\rangle$.

We claim that $\{\bar{c}^k : k\in\omega\}$ is an infinite pairwise dissimilar set of solutions to $A(\bar{x})$ in $\mathcal{A}$. There are three things to check:

(i) If $k\neq k'$, then $\bar{c}^k$ and $\bar{c}^{k'}$ are dissimilar;

(ii) For any k, j, and M, $\mathcal{A}/M\models B_j(\underline{\bar{c}^k(M)})$;

(iii) for any k and j there exists an M such that $\mathcal{A}/M \models C_j(\underline{\bar{c}^k(M)})$.

To show (i), let $i \leq n$ and choose M so that $e_i^k(M) \neq e_i^{k'}(M)$; say $e_i^k(M) = 1$. Then $c_i^k(M) = b_i(M)$ and $c_i^{k'}(M) = a_i(M)$. But $e_i(M) = 1$ since $e_i \supset e_i^k$ and $[a_i \neq b_i] = 1$ since $[a_i \neq b_i] \supset e_i$; hence $a_i(M) \neq b_i(M)$. Thus $c_i^k \neq c_i^{k'}$ for each i , as required.

(ii) is clear since for any k and M , either $\bar{c}^k(M) = \bar{a}(M)$ or $\bar{c}^k(M) = \bar{b}(M)$.

Lastly, note that $f^k \cap e = 0$ so $[\bar{a} = \bar{c}^k] \supset e$, but $e \cap [C_j(\bar{a})] \neq 0$ and thus $[C_j(\bar{c}^k)] \supset [\bar{a} = \bar{b}] \cap [C_j(\bar{a})] \neq 0$; (iii) follows, and the theorem is proved.

§4

In this section T will be a countable, model-complete, algebraically bounded theory; T^+ , a Skolem expansion of T ; T^* , the model-completion of T^+ . The proof of Theorem 2, §1, shows that T^* can be axiomatized by the axioms of T^+ together with a set Φ of sentences $\forall v_1 \ldots v_k C(v_1 \ldots v_k)$, where:

(i) $C(v_1 \ldots v_k) = \exists^{>N} v_{k+1} \ldots v_n (A(\bar{v}) \wedge B'(\bar{v})) \rightarrow \exists v_{k+1} \ldots v_n (A(\bar{v}) \wedge B(\bar{v}))$;

(ii) $A(\bar{v})$ is a quantifier-free formula of $L(T)$;

(iii) $B(\bar{v})$ is a (uniform) configuration, i.e. a conjunction of equalities of the sort $x_0 = f(x_1 \ldots x_m)$ involving Skolem functions;

(iv) $B'(\bar{v})$ is a formula of $L(T)$ which codes "eligibility" for the configuration $B(\bar{v})$, so that the following <u>alteration property</u> holds: if $a_1 \ldots a_n$ are elements of a model $\mathcal{A}^+$ of T^+ such that $\mathcal{A}^+ \models B'(\bar{a})$, then the result of altering the Skolem structure of $\mathcal{A}^+$ precisely so as to satisfy $B(\bar{a})$ is again a model of T^+.

(v) N is an algebraic bound for $A(\bar{v}) \wedge B'(\bar{v})$ in $v_{k+1} \ldots v_n$, i.e. the quantifier "$\exists^{>N}$" in (i) is equivalent to "$\exists^{\infty}$".

We now ask the question: when can a model of T be expanded to a model of T^*? To see that the answer is not "always", let T be the deductive closure of $\{\exists^{>M} xy(R(x) \wedge \neg R(y)) : M \in \omega\}$ and let T^+ be formed from T by adding to $L(T)$ a single unrestricted unary Skolem function f. Then the following is a theorem of T^*:

$$\forall x \exists y (R(y) \wedge f(y) = x) \quad .$$

But if $\mathcal{A}$ is an uncountable model of T such that $\mathrm{card}(R^{\mathcal{A}}(x)) = \aleph_0$, then no expansion of $\mathcal{A}$ can possibly satisfy this. It turns out, however, that two-cardinal models are the only obstruction.

A structure $\mathcal{A} \models T$ will be termed a <u>one-cardinal model</u> if every infinite definable (with parameters) subset of $|\mathcal{A}|$ has the same cardinality. Thus, any finite or countable model is a one-cardinal model; if T admits no Vaughtian pairs, then every model of T is a one-cardinal model.

LEMMA 6: Let $\mathcal{A}$ be a one-cardinal model of an algebraically bounded theory; say card $|\mathcal{A}| = \kappa$. Then any definable d-infinite set of n-tuples from $\mathcal{A}$ has a pairwise dissimilar subset of cardinality κ.

Proof: Assume the lemma is true for k-tuples, $k \leq n$, and suppose that $\mathcal{A} \models \exists^{\infty} x y_1 \ldots y_n A(x y_1 \ldots y_n)$ where A is a formula of $L(\mathcal{A})$. We wish to find a pairwise dissimilar subset of $A^{\mathcal{A}}(x y_1 \ldots y_n)$ having cardinality κ.

Let D be the domain of A, i.e. $D = \{a \in |\mathcal{A}| : \mathcal{A} \models \exists \bar{y} A(\underline{a}\bar{y})\}$, and let $D' = \{a \in D : \mathcal{A} \models \exists^{\infty} \bar{y} A(\underline{a}\bar{y})\}$. If D' is infinite it has cardinality κ, since, on account of algebraic boundedness, D' is definable. In that case let $D' = \{a_\alpha : \alpha < \kappa\}$.

Define a function $f : \kappa \longrightarrow |\mathcal{A}|^n$ by induction, as follows: let $f(\alpha)$ be any n-tuple $\bar{b}$ such that $\mathcal{A} \models A(\underline{a_\alpha}\underline{\bar{b}})$ and such that for any $\beta < \alpha$, $\bar{b}$ is dissimilar to $f(\beta)$. This can never run into trouble, since for any $\alpha < \kappa$, $\{\bar{b} : \mathcal{A} \models A(\underline{a_\alpha}\underline{\bar{b}})\}$ is definable and d-infinite, and thus contains a pairwise dissimilar subset of cardinality κ; and by stage α fewer than κ-many of the members of this subset have been ruled out. The set $\{\langle a_\alpha b_1 \ldots b_n \rangle \in |\mathcal{A}|^{n+1} : \alpha < \kappa,\ \bar{b} = f(\alpha)\}$ fulfills the requirements of the lemma.

Next suppose that D' is not infinite and let $D' = \{a_1 \ldots a_m\}$. The set $\{\langle a b_1 \ldots b_n \rangle : \mathcal{A} \models A(\underline{a}\underline{\bar{b}})$ and $a \notin D'\}$ is still d-infinite, since at most m of infinitely many pairwise dissimilar solutions to $A(x\bar{y})$ have been eliminated; thus its range

$$R = \{\bar{b} \in |\mathcal{A}|^n : \mathcal{A} \models \exists x (A(x\underline{\bar{b}}) \wedge \bigwedge_{1 \leq i \leq m} x \neq \underline{a_i})$$

is also d-infinite. By the induction assumption R contains a pairwise dissimilar subset R' of cardinality κ ; let $J = \{<ab_1 \ldots b_n> : \bar{b} \in R', \mathfrak{A} \models A(\underline{a}\bar{\underline{b}})$ and $a \notin D'\}$, so that $|J| = \kappa$ also. But any given first coordinate can appear only finitely many times in J, by definition of D' ; thus J has a pairwise dissimilar subset of cardinality κ, proving the lemma.

<u>THEOREM 10</u>: Let T be a countable, model-complete and algebraically bounded theory, let T^+ be any Skolem expansion of T, and let T^* be the model-completion of T^+. Then any one-cardinal model of T can be expanded to a model of T^*.

<u>Proof</u>: Let $\mathfrak{A}$ be a one-cardinal model of T, say of cardinality κ, and let the set Φ of axioms for T^* be as described above. Let $\Phi(\mathfrak{A})$ be the set of instances in $\mathfrak{A}$ of formulas in Φ, i.e.

$$\Phi(\mathfrak{A}) = \{C(\underline{a}_1 \ldots \underline{a}_k) : a_1 \ldots a_k \in |\mathfrak{A}|, \text{ and } "\forall v_1 \ldots v_k C(v_1 \ldots v_k)" \in \Phi\} .$$

It will thus suffice to expand $\mathfrak{A}$ to $\mathfrak{A}^+ \models T^+ \cup \Phi(\mathfrak{A})$.

Let Θ be a set of sentences of the form $\underline{a}_0 = f(\underline{a}_1 \ldots \underline{a}_m)$, for various Skolem functions f and members $a_0 \ldots a_m$ of $|\mathfrak{A}|$. Θ will be called a <u>partial expansion</u> of $\mathfrak{A}$ if $|\Theta| < \kappa$ and $\Theta \cup T^+ \cup D\mathfrak{A}$ is consistent, where $D\mathfrak{A}$ is the complete diagram of $\mathfrak{A}$. We will define by induction an increasing chain $\{\Theta_\alpha : \alpha < \kappa\}$ of partial expansions of $\mathfrak{A}$.

First, fix a κ-enumeration of $\Phi(\mathfrak{A})$. Set $\Theta_0 = \emptyset$, and if β is a limit ordinal, set $\Theta_\beta = \bigcup_{\alpha<\beta} \Theta_\alpha$. At stage $\alpha+1$, let $C(\underline{a}_1 \ldots \underline{a}_k)$ be the αth member of $\Phi(\mathfrak{A})$; then we may assume that C is in the form given in (i) through (v) above, so that

$$C(\underline{a}_1\ldots\underline{a}_k) =$$

$$\exists^{>N} v_{k+1}\ldots v_n(A(\underline{a}_1\ldots\underline{a}_k v_{k+1}\ldots v_n)\wedge B'(\underline{a}_1\ldots\underline{a}_k v_{k+1}\ldots v_n))$$
$$\rightarrow \exists v_{k+1}\ldots v_n(A(\underline{a}_1\ldots\underline{a}_k v_{k+1}\ldots v_n)\wedge B(\underline{a}_1\ldots\underline{a}_k v_{k+1}\ldots v_n)) .$$

Now if $\mathcal{A}\models\exists^{\leq N} v_{k+1}\ldots v_n(A(\underline{a}_1\ldots v_n)\wedge B'(\underline{a}_1\ldots v_n))$, set $\theta_{\alpha+1} = \theta_\alpha$. Otherwise, $\mathcal{A}\models\exists^{\infty} v_{k+1}\ldots v_n(A(\underline{a}_1\ldots v_n)\wedge B'(\underline{a}_1\ldots v_n))$, so by Lemma 6 there is a pairwise dissimilar set G of witnesses to $v_{k+1}\ldots v_n$ with $|G| = \kappa$. Since $|\theta_\alpha| < \kappa$, there is an (n-k)-tuple $\langle a_{k+1}\ldots a_n\rangle \in G$ none of whose coordinates is mentioned in θ_α ; let $\theta_{\alpha+1}$ be the union of θ_α and the set of conjuncts of $B(\underline{a}_1\ldots\underline{a}_k\underline{a}_{k+1}\ldots\underline{a}_n)$. It follows from the alteration property that $T^+\cup D\mathcal{A}\cup\theta_\alpha\cup\{B(\underline{a}_1\ldots\underline{a}_n)\}$ is consistent, so $\theta_{\alpha+1}$ is again a partial expansion of $\mathcal{A}$.

Finally, set $\theta = \bigcup_{\alpha<\kappa}\theta_\alpha$. Since $T^+\cup D\mathcal{A}\cup\theta$ is still consistent, it is clear that we can complete the Skolem expansion of $\mathcal{A}$ in the usual manner, and so obtain an expansion $\mathcal{A}^+$ which is a model of $T^+\cup\theta$; but then by construction, $\mathcal{A}^+\models\Phi(\mathcal{A})$ and the theorem is proved.

<u>COROLLARY</u>: If T is as above, and in addition admits no Vaughtian pairs, then any model of T can be expanded to a model of T* .

The corollary applies to both cases of original interest to Robinson, i.e. the theories of algebraically closed fields and real-closed fields.

In the next section Theorem 10, and the technique used in its proof, will be applied to a particular problem requiring the simultaneous expansion of two structures.

§5

The following question was asked by Professor H. Jerome Keisler of the University of Wisconsin, at the UCLA Logic Year ('67-'68) :

> Let $\mathcal{A}$ and $\mathcal{B}$ be L-structures with $\mathcal{A} \prec \mathcal{B}$, and let $S(x_1 \ldots x_m y)$ be a formula of L. Can a Skolem function $f\colon |\mathcal{B}|^m \to |\mathcal{B}|$ for S be found so as to preserve the elementary embedding?

A pair $\mathcal{A} \prec \mathcal{B}$ for which this holds, for all formulas S, will be said to have Keisler's Property (KP). Not all pairs have KP; uncountable counterexamples have been found by Payne [15] and Knight [10]. However, in the countable case ($\mathcal{A}$, $\mathcal{B}$ and L countable) the question is still open.

Positive results for this problem have proved to be surprisingly difficult to attain, even in the case of countable dense linear orderings (CDLO's, for short.) Keisler himself proved KP for the case where $\mathcal{A}$ is a CDLO and $\mathcal{B}$ a one-point extension; Julia Knight used weak second-order forcing to extend this result to the case where $|\mathcal{B}| - |\mathcal{A}|$ consists entirely of isolated points of $\mathcal{B}$ (see [11]). One consequence of this section will be the generalization of these results to the case where $|\mathcal{A}|$ is dense in $\mathcal{B}$ and the case where $|\mathcal{B}| - |\mathcal{A}|$ is dense in $\mathcal{B}$.

Our method is roughly as follows: suppose $\mathcal{A}$ and $\mathcal{B}$ are models of an algebraically bounded theory T. We may (and do) assume, by "Morleyizing", that T admits elimination of quantifiers and thus in particular is model-complete. Let T^+ be the Skolem expansion of T for the formula S, i.e.

$$T^+ = T \cup \{\exists y S(x_1 \ldots x_m y) \rightarrow S(x_1 \ldots x_m f(x_1 \ldots x_m))\} ;$$

then by Theorem 2, T^+ has a model-completion T^*. By Theorem 10, $\mathcal{A}$ and $\mathcal{B}$ can each be expanded to models $\mathcal{A}^+$ and $\mathcal{B}^+$ of T^*. We would then have $\mathcal{A}^+ \prec \mathcal{B}^+$ as sought, provided the two expansions are compatible (i.e. for any $a_1 \ldots a_m \in |\mathcal{A}|$, $f^{\mathcal{A}^+}(a_1 \ldots a_m) = f^{\mathcal{B}^+}(a_1 \ldots a_m)$), but it is not always possible to guarantee this. Alternatively we could expand $\mathcal{B}$ and let $\mathcal{A}$ take care of itself, but then we have to worry about whether the Skolem function leads out of $\mathcal{A}$ (i.e. for some $a_1 \ldots a_m \in |\mathcal{A}|$, $f^{\mathcal{B}^+}(a_1 \ldots a_m) \notin |\mathcal{A}|$.) Curiously, complementary assumptions are needed for the two procedures.

In the following definitions, X is a subset of $|\mathcal{B}|$, and by "definable" we mean "definable with parameters in $\mathcal{B}$."

X is <u>dense</u> (respectively <u>codense</u>) in $\mathcal{B}$ if every infinite definable subset of $|\mathcal{B}|$ intersects X (resp. $|\mathcal{B}| - X$).

X is <u>d-dense</u> (resp. <u>d-codense</u>) in $\mathcal{B}$ if, for any $n \geq 1$, every d-infinite definable subset of $|\mathcal{B}|^n$ intersects X^n (resp. $(|\mathcal{B}| - X)^n$).

Correspondingly an embedding $\mathcal{A} < \mathcal{B}$ will be termed dense (codense, etc.) if $|\mathcal{A}|$ is dense (etc.) in $\mathcal{B}$; note that codensity in this sense is strictly stronger than not being a Vaughtian pair. This notion of density coincides with that used elsewhere in this volume by Macintyre [14] but d-density is unfortunately somewhat stronger. Proper embeddings of algebraically closed fields, for example, are dense but not d-dense (if b is an element not in the small field, then the formula $x_1 + x_2 = \underline{b}$ defies d-density.) The following example, suggested by Walter Baur, shows that also codensity and d-codensity are not equivalent: Let G_2 be the two-element group, set $\mathcal{A} = \bigoplus_{\omega} G_2$, $\mathcal{B} = \mathcal{A} \oplus G_2$. An automorphism argument shows $\mathcal{A}$ is a codense elementary substructure of $\mathcal{B}$, but the formula $x_1+x_2+x_3=0$ has no solutions in $(|\mathcal{B}|-|\mathcal{A}|)^3$.

In the case of CDLO's it is easily seen that density, d-density and ordinary order-density coincide; similarly for codensity.

<u>THEOREM 11</u>: Let T be a countable, algebraically bounded theory. Then any d-codense embedding of countable models of T has KP.

<u>Proof</u>: Suppose that $\mathcal{A}$, $\mathcal{B}$ and T satisfy the hypotheses and let $S(x_1 \ldots x_m y)$ be an arbitrary formula of L(T). Let T^+ be the Skolem expansion of T for S, T* its model-completion (recall that we can assume T is model-complete). Expand $\mathcal{A}$ to a model $\mathcal{A}^+$ of T* via Theorem 10.

We now expand the rest of $\mathcal{B}$ by constructing a chain $<\theta_\alpha : \alpha < \omega>$ of partial expansions of $\mathcal{B}$; we use the notation in the proof of Theorem 10.

Set $\theta_0 = \emptyset$. At stage $\alpha+1$ let the αth member $C(a_1 \ldots a_k)$ of $\Phi(\mathcal{B})$ be in the usual "canonical form"; if it is not the case that $\mathcal{B} \models \exists^\infty v_{k+1} \ldots v_n (A(\underline{a}_1 \ldots v_n) \wedge B'(\underline{a}_1 \ldots v_n))$, let $\theta_{\alpha+1} = \theta_\alpha$. Otherwise,

$$\mathcal{B} \models \exists^\infty v_{k+1} \ldots v_n (A(\underline{a}_1 \ldots v_n) \wedge B'(\underline{a}_1 \ldots v_n) \wedge \bigwedge_{\substack{1 \le j \le p \\ k < i \le n}} v_i \ne \underline{b}_j)$$

where $\{b_j : 1 \le j \le p\}$ is a list of the elements of $|\mathcal{B}| - |\mathcal{A}|$ whose names appear in θ_α. In view of d-codensity this formula has a solution $<a_{k+1} \ldots a_n> \in (|\mathcal{B}| - |\mathcal{A}|)^{n-k}$, and we can then let

$\Theta_{\alpha+1}$ be the union of Θ_α and the set of conjuncts of $B(\underline{a}_1 \ldots \underline{a}_k \underline{a}_{k+1} \ldots \underline{a}_n)$. Finally set $\Theta = \bigcup_{\alpha \in \omega} \Theta_\alpha$ and let $\mathcal{B}^+$ be any expansion of $\mathcal{B}$ consistent with Θ, T^+ and the diagram of $\mathcal{A}^+$. Then $\mathcal{A}^+ \subset \mathcal{B}^+ \models T^*$, so $\mathcal{A}^+ \prec \mathcal{B}^+$ and the theorem is proved.

Before going on to examples and corollaries, we remark on the limitations of this method in proving KP. First, there are embeddings of countable models which cannot be compatibly expanded to models of T^* ; for example let $\mathcal{B}$ be an end-extension of a CDLO $\mathcal{A}$, e.g. take $\mathcal{A}$ to be the negative rationals, $\mathcal{B}$ all the rationals. If $S(xy)$ is $x < y$, then $T^* \vdash \forall xy \exists z\,(z < x \wedge f(z) > y)$ but if $x \in |\mathcal{A}|$ and $y \notin |\mathcal{A}|$ this cannot hold in $\mathcal{B}$ unless the Skolem function f leads out of $\mathcal{A}$. (It does happen that KP holds in this case; it is an amusing exercise to prove it.)

It might seem that the algebraic boundedness condition is stronger than necessary, since we do not use anywhere near the full strength of the fact that T^* is the model-completion of T^+ . The proof of Theorem 2, in the absence of algebraic boundedness, gives an axiomatization for the existentially closed models of T^+ in the language having the quantifier $\exists^\infty$; the proof of Theorem 10 then enables us to expand any countable model to an existentially closed model of T^+ . Thus d-codensity allows us to expand $\mathcal{A}$ and $\mathcal{B}$ compatibly to existentially closed models; if these were <u>generic</u> models, in the sense of infinite forcing (see, e.g., [17]) we would have the desired elementary extension. It sometimes happens that $\underline{E} = \underline{G}$, where $\underline{G}$ is the class of generic models, even when there is no model-companion; but, alas, this is not generally the case for T^+ . As an example let T be the theory of the natural numbers with constant 0 , order $<$, and successor function s . Let $\mathcal{A}$ be the prime model of T , and let $\mathcal{B}$ be a countable saturated elementary extension; it is easily checked that this embedding is d-codense. Let S(xy) be $x<y$ and expand $\mathcal{A}$ and $\mathcal{B}$ compatibly to existentially closed models of T^+ , using the proof of Theorem 11. Now, since $\mathcal{B}^+ \models \exists x(\exists^\infty y(y<x))$, $\mathcal{B}^+$ must satisfy

$$\exists x \forall z \exists y(z>x \longrightarrow y<x \wedge f(y)=z)$$

but this clearly cannot hold in $\mathcal{A}^+$, so $\mathcal{A}^+ \not\prec \mathcal{B}^+$ and consequently $\underline{E} \neq \underline{G}$.

We now exhibit a few consequences of Theorem 11.

<u>COROLLARY 1</u>: Any codense embedding of CDLO's has KP.

Here, as remarked above, we can take codense to mean simply that $|\mathcal{B}| - |\mathcal{A}|$ is order-dense in $\mathcal{B}$.

<u>COROLLARY 2</u>: Any embedding of countable algebraically closed fields has KP.

<u>Proof</u>: Let T_2 be the theory of pairs of algebraically closed fields, i.e. the theory of algebraically closed fields with a distinguished proper algebraically closed subfield. By a theorem of Robinson [19] T_2 is complete. Since the ordinary theory is algebraically bounded, d-codensity is expressible in the language of T_2 ; thus if any embedding of algebraically closed fields is d-codense, they all are.

But, any infinite structure $\mathcal{A}$ has an elementary extension $\mathcal{B}$ in which it is d-codense; e.g., take $\mathcal{B}$ to be $(\mathrm{card}\,|\mathcal{A}|)^+$-saturated. If $\mathcal{B} \models \exists^\infty v_1 \ldots v_k C(v_1 \ldots v_k \underline{b}_{k+1} \ldots \underline{b}_n)$ then

$$\{C(v_1 \ldots v_k \underline{b}_{k+1} \ldots \underline{b}_n)\} \cup \{v_i \neq \underline{a} \,:\, 1 \leq i \leq k \,,\, a \in |\mathcal{A}|\}$$

can be extended to a consistent k-type over $|\mathcal{A}| \cup \{b_{k+1} \ldots b_n\}$, which must then be satisfied in $\mathcal{B}$.

With slightly more effort one can remove the countability assumption from the statement of the corollary.

Corollary 3 below is actually a strong form of the theorem. In it we use the symbol $T^{\mathcal{S}}$ for the full iterated Skolem expansion of T, i.e. $T^{\mathcal{S}} = \bigcup_n T^n$ where $T^0 = T$ and T^{n+1} is the Skolem expansion of T^n with respect to all formulas of $L(T^n)$.

COROLLARY 3: Let $\mathcal{A} \prec \mathcal{B}$ be a d-codense embedding of countable models of a countable, algebraically bounded theory T. Then $\mathcal{A}$ and $\mathcal{B}$ can be expanded compatibly to models of $T^{\mathcal{S}}$.

Proof: Clearly, this follows by repeated applications of Theorem 11, provided we can show that the hypotheses are preserved; i.e. that T^* is algebraically bounded and the new embedding $\mathcal{A}^+ \prec \mathcal{B}^+$ is still d-codense. The former is just the statement of Corollary 3 to Theorem 4, back in §1; the latter requires some work.

Accordingly, let $D(v_1 \ldots v_p \underline{b}_1 \ldots \underline{b}_q)$ be an arbitrary formula of $L(\mathcal{B}^+)$ and suppose that $\mathcal{B} \models \exists^\infty v_1 \ldots v_p D(v_1 \ldots v_p \underline{b}_1 \ldots \underline{b}_q)$. Since T^* is model-complete, D is equivalent to an existential formula $\exists v_{p+1} \ldots v_r E(v_1 \ldots v_r \underline{b}_1 \ldots \underline{b}_q)$; and we may assume that E has no disjunction symbols, since if we write E in disjunctive form, $\exists v_{p+1} \ldots v_r E_i(v_1 \ldots \underline{b}_q)$ must have d-infinitely many solutions for at least one of the disjuncts E_i.

Now, using the techniques of the first section, we may put everything into splintered form so that $D(v_1 \ldots v_p \underline{b}_1 \ldots \underline{b}_q)$ is now equivalent to

$$\exists v_{p+1} \ldots v_n (A(v_1 \ldots v_n \underline{b}_1 \ldots \underline{b}_q) \wedge B(v_1 \ldots v_n \underline{b}_1 \ldots \underline{b}_q))$$

where A is free of both quantifiers and Skolem functions, and B is a configuration. If we let $B'(v_1 \ldots v_n \underline{b}_1 \ldots \underline{b}_q)$ be the eligibility criterion for B, then we have that $\mathcal{B}^+ \models$

$$\exists^\infty v_1 \ldots v_p \exists v_{p+1} \ldots v_n (A(v_1 \ldots \underline{b}_q) \wedge B'(v_1 \ldots \underline{b}_q)) .$$

Now some sleight-of-hand is necessary to dispose of the inner quantifier. Choose any infinite pairwise-dissimilar set of p-tuples witnessing $v_1 \ldots v_p$, and for each of these p-tuples a corresponding (n-p)-tuple witnessing $v_{p+1} \ldots v_n$. Check to see whether v_{p+1} takes on infinitely many different values in this list. If so, we can commute v_{p+1} with the existential quantifier, i.e. $\mathcal{B}^+ \models$

$$\exists^\infty v_1 \ldots v_{p+1} \exists v_{p+2} \ldots v_n (A(\text{---}) \wedge B'(\text{---})) .$$

If not, there is some value a_{p+1} for v_{p+1} which shows up infinitely often, and thus $\mathcal{B}^+ \models$

$$\exists^\infty v_1 \ldots v_p \exists v_{p+2} \ldots v_n (A(v_1 \ldots v_p \underline{a}_{p+1} v_{p+2} \ldots \underline{b}_q) \wedge B'(\text{ditto})) .$$

We now repeat this procedure for v_{p+2}, etc. until, after rearranging subscripts between p+1 and n, we have $\mathcal{B}^+ \models$

$$\exists^\infty v_1 \ldots v_k (A(v_1 \ldots v_k \underline{a}_{k+1} \ldots \underline{a}_n \underline{b}_1 \ldots \underline{b}_q) \wedge B'(\text{ditto})) .$$

Now, finally, what we have is precisely the "if" clause of a sentence of $\Phi(\mathcal{B})$, say the αth sentence. At stage $\alpha+1$, in the proof of Theorem 11, witnesses $a_1 \ldots a_k$ <u>from outside $|\mathcal{A}|$</u> were chosen for the formula above, and $\mathcal{B}^+$ was made to satisfy $B(\underline{a}_1 \ldots \underline{a}_k \underline{a}_{k+1} \ldots \underline{a}_n \underline{b}_1 \ldots \underline{b}_q)$. It follows that the p-tuple $\langle a_1 \ldots a_p \rangle \in (|\mathcal{B}| - |\mathcal{A}|)^p$ satisfies the original formula D, and thus the embedding is d-codense and the proof complete.

If we try to prove KP by expanding $\mathcal{A}$ and $\mathcal{B}$ at the same time, we run into difficulties with the Skolem function leading out of $|\mathcal{A}|$. One way to circumvent them is to assume that the Skolem formula itself does not lead out of $|\mathcal{A}|$; that is, if $\mathcal{B} \models S(\underline{b}_1 \ldots \underline{b}_m \underline{c})$ with $b_1 \ldots b_m \in |\mathcal{A}|$, then $c \in |\mathcal{A}|$ also. It is precisely under this assumption that Julia Knight [12] was able to prove KP for weak second-order elementary extensions. Here, we have:

<u>LEMMA 7</u>: Let $\mathcal{A} \prec \mathcal{B}$ be one-cardinal models of an algebraically bounded theory T, and suppose that whenever $\mathcal{B} \models S(\underline{b}_1 \ldots \underline{b}_m \underline{c})$, and $b_1 \ldots b_m \in |\mathcal{A}|$, then $c \in |\mathcal{A}|$. Then KP holds for $\mathcal{A}$ and $\mathcal{B}$ with respect to the formula $S(x_1 \ldots x_m y)$.

<u>Proof</u>: We find a Skolem function f not for $S(x_1 \ldots x_m y)$ but for the formula $S'(x_1 \ldots x_m y)$, where, for some fixed element a of $|\mathcal{A}|$, $S'(x_1 \ldots x_m y)$ is

$$S(x_1 \ldots x_m y) \vee (\neg \exists z S(x_1 \ldots x_m z) \wedge y = \underline{a}) .$$

This requires adding a new constant symbol to $L(T)$, but algebraic boundedness and one-cardinality are unaffected. The Skolem function f will also serve for the formula S; and f cannot possibly lead out of $\mathcal{A}$.

We may assume that $\text{card}\,|\mathcal{A}| = \text{card}\,|\mathcal{B}| = \kappa$, for if the models were of different cardinalities, the one-cardinality of the larger model would force the embedding to be d-codense, whereupon Theorem 11 would apply.

We now produce a chain $< \theta_\alpha : \alpha < \kappa >$ of partial expansions of $\mathcal{B}$ as in the proof of Theorem 10, with the following slight restriction at stage $\alpha+1$: if it happens that the parameters $a_1 \ldots a_k$ in the αth member $C(\underline{a}_1 \ldots \underline{a}_k)$ of $\Phi(\mathcal{B})$ are all in $|\mathcal{A}|$, i.e. $C(\underline{a}_1 \ldots \underline{a}_k) \in \Phi(\mathcal{A})$, then we make sure that the chosen (n-k)-tuple $\langle a_{k+1} \ldots a_n \rangle$ is in $|\mathcal{A}|^{n-k}$. This is always possible, since $\exists^{\infty} v_{k+1} \ldots v_n (A(\underline{a}_1 \ldots \underline{a}_k v_{k+1} \ldots v_n) \wedge B'(\underline{a}_1 \ldots v_n))$ is satisfied in $\mathcal{A}$ if it is satisfied in $\mathcal{B}$; thus Lemma 6 enables us to find witnesses in $|\mathcal{A}|$ unmentioned in Θ_α, which has cardinality less than κ.

As a result, when we form the union $\Theta = \bigcup_{\alpha<\kappa} \Theta_\alpha$, we have not only $\mathcal{B} \models \Theta \cup T^+ \rightarrow \Phi(\mathcal{B})$ but also $\mathcal{A} \models \Theta \cup T^+ \rightarrow \Phi(\mathcal{A})$. We expand $\mathcal{B}$ to a model $\mathcal{B}^+$ of T^+ consistent with Θ, and the lemma follows.

In the next (and final) lemma we make a brief excursion into the realm of relational expansions. Let T be an algebraically bounded theory (assumed model-complete, as usual) and set $L_1 = L(T) \cup \{R\}$ where R is a new unary relation symbol. Let $T_1 = T(L_1)$ (i.e. the trivial expansion of T to the new language) and let T_1^* be the model-completion of T_1 (existence guaranteed by Theorem 3).

<u>LEMMA 8</u>: Let $\mathcal{A} \prec \mathcal{B}$ be a d-dense embedding of one-cardinal models of the above theory T. Then $\mathcal{A}$ and $\mathcal{B}$ can be expanded to models $\mathcal{A}_1$ and $\mathcal{B}_1$ of T_1^* such that $\mathcal{A}_1 \prec \mathcal{B}_1$ and $R^{\mathcal{B}_1} = R^{\mathcal{A}_1}$.

<u>Proof</u>: The axiom-set $\Phi(\mathcal{B})$ in this case is similar to, but simpler than, the Skolem expansion case. A configuration is now just a conjunction of formulas of the form $R(v_i)$ or $\neg R(v_i)$; the eligibility criterion for a configuration $B(v_1 \ldots v_n)$ is merely the conjunction of the formulas $v_i \neq v_j$ where $R(v_i)$ and

$\neg R(v_j)$ both appear in $B(v_1 \ldots v_n)$. We build partial expansions as usual, but using d-density we can choose all witnesses from $|\mathcal{A}|$. Finally set $R^{\mathcal{B}} = \{a \in |\mathcal{B}| : \theta \vdash R(\underline{a})\}$, and we are done. There is no problem, of course, with a relation "leading out of $|\mathcal{A}|$".

Note that $R^{\mathcal{B}}$ is perforce d-dense and d-codense in both $\mathcal{A}$ and $\mathcal{B}$, on account of members of $\Phi(\mathcal{B})$ in which $B(v_1 \ldots v_n) = \bigwedge_{1 \le i \le n} R(v_i)$ or in which $B(v_1 \ldots v_n) = \bigwedge_{1 \le i \le n} \neg R(v_i)$.

THEOREM 12: Let $\mathcal{A} \prec \mathcal{B}$ be a d-dense embedding of countable models of a countable algebraically bounded theory T, and suppose that there is a linear order "<" on $|\mathcal{A}|$ definable in $\mathcal{A}$. Then $\mathcal{A}$ and $\mathcal{B}$ have KP.

Proof: Let $\mathcal{A}_1$ and $\mathcal{B}_1$ be as in the conclusion of Lemma 8, and let $S(x_1 \ldots x_m y)$ be an arbitrary formula of $L(T)$. Set

$$S'(x_1 \ldots x_m y) = S(x_1 \ldots x_m y) \wedge R(y) .$$

Since $R^{\mathcal{B}} \subset |\mathcal{A}|$, Lemma 7 allows us to assign a Skolem function f' for S' such that $\mathcal{A}_1^+ \prec \mathcal{B}_1^+$. Now we define a function f: $|\mathcal{B}|^m \longrightarrow |\mathcal{B}|$ as follows:

$$f(a_1 \ldots a_m) = \begin{cases} f'(a_1 \ldots a_m), & \text{if } \mathcal{B} \models \neg \exists y S(\underline{a}_1 \ldots \underline{a}_m y); \\ f'(a_1 \ldots a_m), & \text{if } \mathcal{B} \models \exists^{\infty} y S(\underline{a}_1 \ldots \underline{a}_m y); \\ b & \text{otherwise, where } b \text{ is the unique element of } |\mathcal{B}| \text{ such that } \mathcal{B} \models S(\underline{a}_1 \ldots \underline{a}_m \underline{b}) \wedge \forall x (x < \underline{b} \rightarrow \neg S(\bar{\underline{a}} x)) . \end{cases}$$

This generalizes the method used by Knight [11] to obtain her result on CDLO's. The definition is elementary, by Corollaries 1 and 3 to Theorem 4, which together imply that the model-completion of $(T_1^*)^+$ is algebraically bounded. Thus the introduction of a symbol for f preserves the elementary embedding. To see that f serves as a Skolem function for S we need only note that $\mathcal{B} \models \exists^{\infty} y S(x_1 \ldots x_m y) \longrightarrow \exists y (S(x_1 \ldots x_m y) \wedge R(y))$ since $R^{\mathcal{B}}$ is d-dense in $\mathcal{B}$.

COROLLARY: Any dense embedding of CDLO's has KP.

KP can be proved for many special cases of embeddings of CDLO's which are neither dense nor codense, using hammer-and-tongs methods, but as far as the author knows the general case for CDLO's remains open.

REFERENCES

[1] P. Bacsich, "The strong amalgamation property," to appear.

[2] J.T. Baldwin and A.H. Lachlan, "On strongly minimal sets," J. Symb. Logic 36 (1971) #1, pp 79-96.

[3] J. Barwise and A. Robinson, "Completing theories by forcing," Ann. Math. Logic 2 (1970), pp 119-142.

[4] O.B. Belegradek and B.I. Zil'ber, "The model companion of an $\aleph_1$-categorical theory," Third All-Union Conference on Mathematical Logic, Институт Математики СО АН СССР, Novosibirsk June 1974, p 10.

[5] A.B. Carson, "The model completion of the theory of commutative regular rings," J. of Algebra 27 (1973) #1, pp 136-146.

[6] G. Cherlin, "Algebraically closed commutative rings," J. Symb. Logic 38 (1973) #3, pp 493-499.

[7] P. Eklof, "Algebraic closure operators and strong amalgamation bases," preprint.

[8] P. Eklof and G. Sabbagh, "Model completions and modules," Ann. Math. Logic 2 (1971) #3, pp 251-255.

[9] P. Henrard, "Classes of cotheories," Proceedings of the Bertrand Russell Memorial Logic Conference, Uldum, Denmark 1971; Leeds 1973.

[10] J. Knight, "An example involving Skolem functions and elementary embeddings," Notices Am.Math. Soc. 17 (1970) p 964.

[11] J. Knight, "Generic expansions of structures," J. Symb. Logic 38 (1973) #4, pp 561-570.

[12] J. Knight, "Some problems in model theory," Doctoral dissertation, University of California at Berkeley 1972.

[13] L. Lipshitz and D. Saracino, "The model companion of the theory of commutative rings without nilpotent elements," Proc. Am. Math. Soc. 38 (1973) pp 381-387.

[14] A. Macintyre, "Dense embeddings," this volume.

[15] T.H. Payne, "An elementary submodel never preserved by Skolem expansions," Zeitschrift für Mathematische Logik und Grundlagen der Mathematik 15 (1969) pp 435-436.

[16] A. Robinson, Complete Theories, North-Holland, Amserdam 1956.

[17] A. Robinson, "Infinite forcing in model theory," Proceedings of the Second Scandinavian Logic Symposium (J.E. Fenstad, ed), North-Holland, Amsterdam 1971; pp 317-340.

[18] A. Robinson, Introduction to Model Theory and to the Metamathematics of Algebra, North-Holland, Amsterdam 1963.

[19] A. Robinson, "Solution of a problem of Tarski," Fund. Math. 47 (1959) pp 179-204.

[20] G.E. Sacks, Saturated Model Theory, W.A. Benjamin Inc., Reading, Mass. 1972.

[21] D. Saracino, "A counterexample in the theory of model companions," Yale U. preprint, Oct 1973; also Notices Am. Math. Soc. 21 (1974) #2, abstract 74T-E9.

[22] D. Saracino, "Existentially complete nilpotent groups," to appear.

[23] D. Saracino, "Model companions for $\aleph_0$-categorical theories," Proc. Am. Math. Soc. 39 (1973) #3, pp 591-598.

[24] D. Saracino, "Wreath products and existentially complete solvable groups," Yale U. preprint June 1973.

[25] J.R. Shoenfield, Mathematical Logic, Addison-Wesley, Reading, Mass. 1967.

Vol. 342: Algebraic K-Theory II, "Classical" Algebraic K-Theory, and Connections with Arithmetic. Edited by H. Bass. XV, 527 pages. 1973.

Vol. 343: Algebraic K-Theory III, Hermitian K-Theory and Geometric Applications. Edited by H. Bass. XV, 572 pages. 1973.

Vol. 344: A. S. Troelstra (Editor), Metamathematical Investigation of Intuitionistic Arithmetic and Analysis. XVII, 485 pages. 1973.

Vol. 345: Proceedings of a Conference on Operator Theory. Edited by P. A. Fillmore. VI, 228 pages. 1973.

Vol. 346: Fučik et al., Spectral Analysis of Nonlinear Operators. II, 287 pages. 1973.

Vol. 347: J. M. Boardman and R. M. Vogt, Homotopy Invariant Algebraic Structures on Topological Spaces. X, 257 pages. 1973.

Vol. 348: A. M. Mathai and R. K. Saxena, Generalized Hypergeometric Functions with Applications in Statistics and Physical Sciences. VII, 314 pages. 1973.

Vol. 349: Modular Functions of One Variable II. Edited by W. Kuyk and P. Deligne. V, 598 pages. 1973.

Vol. 350: Modular Functions of One Variable III. Edited by W. Kuyk and J.-P. Serre. V, 350 pages. 1973.

Vol. 351: H. Tachikawa, Quasi-Frobenius Rings and Generalizations. XI, 172 pages. 1973.

Vol. 352: J. D. Fay, Theta Functions on Riemann Surfaces. V, 137 pages. 1973.

Vol. 353: Proceedings of the Conference. on Orders, Group Rings and Related Topics. Organized by J. S. Hsia, M. L. Madan and T. G. Ralley. X, 224 pages. 1973.

Vol. 354: K. J. Devlin, Aspects of Constructibility. XII, 240 pages. 1973.

Vol. 355: M. Sion, A Theory of Semigroup Valued Measures. V, 140 pages. 1973.

Vol. 356: W. L. J. van der Kallen, Infinitesimally Central Extensions of Chevalley Groups. VII, 147 pages. 1973.

Vol. 357: W. Borho, P. Gabriel und R. Rentschler, Primideale in Einhüllenden auflösbarer Lie-Algebren. V, 182 Seiten. 1973.

Vol. 358: F. L. Williams, Tensor Products of Principal Series Representations. VI, 132 pages. 1973.

Vol. 359: U. Stammbach, Homology in Group Theory. VIII, 183 pages. 1973.

Vol. 360: W. J. Padgett and R. L. Taylor, Laws of Large Numbers for Normed Linear Spaces and Certain Fréchet Spaces. VI, 111 pages. 1973.

Vol. 361: J. W. Schutz, Foundations of Special Relativity: Kinematic Axioms for Minkowski Space-Time. XX, 314 pages. 1973.

Vol. 362: Proceedings of the Conference on Numerical Solution of Ordinary Differential Equations. Edited by D.G. Bettis. VIII, 490 pages. 1974.

Vol. 363: Conference on the Numerical Solution of Differential Equations. Edited by G. A. Watson. IX, 221 pages. 1974.

Vol. 364: Proceedings on Infinite Dimensional Holomorphy. Edited by T. L. Hayden and T. J. Suffridge. VII, 212 pages. 1974.

Vol. 365: R. P. Gilbert, Constructive Methods for Elliptic Equations. VII, 397 pages. 1974.

Vol. 366: R. Steinberg, Conjugacy Classes in Algebraic Groups (Notes by V. V. Deodhar). VI, 159 pages. 1974.

Vol. 367: K. Langmann und W. Lütkebohmert, Cousinverteilungen und Fortsetzungssätze. VI, 151 Seiten. 1974.

Vol. 368: R. J. Milgram, Unstable Homotopy from the Stable Point of View. V, 109 pages. 1974.

Vol. 369: Victoria Symposium on Nonstandard Analysis. Edited by A. Hurd and P. Loeb. XVIII, 339 pages. 1974.

Vol. 370: B. Mazur and W. Messing, Universal Extensions and One Dimensional Crystalline Cohomology. VII, 134 pages. 1974.

Vol. 371: V. Poenaru, Analyse Différentielle. V, 228 pages. 1974.

Vol. 372: Proceedings of the Second International Conference on the Theory of Groups 1973. Edited by M. F. Newman. VII, 740 pages. 1974.

Vol. 373: A. E. R. Woodcock and T. Poston, A Geometrical Study of the Elementary Catastrophes. V, 257 pages. 1974.

Vol. 374: S. Yamamuro, Differential Calculus in Topological Linear Spaces. IV, 179 pages. 1974.

Vol. 375: Topology Conference. Edited by R. F. Dickman Jr. and P. Fletcher. X, 283 pages 1974.

Vol. 376: I. J. Good and D. B. Osteyee, Information, Weight of Evidence. The Singularity between Probability Measures and Signal Detection. XI, 156 pages. 1974.

Vol. 377: A. M. Fink, Almost Periodic Differential Equations. VIII, 336 pages. 1974.

Vol. 378: TOPO 72 - General Topology and its Applications. Proceedings 1972. Edited by R. A. Alò, R. W. Heath and J. Nagata. XIV, 651 pages. 1974.

Vol. 379: A. Badrikian et S. Chevet, Mesures Cylindriques, Espaces de Wiener et Fonctions Aléatoires Gaussiennes. X, 383 pages. 1974.

Vol. 380: M. Petrich, Rings and Semigroups. VIII, 182 pages. 1974.

Vol. 381: Séminaire de Probabilités VIII. Edité par P. A. Meyer. IX, 354 pages. 1974.

Vol. 382: J. H. van Lint, Combinatorial Theory Seminar Eindhoven University of Technology. VI, 131 pages. 1974.

Vol. 383: Séminaire Bourbaki - vol. 1972/73. Exposés 418-435. IV, 334 pages. 1974.

Vol. 384: Functional Analysis and Applications, Proceedings 1972. Edited by L. Nachbin. V, 270 pages. 1974.

Vol. 385: J. Douglas Jr. and T. Dupont, Collocation Methods for Parabolic Equations in a Single Space Variable (Based on C^1-Piecewise-Polynomial Spaces). V, 147 pages. 1974.

Vol. 386: J. Tits, Buildings of Spherical Type and Finite BN-Pairs. X, 299 pages. 1974.

Vol. 387: C. P. Bruter, Eléments de la Théorie des Matroïdes. V, 138 pages. 1974.

Vol. 388: R. L. Lipsman, Group Representations. X, 166 pages. 1974.

Vol. 389: M.-A. Knus et M. Ojanguren, Théorie de la Descente et Algèbres d' Azumaya. IV, 163 pages. 1974.

Vol. 390: P. A. Meyer, P. Priouret et F. Spitzer, Ecole d'Eté de Probabilités de Saint-Flour III - 1973. Edité par A. Badrikian et P.-L. Hennequin. VIII, 189 pages. 1974.

Vol. 391: J. W. Gray, Formal Category Theory: Adjointness for 2-Categories. XII, 282 pages. 1974.

Vol. 392: Géométrie Différentielle, Colloque, Santiago de Compostela, Espagne 1972. Edité par E. Vidal. VI, 225 pages. 1974.

Vol. 393: G. Wassermann, Stability of Unfoldings. IX, 164 pages. 1974.

Vol. 394: W. M. Patterson, 3rd, Iterative Methods for the Solution of a Linear Operator Equation in Hilbert Space - A Survey. III, 183 pages. 1974.

Vol. 395: Numerische Behandlung nichtlinearer Integrodifferential- und Differentialgleichungen. Tagung 1973. Herausgegeben von R. Ansorge und W. Törnig. VII, 313 Seiten. 1974.

Vol. 396: K. H. Hofmann, M. Mislove and A. Stralka, The Pontryagin Duality of Compact O-Dimensional Semilattices and its Applications. XVI, 122 pages. 1974.

Vol. 397: T. Yamada, The Schur Subgroup of the Brauer Group. V, 159 pages. 1974.

Vol. 398: Théories de l'Information, Actes des Rencontres de Marseille-Luminy, 1973. Edité par J. Kampé de Fériet et C.-F. Picard. XII, 201 pages. 1974.

Vol. 399: Functional Analysis and its Applications. Proceedings 1973. Edited by H. G. Garnir, K. R. Unni and J. H. Williamson. II, 584 pages. 1974.

Vol. 400: A Crash Course on Kleinian Groups. Proceedings 1974. Edited by L. Bers and I. Kra. VII, 130 pages. 1974.

Vol. 401: M. F. Atiyah, Elliptic Operators and Compact Groups. V, 93 pages. 1974.

Vol. 402: M. Waldschmidt, Nombres Transcendants. VIII, 277 pages. 1974.

Vol. 403: Combinatorial Mathematics. Proceedings 1972. Edited by D. A. Holton. VIII, 148 pages. 1974.

Vol. 404: Théorie du Potentiel et Analyse Harmonique. Edité par J. Faraut. V, 245 pages. 1974.

Vol. 405: K. J. Devlin and H. Johnsbråten, The Souslin Problem. VIII, 132 pages. 1974.

Vol. 406: Graphs and Combinatorics. Proceedings 1973. Edited by R. A. Bari and F. Harary. VIII, 355 pages. 1974.

Vol. 407: P. Berthelot, Cohomologie Cristalline des Schémas de Caracteristique p > o. II, 604 pages. 1974.

Vol. 408: J. Wermer, Potential Theory. VIII, 146 pages. 1974.

Vol. 409: Fonctions de Plusieurs Variables Complexes, Séminaire François Norguet 1970–1973. XIII, 612 pages. 1974.

Vol. 410: Séminaire Pierre Lelong (Analyse) Année 1972–1973. VI, 181 pages. 1974.

Vol. 411: Hypergraph Seminar. Ohio State University, 1972. Edited by C. Berge and D. Ray-Chaudhuri. IX, 287 pages. 1974.

Vol. 412: Classification of Algebraic Varieties and Compact Complex Manifolds. Proceedings 1974. Edited by H. Popp. V, 333 pages. 1974.

Vol. 413: M. Bruneau, Variation Totale d'une Fonction. XIV, 332 pages. 1974.

Vol. 414: T. Kambayashi, M. Miyanishi and M. Takeuchi, Unipotent Algebraic Groups. VI, 165 pages. 1974.

Vol. 415: Ordinary and Partial Differential Equations. Proceedings 1974. XVII, 447 pages. 1974.

Vol. 416: M. E. Taylor, Pseudo Differential Operators. IV, 155 pages. 1974.

Vol. 417: H. H. Keller, Differential Calculus in Locally Convex Spaces. XVI, 131 pages. 1974.

Vol. 418: Localization in Group Theory and Homotopy Theory and Related Topics. Battelle Seattle 1974 Seminar. Edited by P. J. Hilton. VI, 172 pages 1974.

Vol. 419: Topics in Analysis. Proceedings 1970. Edited by O. E. Lehto, I. S. Louhivaara, and R. H. Nevanlinna. XIII, 392 pages. 1974.

Vol. 420: Category Seminar. Proceedings 1972/73. Edited by G. M. Kelly. VI, 375 pages. 1974.

Vol. 421: V. Poénaru, Groupes Discrets. VI, 216 pages. 1974.

Vol. 422: J.-M. Lemaire, Algèbres Connexes et Homologie des Espaces de Lacets. XIV, 133 pages. 1974.

Vol. 423: S. S. Abhyankar and A. M. Sathaye, Geometric Theory of Algebraic Space Curves. XIV, 302 pages. 1974.

Vol. 424: L. Weiss and J. Wolfowitz, Maximum Probability Estimators and Related Topics. V, 106 pages. 1974.

Vol. 425: P. R. Chernoff and J. E. Marsden, Properties of Infinite Dimensional Hamiltonian Systems. IV, 160 pages. 1974.

Vol. 426: M. L. Silverstein, Symmetric Markov Processes. X, 287 pages. 1974.

Vol. 427: H. Omori, Infinite Dimensional Lie Transformation Groups. XII, 149 pages. 1974.

Vol. 428: Algebraic and Geometrical Methods in Topology, Proceedings 1973. Edited by L. F. McAuley. XI, 280 pages. 1974.

Vol. 429: L. Cohn, Analytic Theory of the Harish-Chandra C-Function. III, 154 pages. 1974.

Vol. 430: Constructive and Computational Methods for Differential and Integral Equations. Proceedings 1974. Edited by D. L. Colton and R. P. Gilbert. VII, 476 pages. 1974.

Vol. 431: Séminaire Bourbaki – vol. 1973/74. Exposés 436–452. IV, 347 pages. 1975.

Vol. 432: R. P. Pflug, Holomorphiegebiete, pseudokonvexe Gebiete und das Levi-Problem. VI, 210 Seiten. 1975.

Vol. 433: W. G. Faris, Self-Adjoint Operators. VII, 115 pages. 1975.

Vol. 434: P. Brenner, V. Thomée, and L. B. Wahlbin, Besov Spaces and Applications to Difference Methods for Initial Value Problems. II, 154 pages. 1975.

Vol. 435: C. F. Dunkl and D. E. Ramirez, Representations of Commutative Semitopological Semigroups. VI, 181 pages. 1975.

Vol. 436: L. Auslander and R. Tolimieri, Abelian Harmonic Analysis, Theta Functions and Function Algebras on a Nilmanifold. V, 99 pages. 1975.

Vol. 437: D. W. Masser, Elliptic Functions and Transcendence. XIV, 143 pages. 1975.

Vol. 438: Geometric Topology. Proceedings 1974. Edited by L. C. Glaser and T. B. Rushing. X, 459 pages. 1975.

Vol. 439: K. Ueno, Classification Theory of Algebraic Varieties and Compact Complex Spaces. XIX, 278 pages. 1975

Vol. 440: R. K. Getoor, Markov Processes: Ray Processes and Right Processes. V, 118 pages. 1975.

Vol. 441: N. Jacobson, PI-Algebras. An Introduction. V, 115 pages. 1975.

Vol. 442: C. H. Wilcox, Scattering Theory for the d'Alembert Equation in Exterior Domains. III, 184 pages. 1975.

Vol. 443: M. Lazard, Commutative Formal Groups. II, 236 pages. 1975.

Vol. 444: F. van Oystaeyen, Prime Spectra in Non-Commutative Algebra. V, 128 pages. 1975.

Vol. 445: Model Theory and Topoi. Edited by F. W. Lawvere, C. Maurer, and G. C. Wraith. III, 354 pages. 1975.

Vol. 446: Partial Differential Equations and Related Topics. Proceedings 1974. Edited by J. A. Goldstein. IV, 389 pages. 1975.

Vol. 447: S. Toledo, Tableau Systems for First Order Number Theory and Certain Higher Order Theories. III, 339 pages. 1975.

Vol. 448: Spectral Theory and Differential Equations. Proceedings 1974. Edited by W. N. Everitt. XII, 321 pages. 1975.

Vol. 449: Hyperfunctions and Theoretical Physics. Proceedings 1973. Edited by F. Pham. IV, 218 pages. 1975.

Vol. 450: Algebra and Logic. Proceedings 1974. Edited by J. N. Crossley. VIII, 307 pages. 1975.

Vol. 451: Probabilistic Methods in Differential Equations. Proceedings 1974. Edited by M. A. Pinsky. VII, 162 pages. 1975.

Vol. 452: Combinatorial Mathematics III. Proceedings 1974. Edited by Anne Penfold Street and W. D. Wallis. IX, 233 pages. 1975.

Vol. 453: Logic Colloquium. Symposium on Logic Held at Boston, 1972–73. Edited by R. Parikh. IV, 251 pages. 1975.

Vol. 454: J. Hirschfeld and W. H. Wheeler, Forcing, Arithmetic, Division Rings. VII, 266 pages. 1975.

Vol. 455: H. Kraft, Kommutative algebraische Gruppen und Ringe. III, 163 Seiten. 1975.

Vol. 456: R. M. Fossum, P. A. Griffith, and I. Reiten, Trivial Extensions of Abelian Categories. Homological Algebra of Trivial Extensions of Abelian Categories with Applications to Ring Theory. XI, 122 pages. 1975.

Vol. 457: Fractional Calculus and Its Applications. Proceedings 1974. Edited by B. Ross. VI, 381 pages. 1975.

Vol. 458: P. Walters, Ergodic Theory – Introductory Lectures. VI, 198 pages. 1975.

Vol. 459: Fourier Integral Operators and Partial Differential Equations. Proceedings 1974. Edited by J. Chazarain. VI, 372 pages. 1975.

Vol. 460: O. Loos, Jordan Pairs. XVI, 218 pages. 1975.

Vol. 461: Computational Mechanics. Proceedings 1974. Edited by J. T. Oden. VII, 328 pages. 1975.

Vol. 462: P. Gérardin, Construction de Séries Discrètes p-adiques. »Sur les séries discrètes non ramifiées des groupes réductifs déployés p-adiques«. III, 180 pages. 1975.

Vol. 463: H.-H. Kuo, Gaussian Measures in Banach Spaces. VI, 224 pages. 1975.

Vol. 464: C. Rockland, Hypoellipticity and Eigenvalue Asymptotics. III, 171 pages. 1975.

Vol. 465: Séminaire de Probabilités IX. Proceedings 1973/74. Edité par P. A. Meyer. IV, 589 pages. 1975.

Vol. 466: Non-Commutative Harmonic Analysis. Proceedings 1974. Edited by J. Carmona, J. Dixmier and M. Vergne. VI, 231 pages. 1975.

Vol. 467: M. R. Essén, The Cos $\pi\lambda$ Theorem. With a paper by Christer Borell. VII, 112 pages. 1975.

Vol. 468: Dynamical Systems – Warwick 1974. Proceedings 1973/74. Edited by A. Manning. X, 405 pages. 1975.

Vol. 469: E. Binz, Continuous Convergence on C(X). IX, 140 pages. 1975.

Vol. 470: R. Bowen, Equilibrium States and the Ergodic Theory of Anosov Diffeomorphisms. III, 108 pages. 1975.

Vol. 471: R. S. Hamilton, Harmonic Maps of Manifolds with Boundary. III, 168 pages. 1975.

Vol. 472: Probability-Winter School. Proceedings 1975. Edited by Z. Ciesielski, K. Urbanik, and W. A. Woyczyński. VI, 283 pages. 1975.

Vol. 473: D. Burghelea, R. Lashof, and M. Rothenberg, Groups of Automorphisms of Manifolds. (with an appendix by E. Pedersen) VII, 156 pages. 1975.

Vol. 474: Séminaire Pierre Lelong (Analyse) Année 1973/74. Edité par P. Lelong. VI, 182 pages. 1975.

Vol. 475: Répartition Modulo 1. Actes du Colloque de Marseille-Luminy, 4 au 7 Juin 1974. Edité par G. Rauzy. V, 258 pages. 1975. 1975.

Vol. 476: Modular Functions of One Variable IV. Proceedings 1972. Edited by B. J. Birch and W. Kuyk. V, 151 pages. 1975.

Vol. 477: Optimization and Optimal Control. Proceedings 1974. Edited by R. Bulirsch, W. Oettli, and J. Stoer. VII, 294 pages. 1975.

Vol. 478: G. Schober, Univalent Functions – Selected Topics. V, 200 pages. 1975.

Vol. 479: S. D. Fisher and J. W. Jerome, Minimum Norm Extremals in Function Spaces. With Applications to Classical and Modern Analysis. VIII, 209 pages. 1975.

Vol. 480: X. M. Fernique, J. P. Conze et J. Gani, Ecole d'Eté de Probabilités de Saint-Flour IV-1974. Edité par P.-L. Hennequin. XI, 293 pages. 1975.

Vol. 481: M. de Guzmán, Differentiation of Integrals in R^n. XII, 226 pages. 1975.

Vol. 482: Fonctions de Plusieurs Variables Complexes II. Séminaire François Norguet 1974–1975. IX, 367 pages. 1975.

Vol. 483: R. D. M. Accola, Riemann Surfaces, Theta Functions, and Abelian Automorphisms Groups. III, 105 pages. 1975.

Vol. 484: Differential Topology and Geometry. Proceedings 1974. Edited by G. P. Joubert, R. P. Moussu, and R. H. Roussarie. IX, 287 pages. 1975.

Vol. 485: J. Diestel, Geometry of Banach Spaces – Selected Topics. XI, 282 pages. 1975.

Vol. 486: S. Stratila and D. Voiculescu, Representations of AF-Algebras and of the Group U (∞). IX, 169 pages. 1975.

Vol. 487: H. M. Reimann und T. Rychener, Funktionen beschränkter mittlerer Oszillation. VI, 141 Seiten. 1975.

Vol. 488: Representations of Algebras, Ottawa 1974. Proceedings 1974. Edited by V. Dlab and P. Gabriel. XII, 378 pages. 1975.

Vol. 489: J. Bair and R. Fourneau, Etude Géométrique des Espaces Vectoriels. Une Introduction. VII, 185 pages. 1975.

Vol. 490: The Geometry of Metric and Linear Spaces. Proceedings 1974. Edited by L. M. Kelly. X, 244 pages. 1975.

Vol. 491: K. A. Broughan, Invariants for Real-Generated Uniform Topological and Algebraic Categories. X, 197 pages. 1975.

Vol. 492: Infinitary Logic: In Memoriam Carol Karp. Edited by D. W. Kueker. VI, 206 pages. 1975.

Vol. 493: F. W. Kamber and P. Tondeur, Foliated Bundles and Characteristic Classes. XIII, 208 pages. 1975.

Vol. 494: A Cornea and G. Licea. Order and Potential Resolvent Families of Kernels. IV, 154 pages. 1975.

Vol. 495: A. Kerber, Representations of Permutation Groups II. V, 175 pages. 1975.

Vol. 496: L. H. Hodgkin and V. P. Snaith, Topics in K-Theory. Two Independent Contributions. III, 294 pages. 1975.

Vol. 497: Analyse Harmonique sur les Groupes de Lie. Proceedings 1973–75. Edité par P. Eymard et al. VI, 710 pages. 1975.

Vol. 498: Model Theory and Algebra. A Memorial Tribute to Abraham Robinson. Edited by D. H. Saracino and V. B. Weispfenning. X, 463 pages. 1975.